Springer Proceedings in Mathematics & Statistics

Volume 201

Springer Proceedings in Mathematics & Statistics

This book series features volumes composed of selected contributions from workshops and conferences in all areas of current research in mathematics and statistics, including operation research and optimization. In addition to an overall evaluation of the interest, scientific quality, and timeliness of each proposal at the hands of the publisher, individual contributions are all refereed to the high quality standards of leading journals in the field. Thus, this series provides the research community with well-edited, authoritative reports on developments in the most exciting areas of mathematical and statistical research today.

More information about this series at http://www.springer.com/series/10533

Gregory E. Fasshauer · Larry L. Schumaker
Editors

Approximation Theory XV: San Antonio 2016

Springer

Editors
Gregory E. Fasshauer
Department of Applied Mathematics and Statistics
Colorado School of Mines
Golden, CO
USA

Larry L. Schumaker
Department of Mathematics
Vanderbilt University
Nashville, TN
USA

ISSN 2194-1009 ISSN 2194-1017 (electronic)
Springer Proceedings in Mathematics & Statistics
ISBN 978-3-319-59911-3 ISBN 978-3-319-59912-0 (eBook)
DOI 10.1007/978-3-319-59912-0

Library of Congress Control Number: 2017943223

Mathematics Subject Classification (2010): 41-XX, 42-XX, 65-XX, 42C05, 65D07, 94A12

Printed on acid-free paper

This Springer imprint is published by Springer Nature
The registered company is Springer International Publishing AG
The registered company address is: Gewerbestrasse 11, 6330 Cham, Switzerland

Preface

These proceedings are based on papers presented at the international conference *Approximation Theory XV*, which was held May 22–25, 2016 in San Antonio, Texas. The conference was the fifteenth in a series of meetings in Approximation Theory held at various locations in the USA and was attended by 146 participants. Previous conferences in the series were held in Austin, Texas (1973, 1976, 1980, 1992); College Station, Texas (1983, 1986, 1989, 1995); Nashville, Tennessee (1998); St. Louis, Missouri (2001); Gatlinburg, Tennessee (2004); and San Antonio, Texas (2007, 2010, 2013).

We are particularly indebted to our plenary speakers: Josef Dick (Univ. New South Wales), Simon Foucart (Texas A&M), Elisabeth Larsson (Uppsala Univ.), Doron Lubinsky (Georgia Tech), Carla Manni (Univ. Rome—Tor Vergata), Mike Neamtu (Vanderbilt Univ.), and Ulrich Reif (Tech. Univ. Darmstadt). The eighth Vasil A. Popov Prize in Approximation Theory was awarded to Jean-Marie Mirebeau (CNRS, Université Paris-Sud), who also presented a plenary lecture. Thanks are also due to the presenters of contributed papers, as well as everyone who attended for making the conference a success.

We are especially grateful to the National Science Foundation for financial support, and also to the Department of Mathematics at Vanderbilt University for its logistical support.

We would also like to express our sincere gratitude to the reviewers who helped select articles for inclusion in this proceedings volume, and also for their suggestions to the authors for improving their papers.

Golden, CO, USA — Gregory E. Fasshauer
Nashville, TN, USA — Larry L. Schumaker

Contents

Contributors

Jean-Paul Berrut Department of Mathematics, University of Fribourg, Fribourg, Switzerland

Oleg Davydov Department of Mathematics, University of Giessen, Giessen, Germany

Stefano De Marchi Department of Mathematics “Tullio Levi-Civita”, University of Padova, Padova, Italy

Simon Foucart Department of Mathematics, Texas A&M University, College Station, TX, USA

Daan Huybrechs KU Leuven, Leuven, Belgium

Andrea Idda Banco Popolare, Verona, Italy

Ming-Jun Lai Department of Mathematics, University of Georgia, Athens, GA, USA

James Lanterman Department of Mathematics, University of Georgia, Athens, GA, USA

Wayne Lawton School of Mathematics & Statistics, University of Western Australia, Perth, Australia

Jeff Ledford Virginia Commonwealth University, Richmond, VA, USA

Shiying Li Department of Mathematics, Vanderbilt University, Nashville, TN, USA

D.S. Lubinsky School of Mathematics, Georgia Institute of Technology, Atlanta, GA, USA

Carla Manni Department of Mathematics, University of Rome “Tor Vergata”, Rome, Italy

Roel Matthysen KU Leuven, Leuven, Belgium

Tom McKinley University of South Florida, Tampa, FL, USA

Clayton Mersmann Department of Mathematics, University of Georgia, Athens, GA, USA

Deanna Needell Claremont McKenna College, Claremont, CA, USA

Tomoaki Okayama Department of Systems Engineering, Graduate School of Information Sciences, Hiroshima City University, Asaminami-ku, Hiroshima, Japan

Laura Pezza Dipartimento SBAI, Università di Roma “La Sapienza”, Roma, Italy

Francesca Pitolli Dipartimento SBAI, Università di Roma “La Sapienza”, Roma, Italy

David W. Roach Murray State University, Murray, KY, USA

Fabio Roman Department of Mathematics, University of Turin, Turin, Italy

Gabriele Santin IANS University of Stuttgart, Stuttgart, Germany

Larry L. Schumaker Department of Mathematics, Vanderbilt University, Nashville, TN, USA

Boris Shekhtman University of South Florida, Tampa, FL, USA

Hendrik Speleers Department of Mathematics, University of Rome “Tor Vergata”, Rome, Italy

Masaaki Sugihara Department of Physics and Mathematics, College of Science and Engineering, Aoyama Gakuin University, Chuo-ku, Sagamihara-shi, Kanagawa, Japan

Ken’ichiro Tanaka Department of Mathematical Informatics, Graduate School of Information Science and Technology, The University of Tokyo, Bunkyo-ku, Tokyo, Japan

Vesselin Vatchev School of Mathematical and Statistical Sciences, The University of Texas Rio Grande Valley, Brownsville, TX, USA

Rachel Ward University of Texas, Austin, TX, USA

Qi Ye School of Mathematical Sciences, South China Normal University, Guangzhou, Guangdong, China

Wee Ping Yeo Faculty of Science, Universiti Brunei Darussalam, Gadong, Brunei Darussalam

Linear Barycentric Rational Interpolation with Guaranteed Degree of Exactness

Jean-Paul Berrut

Abstract In recent years, linear barycentric rational interpolants, introduced in 1988 and improved in 2007 by Floater and Hormann, have turned out to be among the most efficient infinitely smooth interpolants, in particular with equispaced points. In the present contribution, we introduce a new way of obtaining linear barycentric rational interpolants with relatively high orders of convergence. The basic idea is to modify the interpolant with equal weights of 1988 to force it to interpolate exactly the monomials up to a certain degree. This is obtained by modifying a few weights at each extremity of the interval of interpolation. Numerical experience demonstrates that the method is indeed able to interpolate with much higher orders than the original 1988 interpolant and in a very stable way.

Keywords Barycentric interpolation · Linear rational interpolation · Degree of exactness · Exactness conditions · Degree diminishing conditions

1 Introduction

The present work is concerned with linear barycentric rational interpolation (LBRI) on the interval (for the circle case, see [2, 6]). Let $x_0 < x_1 < \ldots < x_n$ be distinct and ordered abscissae in an interval $[a, b] \in \mathbb{R}$, f a function $[a, b] \mapsto \mathbb{R}$, and let $f_j := f(x_j)$, $j = 0(1)n$. The (true [35, p. 106]) barycentric Lagrange formula for the polynomial of degree at most n that interpolates f at the x_j's is

$$p_n(x) = \sum_{j=0}^{n} \frac{\lambda_j}{x - x_j} f_j \Big/ \sum_{j=0}^{n} \frac{\lambda_j}{x - x_j}, \qquad \lambda_j := 1 \Big/ \prod_{\substack{i=0 \\ i \neq j}} (x_j - x_i). \tag{1}$$

J.-P. Berrut (✉)
Department of Mathematics, University of Fribourg, Pérolles,
1700 Fribourg, Switzerland
e-mail: jean-paul.berrut@unifr.ch

G.E. Fasshauer and L.L. Schumaker (eds.), *Approximation Theory XV: San Antonio 2016*, Springer Proceedings in Mathematics & Statistics 201,
DOI 10.1007/978-3-319-59912-0_1

The λ_j's are called the *weights* of the formula.

p_n is well-known to be useless for large numbers of arbitrary nodes, the most striking example being that of equispaced points: p_n is ill-conditioned for large n and even divergent for many functions as $n \to \infty$, e.g., when the complex singularities of a meromorphic f come close to the center of $[x_0, x_n]$, as with the notorious Runge example [42, Chap. 13].

Thirty years ago, Werner [45] noticed that the interpolation property is preserved in (1), when one changes the λ_j's to any set of non-vanishing numbers. As a way of avoiding Runge's phenomenon, we have suggested in [6] to replace the λ_j's by a set of well chosen weights β_j, $j = 0(1)n$. Multiplying numerator and denominator of p_n in (1) by

$$\ell(x) := \prod_{i=0}^{n} (x - x_i),$$

one sees that it then becomes a rational function, more precisely the quotient of two polynomials of degree at most n. In the classic approach to the rational interpolation problem, numerator and denominator degrees are forced to sum up to n in order to guarantee the unicity of the interpolant (when it exists, i.e., in the absence of so-called unattainable points) [10, 37, 38]; consequently, the denominator depends on the f_j's, and the resulting interpolant looses the linearity of the dependence on f [15]. The suggestion in [6] was to choose instead a set of β_j's that just depend on the nodes, not on the f_j's, abstaining from imposing a degree condition on the denominator. Notice that any choice of the β_j in

$$r_n(x) := \sum_{j=0}^{n} \frac{\gamma_j f_j}{x - x_j} \Big/ \sum_{j=0}^{n} \frac{\beta_j}{x - x_j}$$

fully determines r_n, as γ_j has to equal β_j for interpolation.

In order to avoid the presence of poles for some f, the signs of the β_j should alternate.

Lemma 1 *If the linear interpolant*

$$r_n(x) \equiv r_n[f](x) := \sum_{j=0}^{n} \frac{\beta_j}{x - x_j} f_j \Big/ \sum_{j=0}^{n} \frac{\beta_j}{x - x_j}, \qquad \beta_j \neq 0 \,\forall\, j, \tag{2}$$

is to be free of poles for all $f \in C[x_0, x_n]$, *then* $\operatorname{sign}(\beta_{j+1}) = -\operatorname{sign}(\beta_j)$, $j = 0(1)n - 1$.

Proof Write $r_n = p/q$ with $p(x) := \ell(x) \sum_{j=0}^{n} \frac{\beta_j}{x - x_j} f_j$ and $q(x) := \ell(x) \sum_{j=0}^{n} \frac{\beta_j}{x - x_j}$. q does not vanish at any node, as $\lim_{x \to x_k} q(x) = \beta_k \prod_{i \neq k} (x_k - x_i)$. But it does not vanish at any other x^* either: Suppose $q(x^*) = 0$ and let (x_k, x_{k+1}) be the interval containing x^*. Choose a continuous function f with $f_j = 0$ for $j \neq k$ and

$f_k \neq 0$: Then, $p(x^*) = \ell(x^*)\beta_k f_k/(x^* - x_k) \neq 0$, and r_n has a pole at x^*, which contradicts the hypothesis.

Now, consider an interval (x_k, x_{k+1}); on it, $\ell(x)$ has constant sign. On its very left (i.e., close to x_k), $\sum_{j=0}^{n} \beta_j/(x - x_j)$ is dominated by $\beta_k/(x - x_k)$, on the very right by $\beta_{k+1}/(x - x_{k+1})$. But in (x_k, x_{k+1}), $\frac{1}{x-x_k}$ is positive and $\frac{1}{x-x_{k+1}}$ negative. The constant sign of q over (x_k, x_{k+1}) therefore is merely possible if $\text{sign}(\beta_{k+1}) = -\text{sign}(\beta_k)$. ■

Notice that Proposition 8 of [37] does not apply to this case, but merely to the classic one, as it assumes that r_n is reduced. Counterexamples are easy to find in case this condition is not satisfied [5, 29]. In [29, p. 199], interpolants which may be written in a simple closed-form expression such as (2) and reproduce linear functions are called *barycentric*.

In [6], starting from the observation that p_n has very regular λ_j's when the polynomial problem is well conditioned, we began with $\beta_j = (-1)^j$ and proved that the resulting denominator of r_n in (2) does not have real zeros. (It is shown in [8] that the Lebesgue constant Λ_n of the linear operator, which to every $f \in C[a, b]$ associates its polynomial p_n in (1), grows at least as rapidly as

$$\Lambda_n \geq \frac{1}{2n^2} \frac{\max |\lambda_j|}{\min |\lambda_j|};$$

weights with similar absolute values seem therefore a natural thing to try.) Considerable—and at that time very costly—numerical evidence led in [6] to the conjecture that the corresponding interpolant, say R_0, converges to f as $\mathcal{O}(h)$, where $h := \max_{0 \leq i \leq n-1}(x_{i+1} - x_i)$. This has been proved by Floater and Hormann in [24] under a "local mesh ratio condition" on the manner h tends to 0. In the meantime, much is known about R_0, as testified by the references in [29].

In [6], a trigonometric version of the interpolant was also suggested; its transplantation to the interval by the map $x = \cos\phi$ suggested to change the extremal weights (i.e., β_0 and β_n) to 1/2, resp. $(1/2)(-1)^n$, as the corresponding nodes are used (counted) twice. With this change, the interpolant, which we shall denote by R_1, still is free of poles in $[x_0, x_n]$. Numerical results showed R_1 to be much more accurate than R_0; for equispaced nodes, its $\mathcal{O}(h^2)$ convergence was later conjectured in [3] and proved in [24]. For arbitrary nodes, this holds under some condition on these [14]. Note that this interpolant converges exponentially, when the x_j's are conformal maps of Chebyshev nodes [3].

A remarkable step was consequently made by Floater and Hormann, who in [24] suggested a set of β_j's that guarantee any order of convergence prescribed in advance. More precisely, for every $d \in \mathbb{N}$, these authors construct a linear barycentric rational interpolant as a rational blend of the $n - d + 1$ polynomials interpolating at $d + 1$ consecutive nodes, which converges as $\mathcal{O}(h^{d+1})$; its denominator does not have any zeros in $[x_0, x_n]$. They also provide a simple formula for the corresponding β_j's, which merely depend on the x_j's and d, and alternate in sign. A detailed study of the convergence of this family of interpolants for meromorphic functions, leading to a suggestion for choosing d, has been performed in [27].

For *equispaced nodes*, this yields again $\beta_j = (-1)^j$, up to a "boundary correction" of d nodes that decrease from 1 to 2^{-d} at each extremity. For $d = 1$, $|\beta_0| = |\beta_n| = 1/2$, so that the interpolant coincides with R_1, which proved its $\mathcal{O}(h^2)$-convergence. Note that exponential convergence is not possible for equispaced nodes, if the scheme has to be reasonably conditioned [34]: Algebraic convergence may well be close to best possible.

For *arbitrary nodes*, however, the Floater–Hormann β_j's depend on d nodes about x_j and for $d = 1$ therefore do not coincide with those of R_1 [29]. The Floater–Hormann (FH) interpolant has already led to several applications in numerical analysis [13, 16]. In view of what follows, we notice that it reproduces every polynomial of degree up to d if $n - d$ is even and of degree up to $d + 1$ if $n - d$ is odd [32, p. 199].

In the present work, we intend to increase the rate of convergence of our interpolants above the value of two by applying an "endpoint correction" at d_ℓ nodes on the left-hand side and d_r on the right-hand side, while keeping the interior nodes at $\beta_j = (-1)^j$, independently of the particular nodes.

Our basic observation is that there seems to be a connection between the number of monomials of consecutive degrees that are interpolated exactly and the order of convergence of the corresponding LBRI; for instance, as just mentioned, the FH interpolant reproduces monomials up to degree d and converges with order $d + 1$; this does not seem fortuitous.

2 The Degree of Exactness of a Linear Barycentric Rational Interpolant

Here, we borrow a concept from numerical quadrature, already introduced for interpolation in [20] and maybe earlier. Recall that the degree of exactness of a quadrature rule Q is the largest $\rho \in \mathbb{N}^0$ such that Q integrates all polynomials of degree $\leq \rho$ exactly. In view of the linearity, this is the case iff $1, x, x^2, \ldots, x^\rho$ are integrated exactly and $x^{\rho+1}$ is not [36, p. 396]. (Polynomial) interpolating (so-called Newton–Cotes) quadrature formulae for $n + 1$ equispaced points reach a degree of exactness n or $n + 1$, depending on the parity of n [36, p. 398], while symmetric linear barycentric rational quadrature based on the FH interpolant reaches d to $d + 2$, depending on the parity of n and d [32, p. 419].

Definition 1 By the degree of exactness of a (linear) interpolant $f \mapsto If$, we shall mean the largest $\rho \in \mathbb{N}^0$ such that $Iq = q$ for every polynomial of degree $\leq \rho$ and such that there exists a polynomial q with $\deg q = \rho + 1$ and $Iq \neq q$.

In the present work, we shall modify the first d_ℓ and last d_r weights of our original interpolant determined by $\beta_j = (-1)^j$ to obtain an r_n in (2) with a guaranteed degree of exactness. We shall also give a formula for computing the

$$d := d_\ell + d_r$$

weights to be modified. The following equivalence result, already implicitly contained in [9], is our key ingredient.

Theorem 1 *If the denominator* $\ell(x)\sum_{j=0}^{n}\frac{\beta_j}{x-x_j}$, $\beta_j \neq 0$, *of* r_n *does not have any real zero, then the following properties are equivalent:*

(a) r_n *reproduces all polynomials of degree* $\leq d$;
(b) the degree of the denominator of r_n *is* $\leq n-d$.

Proof We start with the proof that (b) implies (a). Multiplying numerator and denominator of (2) by ℓ yields the representation

$$r_n[f](x) = \sum_{j=0}^{n} \beta_j f_j s_j(x) \Big/ \sum_{j=0}^{n} \beta_j s_j(x),$$

with $s_j(x) = \prod_{k\neq j}(x - x_k)$. Let now y be a polynomial of degree $\leq d$: $y(x) = \sum_{i=0}^{d} a_i x^i$ for some a_i; let $y_j := y(x_j) = \sum_{i=0}^{d} a_i x_j^i$. Then

$$\begin{aligned} r_n[y](x) - y(x) &= \sum_{j=0}^{n} \beta_j [y_j - y(x)] s_j(x) \Big/ \sum_{j=0}^{n} \beta_j s_j(x), \qquad (3)\\ &= \sum_{j=0}^{n} \beta_j s_j(x) \sum_{i=0}^{d} [a_i(x_j^i - x^i)] \Big/ \sum_{j=0}^{n} \beta_j s_j(x), \end{aligned}$$

where by (b) $\deg\left(\sum \beta_j s_j(x)\right) \leq n-d$. $r_n[y]-y$ vanishes at the $n+1$ nodes x_j; its numerator reads

$$\sum_{i=0}^{d} a_i \sum_{j} x_j^i \beta_j s_j(x) - \sum_{i=0}^{d} a_i x^i \sum_{j} \beta_j s_j(x).$$

Both terms are of degree at most n, the first by definition of the s_j's and second by assumption (b). Altogether, it is a polynomial of degree at most n with $n+1$ distinct zeros. Thus, $r[y] \equiv y$, which is (a).

To show the other direction of the equivalence, we note that, if r_n reproduces all polynomials of degree up to d, then the numerator of (3) vanishes, which, if the denominator does not have real zeros, implies

$$\sum_{j=0}^{n} \beta_j y_j s_j(x) = y(x) \sum_{j=0}^{n} \beta_j s_j(x)$$

for all $y(x)$, i.e., for all sets of coefficients a_i, $i = 0(1)d$. As the left-hand side is a polynomial of degree $\leq n$, the right-hand side must be one as well. Since some sets of a_i lead to polynomials of degree d, the degree p of the denominator $\sum_{j=0}^{n}\beta_j s_j(x)$ is at most $n-d$, which was to be proved. ■

An increase by one unit of the guaranteed minimal degree of exactness can thus be obtained through a decrease of the denominator degree by one unit. The following set of linear equalities to impose to the β_j's in order to lower the denominator degree from n to $n-d$ has been given in [10].

Theorem 2 *The denominator* $\ell(x)\sum_{j=0}^{n}\frac{\beta_j}{x-x_j}$ *of* r_n *in (2) has degree* $\leq n-d$ *iff*

$$\sum_{j=0}^{n} x_j^k \beta_j = 0, \qquad k = 0(1)d-1. \tag{4}$$

The hypothesis in Theorem 1 that the denominator of r_n does not vanish is not necessary for exactness.

Theorem 3 r_n *reproduces all polynomials of degree* $\leq d$ *iff its weights satisfy the relations (4).*

Proof The result obviously is true for $d=0$—any rational interpolant in barycentric form reproduces constants. It holds for $d=1$ [29]. Assume that it is true for every degree up to d. By linearity, r_n reproduces polynomials of degree $d+1$ iff it reproduces the monomial x^{d+1}, i.e., iff

$$\sum_{j=0}^{n}\frac{\beta_j}{x-x_j}x_j^{d+1}\Big/\sum_{j=0}^{n}\frac{\beta_j}{x-x_j} = x^{d+1} \quad \forall\, x$$

or

$$\sum_{j=0}^{n}\frac{\beta_j}{x-x_j}(x^{d+1}-x_j^{d+1})\Big/\sum_{j=0}^{n}\frac{\beta_j}{x-x_j} \equiv 0 \quad \forall\, x \neq x_j,\ \text{all } j.$$

The expression on the left of the slash equals

$$\sum_{j=0}^{n}\beta_j\sum_{k=0}^{d}x^{d-k}x_j^k = \sum_{k=0}^{d}x^{d-k}\sum_{j=0}^{n}\beta_j x_j^k.$$

By induction, $\sum_{j=0}^{n}\beta_j x_j^k = 0$, $k=0(1)d-1$, and the above expression vanishes for all $x \neq x_j$, all j, iff $\sum_{j=0}^{n}\beta_j x_j^d = 0$. ■

Corollary 1 *The Floater–Hormann weights with parameter d satisfy (4).*

Proof The rational interpolant with parameter d reproduces all polynomials of degree up to d by construction. ■

Equation (4) is a system of d homogeneous linear equations in the $n+1$ unknowns β_j. We suggested in [10] to bring it analytically to Gauss–Jordan form. Here, we shall use the latter to solve for the d unknown weights by back substitution, once the others have been fixed. The result is an upper triangular system with $n+d-2$ diagonals: For $i = 0(1)d-s$,

$$\sum_{j=i}^{n-d+i+1} a_{ij}\beta_j = 0, \qquad a_{ij} := \prod_{k=0}^{i-1}(x_j - x_k) \prod_{k=n-d+i+2}^{n} (x_j - x_k). \tag{5}$$

All these a_{ij} contain the same number $d+1$ of factors (empty products $\prod_{k=r}^{s}$ with $s < r$ are set to 1), which do not vanish since the x_j's are distinct. The matrix (a_{ij}) therefore has maximal rank. *Our method consists in choosing d free nodes and to determine their weights by* (5) *from* $n-d$ *fixed ones set to* $\beta_j = (-1)^j$, *i.e., so that they guarantee a degree of exactness at least d with the fixed weights.* This linear barycentric rational interpolation with guaranteed degree of exactness will be denoted by LBRIGDE, or simply GDE, in the sequel.

Applying back substitution to (5) yields

$$\beta_i = \Big(\sum_{j=i+1}^{n-d+i+1} a_{ij}\beta_j\Big)\Big/a_{ii}, \qquad i = d-1(-1)0. \tag{6}$$

This however requires that the known weights are located at the tail of the vector, which is not the case here, as d_r of those to be determined lie at the end. Note that modifying the system in order to use forward substitution would not help, as the d_ℓ first unknowns would then appear at the beginning, no favorable location either. For that reason, we exchange the nodes $x_{n-d_r+1}, \ldots, x_n$ with x_{d_ℓ} to $x_{d_\ell+d_r-1}$ before using formula (6)—the order of the nodes in (4) is irrelevant.

In fact, we do even more in practical computations. For large n, formula (5) involves many differences of x_j's, whose computation is affected by cancellation when the latter come close to one another. The points should be reordered [23, 40], what we do as in [10] with the van der Corput sequence prior to any computation. Since the last d_r elements are then far from each other, bringing them to position d_ℓ to $d_\ell + d_r - 1$ did not cause any difficulty in our examples, as documented in §5.

We conclude this section with two remarks:

(a) Equation (6) is a simple closed formula for the weights β_j. It takes into account the full information at the nodes, unlike some alternate strategies such as those using mock–Chebyshev nodes, see [22] for a recent reference. Moreover, the method does not require solving a least squares problem, in contrast, e.g., with those recently introduced in [1, 22] or [33].

(b) The cost of the method may be markedly lowered in symmetric cases. For instance, when the nodes lie symmetrically with respect to their center $(x_0 + x_n)/2$ and $d_\ell = d_r$, about half of the equations are automatically satisfied, and the number of unknowns of the remaining ones is reduced by half. We did not program this particular case with the van der Corput reordering of the nodes, though, and thus could not compute with large n.

3 The Lebesgue Constant of LBRIGDE

The norm of a linear approximation operator is its Lebesgue constant. When a Lagrange basis $\{b_j(x)\}$, i.e., a basis with the Lagrange property $\{b_j(x_i)\} = \delta_{ij}$, of the subspace of the approximants is known, the Lebesgue constant is the maximum of the Lebesgue function $\sum_{j=0}^{n} |b_j(x)|$, here [8]

$$\Lambda_n = \max_{a \le x \le b} \sum_{j=0}^{n} \left| \frac{\beta_j}{x - x_j} \right| \Big/ \left| \sum_{j=0}^{n} \frac{\beta_j}{x - x_j} \right|.$$

Bos, De Marchi, and Hormann [17] have shown that, with the original interpolant R_0 with weights $\beta_j = (-1)^j$, Λ_n grows merely logarithmically with n for equispaced nodes, as that of the polynomial interpolant at nearly optimal nodes. This holds also for most nodes used in practice [19], while with the FH weights and equispaced nodes an extra exponential factor 2^{d-1} appears in Λ_n [18].

Our computations with equispaced nodes given in Sect. 5.4 show the Lebesgue constant of GDE to be very close to that of FH interpolation for such points.

4 Accuracy of the Scheme

By itself, a high degree of exactness ρ is no guarantee for good convergence: The interpolating polynomial p_n from (1) has $\rho = n$ for any set of $n + 1$ nodes, but by Runge, it does not converge for all $f \in C^{\infty}[a, b]$. Good condition, as expressed by the Lebesgue constant, seems necessary.

We could not prove the convergence of GDE. In [9], it is shown that if the degree of the denominator is $n - d$, then

$$\|f - r_n\| \le (1 + \Lambda_n)\|f - r_d^*\|,$$

where r_d^* is the best approximation of f in the subspace of all linear barycentric rational interpolants with weights β, and also that (Theorem 2.1 of [9])

$$\|f - r_d^*\| \le \|f - p_d^*\|,$$

where p_d^* is the best polynomial approximation of degree at most d of f on $[a, b]$.

Since by Theorem 1, a degree of exactness d is equivalent to a denominator degree $n-d$, and since by Jackson's theorem V $\|f - p_d^*\| = \mathscr{O}(d^{-s})$ for $f \in C^s[a,b]$, one has $\|f - r_n\| \le (1+\Lambda_n)\mathscr{O}(d^{-s})$. But this does not lead to convergence, even with increasing d, as Λ_n may grow exponentially with d, as in the FH case for equispaced nodes.

Here, we are interested in a fixed d and increasing n. Our numerical tests lead us to the following conjecture.

Conjecture 1 *Let $f \in C[a,b]$. If the weights β_j in (2) are given by*

$$\beta_j = (-1)^j, \qquad d_\ell \le j \le n - d_r + 1,$$

and as the solution of (5) for $j = 0(1)d_\ell - 1$ and $j = n - d_r + 1(1)n$, then

$$|r_n(x) - f(x)| \le C \cdot h^{\min\{d_\ell, d_r\}+1}.$$

The correctness of this bound would mean that the number of weights devoted at each extremity to the endpoint correction, i.e., to the descent of the weights from the (absolute) value of 1, is decisive for the convergence order.

5 Numerical Results

We have tested the idea on several examples. Up to that in Sect. 5.5 with the non-symmetric nodes, the function we thereby used was $f(x) = \sin x + 1/(1+25x^2)$ on $[-1,1]$, i.e., a Runge function modified to make it not even. We first considered the classic, symmetric arrays of nodes.

5.1 Chebyshev Points of the Second Kind (Chebyshev–Gauss–Lobatto nodes)

We started with these nodes, as with them the polynomial weights λ_j then already are all $(-1)^j$ but the two extremal ones, which are $\lambda_0 = 1/2$ and $\lambda_n = (-1)^n/2$ (see the discussion of R_1 in the Introduction). As the nodes lie symmetrically about their center, we put $d_\ell = d_r$, i.e., $d = 2d_\ell = 2d_r$. Our results are displayed in Table 1.

The first column gives the number of interpolation nodes (which were taken odd in order to have one node at 0, the maximum of f) and the second, the error with the interpolating polynomial (which decays exponentially as to be expected [42, p. 57]). The errors were all computed as an approximation of the infinite norm, with the maximum taken among 999 values of the absolute error at equispaced evaluation points. The three pairs of columns each give, for the value of d displayed on top, the interpolation error as just explained and the infinite norm of the difference between

Table 1 Approximation results for Chebyshev points of the second kind

$n+1$	Polyn.	$d=2$		$d=4$		$d=10$	
	$\|p_n - f\|$	$\|r_n - f\|$	$\|\beta - \lambda\|$	$\|r_n - f\|$	$\|\beta - \lambda\|$	$\|r_n - f\|$	$\|\beta - \lambda\|$
11	$1.3e-01$	$1.3e-01$	$4.4e-16$	$1.3e-01$	$4.4e-15$	$1.3e-01$	$5.8e-15$
21	$1.8e-02$	$1.8e-02$	$1.8e-15$	$1.8e-02$	$3.2e-14$	$1.8e-02$	$3.2e-12$
41	$3.4e-04$	$3.4e-04$	$4.4e-15$	$3.4e-04$	$6.0e-13$	$3.4e-04$	$2.7e-10$
81	$1.2e-07$	$1.2e-07$	$2.0e-14$	$1.2e-07$	$2.7e-12$	$1.2e-07$	$8.1e-07$
161	$1.5e-14$	$8.9e-15$	$8.8e-13$	$1.6e-14$	$9.4e-11$	$1.5e-14$	$7.2e-04$

the vectors of the polynomial weights $\lambda := [\lambda_0, \lambda_1, \ldots, \lambda_n]^T$ and the rational weights $\beta := [\beta_0, \beta_1, \ldots, \beta_n]^T$.

The accuracy basically is the same for all d's and still exponential, therefore much faster than the expected $h^{d/2+1}$. Noticeable is the fact that for $d = 2$ the weights β of GDE still are about the same as the polynomial ones, but with increasing d they drift away from λ. This seems to indicate that there are, numerically, a large number of vectors of weights delivering a certain degree of exactness.

5.2 *Chebyshev Points of the First Kind (Chebyshev–Gauss Nodes)*

Here, the nodes are $x_j = \cos\phi_j$, $\phi_j := \left(\frac{2j+1}{n+1}\frac{\pi}{2}\right)$, and the (simplified) polynomial weights are given by $\lambda_j = (-1)^j \sin\phi_j$ [28, p. 249]. In a sense, they are best for polynomial interpolation, as they solve Chebyshev's famous min–max problem [26, p. 91]. Approximation results are given in Table 2, where, again, the second column gives the polynomial error and the next three the GDE errors with $d_\ell = d_r = 1, 2$ and 5, thus $d = 2, 4, 10$.

The polynomial error again decays exponentially, as expected from the theory [26, p. 92]. For $d = 2$, we obtain the same weights as those of the Chebyshev points of the second kind, which are also those of R_1. It is interesting to note that up to $n + 1 = 41$, the error does not improve by increasing d and also that for $d = 10$, the

Table 2 Approximation results for the Chebyshev points of the first kind

$n+1$	Polyn.	$d=2$	$d=4$	$d=10$
	$\|p_n - f\|$	$\|r_n - f\|$	$\|r_n - f\|$	$\|r_n - f\|$
11	$1.1e-01$	$1.1e-01$	$1.1e-01$	$1.1e-01$
21	$1.5e-02$	$1.5e-02$	$1.5e-02$	$1.5e-02$
41	$2.9e-04$	$2.8e-04$	$2.8e-04$	$2.8e-04$
81	$1.0e-07$	$2.1e-06$	$1.0e-07$	$9.8e-08$
161	$1.4e-14$	$2.8e-07$	$1.3e-10$	$1.3e-14$

convergence for large n is much faster than the $\mathcal{O}(h^6)$ to be expected: In fact, the accuracy is even slightly better than with the polynomial and the same number of nodes, which, let us recall, is exponential! One reason becomes apparent by checking the empirical degree of exactness: The first not guaranteed condition, the 11th, is still satisfied with an accuracy which is smaller than the maximum error. The practical degree of exactness therefore is much larger than the value of 10 guaranteed by formula (6).

Keep in mind that only d weights are different from ± 1 and need to be computed. All factors $\beta_j/(x - x_j)$ but those corresponding to the d modified weights therefore merely require the computation of the reciprocal of a number, in place of a division.

5.3 *Legendre Points*

Legendre nodes are very much in use as well, e.g., for the solution of fluid flow problems [25]. Here, we have taken a fixed value of d ($d_\ell = d_r = 2$, $d = 4$), but added some comparisons to the results with the FH weights. Table 3 summarizes our findings.

The second column gives the results for the polynomial, demonstrating the exponential convergence for such nodes [44, p. 868]: Every doubling of n squares the error. We could not evaluate the polynomial errors for n larger than about 1,000 with MATLAB on our computer (a MacBook Pro 2 GHz Intel Core i7), as we did not use recent methods for computing the Legendre nodes in $\mathcal{O}(n)$ operations incorporated in Chebfun [43], but this does not have any impact on our findings. We have determined numerically that R_1 does *not* yield exponential convergence with these nodes. The next three columns give the results with the FH weights for $d = 4$: first the error $\|r_n^{(4)} - f\|$, second the empirical degree of exactness, and third the empirical order of convergence; the last three columns give the corresponding values for GDE. The empirical orders of convergence are computed in the usual way [21, p. 18].

Table 3 Approximation results for Legendre points and $d = 4$

$n+1$	Polyn.	Floater–Hormann			LBRIGDE		
	$\|p_n - f\|$	$\|r_n^{(4)} - f\|$	ρ	Order	$\|r_n - f\|$	ρ	Order
11	$9.9e-02$	$8.0e-02$	4		$9.4e-02$	8	
21	$1.4e-02$	$1.1e-02$	4	3.06	$1.3e-02$	8	3.02
41	$3.2e-04$	$2.2e-04$	4	5.88	$2.5e-04$	8	5.93
81	$9.7e-08$	$9.2e-08$	4	11.4	$8.9e-08$	8	11.7
161	$1.7e-14$	$4.6e-10$	4	7.68	$1.5e-14$	10	22.7
321	$3.1e-15$	$1.6e-11$	8	4.88	$2.7e-15$	20	2.48
641	$4.0e-15$	$2.7e-11$	>20	0.12	$3.6e-15$	>20	5.00
1281	$8.0e-15$	$2.3e-10$	>20	−2.79	$5.0e-15$	>20	−0.49
2561	$1.6e-14$	$2.2e-09$	>20	−3.73	$8.4e-15$	>20	−0.76

Very interesting is the fact that all three interpolants behave in about the same manner until $n = 100$: The orders of convergence manifest themselves merely beyond this value. At least for this f, the FH weights give the best results until n about 60. Strangely, GDE is even better than the polynomial for all values of n. And after about $n = 60$, it is the best of the three, the reason being that the FH family becomes unstable. We think that this is a consequence of the accumulation of the nodes near the extremities of $[-1, 1]$: The FH formula for the weights suffers from cancellation; no reordering of the nodes seems possible, as they are determined by the interpolating polynomials corresponding to $d + 1$ consecutive nodes. The top picture in Fig. 1 displays the difference between the three interpolants in a graphical manner.

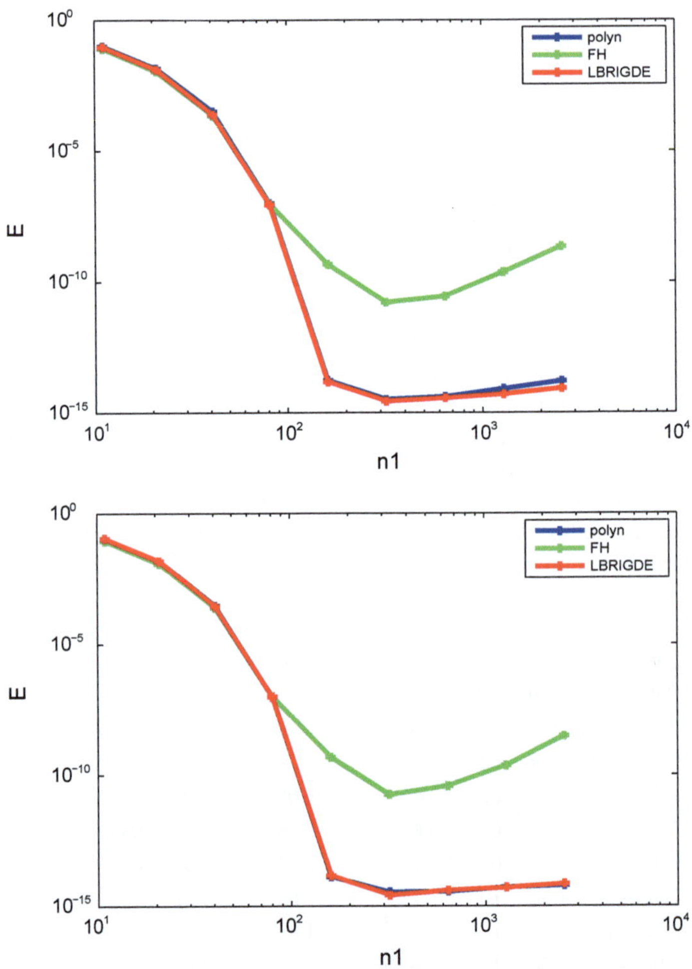

Fig. 1 Errors with various interpolants, *top* for Legendre points, *bottom* for Chebyshev points of the first kind

The labels are self-explanatory. A similar behavior of the interpolation with the FH weights may be observed for Chebyshev points of the first kind, see the bottom image in Fig. 1. (The Legendre points are cosines of angles close to equispaced ones, see [39, p. 118].)

Again, only four weights must be stored, those at both ends of β: All others are ± 1; most divisions in $\beta_j/(x-x_j)$ are replaced with reciprocals.

5.4 *Equispaced Points*

With all sets of nodes considered up to this point, the polynomial interpolant converges exponentially, so that linear barycentric rational interpolation is unnecessary in most applications (recall, however, that almost all weights of GDE are ± 1). The most important nodes with which LBRI brings a big improvement are the equispaced ones, for which the interpolating polynomial is ill-conditioned on large intervals and often diverges, see the Sect. 1.

Our results with equispaced nodes for the same example $f(x) = \sin x + \frac{1}{1+25x^2}$ as above are summarized in Table 4.

We have chosen FH with $d = 4$ and, in view of the symmetry of the nodes, $d_\ell = d_r = 4$. Since in some cases we have computed with more than 1,000 equispaced nodes, which could be close to the 999 equispaced evaluation points, we have increased the number of the latter to 9999. We have computed the FH weights with the special formula for equispaced points.

Here, FH is slightly better than GDE, despite the fact that the empirical degree of exactness is half as large. When n is large enough, but not so large that the error approaches machine precision, the empirical orders of convergence are quite precisely 5, which indeed corresponds to $d+1$ for FH and $d_\ell + 1 = d_r + 1$ for GDE.

Table 4 Approximation results for equispaced points

$n+1$	Floater–Hormann			LBRIGDE			$\Vert\beta^{\text{GDE}} - \beta^{\text{FH}}\Vert$
	$\Vert r_n^{(4)} - f\Vert$	ρ	Order	$\Vert r_n - f\Vert$	ρ	Order	
11	$1.26e-01$	4		$2.84e-01$	8		140.0
21	$5.30e-03$	4	4.90	$7.32e-03$	8	5.66	232.9
41	$9.96e-06$	4	9.38	$1.22e-05$	8	9.56	237.1
81	$6.22e-09$	4	10.8	$2.25e-08$	8	9.24	238.7
161	$1.74e-10$	4	5.21	$6.06e-10$	8	5.26	239.4
321	$5.15e-12$	4	5.10	$1.76e-11$	8	5.13	239.7
641	$1.58e-13$	4	5.03	$5.31e-13$	8	5.06	239.8
1281	$7.44e-15$	4	4.42	$1.78e-14$	10	4.91	239.9
2561	$8.99e-15$	4	−0.27	$9.44e-15$	18	0.91	240.0

Table 5 The Lebesgue constant Λ_n for equispaced nodes

$n+1$	Floater–Hormann		GDE	
	Λ_n^{FH}	$\dfrac{\Lambda_n^{\mathrm{FH}}}{\frac{2}{\pi}\ln(2n+2)}$	Λ_n^{GDE}	$\dfrac{\Lambda_n^{\mathrm{GDE}}}{\frac{2}{\pi}\ln(2n+2)}$
10	5.0795	2.6634	17.849	7.8469
20	7.0739	3.0122	11.026	4.0598
40	8.8246	3.1633	10.775	3.4129
80	10.479	3.2434	11.539	3.2067
160	12.090	3.2924	12.682	3.1393
320	13.681	3.3259	14.012	3.4064
640	15.261	3.3506	15.446	3.3912
1280	16.837	3.3700	16.940	3.3906
2560	18.400	3.3840	18.458	3.3946
5120	19.969	3.3969	19.943	3.3925
10240	21.375	3.3822	25.019	3.9588

Interesting, but not understood, is the fact that the infinite norm of the difference of the weights of FH and GDE (last column) seems to converge toward a value of about 240, despite the fact that the number of weights doubles from one row of the table to the next. The results show experimentally again that, for a given set of nodes, there are various sets of weights which deliver about the same order of convergence.

For such equispaced nodes, we have also compared the *Lebesgue constant* of GDE with that of the FH interpolant, for which theoretical results are known. Our results are given in Table 5. To demonstrate that the efficiency of the interpolants does not depend on the parity of n, we have this time chosen even numbers of nodes. The second and third columns give our results with the FH interpolant and $d = 4$: the second the value of Λ_n and the third the same value divided by the main term $\alpha_n := \dfrac{2}{\pi}\ln(2n+2)$ of its growth formula according to current knowledge [30], with the same d. The last two columns provide the corresponding numbers for GDE.

The Floater–Hormann constants Λ_n^{FH} are clearly smaller for small values of n, but the parallelism of the growth of Λ_n^{FH} and Λ_n^{GDE} with increasing n is baffling: For $n+1 = 5120$, $\Lambda_n^{\mathrm{FH}}/\alpha_n = 3.3969$ and $\Lambda_n^{\mathrm{GDE}}/\alpha_n = 3.3925$! Recall that this asymptotic behavior for equispaced points is about the same as that of polynomial interpolation with Chebyshev nodes [30].

Table 6 Approximation results for equidistant nodes and different values of d_ℓ and d_r

$n+1$	(5, 3)	Order	(6, 2)	Order	Order2
21	$4.31e-03$		$6.25e-03$		
41	$2.79e-05$	7.53	$2.55e-02$	−2.10	−2.10
81	$1.44e-06$	4.36	$5.22e-03$	2.33	0.13
161	$8.33e-08$	4.14	$1.12e-04$	5.60	1.98
321	$5.02e-09$	4.07	$3.87e-05$	1.53	1.86
641	$3.09e-10$	4.04	$3.25e-07$	6.92	2.89
1281	$1.91e-11$	4.02	$1.16e-07$	1.48	2.66
2561	$1.19e-12$	4.01	$2.71e-09$	5.43	3.04
5121	$7.85e-14$	3.92	$5.70e-08$	−4.40	2.23

5.5 An Example with $d_r \neq d_\ell$

In the above examples, we have always taken $d_r = d_\ell$. However, all one needs are d equations in d unknown weights, and one obtains such a system with any breakdown of d into d_ℓ and d_r. To test Conjecture 1, it is therefore natural to try other splits of d. We performed corresponding tests with equidistant nodes and $d = 8$, as in Table 4. More precisely, we successively increased d_ℓ and decreased d_r by one unit, thus keeping the sum at 8. Results are given in Table 6, where the combinations of d_ℓ and d_r are denoted by (d_ℓ, d_r).

The numbers should be compared with the same for the pair (4,4) and GDE in Table 4. The order of convergence for (5,3) clearly is four, indeed $\min\{d_\ell, d_r\} + 1$, in accordance with Conjecture 1. The results are the same with (3,5). The situation becomes more intricate for (6,2): The convergence is more erratic and so is its empirical order, as documented in the second last column. For that reason, we have estimated it a second time, by taking the case $n + 1 = 21$, that of the smallest number of points used, as a reference: The results appear as Order2 in the last column. Though not perfectly convincing, they are quite consistent with the order $d_r + 1 = 3$ forecast in Conjecture 1.

5.6 A Non-symmetric Example

Although symmetric nodes certainly are the most important, non-symmetric ones arise in practice; for instance, when their concentration in the vicinity of a front allows for a much better approximation of the solution of differential equations, see [4, 12] and [41]. With a conformal image of Chebyshev nodes of the second kind, R_1 retains the exponential convergence of the polynomial [3] and is thus much faster than FH and GDE; however, it maintains a strong concentration of nodes close to the extremities of the interval, which results in a bad conditioning of the derivatives there. For that reason, we have computed with other distributions, where the nodes are conformal maps of *equispaced* ones.

The map we chose is that of Bayliss and Turkel, with which we have made good experiences in [11]:

$$x = g(y) = \frac{1}{\alpha}\tan[\lambda(y-\mu)] + \beta, \tag{7}$$

where

$$\lambda = \frac{\gamma+\delta}{2}, \qquad \mu = \frac{\gamma-\delta}{\gamma+\delta}$$

with

$$\gamma = \arctan[\alpha(1+\beta)], \qquad \delta = \arctan[\alpha(1-\beta)];$$

in (5.1), y denotes the variable with the equispaced nodes $y_j = a + j\frac{b-a}{n}$. The computational nodes then are $x_j := g(y_j)$. For other and more recent maps, see, e.g., the examples and references in [31].

In our specific example, we have interpolated the function

$$f(x) = \frac{1}{1+200(x-(\sqrt{2}-1))^2} + \tanh\left(50(x-(\sqrt{2}-1))^2\right),$$

still between -1 and 1. The function is shown in Fig. 2. Since β stands for the location of the front in physical space, we naturally chose it as $\sqrt{2}-1$; α on its part was arbitrarily taken as 3 after a short trial and error.

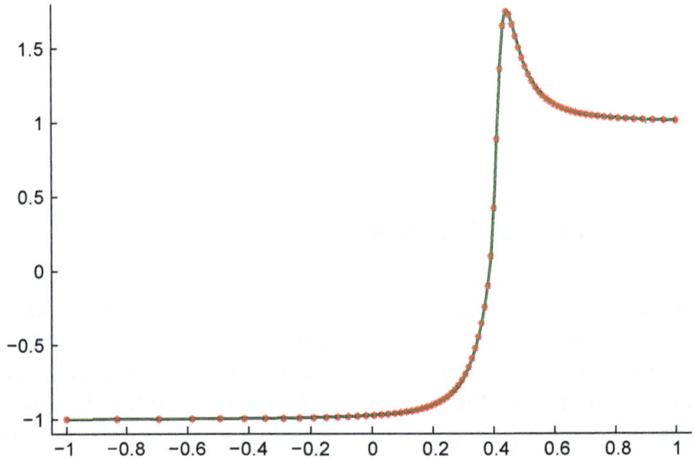

Fig. 2 Function of Example 5.6 with 80 shifted nodes and the corresponding GDE–approximation with $d_\ell = d_r = 4$

Table 7 Approximation results for shifted equispaced nodes

$n+1$	R_1	$\widehat{r}^{(8)}$	Floater–Hormann		GDE		
	$\|R_1 - f\|$	$\|\widehat{r}_n^{(8)} - f\|$	$\|r_n^{(8)} - f\|$	$\left\|\sum \beta_j^{(8)}\right\|$	$\|r_n - f\|$	$\left\|\sum \beta_j\right\|$	$\mathcal{O}$
10	$5.5e-01$	2.2	$1.5e+04$	$1.7e-10$	$4.9e+03$	$1.4e-11$	
20	$1.4e-01$	4.9	$6.4e+06$	$3.3e-08$	$6.7e+01$	$2.3e-11$	6.19
40	$1.8e-02$	2.0	$1.3e+07$	$1.6e-06$	1.4	$1.4e-12$	5.60
80	$3.8e-04$	$1.0e-01$	$1.5e+06$	$2.3e-05$	$2.7e-02$	$6.9e-13$	5.71
160	$1.0e-04$	$5.0e-03$	$1.8e+01$	$6.0e-04$	$4.4e-04$	$8.8e-13$	5.93
320	$2.5e-05$	$1.8e-06$	$2.5e-05$	$2.5e-03$	$2.1e-06$	$2.0e-12$	7.71
640	$6.3e-06$	$6.0e-14$	$2.1e-05$	$2.0e-03$	$5.3e-07$	$5.6e-12$	1.98
1280	$1.6e-06$	$1.1e-14$	$1.6e-05$	$3.3e-03$	$3.6e-08$	$1.3e-11$	3.89
2560	$3.9e-07$	$1.2e-14$	$6.1e-06$	$2.1e-03$	$4.7e-11$	$2.0e-11$	9.57
5120	$9.6e-08$	$1.8e-14$	$1.7e-05$	$2.5e-02$	$1.4e-12$	$1.1e-10$	5.06

The interpolating polynomial is useless with such nodes, in view of Runge's phenomenon. We have therefore compared GDE and FH with R_1. The errors with the latter are given in the second column of Table 7. Here, we may as well take an even number of nodes: The abscissa of maximal curvature is not in the center any longer. One clearly sees the error decay according to $\mathcal{O}(h^2)$, as the error is asymptotically divided by four with each doubling of the number of nodes.

The third column provides the results with FH interpolation at *(unshifted) equispaced nodes*, for the sake of comparison.

The other columns display the results with $d = 8$ for FH and $d_\ell = d_r = 4$ for GDE. For each of the two, the first column gives the maximum error and the second, the accuracy with which the first degree condition, the worst of the first eight, is satisfied (in fact, as $n + 1$ is even, FH mathematically interpolates exactly all polynomials of degree 9, see the Introduction). The last column, finally, gives the order of GDE.

Up to about 200 nodes, R_1 is the best. With these shifted nodes, GDE is always better than FH. Its faster convergence than that of R_1 shows after 200. FH does not like scarce nodes close to the edges. Its jump in accuracy at about 200 nodes is impressive; numerically, however, it does not improve below about $5 \cdot 10^{-5}$, one reason being that its weights are massively different in size and another that the order conditions are hardly satisfied, as exemplified by the fifth column of the table for the condition for degree $\leq n - 1$; with Corollary 1, this shows that the computation of the FH weights is unstable. GDE does not suffer from this kind of weakness. With the present example, it is the only one which brings the error below 10^{-8} with these shifted nodes and $n + 1 \leq 5120$. The above findings are illustrated in Fig. 3.

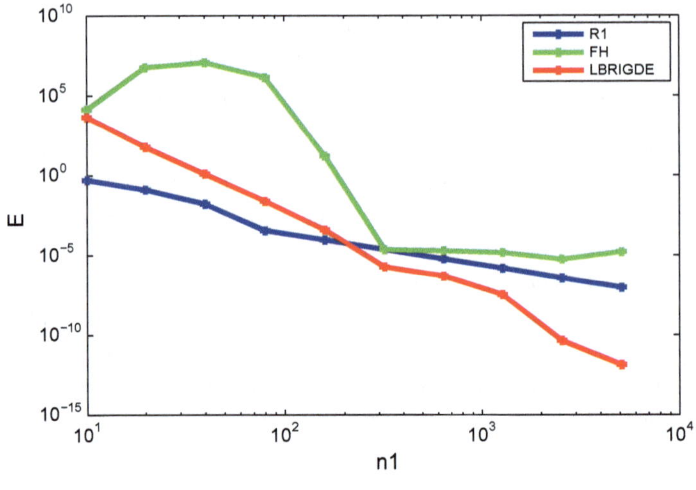

Fig. 3 Errors with Example 5.6 and various interpolants

6 Conclusions

In this work, we have demonstrated how solving a homogeneous system of equations for a few weights can massively improve upon the accuracy of the linear rational interpolants introduced in [6]. Complemented with a reordering of the nodes, the computation of the weights turns out to be impressively stable. In many examples, the conditions guaranteeing a certain degree of exactness do not require to be fulfilled to machine precision to yield an interpolant with that accuracy.

Once the weights are computed, a significant advantage of this family of interpolants is the fact that all weights are ± 1 besides the first and last: There is no need for memory to store the weights, and every term $\beta_j/(x - x_j)$ merely requires the computation of a reciproqual, no division. This is rarely crucial in one dimension, but will become important when going over to higher dimensions [7]. Moreover, these weights could improve the conditioning of the derivatives.

Acknowledgements The author is indebted to Nick Trefethen for an interesting discussion after a talk on this research at the 25th Numerical Analysis Conference in Glasgow.

References

1. B. Adcock, R.B. Platte, A mapped polynomial method for high-accuracy approximations on arbitrary grids. SIAM. J. Numer. Anal. **54**, 2256–2281 (2016)
2. R. Baltensperger, Some results on linear rational trigonometric interpolation. Comput. Math. Appl. **43**, 737–746 (2002)

3. R. Baltensperger, J.-P. Berrut, B. Noël, Exponential convergence of a linear rational interpolant between transformed Chebyshev points. Math. Comp. **68**, 1109–1120 (1999)
4. A. Bayliss, E. Turkel, Mappings and accuracy for Chebyshev pseudo-spectral approximations. J. Comput. Phys. **101**, 349–359 (1992)
5. J.-P. Berrut, Linear rational interpolation of continuous functions over an interval, in *Mathematics of Computation 1943–1993: A Half-Century of Computational Mathematics*, ed. by W. Gautschi. Proceedings of Symposia in Applied Mathematics (American Mathematical Society, Providence, 1994), pp. 261–264
6. J.-P. Berrut, Rational functions for guaranteed and experimentally well-conditioned global interpolation. Comput. Math. Appl. **15**, 1–16 (1988)
7. J.-P. Berrut, Equispaced linear barycentric rational interpolation in several dimensions, in preparation
8. J.-P. Berrut, H. Mittelmann, Lebesgue constant minimizing linear rational interpolation of continuous functions over the interval. Comput. Math. Appl. **33**, 77–86 (1997)
9. J.-P. Berrut, H. Mittelmann, Exponentially convergent linear rational interpolation between equidistant (and other) points. Meth. Appl. Anal. **4**, 67–76 (1997)
10. J.-P. Berrut, H. Mittelmann, Matrices for the direct determination of the barycentric weights of rational interpolation. J. Comput. Appl. Math. **78**, 355–370 (1997)
11. J.-P. Berrut, H.D. Mittelmann, Adaptive point shifts in rational approximation with optimized denominator. J. Comput. Appl. Math. **164–165**, 81–92 (2004)
12. J.-P. Berrut, H.D. Mittelmann, Optimized point shifts and poles in the linear rational pseudospectral method for boundary value problems. J. Comput. Phys. **204**, 292–301 (2005)
13. J.-P. Berrut, G. Klein, Recent advances in linear barycentric rational interpolation. J. Comput. Appl. Math. **259**, 95–107 (2014)
14. J.-P. Berrut, M.S. Floater, title not finalized, in preparation
15. J.-P. Berrut, R. Baltensperger, H.D. Mittelmann, Recent developments in barycentric rational interpolation, in *Trends and Applications in Constructive Approximation*, eds. by M.G. de Bruin, D.H. Mache, J. Szabados. International Series of Numerical Mathematics ISNM, vol. 151 (Birkhäuser, Basel, 2005), pp. 27–51
16. J.-P. Berrut, S.A. Hosseini, G. Klein, The linear barycentric rational quadrature method for Volterra integral equations. SIAM. J. Sci. Comput. **36**, A105–A123 (2014)
17. L. Bos, S. De Marchi, K. Hormann, On the Lebesgue constant of Berrut's rational interpolant at equidistant nodes. J. Comput. Appl. Math. **236**, 504–510 (2011)
18. L. Bos, S. De Marchi, K. Hormann, G. Klein, On the Lebesgue constant of barycentric rational interpolation at equidistant nodes. Numer. Math. **121**, 461–471 (2012)
19. L. Bos, S. De Marchi, K. Hormann, J. Sidon, Bounding the Lebesgue constant for Berrut's rational interpolant at general nodes. J. Approx. Theory **169**, 7–22 (2013)
20. G. Coman, R.T. Trimbitas, Combined Shepard univariate operators. East J. Approx. **7**, 471–483 (2001)
21. C. de Boor, *A Practical Guide to Splines*, Revised Edition. Applied Mathematical Sciences, vol. 27 (Springer, New York, 2001)
22. S. De Marchi, F. Dell'Accio, M. Mazza, On the constrained Mock–Chebyshev least squares. J. Comput. Appl. Math. **280**, 94–109 (2015)
23. B. Fischer, L. Reichel, Newton interpolation in Fejér and Chebyshev points. Math. Comp. **53**, 265–278 (1989)
24. M.S. Floater, K. Hormann, Barycentric rational interpolation with no poles and high rates of approximation. Numer. Math. **107**, 315–331 (2007)
25. D. Funaro, *Spectral Elements for Transport-Dominated Equations*. Lecture Notes in Computational Science and Engineering (Springer, Berlin, 1997)
26. W. Gautschi, *Numerical Analysis: An Introduction* (Birkhäuser, Boston, 1997)
27. S. Güttel, G. Klein, Convergence of linear barycentric rational interpolation for analytic functions. SIAM. J. Numer. Anal. **50**, 2560–2580 (2012)
28. P. Henrici, *Essentials of Numerical Analysis with Pocket Calculator Demonstrations* (Wiley, New York, 1982)

29. K. Hormann, Barycentric interpolation, in *Approximation Theory XIV: San Antonio 2013*, ed. by G.E. Fasshauer, L.L. Schumaker. Springer Proceedings in Mathematics & Statistics, vol. 83 (Springer, New York, 2014), pp. 197–218
30. B.A. Ibrahimoglu, A. Cuyt, Sharp bounds for Lebesgue constants of barycentric rational interpolation at equidistant points. Exp. Math. **25**, 347–354 (2016)
31. H.A. Jafari-Varzaneh, S.M. Hosseini, A new map for the Chebyshev pseudospectral solution of differential equations with large gradients. Numer. Algor. **69**, 95–108 (2015)
32. G. Klein, J.-P. Berrut, Linear barycentric rational quadrature. BIT **52**, 407–424 (2012)
33. Y. Nakatsukasa, O. Sète, L.N. Trefethen, The AAA algorithm for rational approximation, submitted
34. R.B. Platte, L.N. Trefethen, A.B.J. Kuijlaars, Impossibility of fast stable approximation of analytic functions from equispaced samples. SIAM Rev. **53**, 308–318 (2011)
35. H. Rutishauser, *Vorlesungen über Numerische Mathematik*, vol. 1 (Birkhäuser, Basel-Stuttgart, 1976)
36. H.R. Schwarz, *Numerische Mathematik* 4te Aufl., Teubner, 1997; English translation of the second edition: *Numerical Analysis. A Comprehensive Introduction* (Wiley, New York, 1989)
37. C. Schneider, W. Werner, Some new aspects of rational interpolation. Math. Comp. **47**, 285–299 (1986)
38. J. Stoer, *Einführung in die Numerische Mathematik I.4.Aufl* (Springer, Berlin, 1983)
39. G. Szegö, *Orthogonal Polynomials*, vol. XXIII (AMS Colloquium Publications, Providence, 1939) (1978)
40. H. Tal-Ezer, High degree polynomial interpolation in Newton form. SIAM. J. Sci. Stat. Comput. **12**, 648–667 (1991)
41. T.W. Tee, L.N. Trefethen, A rational spectral collocation method with adaptively transformed Chebyshev grid points. SIAM. J. Sci. Comput. **28**, 1798–1811 (2006)
42. L.N. Trefethen, *Approximation Theory and Approximation Practice* (SIAM, Philadelphia, 2013)
43. L.N. Trefethen, others, Chebfun Version 4.2 The Chebfun Development Team (2011), http://www.maths.ox.ac.uk/chebfun/
44. H. Wang, S. Xiang, On the convergence rates of Legendre approximation. Math. Comp. **81**, 861–877 (2012)
45. W. Werner, Polynomial interpolation: Lagrange versus Newton. Math. Comp. **43**, 205–217 (1984)

Approximation by C^1 Splines on Piecewise Conic Domains

Oleg Davydov and Wee Ping Yeo

Abstract We develop a Hermite interpolation scheme and prove error bounds for C^1 bivariate piecewise polynomial spaces of Argyris type vanishing on the boundary of curved domains enclosed by piecewise conics.

Keywords Curved elements · C^1 elements · Bivariate splines

1 Introduction

Spaces of piecewise polynomials defined on domains bounded by piecewise algebraic curves and vanishing on parts of the boundary can be used in the finite element method as an alternative to the classical mapped curved elements [10, 12]. Since implicit algebraic curves and surfaces provide a well-known modeling tool in CAGD [1], these methods are inherently isogeometric in the sense of [14]. Moreover, this approach does not suffer from the usual difficulties of building a globally C^1 or smoother space of functions on curved domains (see [4, Sect. 4.7]) shared by the classical curved finite elements and the B-spline-based isogeometric analysis.

In particular, a space of C^1 piecewise polynomials on domains enclosed by piecewise conic sections has been studied in [10] and applied to the numerical solution of fully nonlinear elliptic equations. These piecewise polynomials are quintic on the interior triangles of a triangulation of the domain and sextics on the boundary triangles (pie-shaped triangles with one side represented by a conic section as well as those triangles that share with them an interior edge with one endpoint on the boundary) and generalize the well-known Argyris finite element. Although local bases for

O. Davydov (✉)
Department of Mathematics, University of Giessen, Arndtstrasse 2,
35392 Giessen, Germany
e-mail: oleg.davydov@math.uni-giessen.de

W.P. Yeo
Faculty of Science, Universiti Brunei Darussalam, Gadong BE1410,
Brunei Darussalam
e-mail: weeping.yeo@ubd.edu.bn

G.E. Fasshauer and L.L. Schumaker (eds.), *Approximation Theory XV: San Antonio 2016*, Springer Proceedings in Mathematics & Statistics 201,
DOI 10.1007/978-3-319-59912-0_2

these spaces have been constructed in [10] and numerical examples demonstrated the convergence orders expected from a piecewise quintic finite element, no error bounds have been provided.

In this paper, we study the approximation properties of the spaces introduced in [10]. We define a Hermite-type interpolation operator and prove an error bound that shows the convergence order $\mathcal{O}(h^6)$ of the residual in L_2-norm and order $\mathcal{O}(h^{6-k})$ in Sobolev spaces $H^k(\Omega)$. This extends the techniques used in [12] for C^0 splines to Hermite interpolation.

The paper is organized as follows. We introduce in Sect. 2 the spaces $\mathbb{S}^{1,2}_{d,0}(\Delta)$ of C^1 piecewise polynomials on domains bounded by a number of conic sections, with homogeneous boundary conditions, define in Sect. 3 our interpolation operator in the case $d = 5$, and investigate in Sect. 4 its approximation error for functions in Sobolev spaces $H^m(\Omega)$, $m = 5, 6$, vanishing on the boundary.

2 C^1 Piecewise Polynomials on Piecewise Conic Domains

We make the same assumptions on the domain and its triangulation as in [10, 12], as outlined below.

Let $\Omega \subset \mathbb{R}^2$ be a bounded curvilinear polygonal domain with $\Gamma = \partial\Omega = \bigcup_{j=1}^{n} \overline{\Gamma}_j$, where each Γ_j is an open arc of an algebraic curve of at most second order (i.e., either a straight line or a conic). For simplicity, we assume that Ω is simply connected, so that its boundary Γ is a closed curve without self-intersections. Let $Z = \{z_1, \ldots, z_n\}$ be the set of the endpoints of all arcs numbered counterclockwise such that z_j, z_{j+1} are the endpoints of Γ_j, $j = 1, \ldots, n$, with $z_{j+n} = z_j$. Furthermore, for each j, we denote by ω_j the internal angle between the tangents τ_j^+ and τ_j^- to Γ_j and Γ_{j-1}, respectively, at z_j. We assume that $\omega_j \in (0, 2\pi)$ for all j. Hence, Ω is a Lipschitz domain.

Let Δ be a *triangulation* of Ω, i.e., a subdivision of Ω into triangles, where each triangle $T \in \Delta$ has at most one edge replaced with a curved segment of the boundary $\partial\Omega$, and the intersection of any pair of the triangles is either a common vertex or a common (straight) edge if it is non-empty. The triangles with a curved edge are said to be *pie-shaped*. Any triangle $T \in \Delta$ that shares at least one edge with a pie-shaped triangle is called a *buffer* triangle, and the remaining triangles are *ordinary*. We denote by Δ_0, Δ_B, and Δ_P the sets of all ordinary, buffer, and pie-shaped triangles of Δ, respectively, such that $\Delta = \Delta_0 \cup \Delta_B \cup \Delta_P$ is a disjoint union, see Fig. 1. Let $V, E, V_I, E_I, V_\partial, E_\partial$ denote the set of all vertices, all edges, interior vertices, interior edges, boundary vertices, and boundary edges, respectively.

For each $j = 1, \ldots, n$, let $q_j \in \mathbb{P}_2$ be a polynomial such that $\Gamma_j \subset \{x \in \mathbb{R}^2 : q_j(x) = 0\}$, where $\mathbb{P}_d$ denotes the space of all bivariate polynomials of total degree at most d. By changing the sign of q_j if needed, we ensure that $q_j(x)$ is positive for points in Ω near the boundary segment Γ_j. For simplicity, we assume in this paper that all boundary segments Γ_j are curved. Hence, each q_j is an irreducible quadratic polynomial and

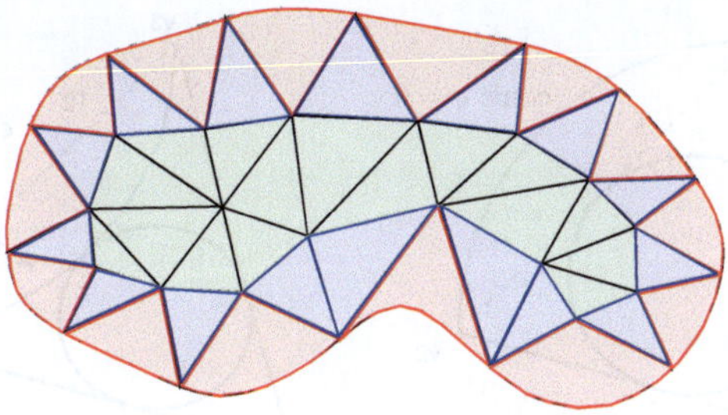

Fig. 1 A triangulation of a curved domain with ordinary triangles (*green*), pie-shaped triangles (*pink*), and buffer triangles (*blue*)

$$\nabla q_j(x) \neq 0 \quad \text{if} \quad x \in \Gamma_j. \tag{1}$$

We assume that $\triangle$ satisfies the following conditions:

(A) $Z = \{z_1, \ldots, z_n\} \subset V_\partial$.
(B) No interior edge has both endpoints on the boundary.
(C) No pair of pie-shaped triangles shares an edge.
(D) Every $T \in \triangle_P$ is star-shaped with respect to its interior vertex v.
(E) For any $T \in \triangle_P$ with its curved side on Γ_j, $q_j(z) > 0$ for all $z \in T \setminus \Gamma_j$.
(F) No pair of buffer triangles shares an edge.

It can be easily seen that (B) and (C) are achievable by a slight modification of a given triangulation, while (D) and (E) hold for sufficiently fine triangulations. The assumption (F) is made for the sake of simplicity of the analysis. Note that the triangulation shown in Fig. 1 does not satisfy (F).

For any $T \in \triangle$, let h_T denote the diameter of T, and let ρ_T be the radius of the disk B_T inscribed in T if $T \in \triangle_0 \cup \triangle_B$ or in $T \cap T^*$ if $T \in \triangle_P$, where T^* denotes the triangle obtained by joining the boundary vertices of T by a straight line, see Fig. 2. Note that every triangle $T \in \triangle$ is star-shaped with respect to B_T. In particular, for $T \in \triangle_P$, this follows from Condition (D) and the fact that the conics do not possess inflection points.

We define the *shape regularity constant* of $\triangle$ by

$$R = \max_{T \in \triangle} \frac{h_T}{\rho_T}. \tag{2}$$

For any $d \geq 1$, we set

$$\mathbb{S}^1_d(\triangle) := \{s \in C^1(\Omega) \,:\, s|_T \in \mathbb{P}_d,\ T \in \triangle_0, \ \text{ and } s|_T \in \mathbb{P}_{d+1},\ T \in \triangle_P \cup \triangle_B\},$$
$$\mathbb{S}^{1,2}_{d,I}(\triangle) := \{s \in \mathbb{S}^1_d(\triangle) \,:\, s \text{ is twice differentiable at any } v \in V_I\},$$
$$\mathbb{S}^{1,2}_{d,0}(\triangle) := \{s \in \mathbb{S}^{1,2}_{d,I}(\triangle) \,:\, s|_\Gamma = 0\}.$$

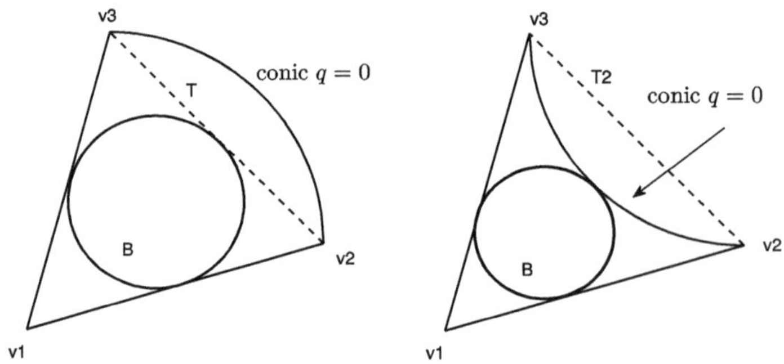

Fig. 2 A pie-shaped triangle with a curved edge and the associated triangle T^* with straight sides and vertices v_1, v_2, v_3. The curved edge can be either outside (*left*) or inside T^* (*right*)

We refer to [10] for the construction of a local basis for the space $\mathbb{S}^{1,2}_{5,0}(\triangle)$ and its applications in the finite element method.

Our goal is to obtain an error bound for the approximation of functions vanishing on the boundary by splines in $\mathbb{S}^{1,2}_{5,0}(\triangle)$. This is done through the construction of an interpolation operator of Hermite type. Note that a method of *stable splitting* was employed in [6–8] to estimate the approximation power of C^1 splines vanishing on the boundary of a polygonal domain. C^1 finite element spaces with a stable splitting are also required in Böhmer's proofs of the error bounds for his method of numerical solution of fully nonlinear elliptic equations [2]. A stable splitting of the space $\mathbb{S}^{1,2}_{5,I}(\triangle)$ will be obtained if a stable local basis for a stable complement of $\mathbb{S}^{1,2}_{5,0}(\triangle)$ in $\mathbb{S}^{1,2}_{5,I}(\triangle)$ is constructed, which we leave to a future work.

3 Interpolation Operator

We denote by $\partial^\alpha f$, $\alpha \in \mathbb{Z}^2_+$, the partial derivatives of f and consider the usual Sobolev spaces $H^m(\Omega)$ with the seminorm and norm defined by

$$|f|^2_{H^m(\Omega)} = \sum_{|\alpha|=m} \|\partial^\alpha f\|^2_{L^2(\Omega)}, \quad \|f\|^2_{H^m(\Omega)} = \sum_{k=0}^{m} |f|^2_{H^k(\Omega)} \quad (H^0(\Omega) = L^2(\Omega)),$$

where $|\alpha| := \alpha_1 + \alpha_2$. We set $H^1_0(\Omega) = \{f \in H^1(\Omega) : f|_{\partial\Omega} = 0\}$.

In this section, we construct an interpolation operator $I_\triangle : H^5(\Omega) \cap H^1_0(\Omega) \to \mathbb{S}^{1,2}_{5,0}(\triangle)$ and estimate its error for the functions in $H^m(\Omega) \cap H^1_0(\Omega)$, $m = 5, 6$, in the next section.

As in [12], we choose domains $\Omega_j \subset \Omega$, $j = 1, \dots, n$, with Lipschitz boundary such that

(a) $\partial\Omega_j \cap \partial\Omega = \Gamma_j$,

(b) $\partial\Omega_j \setminus \partial\Omega$ is composed of a finite number of straight line segments,
(c) $q_j(x) > 0$ for all $x \in \overline{\Omega}_j \setminus \Gamma_j$, and
(d) $\Omega_j \cap \Omega_k = \emptyset$ for all $j \neq k$.

In addition, we assume that the triangulation $\triangle$ is such that

(e) $\overline{\Omega}_j$ contains every triangle $T \in \triangle_P$ whose curved edge is part of Γ_j

and that q_j satisfy (without loss of generality)

(f) $\max_{x\in\overline{\Omega}_j} \|\nabla q_j(x)\|_2 \leq 1$ and $\|\nabla^2 q_j\|_2 \leq 1$, for all $j = 1, \ldots, n$,

where $\nabla^2 q_j$ denotes the (constant) Hessian matrix of q_j.

Note that (e) will hold with the same set $\{\Omega_j : j = 1, \ldots, n\}$ for any triangulations obtained by subdividing the triangles of $\triangle$.

The following lemma can be shown following the lines of the proof of [13, Theorem 6.1], see also [12, Theorem 3.1].

Lemma 1 *There is a constant K depending only on Ω, the choice of Ω_j, $j = 1, \ldots, n$, and $m \geq 1$, such that for all j and $u \in H^m(\Omega) \cap H_0^1(\Omega)$,*

$$|u/q_j|_{H^{m-1}(\Omega_j)} \leq K \|u\|_{H^m(\Omega_j)}. \tag{3}$$

Given a unit vector $\tau = (\tau_x, \tau_y)$ in the plane, we denote by D_τ the directional derivative operator in the direction of τ in the plane, so that

$$D_\tau f := \tau_x D_x f + \tau_y D_y f, \quad D_x f := \partial f/\partial x, \quad D_y f := \partial f/\partial y.$$

Given $f \in C^{\alpha+\beta}(\triangle)$, $\alpha, \beta \geq 0$, any number

$$\eta f = D_{\tau_1}^{\alpha} D_{\tau_2}^{\beta} (f|_T)(z),$$

where $T \in \triangle$, $z \in T$, and τ_1, τ_2 are some unit vectors in the plane, is said to be a *nodal value* of f, and the linear functional $\eta : C^{\alpha+\beta}(\triangle) \to \mathbb{R}$ is a *nodal functional*, with $d(\eta) := \alpha + \beta$ being the *degree* of η.

For some special choices of z, τ_1, τ_2, we use the following notation:

- If v is a vertex of $\triangle$ and e is an edge attached to v, we set

$$D_e^{\alpha} f(v) := D_\tau^{\alpha} (f|_T)(v), \quad \alpha \geq 1,$$

  where τ is the unit vector in the direction of e away from v, and $T \in \triangle$ is one of the triangles with edge e.
- If v is a vertex of $\triangle$ and e_1, e_2 are two consecutive edges attached to v, we set

$$D_{e_1}^{\alpha} D_{e_2}^{\beta} f(v) := D_{\tau_1}^{\alpha} D_{\tau_2}^{\beta} (f|_T)(v), \quad \alpha, \beta \geq 1,$$

where $T \in \Delta$ is the triangle with vertex v and edges e_1, e_2, and τ_i is the unit vector in the e_i direction away from v.

- For every edge e of the triangulation Δ, we choose a unit vector $\tau^\perp$ (one of two possible) orthogonal to e and set

$$D^\alpha_{e^\perp} f(z) := D^\alpha_{\tau^\perp} f(z), \quad z \in e, \quad \alpha \geq 1,$$

provided $f \in C^\alpha(z)$.

On every edge e of Δ, with vertices v' and v'', we define three points on e by

$$z_e^j := v' + \frac{j}{4}(v'' - v'), \quad j = 1, 2, 3.$$

For every triangle $T \in \Delta_0$ with vertices v_1, v_2, v_3 and edges e_1, e_2, e_3, we define $\mathcal{N}_T^0$ to be the set of nodal functionals corresponding to the nodal values

$$D_x^\alpha D_y^\beta f(v_i), \quad 0 \leq \alpha + \beta \leq 2, \quad i = 1, 2, 3,$$

$$D_{e_i^\perp} f(z_{e_i}^2), \quad i = 1, 2, 3,$$

see Fig. 3 (left), where the nodal functionals are depicted in the usual way by dots, segments, and circles as, for example, in [5].

Let $T \in \Delta_P$. We define $\mathcal{N}_T^P$ to be the set of nodal functionals corresponding to the nodal values

$$D_x^\alpha D_y^\beta f(v_1), \quad 0 \leq \alpha + \beta \leq 2,$$

$$D_x^\alpha D_y^\beta f(v_i), \quad 0 \leq \alpha + \beta \leq 1, \quad i = 2, 3,$$

$$D_x^\alpha D_y^\beta f(c_T), \quad 0 \leq \alpha + \beta \leq 1,$$

where v_1 is the interior vertex of T, v_2, v_3 are boundary vertices, and c_T is the center of the disk B_T, see Fig. 4.

Let $T \in \Delta_B$ with vertices v_1, v_2, v_3. We define $\mathcal{N}_T^{B,1}$ to be the set of nodal functionals corresponding to the nodal value

$$f(c_T), \quad c_T := (v_1 + v_2 + v_3)/3.$$

Also, we define $\mathcal{N}_T^{B,2}$ to be the set of nodal functionals corresponding to the nodal values

$$f(z_{e_i}^2), \quad i = 1, 2, 3,$$

$$D_x^\alpha D_y^\beta f(v_i), \quad 0 \leq \alpha + \beta \leq 2, \quad i = 1, 2, 3,$$

$$D_{e_i^\perp} f(z_{e_i}^j), \quad j = 1, 3, \quad i = 1, 2, 3,$$

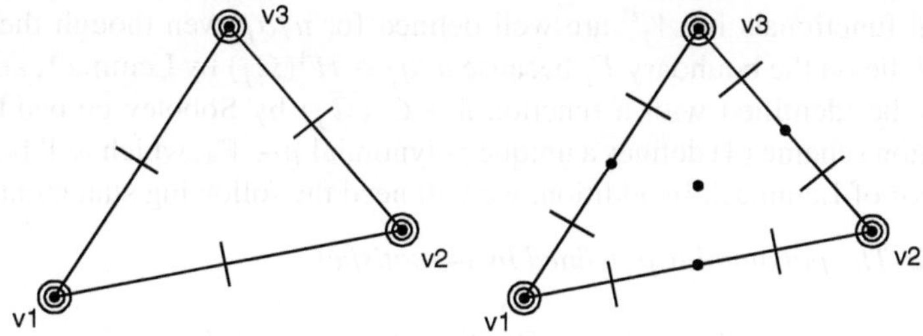

Fig. 3 Nodal functionals corresponding to $\mathcal{N}_T^0$ (*left*) and $\mathcal{N}_T^B$ (*right*)

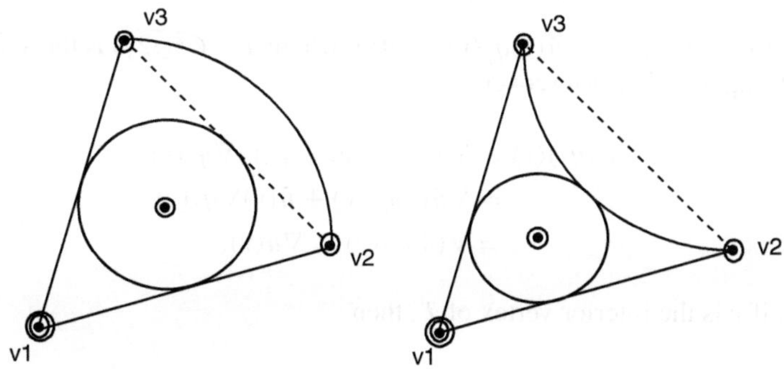

Fig. 4 Nodal functionals corresponding to $\mathcal{N}_T^P$

where v_1 is the boundary vertex, and v_2, v_3 are the interior vertices of T. We set

$$\mathcal{N}_T^B := \mathcal{N}_T^{B,1} \cup \mathcal{N}_T^{B,2},$$

see Fig. 3 (right).

We define an operator $I_\triangle : H^5(\Omega) \cap H_0^1(\Omega) \to \mathbb{S}_{5,0}^{1,2}(\triangle)$ of interpolatory type. Let $u \in H^5(\Omega) \cap H_0^1(\Omega)$. By Sobolev embedding, we assume without loss of generality that $u \in C^3(\overline{\Omega})$. For all $T \in \triangle_0 \cup \triangle_P$, we set $I_\triangle u|_T = I_T(u|_T)$, with the local operators I_T defined as follows.

If $T \in \triangle_0$, then $p := I_T u$ is the polynomial of degree 5 that satisfies the following interpolation conditions:

$$\eta p = \eta u, \quad \text{for all } \eta \in \mathcal{N}_T^0.$$

This is a well-known Argyris interpolation scheme, see, e.g., [15, Sect. 6.1], which guarantees the existence and uniqueness of the polynomial p.

Let $T \in \triangle_P$ with the curved edge on Γ_j. Then, $I_T u := p q_j$, where $p \in \mathbb{P}_4$ satisfies the following interpolation condition:

$$\eta p = \eta(u/q_j), \quad \text{for all } \eta \in \mathcal{N}_T^P. \tag{4}$$

The nodal functionals in $\mathcal{N}_T^P$ are well defined for u/q_j even though the vertices v_2, v_3 of T lie on the boundary Γ_j because $u/q_j \in H^4(\Omega_j)$ by Lemma 1, and hence, u/q_j may be identified with a function $\tilde{u} \in C^2(\overline{\Omega}_j)$ by Sobolev embedding. The interpolation scheme (4) defines a unique polynomial $p \in \mathbb{P}_4$, which will be justified in the proof of Lemma 3. In addition, we will need the following statement.

Lemma 2 *The polynomial p defined by* (4) *satisfies*

$$D_x^\alpha D_y^\beta (pq_j)(v) = D_x^\alpha D_y^\beta u(v), \quad 0 \le \alpha + \beta \le 2,$$

where v is any vertex of the pie-shaped triangle T.

Proof By (4), $p(v)q_j(v) = \tilde{u}(v)q_j(v) = u(v)$, where $\tilde{u} \in C^2(\overline{\Omega}_j)$ is the above function satisfying $u = \tilde{u}q_j$. Moreover,

$$\begin{aligned}\nabla(pq_j)(v) &= \nabla p(v)q_j(v) + p(v)\nabla q_j(v)\\ &= \nabla\tilde{u}(v)q_j(v) + \tilde{u}(v)\nabla q_j(v)\\ &= \nabla(\tilde{u}q_j)(v) = \nabla u(v).\end{aligned}$$

Similarly, if v is the interior vertex of T, then

$$\begin{aligned}\nabla^2(pq_j)(v) &= \nabla^2 p(v)q_j(v) + \nabla p(v)(\nabla q_j(v))^T + \nabla q_j(v)(\nabla p(v))^T + p(v)\nabla^2 q_j(v)\\ &= \nabla^2\tilde{u}(v)q_j(v) + \nabla\tilde{u}(v)(\nabla q_j(v))^T + \nabla q_j(v)(\nabla\tilde{u}(v))^T + \tilde{u}(v)\nabla^2 q_j(v)\\ &= \nabla^2 u(v).\end{aligned}$$

If v is one of the boundary vertices, then $q_j(v) = 0$, and hence,

$$\begin{aligned}\nabla^2(pq_j)(v) &= \nabla p(v)(\nabla q_j(v))^T + \nabla q_j(v)(\nabla p(v))^T + p(v)\nabla^2 q_j(v)\\ &= \nabla\tilde{u}(v)(\nabla q_j(v))^T + \nabla q_j(v)(\nabla\tilde{u}(v))^T + \tilde{u}(v)\nabla^2 q_j(v)\\ &= \nabla^2 u(v).\end{aligned}$$

□

It is easy to deduce from Lemma 2 that the interpolation conditions for p at the boundary vertices v_2, v_3 of T can be equivalently formulated as follows: For $i = 2, 3$,

$$\begin{gathered} p(v_i) = \frac{\partial u}{\partial n_i}(v_i) \Big/ \frac{\partial q_j}{\partial n_i}(v_i),\\ \frac{\partial p}{\partial n_i}(v_i) = \frac{1}{2}\frac{\partial^2 u}{\partial n_i^2}(v_i) \Big/ \frac{\partial q_j}{\partial n_i}(v_i), \qquad \frac{\partial p}{\partial \tau_i}(v_i) = \frac{\partial^2 u}{\partial n_i \partial \tau}(v_i) \Big/ \frac{\partial q_j}{\partial n_i}(v_i), \end{gathered} \tag{5}$$

where n_i and τ_i are the normal and the tangent unit vectors to the curve $q_j(x) = 0$ at v_i.

Finally, assume that $T \in \triangle_B$ with vertices v_1, v_2, v_3 where v_1 is a boundary vertex. Then, $I_T u = p \in \mathbb{P}_6$ satisfies the following interpolation conditions:

$$\eta p = \eta u, \quad \text{for all } \eta \in \mathcal{N}_T^{B,1},$$

and

$$\eta p = \eta I_{T_i} u, \quad \text{for all } \eta \in \mathcal{N}_i \subset \mathcal{N}_T^{B,2}, \quad i = 1, 2, 3,$$

where T_1 is a triangle in Δ_0 sharing an edge $e_1 = \langle v_2, v_3 \rangle$ with T, and $\mathcal{N}_1$ corresponds to the nodal values

$$f(z_{e_1}^2), \quad D_{e_1^\perp} f(z_{e_1}^i), \quad i = 1, 3,$$

$$D_x^\alpha D_y^\beta f(v_i), \quad 0 \le \alpha + \beta \le 2, \quad i = 2, 3;$$

T_2 is a triangle in Δ_P sharing an edge $e_2 = \langle v_1, v_2 \rangle$ with T, and $\mathcal{N}_2$ corresponds to the nodal values

$$f(z_{e_2}^2), \quad D_{e_2^\perp} f(z_{e_2}^i), \quad i = 1, 3,$$

$$D_x^\alpha D_y^\beta f(v_1), \quad 0 \le \alpha + \beta \le 2;$$

and T_3 is a triangle in Δ_P sharing an edge $e_3 = \langle v_1, v_3 \rangle$ with T, and $\mathcal{N}_3$ corresponds to the nodal values

$$f(z_{e_3}^2), \quad D_{e_3^\perp} f(z_{e_3}^i), \quad i = 1, 3.$$

Since $\mathcal{N}_T^{B,2} = \mathcal{N}_1 \cup \mathcal{N}_2 \cup \mathcal{N}_3$ and $\mathcal{N}_T^B = \mathcal{N}_T^{B,1} \cup \mathcal{N}_T^{B,2}$ are a well posed interpolation scheme [16] for polynomials of degree 6, it follows that p is uniquely defined by the above conditions.

Theorem 1 *Let $u \in H^5(\Omega) \cap H_0^1(\Omega)$. Then $I_\Delta u \in \mathbb{S}_{5,0}^{1,2}(\Delta)$.*

Proof By the above construction, $I_\Delta u$ is a piecewise polynomial of degree 5 on all triangles in Δ_0 and degree 6 on the triangles in $\Delta_P \cup \Delta_B$. Moreover, $I_\Delta u$ vanishes on the boundary of Ω.

To see that $I_\Delta u \in \mathbb{S}_{5,0}^{1,2}(\Delta)$ we thus need to show the C^1 continuity of $I_\Delta u$ across all interior edges of Δ. If e is a common edge of two triangles $T', T'' \in \Delta_0$, then the C^1 continuity follows from the standard argument for C^1 Argyris finite element, see [4, Chap. 3] and [15, Sect. 6.1].

Next, we will show the C^1 continuity of $I_\Delta u$ across edges shared by buffer triangles with either ordinary or pie-shaped triangles. Let $T \in \Delta_B$ and $T' \in \Delta_0 \cup \Delta_P$ with common edge $e' = \langle v', v'' \rangle$, and let $p = I_T u$ and $s = I_{T'} u$. Consider the univariate polynomials $p|_{e'}$ and $s|_{e'}$, and let $q = p|_{e'} - s|_{e'}$. Assuming that the edge e' is parameterized by $t \in [0, 1]$, then q is a univariate polynomial of degree 6 with respect to the parameterization $v' + t(v'' - v')$, $t \in [0, 1]$. Similarly, we consider the orthogonal/normal derivatives $D_{e'^\perp} p|_{e'}$ and $D_{e'^\perp} s|_{e'}$, and let $r = D_{e'^\perp} p|_{e'} - D_{e'^\perp} s|_{e'}$; then, r is a univariate polynomial of degree 5 with respect to the same parameter t. The C^1 continuity will follow if we show that both q and r are zero functions.

If $T' = T_1 \in \Delta_0$, then using the interpolation conditions corresponding to $\mathcal{N}_1 \subset \mathcal{N}_T^{B,2}$, we have

$$q(0) = q'(0) = q''(0) = q(1/2) = q(1) = q'(1) = q''(1) = 0,$$
$$r(0) = r'(0) = r(1/4) = r(3/4) = r(1) = r'(1) = 0,$$

which implies $q \equiv 0$ and $r \equiv 0$.

If $T' = T_2 \in \Delta_P$, then the interpolation conditions corresponding to $\mathcal{N}_2 \subset \mathcal{N}_T^{B,2}$ imply

$$q(0) = q'(0) = q''(0) = q(1/2) = 0,$$
$$r(0) = r'(0) = r(1/4) = r(3/4) = 0.$$

In view of Lemma 2, we have

$$D_x^\alpha D_y^\beta s(v_2) = D_x^\alpha D_y^\beta u(v_2) = D_x^\alpha D_y^\beta p(v_2), \quad 0 \le \alpha + \beta \le 2,$$

which implies

$$q(1) = q'(1) = q''(1) = 0, \quad r(1) = r'(1) = 0,$$

and hence, $q \equiv 0$ and $r \equiv 0$.

If $T' = T_3 \in \Delta_P$, then the interpolation conditions corresponding to $\mathcal{N}_3 \subset \mathcal{N}_T^{B,2}$ imply

$$q(1/2) = 0, \quad r(1/4) = r(3/4) = 0,$$

whereas Lemma 2 gives

$$q(0) = q'(0) = q''(0) = 0, \quad r(0) = r'(0) = 0,$$
$$q(1) = q'(1) = q''(1) = 0, \quad r(1) = r'(1) = 0,$$

which completes the proof. □

It follows from Lemma 2 that $I_\triangle u$ is twice differentiable at the boundary vertices, and thus,

$$I_\triangle u \in \{s \in \mathbb{S}_5^1(\triangle) \;:\; s \text{ is twice differentiable at any vertex and } s|_\Gamma = 0\}.$$

Moreover, $I_\triangle u$ satisfies the following interpolation conditions:

$$D_x^\alpha D_y^\beta I_\triangle u(v) = D_x^\alpha D_y^\beta u(v), \quad 0 \le \alpha + \beta \le 2, \quad \text{for all } v \in V,$$

$$D_{e^\perp} I_\triangle u(z_e^2) = D_{e^\perp} u(z_e^2), \quad \text{for all edges } e \text{ of } \Delta_0,$$

$$D_x^\alpha D_y^\beta I_\triangle u(c_T) = D_x^\alpha D_y^\beta u(c_T), \quad 0 \le \alpha + \beta \le 1, \quad \text{for all } T \in \triangle_P,$$
$$I_\triangle u(c_T) = u(c_T), \quad \text{for all} T \in \triangle,$$

where c_T denotes the center of the disk B_T inscribed into T^* if T is a pie-shaped triangle and the barycenter of T if T is a buffer triangle. In view of (5), $I_\triangle u \in \mathbb{S}_{5,0}^{1,2}(\triangle)$ is uniquely defined by these conditions for any $u \in C^2(\overline{\Omega})$.

4 Error Bounds

In this section, we estimate the error $\|u - I_\triangle u\|_{H^k(\Omega)}$ for functions $u \in H^m(\Omega) \cap H_0^1(\Omega)$, $m = 5, 6$. Similar to [12, Sect. 3], we follow the standard finite element techniques involving the Bramble–Hilbert Lemma (see [4, Chap. 4]) on the ordinary triangles and make use of the estimate (3) on the pie-shaped triangles. Since the spline $I_\triangle u$ on the buffer triangles is constructed in part by interpolation and in part by the smoothness conditions, the estimate of the error on such triangles relies in particular on the estimates of the interpolation error on the neighboring ordinary and buffer triangles.

Lemma 3 *If $p \in \mathbb{P}_4$ and $T \in \triangle_P$, then*

$$\|p|_{T^*}\|_{L^\infty(T^*)} \le \max_{\eta \in \mathcal{N}_T^P} h_{T^*}^{d(\eta)} |\eta p|, \tag{6}$$

where T^* is the triangle obtained by replacing the curved edge of T by the straight line segment, and h_{T^*} is the diameter of T^*. Similarly, if $p \in \mathbb{P}_6$ and $T \in \triangle_B$, then

$$\|p|_T\|_{L^\infty(T)} \le \max_{\eta \in \mathcal{N}_T^B} h_T^{d(\eta)} |\eta p|, \tag{7}$$

where h_T is the diameter of T.

Proof To show the estimate (6) for T^*, we follow the proof of [11, Lemma 3.9]. We note that we only need to show that the interpolation scheme for pie-shaped triangles is a valid scheme, that is, we need to show that $\mathcal{N}_T^P$ is $\mathbb{P}_4$-unisolvent, and the rest of the proof can be done similar to that of [11, Lemma 3.9]. Recall that a set of functionals $\mathcal{N}$ are said to be $\mathbb{P}_d$-*unisolvent* if the only polynomial $p \in \mathbb{P}_d$ satisfying $\eta p = 0$ for $\eta \in \mathcal{N}$ is the zero function.

Let $T^* = \langle v_1, v_2, v_3 \rangle$, where v_1 is the interior vertex. Set $e_1 := \langle v_1, v_2 \rangle$, $e_2 := \langle v_2, v_3 \rangle$, $e_3 := \langle v_3, v_1 \rangle$, see Fig. 4. The interpolation conditions along e_1, e_3 imply that s vanishes on these edges. After splitting out the linear polynomial factors which vanish along the edges e_1, e_3, we obtain a valid interpolation scheme for quadratic polynomials with function values at the three vertices, and function and gradient values at the the barycenter c of $B_T \subset T^*$. The validity of this scheme can be seen by looking at a straight line ℓ through c and any one of the vertices of T^*. Along the line ℓ, a function value is given at the vertex and a function value together with the first derivative is given at the point c, and this set of data are unisolvent for the univariate quadratic polynomials, which means s must vanish along ℓ. After factoring out the respective linear polynomial, we are left with function values at three non-collinear points, which defines a valid interpolation scheme for the remaining linear polynomial factor of s.

To show the estimate (7) for $T \in \triangle_B$, the proof is similar. We need to show the set of functionals $\mathcal{N}_T^B$ is $\mathbb{P}_6$-unisolvent but this follows from the standard scheme of [16] for polynomials of degree six.

We note that the argument of the proof of [11, Lemma 3.9] applies to affine invariant interpolation schemes, that is the schemes that use the edge derivatives. As our scheme relies on the standard derivatives in the direction of the x, y axes, we need to express the edge derivatives as linear combinations of the x, y derivatives as follows. Assume that e_1, e_2 are two edges that emanate from a vertex v. Let $\tau_i = (\tau_{i1}, \tau_{i2})$ be the unit vector in the direction of e_i away from v, $i = 1, 2$. Then, we can easily obtain the following identities

$$D_{e_i} f(v) = \tau_{i1} D_x f(v) + \tau_{i2} D_y f(v),$$

$$D_{e_i}^2 f(v) = \tau_{i1}^2 D_x^2 f(v) + 2\tau_{i1}\tau_{i2} D_x D_y f(v) + \tau_{i2}^2 D_y^2 f(v),$$

$$D_{e_1} D_{e_2} f(v) = \tau_{11}\tau_{21} D_x^2 f(v) + (\tau_{11}\tau_{22} + \tau_{12}\tau_{21}) D_x D_y f(v) + \tau_{12}\tau_{22} D_y^2 f(v).$$

□

Lemma 4 *Let $T \in \triangle_P$ and its curved edge $e \subset \Gamma_j$. Then*

$$\|I_T u\|_{L^\infty(T)} \le C_1 \max_{0\le \ell \le 2} h_T^{\ell+1} |u/q_j|_{W_\infty^\ell(T)} \quad \text{if } u \in H^5(\Omega) \cap H_0^1(\Omega), \tag{8}$$

where C_1 depends only on h_T/ρ_T. Moreover, if $5 \le m \le 6$, then for any $u \in H^m(\Omega) \cap H_0^1(\Omega)$,

$$\|u - I_T u\|_{H^k(T)} \le C_2 h_T^{m-k} |u/q_j|_{H^{m-1}(T)}, \quad k = 0, \ldots, m-1, \tag{9}$$

$$|u - I_T u|_{W_\infty^k(T)} \le C_3 h_T^{m-k-1} |u/q_j|_{H^{m-1}(T)}, \quad k = 0, \ldots, m-2, \tag{10}$$

where C_2, C_3 depend only on h_T/ρ_T.

Proof We will denote by $\tilde{C}$ constants which may depend only on h_T/ρ_T and on Ω. Assume that $u \in H^5(\Omega) \cap H_0^1(\Omega)$ and recall that by definition $I_T u = pq_j$, where $p \in \mathbb{P}_4$ satisfies the interpolation conditions (4). Since $u \in H^5(\Omega_j) \cap H_0^1(\Omega_j)$, it follows that $u/q_j \in H^4(\Omega_j)$ by Lemma 1, and hence, $u/q_j \in C^2(\overline{\Omega}_j)$ by Sobolev embedding. From Lemma 3, we have

$$\|p\|_{L^\infty(T^*)} \le \max_{\eta\in\mathcal{N}_T^P} h_{T^*}^{d(\eta)}|\eta p|, \tag{11}$$

and hence

$$\|p\|_{L^\infty(T^*)} \le \max_{\eta\in\mathcal{N}_T^P} h_{T^*}^{d(\eta)}|\eta(u/q_j)| \le \tilde{C} \max_{0\le\ell\le 2} h_T^\ell |u/q_j|_{W_\infty^\ell(T)}.$$

As in the proof of [12, Theorem 3.2], we can show that for any polynomial of degree at most 6,

$$\|s\|_{L^\infty(T)} \le \tilde{C}\|s\|_{L^\infty(T^*)} \quad \text{and} \quad \|s\|_{L^\infty(T^*)} \le \tilde{C}\|s\|_{L^\infty(T)}. \tag{12}$$

By using (f), it is easy to show that $\|q_j\|_{L^\infty(T)} \le h_T$, and hence,

$$\|I_T u\|_{L^\infty(T)} = \|pq_j\|_{L^\infty(T)} \le h_T\|p\|_{L^\infty(T)},$$

which completes the proof of (8).

Moreover, since the area of T is less than or equal to $\frac{\pi}{4}h_T^2$ and $\partial^\alpha(I_T u) \in \mathbb{P}_{6-k}$ if $|\alpha| = k$, it follows that

$$\|\partial^\alpha(I_T u)\|_{L^2(T)} \le \frac{\sqrt{\pi}}{2} h_T \|\partial^\alpha(I_T u)\|_{L^\infty(T)} \le \tilde{C} h_T \|\partial^\alpha(I_T u)\|_{L^\infty(T^*)}.$$

By Markov inequality (see, e.g., [15, Theorem 1.2]), we get furthermore

$$\|\partial^\alpha(I_T u)\|_{L^\infty(T^*)} \le \tilde{C}\rho_T^{-k}\|I_T u\|_{L^\infty(T^*)},$$

and hence in view of (12)

$$|I_T u|_{H^k(T)} \le \tilde{C} h_T^{1-k}\|I_T u\|_{L^\infty(T)}.$$

In view of (8), we arrive at

$$|I_T u|_{H^k(T)} \le \tilde{C} \max_{0\le\ell\le 2} h_T^{\ell+2-k}|u/q_j|_{W_\infty^\ell(T)}, \quad \text{if } u \in H^5(\Omega)\cap H_0^1(\Omega). \tag{13}$$

Let $m \in \{5, 6\}$, and let $u \in H^m(\Omega) \cap H_0^1(\Omega)$. It follows from Lemma 1 that $u/q_j \in H^{m-1}(T)$. By the results in [4, Chap. 4], there exists a polynomial $\tilde{p} \in \mathbb{P}_{m-2}$ such that

$$\begin{aligned} \|u/q_j - \tilde{p}\|_{H^k(T)} &\le \tilde{C} h_T^{m-k-1}|u/q_j|_{H^{m-1}(T)}, \quad k = 0, \ldots, m-1, \\ |u/q_j - \tilde{p}|_{W_\infty^k(T)} &\le \tilde{C} h_T^{m-k-2}|u/q_j|_{H^{m-1}(T)}, \quad k = 0, \ldots, m-2. \end{aligned} \tag{14}$$

Indeed, a suitable $\tilde{p}$ is given by the *averaged Taylor polynomial* [4, Definition 4.1.3] with respect to the disk B_T, and the inequalities in (14) follow from [4, Lemma 4.3.8]

(Bramble–Hilbert Lemma) and an obvious generalization of [4, Proposition 4.3.2], respectively. It is easy to check by inspecting the proofs in [4] that the quotient h_T/ρ_T can be used in the estimates instead of the chunkiness parameter used there.

Since

$$u - I_T u = (u/q_j - \tilde{p})q_j - I_T(u - \tilde{p}q_j),$$

we have for any norm $\|\cdot\|$,

$$\|u - I_T u\| \le \|(u/q_j - \tilde{p})q_j\| + \|I_T(u - \tilde{p}q_j)\|.$$

In view of (f) and (14), for any $k = 0, \dots, m-2$,

$$\begin{aligned}|(u/q_j - \tilde{p})q_j|_{W^k_\infty(T)} &\le h_T |u/q_j - \tilde{p}|_{W^k_\infty(T)} + \|u/q_j - \tilde{p}\|_{W^{k-1}_\infty(T)}\\ &\le \tilde{C} h_T^{m-k-1} |u/q_j|_{H^{m-1}(T)},\end{aligned}$$

and for any $k = 0, \dots, m-1$,

$$\begin{aligned}\|(u/q_j - \tilde{p})q_j\|_{H^k(T)} &\le \tilde{C} h_T \|u/q_j - \tilde{p}\|_{H^k(T)} + \tilde{C}\|u/q_j - \tilde{p}\|_{H^{k-1}(T)}\\ &\le \tilde{C} h_T^{m-k} |u/q_j|_{H^{m-1}(T)}.\end{aligned}$$

Furthermore, by the Markov inequality, (8), (13), and (14),

$$|I_T(u - \tilde{p}q_j)|_{W^k_\infty(T)} \le \tilde{C} \max_{0\le\ell\le 2} h_T^{\ell+1-k} |u/q_j - \tilde{p}|_{W^\ell_\infty(T)} \le \tilde{C} h_T^{m-k-1} |u/q_j|_{H^{m-1}(T)},$$

$$\|I_T(u - \tilde{p}q_j)\|_{H^k(T)} \le \tilde{C} \max_{0\le\ell\le 2} h_T^{\ell+2-k} |u/q_j - \tilde{p}|_{W^\ell_\infty(T)} \le \tilde{C} h_T^{m-k} |u/q_j|_{H^{m-1}(T)}.$$

By combining the inequalities in the five last displays, we deduce (9) and (10). □

We are ready to formulate and prove our main result.

Theorem 2 *Let $5 \le m \le 6$. For any $u \in H^m(\Omega) \cap H^1_0(\Omega)$,*

$$\Big(\sum_{T\in\triangle} \|u - I_\triangle u\|^2_{H^k(T)}\Big)^{1/2} \le C h^{m-k} \|u\|_{H^m(\Omega)}, \quad k = 0, \dots, m-1, \qquad (15)$$

where h is the maximum diameter of the triangles in $\triangle$, and C is a constant depending only on Ω, the choice of Ω_j, and the shape regularity constant R of $\triangle$.

Proof We estimate the norms of $u - I_T u$ on all triangles $T \in \triangle$. The letter C stands below for various constants depending only on the parameters mentioned in the formulation of the theorem.

If $T \in \triangle_0$, then $s|_T$ is a macroelement as defined in [15, Chap. 6]. Furthermore, by [15, Theorem 6.3], the set of linear functionals $\mathcal{N}^0_T$ give rise to a stable local nodal basis, which is in particular uniformly bounded. Hence, by [9, Theorem 2], we obtain a Jackson estimate in the form

$$\|u - I_T u\|_{H^k(T)} \le C h_T^{m-k} |u|_{H^m(T)}, \quad k = 0, \dots, m, \tag{16}$$

where C depends only on h_T/ρ_T. If $T \in \Delta_P$, with the curved edge $e \subset \Gamma_j$, then the Jackson estimate (9) holds by Lemma 4.

Let $T \in \Delta_B$, $p := I_\Delta u|_T$, and let $\tilde{p} \in \mathbb{P}_6$ be the interpolation polynomial that satisfies $\eta \tilde{p} = \eta u$ for all $\eta \in \mathscr{N}_T^B$. Then

$$\eta(\tilde{p} - p) = \begin{cases} 0 & \text{if } \eta \in \mathscr{N}_T^{B,1}, \\ \eta(u - I_{T'} u) & \text{if } \eta \in \mathscr{N}_T^{B,2}, \end{cases}$$

where $T' = T'_\eta \in \Delta_0 \cup \Delta_P$. Hence, by Markov inequality and (7) of Lemma 3, we conclude that for $k = 0, \dots, m$,

$$\|\tilde{p} - p\|_{H^k(T)} \le C h_T^{1-k} \|\tilde{p} - p\|_{L^\infty(T)},$$

with

$$\|\tilde{p} - p\|_{L^\infty(T)} \le C \max\{h_T^\ell |u - I_{T'} u|_{W^\ell_\infty(T')} : 0 \le \ell \le 2,\ T' \in \Delta_0 \cup \Delta_P,\ T' \cap T \ne \emptyset\},$$

whereas by the same arguments leading to (16) we have

$$\|u - \tilde{p}\|_{H^k(T)} \le C h_T^{m-k} |u|_{H^m(T)},$$

with the constants depending only on h_T/ρ_T. If $T' \in \Delta_0 \cup \Delta_P$, then by (10) and the analogous estimate for $T' \in \Delta_0$, comparing [4, Corollary 4.4.7], we have for $\ell = 0, 1, 2$,

$$|u - I_{T'} u|_{W^\ell_\infty(T')} \le C h_{T'}^{m-\ell-1} \begin{cases} |u|_{H^m(T')} & \text{if } T' \in \Delta_0, \\ |u/q_j|_{H^{m-1}(T')} & \text{if } T' \in \Delta_P, \end{cases}$$

where C depends only on $h_{T'}/\rho_{T'}$. By combining these inequalities, we obtain an estimate of $\|u - I_T u\|_{H^k(T)}$ by $C\tilde{h}^{m-k}$ times the maximum of $|u|_{H^m(T)}$, $|u|_{H^m(T')}$ for $T' \in \Delta_0$ sharing edges with T, and $|u/q_j|_{H^{m-1}(T')}$ for $T' \in \Delta_P$ sharing edges with T. Here, C depends only on the maximum of h_T/ρ_T and $h_{T'}/\rho_{T'}$, and $\tilde{h}$ is the maximum of h_T and all $h_{T'}$ for $T' \in \Delta_0 \cup \Delta_P$ sharing edges with T.

By using (16) on $T \in \Delta_0$, (9) on $T \in \Delta_P$ and the estimate of the last paragraph on $T \in \Delta_B$, we get

$$\sum_{T \in \Delta} \|u - I_\Delta u\|^2_{H^k(T)} \le C h^{2(m-k)} \Big(\sum_{T \in \Delta_0 \cup \Delta_B} |u|^2_{H^m(T)} + \sum_{T \in \Delta_P} |u/q_{j(T)}|^2_{H^{m-1}(T)} \Big),$$

where $j(T)$ is the index of Γ_j containing the curved edge of $T \in \triangle_P$. Clearly,

$$\sum_{T \in \triangle_0 \cup \triangle_B} |u|^2_{H^m(T)} \leq |u|^2_{H^m(\Omega)} \leq \|u\|^2_{H^m(\Omega)},$$

whereas by Lemma 1,

$$\sum_{T \in \triangle_P} |u/q_{j(T)}|^2_{H^{m-1}(T)} \leq \sum_{j=1}^{n} |u/q_j|^2_{H^{m-1}(\Omega_j)} \leq K \|u\|^2_{H^m(\Omega)},$$

where K is the constant of (3) depending only on Ω and the choice of Ω_j. □

Acknowledgements This research has been supported in part by the grant UBD/PNC2/2/RG/1(301) from Universiti Brunei Darussalam.

References

1. J. Bloomenthal et al., *Introduction to Implicit Surfaces* (Morgan-Kaufmann Publishers Inc., San Francisco, 1997)
2. K. Böhmer, On finite element methods for fully nonlinear elliptic equations of second order. SIAM J. Numer. Anal. **46**, 1212–1249 (2008)
3. K. Böhmer, *Numerical Methods for Nonlinear Elliptic Differential Equations: A Synopsis* (Oxford University Press, Oxford, 2010)
4. S.C. Brenner, L.R. Scott, *The Mathematical Theory of Finite Element Methods* (Springer, New York, 1994)
5. P.G. Ciarlet, *The Finite Element Method for Elliptic Problems* (North-Holland, Amsterdam, 1978)
6. O. Davydov, Smooth finite elements and stable splitting, Berichte "Reihe Mathematik" der Philipps-Universität Marburg, 2007-4 (2007). An adapted version has appeared as [3, Sect. 4.2.6]
7. O. Davydov, A. Saeed, Stable splitting of bivariate spline spaces by Bernstein-Bézier methods, in *Curves and Surfaces - 7th International Conference, Avignon, France, June 24–30, 2010*, eds. by J.-D. Boissonnat et al. LNCS, vol. 6920 (Springer, Berlin, 2012), pp. 220–235
8. O. Davydov, A. Saeed, Numerical solution of fully nonlinear elliptic equations by Böhmer's method. J. Comput. Appl. Math. **254**, 43–54 (2013)
9. O. Davydov, W.P. Yeo, Macro-element hierarchical Riesz bases, in *Mathematical Methods for Curves and Surfaces: 8th International Conference, Oslo, 2012*, eds. by M. Floater et al. LNCS, vol. 8177 (Springer, Berlin, 2014), pp. 112–134
10. O. Davydov, A. Saeed, C^1 quintic splines on domains enclosed by piecewise conics and numerical solution of fully nonlinear elliptic equations. Appl. Numer. Math. **116**, 172–183 (2017)
11. O. Davydov, G. Nürnberger, F. Zeilfelder, Bivariate spline interpolation with optimal approximation order. Constr. Approx. **17**, 181–208 (2001)
12. O. Davydov, G. Kostin, A. Saeed, Polynomial finite element method for domains enclosed by piecewise conics. CAGD **45**, 48–72 (2016)
13. K. Höllig, U. Reif, J. Wipper, Weighted extended B-spline approximation of Dirichlet problem. SIAM J. Numer. Anal. **39**(2), 442–462 (2001)

14. T.J.R. Hughes, J.A. Cottrel, Y. Bazilevs, Isogeometric analysis: CAD, finite elements, NURBS, exact geometry and mesh refinement. Comput. Methods Appl. Mech. Eng. **194**, 4135–4195 (2005)
15. M.J. Lai, L.L. Schumaker, *Spline Functions on Triangulations* (Cambridge University Press, Cambridge, 2007)
16. L.L. Schumaker, On super splines and finite elements. SIAM J. Numer. Anal. **26**(4), 997–1005 (1989)

A Rescaled Method for RBF Approximation

Stefano De Marchi, Andrea Idda and Gabriele Santin

Abstract In the recent paper [1], a new method to compute stable kernel-based interpolants has been presented. This *rescaled interpolation* method combines the standard kernel interpolation with a properly defined rescaling operation, which smooths the oscillations of the interpolant. Although promising, this procedure lacks a systematic theoretical investigation. Through our analysis, this novel method can be understood as standard kernel interpolation by means of a properly rescaled kernel. This point of view allows us to consider its error and stability properties.

Keywords Radial basis functions · Rescaled interpolation · Partition of unit method · Lebesgue constant

1 Introduction

In the last decades, radial basis functions have shown to be a flexible mathematical tool to solve scattered data interpolation problems, to model neural networks, a mesh-free method for solving differential equations, parametrizing shape and surfaces and so on. Interested readers can refer, for example, to the comprehensive monographs [2–4], that give the necessary theoretical background and discuss many of the applications here enumerated.

In the paper [1] has been presented an interpolation method for the construction of stable kernel-based interpolants, called *rescaled interpolation*. It is a consistent

S. De Marchi (✉)
Department of Mathematics "Tullio Levi-Civita", University of Padova, Padova, Italy
e-mail: demarchi@math.unipd.it

A. Idda
Banco Popolare, Verona, Italy
e-mail: a.idda1989@gmail.com

G. Santin
IANS University of Stuttgart, Stuttgart, Germany
e-mail: gabriele.santin@mathematik.uni-stuttgart.de

G.E. Fasshauer and L.L. Schumaker (eds.), *Approximation Theory XV: San Antonio 2016*, Springer Proceedings in Mathematics & Statistics 201,
DOI 10.1007/978-3-319-59912-0_3

local method that combines the standard kernel interpolation with a properly defined rescaling operation, which essentially smooths the oscillations of the interpolant. Although promising, the method lacks a systematic theoretical understanding.

After recalling some necessary notation, we present the method and prove that it is an instance of the well-known *Shepard's method*, when certain weight functions are used. In particular, as for the Shepard's one, it reproduces constant functions.

Second, it is possible to define a modified set of cardinal functions strictly related to the ones of the not-rescaled kernel. Through these functions, we define a Lebesgue function for the rescaled interpolation process and study its maximum—the Lebesgue constant—in different settings.

Also, a preliminary theoretical result on the estimation of the interpolation error is presented.

As an application, we couple our method with a partition of unity algorithm. This setting seems to be the most promising, and we illustrate its behavior with some experiments.

We summarize briefly the paper structure. In the next section, we introduce some basic definitions useful to understand the results presented in the paper. Then, in successive Sect. 3, we present the *rescaled localized* RBF interpolant and discuss some of its properties. In particular, in the successive Sect. 3.1, we present the kernel-based approach to the rescaled interpolant, formalizing some results already presented in [1]. In the case of kernels depending on the shape parameter, an interesting property of the method is that the shape parameter can be chosen neither too small or too big. In Sect. 4, we show that this interpolant is indeed a *Shepard's approximant* and so it reproduces constants functions "at glance." Stability results of this construction are detailed in the successive subsection. The most promising application is the partition of unity method (PUM). In Sect. 5, we apply the rescaled interpolant to the PUM, showing then in the numerical experiments its effectiveness. We will show that the combination of PUM with the rescaled interpolant provides a stable method for interpolation.

2 Useful Notations

We start by recalling some notations useful for the sequel and necessary to understand the results that we are going to present.

Given a real Hilbert space $\mathscr{H}$ of functions from $\mathbb{R}^d$ to $\mathbb{R}$ with inner product $\langle\ ,\ \rangle_{\mathscr{H}}$, a function $K : \Omega \times \Omega \longrightarrow \mathbb{R}$ with $\Omega \subset \mathbb{R}^d$ is called *reproducing kernel* for $\mathscr{H}$ if the two properties hold:

(i) $K(\cdot, x) \in \mathscr{H}$ for all $x \in \Omega$;
(ii) $\langle f, K(\cdot, x)\rangle_{\mathscr{H}} = f(x)$ for all $f \in \mathscr{H}$ and $x \in \Omega$.

The kernel is symmetric (by property (ii)) and positive definite by construction (cf. e.g., [5, §2.3] or [4, §13.1]).

Let $\Phi : \mathbb{R}^d \to \mathbb{R}$ be a continuous function. Φ is called *radial* if there exists a continuous function $\varphi : \mathbb{R}_0^+ \to \mathbb{R}$ such that $\Phi(x) = \varphi(\|x\|)$.

We are interested in *radial kernels*, i.e., kernels of the form $K(x, y) = \Phi(x, y) = \varphi(\|x - y\|)$, which means *invariant under translation and rotation*, with $\|\cdot\|$ the Euclidean distance. Radial kernels are often referred to as *Radial Basis Functions* or shortly RBF. Moreover, since φ is a univariate function, we can call it the *basic radial function* and its values are function of a positive variable $r = \|x\|$.

Examples of $\mathscr{C}^\infty$ kernels are given in Table 1 and kernels with finite smoothness in Table 2. An interesting review of nonstandard kernels is the paper [6] which presents many examples of kernels besides the classical ones presented in the Tables 1 and 2. In the univariate case a special family of kernels with compact support was presented in [7]. We have written the kernels by introducing the parameter $\varepsilon > 0$ which is simply a *shape parameter*: for $\varepsilon \to \infty$, the functions become more and more spiky while as $\varepsilon \to 0$ they become flatter (see Fig. 1). In the case of Wendland's kernels, the parameter d denotes the maximal space dimension for which the functions are positive definite. With the symbol $\doteq$, we denote "equal up to some constants factors." The parameter k indicates that they are $\mathscr{C}^{2k}(\mathbb{R}^d)$. For example, for $d = 1$ and $l = 3$, we have $\varphi_{1,1}(r) \doteq (1 - r)_+^4(4r + 1)$, which is the well-known compactly supported Wendland $\mathscr{C}^2$ function. Introducing also for Wendland functions a shape parameter $\varepsilon > 0$, we simply stretch the support to $[0, 1/\varepsilon]$ having $\delta = 1/\varepsilon$ as its diameter.

Now, given a function $f : \Omega \to \mathbb{R}$, a set $X = \{x_1, \ldots, x_N\} \subset \Omega$ of N distinct points, and the values $f_X = (f(x_1), \ldots, f(x_N))^T$ of the function f at the set X, we seek for interpolants of f_X of the form

$$P_f(x) = \sum_{i=1}^{N} c_i K(x, x_i), \; x \in \Omega, \tag{1}$$

for kernels K which are strictly positive definite and radial. This means that P_f is a function in $H_K(X) = \text{span}\{K(\cdot, x_i), \; i = 1, \ldots, N\}$, formed by translates of K at

Table 1 Examples of infinitely smooth kernels

Kernel	Parameter	Support	Name
$\varphi(\varepsilon r) = e^{-\varepsilon^2 r^2}$	$\varepsilon > 0$	$\mathbb{R}_0^+$	Gaussian
$\varphi(\varepsilon r) = (1 + \varepsilon^2 r^2)^{-1}$	$\varepsilon > 0$	$\mathbb{R}_0^+$	Inverse quadrics
$\varphi(\varepsilon r) = (1 + \varepsilon^2 r^2)^{-1/2}$	$\varepsilon > 0$	$\mathbb{R}_0^+$	Inverse multiquadrics

Table 2 Examples of kernels with finite smoothness. M0 and M2 are Matérn kernels, and W0 and W2 are Wendland kernels with smoothness 0 and 2, respectively

Kernel	Parameter	Support	Name
$\varphi(\varepsilon r) = e^{-\varepsilon r}$	$\varepsilon > 0$	$\mathbb{R}_0^+$	M0
$\varphi(\varepsilon r) = (1 + \varepsilon r)e^{-\varepsilon r}$	$\varepsilon > 0$	$\mathbb{R}_0^+$	M2
$\varphi_{d,0}(r) \doteq (1 - r)_+^l$	$l = \lfloor d/2 + k + 1 \rfloor$	[0,1]	W0
$\varphi_{d,1}(r) \doteq (1 - r)_+^{l+1}((l + 1)r + 1)$	$l = \lfloor d/2 + k + 1 \rfloor$	[0,1]	W2

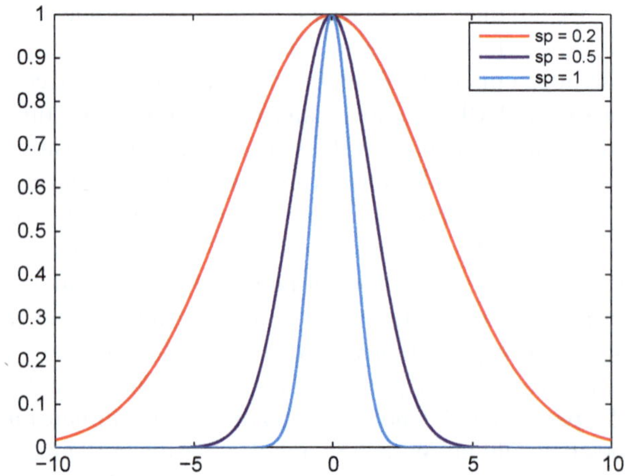

Fig. 1 Effect of changing the shape parameter on the Gaussian kernel plotted in $[-10, 10]$

the point set X. The coefficients in (1) are determined by imposing the interpolation conditions, which is equivalent to find the unique solution of the linear system $Ac = f_X$ with $A_{i,j} = K(x_i, x_j)$. Since the kernel K is assumed to be strictly positive definite, the solution exists and is unique.

The Hilbert space in which the kernel K is reproducing is known as the associate *native space*. We will denote it by $\mathscr{N}_K$, instead of $\mathscr{H}$, to underline the dependence on the kernel. It is equipped with the scalar product $(\cdot, \cdot)_{\mathscr{N}_K}$, from which we get the *native space norm* $\| \ \|_{\mathscr{N}_K}$ (cf. e.g., [4]).

The interpolation process by kernels of functions $f \in \mathscr{N}_K$ gives point-wise errors of the form (see e.g., [5, p. 174] or [6, p. 19])

$$|f(x) - P_f(x)| \le C h_{X,\Omega}^{\beta} \|f\|_{\mathscr{N}_K} \tag{2}$$

for some appropriate exponent β depending on the smoothness of the kernel K, and $h_{X,\Omega}$ that denotes the *mesh-size* (or fill-distance) $h_{X,\Omega} = \max_{x\in\Omega} \min_{x_i \in X} \|x - x_i\|_2$. For functions belonging to bigger spaces than the native space, suitable estimates are based on *sampling inequalities* as discussed e.g., in [5, §9.4].

3 The Rescaled Interpolant

In [1], the authors have proposed a new compactly supported RBF interpolant with the aim of a more accurate interpolation even by using a small diameter for the support. More precisely, on the set of points X, we consider the constant function $g(x) = 1$ $\forall x \in \Omega$, and we denote by $P_g(x)$ the corresponding kernel-based interpolant, that is

$$P_g(x) = \sum_{i=1}^{N} d_i K(x, x_i)\,,$$

whose coefficients $d = (d_1, \ldots, d_N)^T$ can be determined as the solution of the linear system $Ad = 1$, with the vector 1 of ones.

Then, the *rescaled* interpolant is

$$\hat{P}_f(x) = \frac{P_f(x)}{P_g(x)} = \frac{\sum_{i=1}^{N} c_i K(x, x_i)}{\sum_{i=1}^{N} d_i K(x, x_i)}\,. \tag{3}$$

As a simple illustrative example, we want to interpolate $f(x) = x$ on the interval [0,1] by using the W2 function at the point set $X = \{1/6,\, 1/2,\, 5/6\}$ with shape parameter $\varepsilon = 5$. The function, the interpolant, and the errors $|f(x) - P_f(x)|$, $|f(x) - \hat{P}_f(x)|$ are displayed in Fig. 2. The shape parameter has been chosen as the reciprocal of the support radius of the corresponding basis function. Indeed, in this example, the interpolant is the combination of *three* W2 radial functions all having radius of the support $r_j = 1/5,\ j = 1, 2, 3$. Refining the point set, that is, considering the point set $X = \{0,\, 1/6,\, 1/3,\, 1/2,\, 2/3,\, 5/6,\, 1\}$, we get the results shown in Fig. 3 showing the most interesting behavior of the rescaled interpolant, the property to reduce oscillations and so the interpolation error.

We show in Fig. 4 the RMSE (root mean square error) of the stationary interpolation with the rescaled interpolant (3) w.r.t. the classical one (1) on a grid of 25 data points of the square $[0, 1]^2$ at different values of the shape parameter, by using the W2 radial function for the 2D *Franke function*

$$\begin{aligned} f(x, y) = {} & \frac{3}{4} e^{-\frac{1}{4}((9x-2)^2+(9y-2)^2)} + \frac{3}{4} e^{-\frac{1}{49}(9x+1)^2 - \frac{1}{10}(9y+1)} \\ & + \frac{1}{2} e^{-\frac{1}{4}((9x-7)^2+(9y-3)^2)} - \frac{1}{5} e^{-(9x-4)^2-(9y-7)^2}\,. \end{aligned} \tag{4}$$

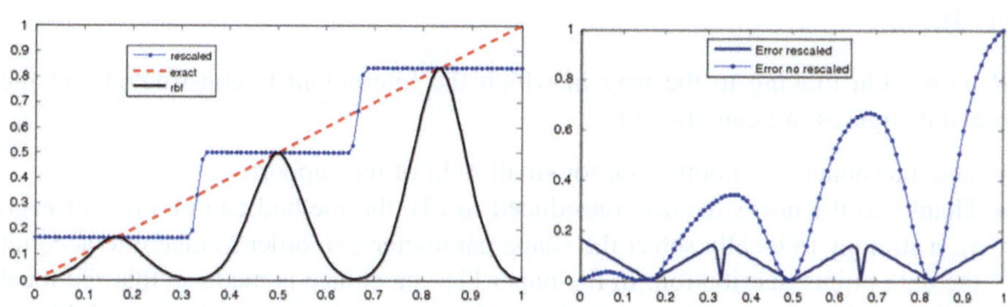

Fig. 2 *Left* the function $f(x) = x$ interpolated with the compactly supported W2 function and the rescaled interpolant at the data set $X = \{1/6,\, 1/2,\, 5/6\}$ and $\varepsilon = 5$. *Right* the corresponding absolute errors

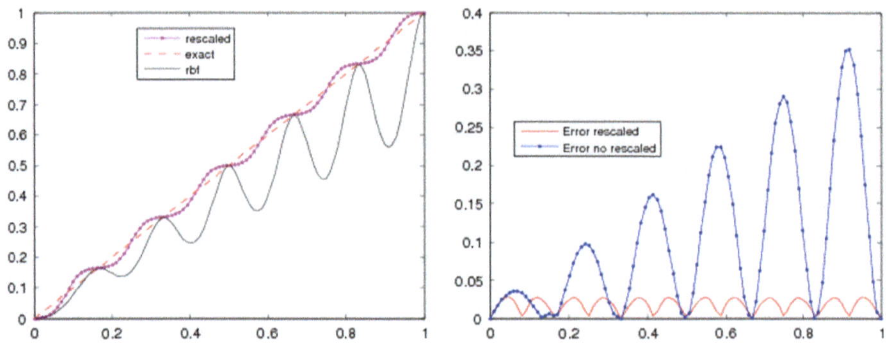

Fig. 3 As in Fig. 2 for the set $X = \{0, 1/6, 1/3, 1/2, 2/3, 5/6, 1\}$ and $\varepsilon = 5$

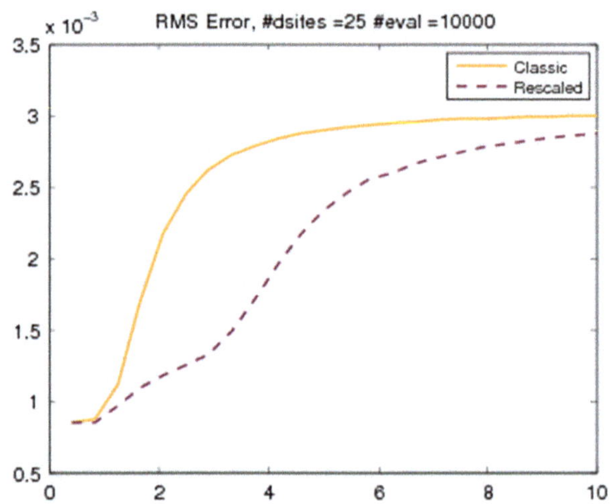

Fig. 4 RMSE behavior at different values of the shape parameter for interpolation of the Franke function with W2 radial function

Similar results can be obtained by using different radial basis functions as studied in [1, 8].

Remarks On looking to the way in which the interpolant is constructed and the previous figures, we can observe

- The interpolant is smooth even for small radii of the support.
- Thanks to the normalization introduced in (3), the method can choose for each x_m a strategy to locally select the shape parameter ε in order to take into account the data points distribution. In the paper [1], the choice is made so that the local radius of the compactly supported kernel gives a constant number of neighbors. This strategy fails when the points are uniformly distributed while gives much better results when the points are not equidistributed. In this second case, we can in fact consider different radii and neighbor points.

3.1 The Rescaled Kernel

Since our interest is the study of the rescaled interpolant as a new kernel which has an associated native space, we start by the following observation. An interpolation process consists in approximating the function f by its interpolant, say P_f, that is,

$$P_f(x) \approx f(x), \quad \forall\, x \in \Omega\,,$$

where equality holds for $x \in X$. Equivalently, we can say that for a constant $k_0 \in \mathbb{R}$,

$$P_f(x) - f(x) \approx k_0, \quad \forall\, x \in \Omega\,.$$

Assuming that $f \neq 0$, then

$$\frac{P_f(x) - f(x)}{f(x)} \approx \frac{k_0}{f(x)} = k_0, \quad \forall\, x \in \Omega\,, \tag{5}$$

the last equality holds when $f \equiv 1$. Hence, assuming that $f = k_1$, then from (5), we get

$$P_{k_1}(x) - k_1 \approx k_0, \quad \forall\, x \in \Omega\,, \tag{6}$$

where we used k_0 to describe the new constant of the right-hand side.

Proposition 1 *The relation $\approx$ induced by (6) is an equivalence relation.*

Proof Let $f, g, h : \Omega \to \mathbb{R}$ and X be a set of distinct points of Ω.

- Reflexivity. Since $f = f$, then $f(\Omega) = f(\Omega)$ implies $f(X) = f(X)$ and so $f \approx f$.
- Symmetry. Let $f \approx g$, then $f(X) = g(X)$ that can be read from right to left and so $g \approx f$.
- Transitivity. Let $f \approx g$ and $g \approx h$. This means $f(X) = g(X)$ and $g(X) = h(X)$. Hence, $f(X) = h(X)$, that is $f \approx h$.

This concludes the proof. □

We can use the transitive property of $\approx$ to combine (5) and (6) to get

$$\begin{aligned}\frac{P_f(x)}{f(x)} - k_1 &\approx P_{k_1}(x) - k_1 \\ \frac{P_f(x)}{f(x)} &\approx P_{k_1}(x) \\ \frac{P_f(x)}{P_{k_1}(x)} &\approx f(x)\,.\end{aligned}$$

Therefore, functions of the form $\dfrac{P_f(x)}{P_{k_1}(x)}$ are rescaled interpolants.

In our setting, we can notice that both $P_f(x)$ and $P_g(x)$ are constructed by using the kernel K with associate native space $\mathcal{N}_K$. In order to identify the native space associated to the rescaled interpolant, we may proceed as follows. Letting

$$\hat{P}_f(x) = \frac{P_f(x)}{P_g(x)}, \tag{7}$$

we may introduce a new kernel, say K_r, associated to the rescaled interpolant, from which we will characterize the associated native space $\mathcal{N}_{K_r}$.

Observing that

$$\hat{P}_f(x) = \frac{P_f(x)}{P_g(x)} = \sum_{j=1}^{N} c_j \frac{K(x, x_j)}{\sum_{i=1}^{N} d_i K(x, x_i)} \tag{8}$$

and recalling that the denominator is the interpolant of the constant function $g(x) = 1$, we have

$$\hat{P}_f(x) = \sum_{j=1}^{N} c_j \left[\frac{K(x, x_j)}{\sum_{i=1}^{N} d_i K(x, x_i) \sum_{i=1}^{N} d_i K(x_j, x_i)} \right]$$

Denoting $q(x) = \sum_{i=1}^{N} d_i K(x, x_i)$, then the square brackets can be rewritten as the set of functions

$$\frac{K(x, x_j)}{q(x) \cdot q(x_j)}, \quad j = 1, \ldots, N$$

which can be interpreted as a *new basis* for the rescaled interpolant.

This theorem finds application in our setting.

Theorem 1 (cf. [9]) *Let $K : \Omega \times \Omega \to \mathbb{R}$ be a (strictly) positive definite kernel. Let $s : \Omega \to \mathbb{R}$ be a continuous and nonvanishing function in Ω. Then*

$$K_s(x, y) = s(x)s(y)K(x, y) \tag{9}$$

is (strictly) positive definite.

In fact, letting P_g the interpolant of the constant $g \equiv 1$ which is by construction continuous in Ω, if it is also nonvanishing, then $s = 1/P_g$. But this property follows from the general error estimation (2), since we can find a set of points X so that $\|P_g(x) - g\|_\infty < g$ that implies $P_g(x) \neq 0, \ \forall x \in \Omega$.

It follows that we can consider $s = 1/P_g$, which is continuous, nonvanishing on Ω and consider the *rescaled kernel*

$$K_r(x, y) = \frac{1}{P_g(x)} \frac{1}{P_g(y)} K(x, y) \tag{10}$$

which turns out to be (strictly) positive definite and we will denote its associate native space by $\mathcal{N}_{K_r}$.

This discussion shows that the rescaled kernel K_r is a kernel approximation process which is well-posed and preserves the properties of the kernel K.

Moreover, by rewriting the rescaled interpolant as

$$\begin{aligned}
\hat{P}_f(x) &= \sum_{j=1}^{N} c_j K_r(x, x_j) \\
&= \sum_{j=1}^{N} c_j \left\{ \frac{1}{\sum_{i=1}^{N} d_i K(x, x_i)} \frac{1}{\sum_{i=1}^{N} d_i K(x_i, x)} \right\} K(x, x_j) \\
&= \sum_{j=1}^{N} c_j \frac{1}{\sum_{i=1}^{N} d_i K(x, x_i)} K(x, x_j) ,
\end{aligned}$$

since $\frac{1}{\sum_{i=1}^{N} d_i K(x,x_i)} = 1$, $\forall x \in \Omega$ is the interpolant of the function $g = 1$. This formulation allows to prove formally that the rescaled interpolant reproduces the constants.

Theorem 2 *Let K be a strictly positive definite kernel, $f : \Omega \to \mathbb{R}\backslash\{0\}$ such that $f(x) = a,\ a \in \mathbb{R}\backslash\{0\}$ and $X = \{x_1, \ldots, x_N\} \subset \Omega$. Then, the associated rescaled interpolant (3) (or equivalently (8)) is such that $\hat{P}_{r,a}(x) = a,\ \ \forall x$.*

Proof The interpolation conditions give the linear system

$$A_r \mathbf{c} = \mathbf{a}$$

where $\mathbf{a} = (a, \ldots, a)^T$ and A_r denotes the collocation matrix w.r.t. the rescaled basis of the function $f(x) = a$. The previous system can be written as

$$\begin{aligned}
a \cdot \left(\frac{1}{a} A_r \mathbf{c} \right) &= a \cdot \mathbf{1} , \\
\frac{1}{a} A_r \mathbf{c} &= \mathbf{1} .
\end{aligned}$$

where $\mathbf{1} = (1, \ldots, 1)^T$. Hence, denoting as $\hat{P}_{r,a}$ the rescaled interpolant of the constant a

$$\hat{P}_{r,a}(\cdot) = a \cdot \hat{P}_g(\cdot) = a \cdot \frac{P_g(\cdot)}{P_g(\cdot)} = a , \tag{11}$$

as required. □

Obviously, the previous results hold for $a = 0$. This comes immediately from (11).

4 Rescaling Is Equivalent to the Shepard's Method

We take into account here a different point of view. We firstly compute the cardinal function form of the interpolant, and then we show the connection with the Shepard's method (see e.g., [4, §23.1]) and provide a stability bound based on the Lebesgue constant. We need to recall the following result (cf e.g., [4, §14.2] or [10]).

Proposition 2 *For any set $X_N = \{x_1, \ldots, x_N\} \subset \Omega$ of pair-wise distinct points, there exists a unique* `cardinal basis` *$U = \{u_j\}_{j=1}^N$ of the span$\{K(\cdot, x), x \in X_N\}$, i.e., a set of functions such that $u_j(x_i) = \delta_{ij}, \quad 1 \le i, j \le N$.*

Using the basis U, the standard interpolant of a function $f \in \mathscr{H}$ can be written in the form $P_f = \sum_{j=1}^N f(x_j)u_j$, and the interpolant of the function $g \equiv 1$ reads as $P_g = \sum_{j=1}^N u_j$. The rescaled interpolant of f then takes the form

$$\hat{P}_f = \frac{\sum_{j=1}^N f(x_j)u_j}{\sum_{k=1}^N u_k} = \sum_{j=1}^N f(x_j)\frac{u_j}{\sum_{k=1}^N u_k} =: \sum_{j=1}^N f(x_j)\hat{u}_j,$$

where we introduced the N functions $\hat{u}_j := u_j / \left(\sum_{k=1}^N u_k\right)$. These functions are still cardinal functions, since $\hat{u}_j(x_i) = \delta_{ij}$, but they do not belong to the subspace span$\{K(\cdot, x), x \in X_N\}$, in general. But they form a partition of unity, in the sense that, for all $x \in \Omega$, we have

$$\sum_{j=1}^N \hat{u}_j(x) = \sum_{j=1}^N \frac{u_j(x)}{\sum_{k=1}^N u_k(x)} = \frac{\sum_{j=1}^N u_j(x)}{\sum_{k=1}^N u_k(x)} = 1.$$

This construction proves the following result.

Proposition 3 *The rescaled interpolation method is a Shepard's method, where the weight functions are defined as $\hat{u}_j = u_j / \left(\sum_{k=1}^N u_k\right)$, $\{u_j\}_j$ being the cardinal basis of span$\{K(\cdot, x), x \in X\}$.*

Remarks

- Looking at Figs. 2 and 3, we notice the typical "flat-spot" behavior of the Shepard's approximation.
- Although this connection allows to relate our work with other existing methods, we remark that it also highlights some possibly severe limitations in the present approach. Indeed, the main reason to consider Shepard's methods relies on the easy computation of the weight functions $\{\hat{u}_j\}_j$, which are usually constructed by solving a small linear system instead of a full interpolation problem. In our case, instead, the computation of the weights requires the computation of the cardinal basis, that is, the solution of the full interpolation problem. In this view, we can not expect good numerical results when the number of points increases. On the other

hand, when the method is coupled with a *partition of unity* procedure, as will be discussed in the next sections, the systems to be solved are of small size, hence a stable computation of the interpolation of $g = 1$ can be expected.

From this construction, we can easily derive stability bounds for the rescaled interpolation process. In fact, as happens in polynomial interpolation by using the cardinal functions, we can define the Lebesgue function $\Lambda_N(x) := \sum_{j=1}^{N} |u_j(x)|$, and its maximum over Ω, $\lambda_N := \|\Lambda_N\|_\infty$, that is, the *Lebesgue constant* which controls the stability of the interpolation process. In fact, for any $x \in \Omega$

$$|P_f(x)| = \left|\sum_{j=1}^{N} f(x_j)u_j(x)\right| \leq \left(\sum_{j=1}^{N} |u_j(x)|\right) \|f\|_{\infty,X} \leq \lambda_N \, \|f\|_{\infty,X}.$$

Extending the setting to our case and by using the rescaled cardinal functions $\{\hat{u}_j\}_j$ instead of the classical cardinals, we can write

$$\hat{\Lambda}_N(x) := \sum_{j=1}^{N} |\hat{u}_j(x)|, \quad \hat{\lambda}_N := \|\hat{\Lambda}_N\|_\infty,$$

which gives the stability bound

$$\|\hat{P}_f\|_\infty \leq \hat{\lambda}_N \|f\|_{\infty,X}.$$

Hence, to quantify the stability gain of the rescaled interpolation process over the standard one, we can simply compare the behavior of $\hat{\lambda}_N$ and λ_N. Numerical experiments showing this comparison are presented in Sect. 6.

5 Application to PUM

Given the domain $\Omega \subset \mathbb{R}^d$, we consider its partition $\{\Omega_k \subset \Omega, \ k = 1, \dots, n\}$ with Ω_k that possibly overlap, such that $\Omega \subseteq \cup_{k=1}^{n} \Omega_k$. We then consider compactly supported functions w_k with $\mathtt{supp}(w_k) \subseteq \Omega_k$, forming a partition of Ω, that is

$$\sum_{k=1}^{n} w_k(x) = 1\,, \quad \forall x \in \Omega\,. \tag{12}$$

Then, we construct a *local interpolant*, p_k, in RBF form

$$p_k(x; X_k) = \sum_{j=1}^{n_k} c_j^{(k)} \Phi_j^{(k)}(x)\,, \tag{13}$$

where X_k is a set of distinct points of Ω_k having $n_k = |X_k|$ as its cardinality and $\Phi^{(k)}$ the RBF kernel at Ω_k. The global interpolant on Ω can be written as

$$P_f(x) = \sum_{k=1}^{n} p_k(x; X_k) w_k(x), \quad x \in \Omega\,. \tag{14}$$

If the local fit interpolates at a given data point, that is, $p_k(x_l) = f(x_l)$, then thanks to the partition of unity property (12), we can conclude that the global fit is also interpolating at the same point

$$P_f(x_l) = \sum_{k=1}^{n} p_k(x_l; X_k) w_k(x_l) = \sum_{k=1}^{n} f(x_l) w_k(x_l) = f(x_l)\,.$$

We can apply the rescaled interpolant to this framework as follows

- by applying the rescaling to the global interpolant (14);
- or by applying the rescaling to every local interpolant (13).

The first approach is equivalent to apply the PUM for interpolating the constant function 1. Hence, it makes sense to rescale *every* local interpolant. The application of the rescaling to every local interpolant of the form (13) gives a global rescaled interpolant of the form

$$P_f(x) = \sum_{k=1}^{n} \hat{R}_k(x; X_k) w_k(x), \quad x \in \Omega\,, \tag{15}$$

with

$$\hat{R}_k(x; X_k) = \sum_{j=1}^{n_k} c_j^{(k)} \frac{\Phi_j^{(k)}(x)}{P_1^{(k)}(x)} = \sum_{j=1}^{n_k} c_j^{(k)} \frac{\Phi_j^{(k)}(x)}{\sum_{l=1}^{n_k} d_l^{(k)} \Phi_l^{(k)}(x)}\,,$$

where the coefficients $d_l^{(k)}$ are chosen so that $\sum_{l=1}^{n_k} d_l^{(k)} \Phi_l^{(k)}(x) = 1, \quad \forall x \in X_k\,.$

6 Numerical Examples

In this section, the rescaled method is tested in different instances with the aim of analyzing its performance and validating our understandings. The first set of experiments, presented in the next subsection, compares the standard and rescaled Lebesgue functions, as defined in Sect. 4. In this case, the tests are limited to a small number N of samples in order to be able to stably compute the cardinal functions, and the results have to be intended as theoretical investigation of the relation between the two Lebesgue functions. In Sect. 6.2, instead, we focus on the comparison between the PUM and the Rescaled-PUM, and we consider a more realistic number of points.

This is possible by limiting the number of points in each subdomain in the partition of unity, so that a stable computation of the rescaled interpolant is numerically feasible.

6.1 Comparison of the Standard and Rescaled Lebesgue Functions

In these experiments, we compare the standard Lebesgue function with that of the rescaled one. Since we need to directly compute the cardinal functions, which is a seriously unstable operation, we keep the example as simple as possible to avoid comparing the effect of the ill-conditioning. To this end, we take $\Omega = [-1, 1]$, a small number of fixed node points (equally spaced) and two classical kernels: the Gaussian kernel (globally supported and $\mathscr{C}^\infty$) and the Wendland W2 kernel, which is compactly supported and $\mathscr{C}^2$. The computation of the standard and rescaled Lebesgue functions has been repeated for $\varepsilon = 0.5,\ 1,\ 4,\ 8$ (Gaussian kernel) and $\varepsilon = 0.5,\ 1,\ 2,\ 4$ (Wendland kernel), as shown in Figs. 5 and 6. In the latter case, the behavior of the Lebesgue function does not change for bigger values of the shape parameter, that is why we stopped at $\varepsilon = 4$. Taking other set of nodes (such as random or Chebyshev points), due to the above observation on the computation of the cardinal functions, we indeed observed essentially a similar behavior for the Gaussian kernel and opposite for the Wendland kernel. See Fig. 7 for a simple example with Chebsyshev points.

Similar behavior can be observed in the two-dimensional setting. In Fig. 8, we show the comparison between the Lebesgue functions for the Wendand function with standard cardinal functions and the rescaled ones on the cardioid contained in $[-1, 1]^2$.

In Fig. 9, we did a similar test on the square. We have chosen $\varepsilon = 3.85$ because we wanted to show that bigger values of ε are meaningless. The reason of this relies on the support of the Wendland function which is $1/\varepsilon$. In the example, we have taken 25 equally spaced points on the square, so that for values of ε bigger than $\varepsilon_M = 2$, the cardinal functions have disjoint support. Therefore, for values of $\varepsilon \geq 2\varepsilon_M$, we can not assume that the interpolant of the function 1 is always not vanishing since some points of the domain fall outside the support of the cardinal functions (giving a value 0 of the interpolant). This explains also why in the one-dimensional case, the cardinal functions with $\varepsilon = 4$ give a Lebesgue function identically equal to 1.

6.2 Comparison Between PUM and Rescaled-PUM

We investigate here the relation between the PUM and the Rescaled-PUM (R-PUM). Two different experiments are analyzed. First, we compare the two methods for a fixed and relatively small number N of samples, in order to highlight the behavior of

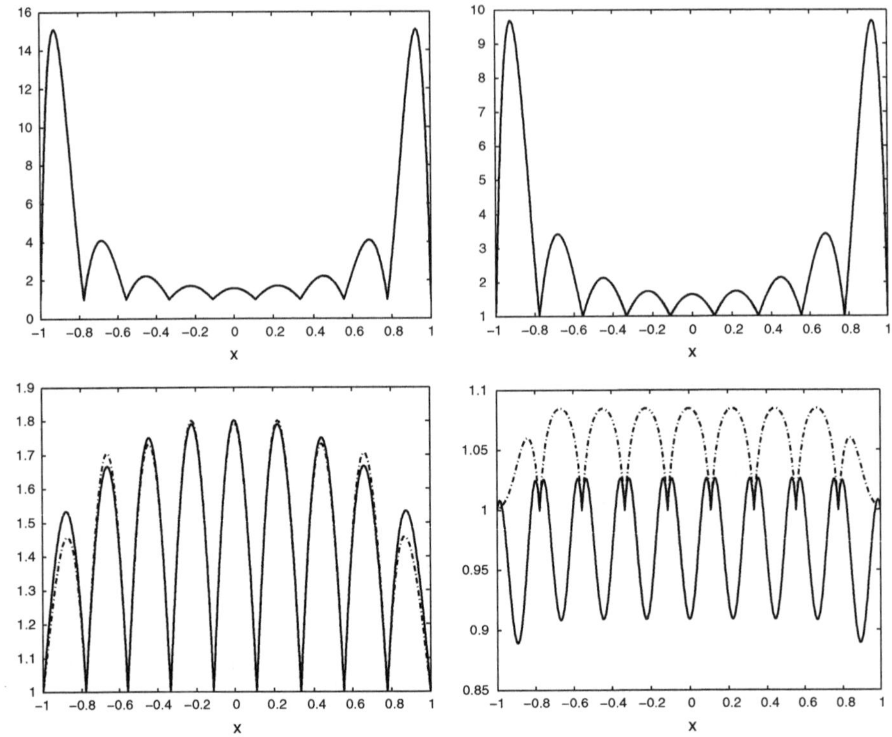

Fig. 5 Standard Lebesgue function (*solid line*) and the rescaled Lebesgue function (*dotted line*) for the Gaussian kernel on 10 equally spaced points of $[-1, 1]$ with different ε. From *top left* to *bottom right*, $\varepsilon = 0.5,\ 1,\ 4,\ 8$

the two methods with respect to changes in the shape parameter of the kernel. Then, we let N increase and we analyze the convergence profile of the two methods, both for a fixed and an optimized shape parameters.

To test the two methods, we use a C^2 Wendland kernel K on bivariate domains $\Omega \subset \mathbb{R}^2$, and a partition of unity defined on a covering of Ω made by balls Ω_j of constant radius and equally spaced centers, such that the resulting number of subdomains is

$$n^2 = \lfloor \frac{N}{4} \rfloor, \tag{16}$$

where $\lfloor \cdot \rfloor$ denotes the integer part. This covering is then equipped with a partition of unity defined by weight functions

$$w_j(x) = \frac{K(x, x_j)}{\sum_{k=1}^{n} K(x, x_k)},$$

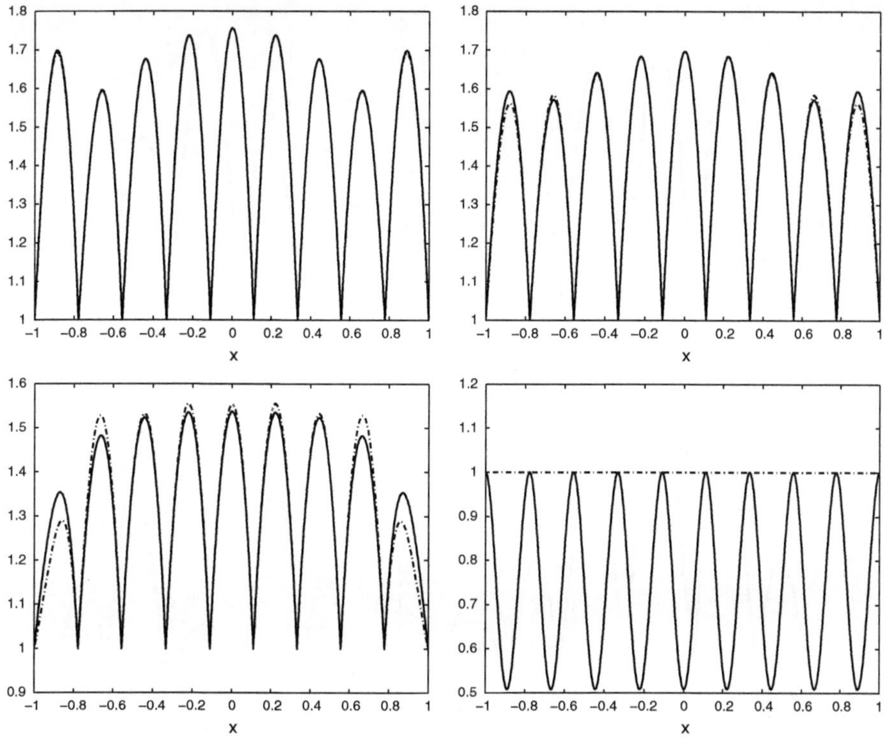

Fig. 6 Comparison between the standard Lebesgue function (*solid line*) and the rescaled Lebesgue function (*dotted line*) for the $\mathscr{C}^2$ Wendland kernel on 10 equally spaced points of $[-1, 1]$ with different ε. From *top left* to *bottom right*, $\varepsilon = 0.5,\ 1,\ 2,\ 4$

K being the Wendland kernel itself.

The error is measured in both cases on a grid of 100×100 uniformly distributed points in the convex hull of the sample set X. We consider both the maximal absolute error and the root mean square error (RMSE).

The algorithm is implemented using a *block-based* partition of unity algorithm presented in [11], which leads to a faster evaluation of the interpolant.

In both experiments, we use as a target function the 2D *Askley's test function* (well-known for testing optimization algorithms [12]), which is defined as

$$f(x, y) = -20\, e^{-0.2\sqrt{0.5(x^2+y^2)}} - e^{-0.5(\cos(2\pi x)+\cos(2\pi y))} + 20 + e. \qquad (17)$$

For the first test, we consider $N = 10^3$ Halton points on the disk Ω of radius 0.5 and centered in $(0.5, 0.5)$. The two methods are tested on 30 values of the shape parameter $\varepsilon \in [0.01, 2]$. The results are shown in the Fig. 10. For both error measurements, the PUM and R-PUM attain the same error for a flat kernel ($\varepsilon \approx 0$), while the R-PUM gives better results for a localized one ($\varepsilon = 2$). These results are easily explained,

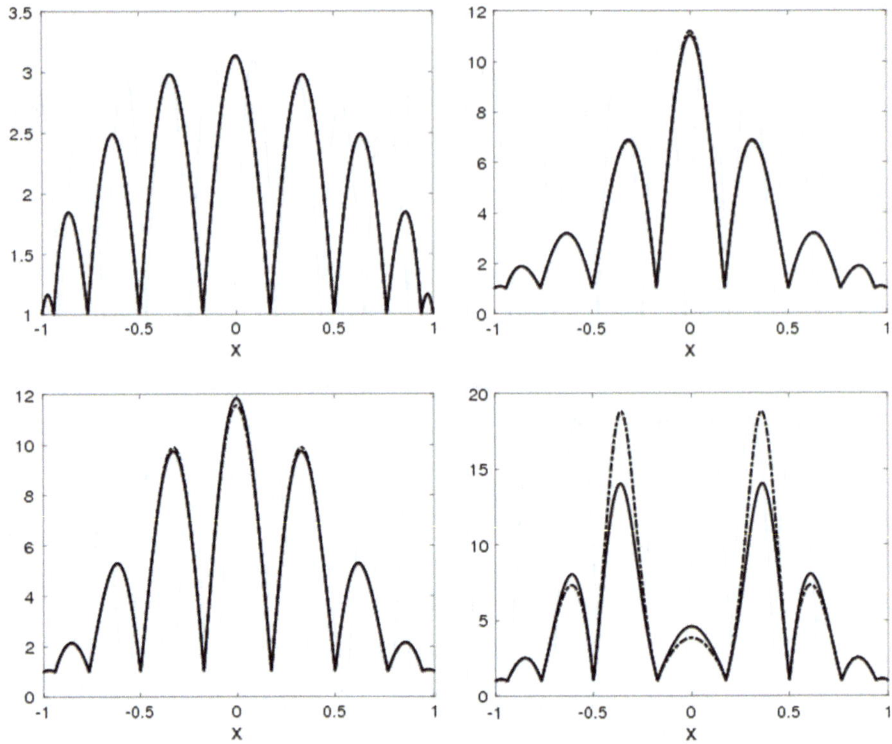

Fig. 7 Standard Lebesgue function (*solid line*) and the rescaled Lebesgue function (*dotted line*) for the Gaussian kernel (*first row*) and Wendland kernel (*second row*) on 10 Chebyshev points of $[-1, 1]$ with $\varepsilon = 0.5$ (*left*) and $\varepsilon = 2$ (*right*)

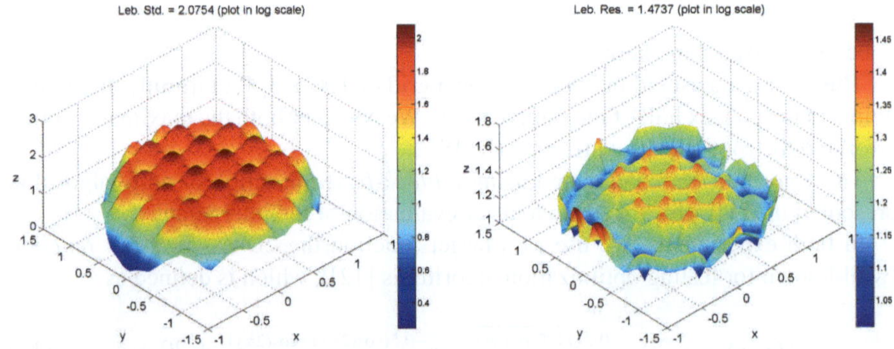

Fig. 8 Comparison between the Lebesgue function with standard basis (*Left*) and the rescaled one (*right*) for the $\mathscr{C}^2$ Wendland kernel on the cardioid with $\varepsilon = 3$

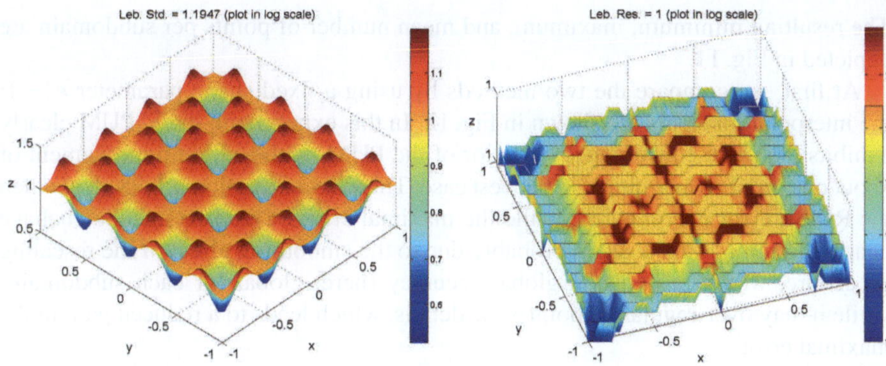

Fig. 9 Comparison between the Lebesgue function with standard basis (*Left*) and the rescaled one (*right*) for the $\mathscr{C}^2$ Wendland kernel on the square with $\varepsilon = 3.85$

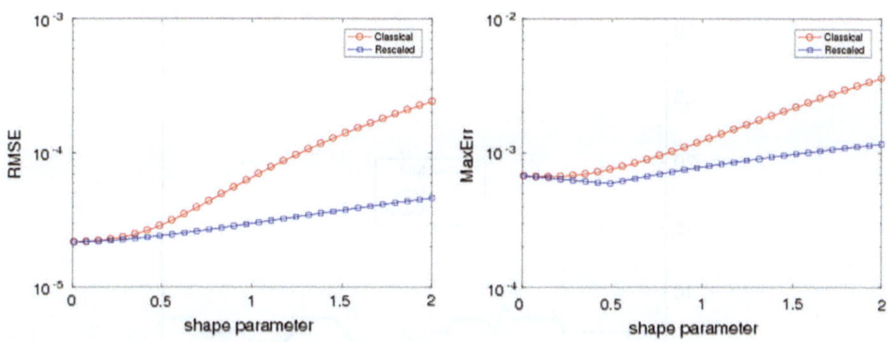

Fig. 10 RMSE (*Left*) and MAXERR (*Right*) for the classical and the rescaled PU for $\varepsilon \in [0.01, 2]$ chosen with Trial and Error

since a localized interpolant is smoothed by the rescaling procedure, and a decreasing of the error can be obtained also in the areas of Ω not well covered by X. The differences between the two methods are more evident by looking at the RMSE, which is a more global measurement than the point-wise maximal error.

We extend now the above comparisons by investigating the convergence behavior as the number N of function samples increases. We let the number n of subdomains to increase along with N (by using the formula (16)), such that each subdomain contains a nearly constant number of points. We remark, as pointed out earlier, that this is the only viable approach to ensure a stable computation of the interpolant of $g = 1$ with the direct method, while an increased size of the local linear system could possibly lead to severe ill-conditioning.

The kernel and partition of unity are as above, and we again address the approximation of the Askley's test function. For simplicity, we use again Halton points on the unit square $[0, 1]^2$. We test the method for $N = 10^3, 2 \cdot 10^3, \ldots, 50 \cdot 10^3$.

The resulting minimum, maximum, and mean number of points per subdomain are depicted in Fig. 11.

At first, we compare the two methods by using a fixed shape parameter $\varepsilon = 1$: the interpolation errors are shown in Fig. 12. In this experiment, the R-PUM clearly exhibits the same convergence behavior of the PUM, with a slight improvement of about one order of magnitude in the best case. This effect is particularly evident when the RMSE error is computed, while the maximal error of the two methods behave nearly the same. This is again probably due to the smoothing effect of the rescaling procedure, which promotes a global accuracy (here, global for each subdomain) while it may over-regularize point-wise details, which leads to a reduced gain in the maximal error.

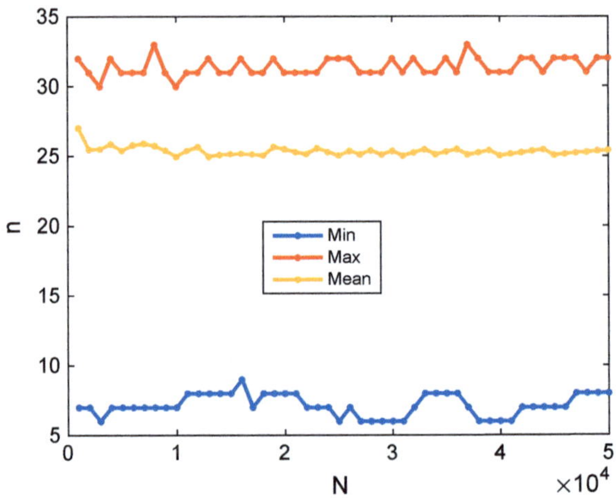

Fig. 11 Minimum, maximum, and mean number of points n per subdomain as the global number N of points increases for the experiments of Sect. 6.2

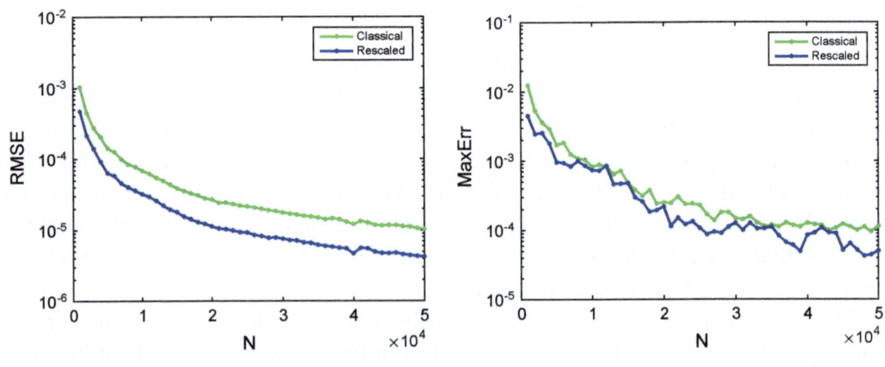

Fig. 12 RMSE (*left*) and maximal error (*right*) for the PUM (*green*) and the R-PUM (*blue*), as functions of the global number N of points, as described in Sect. 6.2

We then repeat the same experiment with shape parameter validation. Namely, the shape parameter is optimized by testing 30 equally spaced values in $[0.01, 2]$, and by picking, for each N, the value which minimize the error. Both root mean square error and maximal error are used, producing two different sequences of parameters for each method. The experiment is restricted in this case to a maximal value $N = 20 \cdot 10^3$.

In Fig. 13, we report the maximal errors and the corresponding selected shape parameters, while in Fig. 14, we show the same results with respect to the RMS error. It is worth noticing that when the RMSE is used as an error indicator, the PUM and R-PUM select exactly the same shape parameter (the green and blue lines overlap), while when the maximal error is employed, the two methods select notably different parameters. The results highlight once more that rescaled method behaves better than the standard one with respect to the RMSE, while the two are essentially equivalent with respect to the maximal error. In both cases, the convergence order for this setting appears to be essentially the same Fig. 11.

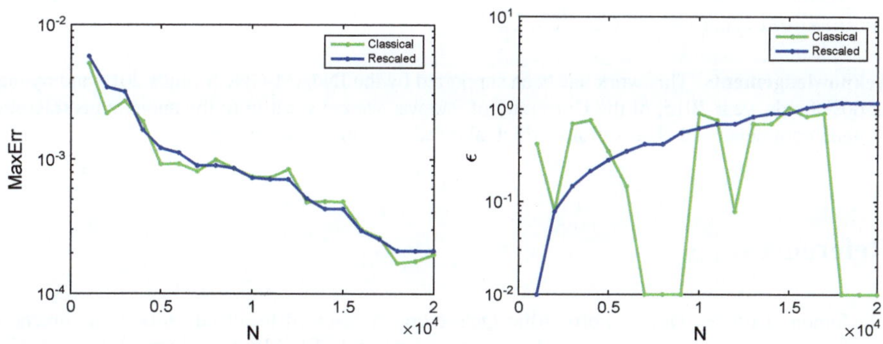

Fig. 13 Maximal error (*left*) and corresponding value of the shape parameter (*right*) for the PUM (*green*) and the R-PUM (*blue*), as functions of the global number N of points, as described in Sect. 6.2

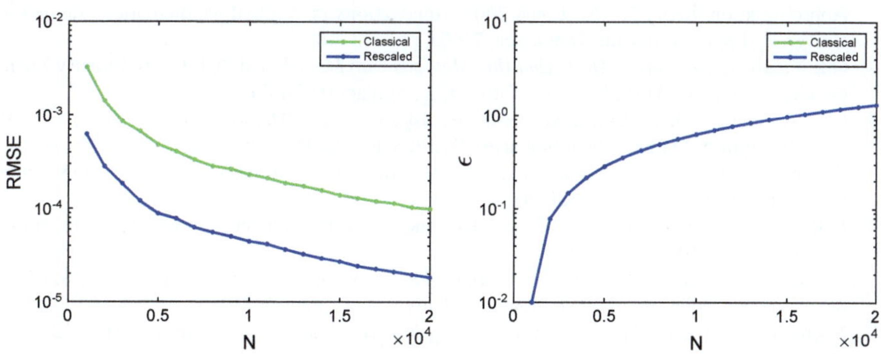

Fig. 14 RMS error (*left*) and corresponding value of the shape parameter (*right*) for the PUM (*green*) and the R-PUM (*blue*), as functions of the global number N of points, as described in Sect. 6.2

As a final remark, we observe that letting T_{RPU}, the evaluation time of the rescaled interpolant, and T_{PU} that of the classical one, we generally saw that $T_{RPU} < c\,T_{PU}$ with $c \approx 1.05$.

7 Conclusions

In this paper, we have studied more deeply the rescaled localized radial basis function interpolant introduced in [1]. We have proved that this interpolant gives a partition of unity method that reproduces constant functions, exactly as does the Shepard's method. One feature is that the shape parameter of the kernel can be chosen in a safe range, not too small and "relatively large," avoiding the classical numerical instability that occurs when the shape parameter is too small, or a severe ill-conditioning in the opposite case.

The numerical experiments on the R-PUM show that the method allows to improve the maximal error over a classical PUM.

Acknowledgements This work has been supported by the INdAM-GNCS funds 2016 and by the ex 60% funds, year 2015, of the University of Padova. We are greatful to the anonymous referees for their comments and observations, which allowed us to improve the paper.

References

1. Simone Deparis, Davide Forti, Alfio Quarteroni, A rescaled localized radial basis function interpolation on non-cartesian and non-conforming grids. SIAM J. Sci. Comp. **36**(6), A2745–A2762 (2014)
2. M. Buhmann, *Radial Basis Functions: Theory and Implementations* (Cambridge University Press, Cambridge, 2003)
3. H. Wendland: Fast Evaluation of Radial Basis Functions: Methods Based on Partition of Unity. Approximation theory X, (St. Louis 2001). Innovations in Applied Mathematics (Vanderbilt University Press, Nashville, Tennessee, 2002), pp. 473–483
4. G.E. Fasshauer, *Meshfree Approximation Methods with Matlab*, vol. 6, Interdisciplinary Mathematical Sciences (World Scientific Publishing, Singapore, 2007)
5. G.E. Fasshauer, M. McCourt, *Kernel-based Approximation Methods using Matlab*, vol. 19, Interdisciplinary Mathematical Sciences (World Scientific Publishing, Singapore, 2015)
6. Stefano De Marchi, Robert Schaback, Nonstandard kernels and their applications. Dolomites Res. Notes Approx. **2**, 16–43 (2009)
7. L. Bos, S. De Marchi, Univariate radial basis functions with compact support cardinal functions east. J. Approx. **14**(1), 69–80 (2008)
8. A. Idda, *A comparison of some RBF interpolation methods: theory and numerics*, Master's degree thesis, Department of Mathematics, University of Padova (2015)
9. Nachman Aronszajn, Theory of reproducing kernels. Trans. Am. Math. Soc. **68**, 337–404 (1950)
10. Z. Wu, R. Schaback, Local error estimates for radial basis function interpolation of scattered data. IMA J. Numer. Anal **13**, 13–27 (1993)

11. R. Cavoretto, A. De Rossi, E. Perracchione, Efficient computation of partition of unity interpolants through a block-based searching technique. Comp. Math. Appl. **71**(2016), 2568–2584 (2016)
12. O. Rody, Many test functions for global optimization algorithms. Matlab Central File Exchange (2015)

Flavors of Compressive Sensing

Simon Foucart

Abstract About a decade ago, a couple of groundbreaking articles revealed the possibility of faithfully recovering high-dimensional signals from some seemingly incomplete information about them. Perhaps more importantly, practical procedures to perform the recovery were also provided. These realizations had a tremendous impact in science and engineering. They gave rise to a field called 'compressive sensing,' which is now in a mature state and whose foundations rely on an elegant mathematical theory. This survey presents an overview of the field, accentuating elements from approximation theory, but also highlighting connections with other disciplines that have enriched the theory, e.g., statistics, sampling theory, probability, optimization, metagenomics, graph theory, frame theory, and Banach space geometry.

Keywords Gelfand width · Random matrices · Restricted isometry property · Iterative hard thresholding · Orthogonal matching pursuit · Basis pursuit

Compressive sensing was born circa 2005 after influential works of Candès–Romberg–Tao [19] and of Donoho [27]. These works triggered rapid developments benefiting from fruitful interactions between mathematics, engineering, computer science, and statistics. The one-sentence summary is as follows:

> Stable and robust recovery of s-sparse vectors in $\mathbb{C}^N$ from an optimal number $m \asymp s\ln(eN/s)$ of measurements is achievable via various efficient algorithms.

This survey will clarify the precise meaning of this sentence and its hidden deep theoretical points, all of which can be found in the book [39]. For complete results, I will mostly refer to [39] instead of original sources, which can be found in the Notes sections of [39]. As a rule, I shall not repeat the full underlying arguments here. Several exceptions will be made when proofs are provided either because some

S. Foucart (✉)
Department of Mathematics, Texas A&M University, College Station, TX, USA
e-mail: foucart@tamu.edu

G.E. Fasshauer and L.L. Schumaker (eds.), *Approximation Theory XV: San Antonio 2016*, Springer Proceedings in Mathematics & Statistics 201,
DOI 10.1007/978-3-319-59912-0_4

assertions are now stronger or because their explanations have been simplified since the publication of [39]. This survey is not intended to be comprehensive, as my first goal is to showcase a view of compressive sensing through an approximation theory lens. In particular, I will not discuss motivating applications such as magnetic resonance imaging, error correction, single-pixel camera, radar, or machine learning, which are all touched upon in [39]. This survey is biased, too, as my second goal is to summarize my own contribution to the field.

1 Stability and Robustness in Compressive Sensing

The scenario considered throughout this survey involves a signal space of very large dimension N. One can only acquire a small number $m \ll N$ of linear measurements about the signals from this space. To enable the recovery of signals from the few available measurements, a structural assumption on the signals must be made. A realistic assumption consists in stipulating that the signals of interest have sparse representations in some basis, and we identify from the onset signals of interest with their coefficient vectors $\boldsymbol{x} \in \mathbb{C}^N$ relative to this basis. We say that $\boldsymbol{x} \in \mathbb{C}^N$ is s-sparse if

$$\|\boldsymbol{x}\|_0 := \mathrm{card}\{j \in [\![1:N]\!] : x_j \neq 0\} \leq s.$$

The standard compressive sensing problem reads, in idealized form[1]:

> Find measurement matrices $\mathsf{A} \in \mathbb{C}^{m \times N}$ and recovery maps $\Delta : \mathbb{C}^m \to \mathbb{C}^N$ such that
>
> $$\Delta(\mathsf{A}\boldsymbol{x}) = \boldsymbol{x} \qquad \text{for all } s\text{-sparse vectors } \boldsymbol{x} \in \mathbb{C}^N. \tag{1}$$

To make the problem solvable, the number of measurements cannot be too small compared to the sparsity level, precisely $m \geq 2s$ is required (see, e.g., [39, p.49]). It turns out that the idealized problem is in fact solvable with $m = 2s$, with the added perks that the matrix A amounts to discrete Fourier measurements and that the map Δ executes an efficient recovery algorithm related to the age-old Prony's method. Its main lines are as follows (see [39, Theorem 2.15] for more details):

- identify $\boldsymbol{x} \in \mathbb{C}^N$ with a function on $[\![0 : N-1]\!]$ and get the $2s$ Fourier coefficients

$$\hat{x}(j) = \sum_{k=0}^{N-1} x(k) \exp(-\mathrm{i}2\pi jk/N), \qquad j \in [\![0 : 2s-1]\!];$$

[1] Although the problem is stated in the complex setting, our account will often be presented in the real setting. There are almost no differences in the theory, but this side step avoids discrepancy with existing literature concerning, e.g., Gelfand widths.

- with $S := \mathrm{supp}(x)$, consider the trigonometric polynomial vanishing on S defined by
$$p(t) := \prod_{k \in S} (1 - \exp(-i2\pi k/N)\exp(i2\pi t/N));$$
- observe that $p \times x = 0$ and deduce by discrete convolution that
$$0 = (\hat{p} * \hat{x})(j) = \sum_{k=0}^{N-1} \hat{p}(k)\hat{x}(j-k), \qquad j \in [\![0 : N-1]\!]; \tag{2}$$
- use the facts that $\hat{p}(0) = 1$ and that $\hat{p}(k) = 0$ when $k > s$ to transform the equations (2) for $j \in [\![s : 2s-1]\!]$ into a (Toeplitz) system with unknowns $\hat{p}(1), \ldots, \hat{p}(s)$;
- solve the system to determine $\hat{p}$, and in turn p, S, and finally x.

However, this algorithm will go astray[2] if the original vector $\boldsymbol{x}$ is not exactly sparse or if the measurements do not exactly equal $\mathsf{A}\boldsymbol{x}$. These two issues are fundamental in compressive sensing: One demands recovery maps that are stable—i.e., they cope well with sparsity defect—and that are robust—i.e., they cope well with measurement error. Let us point out informally that robustness implies stability. Indeed, sparsity defect can be incorporated into the measurement error: If $\boldsymbol{x}$ is not exactly sparse, consider an index set S of s largest absolute entries of $\boldsymbol{x}$, so that the inexact measurement vector $\boldsymbol{y} := \mathsf{A}\boldsymbol{x} + \mathbf{e}$ can be written as $\boldsymbol{y} = \mathsf{A}\boldsymbol{x}_S + \mathbf{e}'$ with $\mathbf{e}' := \mathsf{A}\boldsymbol{x}_{\overline{S}} + \mathbf{e}$. We shall put a special emphasis on stability in this survey. Therefore, with $\sigma_s(\boldsymbol{x})_1$ denoting the error of best s-term approximation to a vector $\boldsymbol{x} \in \mathbb{C}^N$, i.e.,
$$\sigma_s(\boldsymbol{x})_1 := \min\{\|\boldsymbol{x} - \boldsymbol{z}\|_1,\ \boldsymbol{z} \in \mathbb{C}^N \text{is } s\text{-sparse}\},$$
we replace the idealized problem (1) by the following refined version of the standard compressive sensing problem[3]:

> Find measurement matrices $\mathsf{A} \in \mathbb{C}^{m \times N}$ and recovery maps $\Delta : \mathbb{C}^m \to \mathbb{C}^N$ such that
> $$\|\boldsymbol{x} - \Delta(\mathsf{A}\boldsymbol{x})\|_2 \le \frac{C}{\sqrt{s}} \sigma_s(\boldsymbol{x})_1 \qquad \text{for all vectors } \boldsymbol{x} \in \mathbb{C}^N. \tag{3}$$

In the above formulation, $C > 0$ denotes an absolute constant, i.e., a constant that does not depend on any other parameter of the problem. Throughout this survey, the

[2]This is illustrated in the reproducible MATLAB file found on the author's Webpage.

[3]A 'weaker' formulation asks for the estimate $\|\boldsymbol{x} - \Delta(\mathsf{A}\boldsymbol{x})\|_1 \le C\sigma_s(\boldsymbol{x})_1$ for all vectors $\boldsymbol{x} \in \mathbb{C}^N$.

notation $c, c', C, C', d, d', D, D', \ldots$ always stands for positive absolute constants. They are introduced without notice, and they may not represent the same value from line to line. Other pieces of notation used in this survey are quite standard, except possibly $[\![k : \ell]\!]$, which stands for the set $\{k, k+1, \ldots, \ell\}$.

Here is a brief overview of the exposition that follows. Section 2 highlights a connection with Gelfand widths which enables to establish that, due to the stability requirement, the minimal number of measurements increases from $2s$ to linear in s times a logarithmic factor. The choice of suitable measurement matrices is then addressed in Sect. 3, where the popular restricted isometry property (RIP) is introduced and proved for Gaussian random matrices. In Sects. 4 and 5, two recovery procedures, orthogonal matching pursuit (OMP) and basis pursuit (BP), are presented and analyzed under the RIP of the measurement matrix. In Sect. 6, we exploit the notions introduced in Sects. 3–5 to shed a different light on the connection between compressive sensing and Gelfand widths—namely, instead of viewing Gelfand widths as a cornerstone of the compressive sensing theory as in Sect. 2, we derive Gelfand width estimates from pure compressive sensing techniques. In Sect. 7, non-Gaussian random measurement matrices are examined. In Sect. 8, we discuss extensions of the standard compressive sensing problem. In particular, we give new and simplified arguments for the one-bit compressive sensing problem. We close this survey by listing in Sect. 9 a selection of questions left open in the compressive sensing theory.

2 Compressive Sensing: By-product of Gelfand Widths Estimates

This section establishes a connection between compressive sensing and the study of widths. Only one direction of this connection is described for now, namely the impact of Gelfand widths on compressive sensing. We work here in the real setting.

2.1 *Gelfand Widths of the ℓ_1-Ball*

Widths are a well-studied concept in approximation theory, as illustrated, e.g., by the monograph [58]. For a subset K of a real normed space X, the Kolmogorov m-width and Gelfand m-width are defined, respectively, by

$$d_m(K, X) := \inf \left\{ \sup_{x \in K} \inf_{z \in X_m} \|\boldsymbol{x} - \boldsymbol{z}\|, \ X_m \text{ subspace of } X \text{ with } \dim(X_m) \le m \right\},$$

$$d^m(K, X) := \inf \left\{ \sup_{x \in K \cap L^m} \|\boldsymbol{x}\|, \ L^m \text{ subspace of } X \text{ with } \operatorname{codim}(L^m) \le m \right\}.$$

We concentrate on the case of $X = \ell_q^N$ and $K = B_p^N$, i.e., of $\mathbb{R}^N$ equipped with the ℓ_q-norm and the unit ball of ℓ_p^N. In this situation, there is a duality between Kolmogorov and Gelfand widths given by the relation

$$d_m(B_p^N, \ell_q^N) = d^m(B_{q'}^N, \ell_{p'}^N),$$

where $p', q' \geq 1$ are the conjugate exponents of $p, q \geq 1$, i.e., $1/p + 1/p' = 1$ and $1/q + 1/q' = 1$ (see, e.g., [39, Theorem 10.14] for a proof). The behaviors of the widths of B_p^N in ℓ_q^N were almost all known in the mid-1970s, but the Gelfand width of B_1^N in ℓ_2^N was notably missing from the picture. Then, Kashin [46] and Garnaev–Gluskin [40] proved that

$$d^m(B_1^N, \ell_2^N) \asymp \min\left\{1, \sqrt{\frac{\ln(eN/m)}{m}}\right\}.$$

Precisely, Kashin proved the upper estimate $d^m(B_1^N, \ell_2^N) \leq C \min\{1, \sqrt{\ln(eN/m)/m}\}$ (strictly speaking, his estimate featured $\ln(eN/m)^{3/2}$ instead of $\ln(eN/m)^{1/2}$) and Garnaev–Gluskin proved the lower estimate $d^m(B_1^N, \ell_2^N) \geq c \min\{1, \sqrt{\ln(eN/m)/m}\}$. We highlight below the implication that each estimate has for compressive sensing. Note that we shall informally say that a pair (A, Δ) of measurement matrix and recovery map is stable of order s with constant C if (3) holds. This corresponds to the notion of mixed (ℓ_2, ℓ_1)-instance optimality in the terminology of [23].

2.2 *Consequence of the Lower Estimate*

Briefly stated, *the existence of a stable pair* (A, Δ) *of order* s *forces* $m \geq cs \ln(eN/s)$. Indeed, let us consider $\mathbf{v} \in \ker \mathsf{A}$ and let us apply (3) to $-\mathbf{v}_S$ and to $\mathbf{v}_{\overline{S}}$ for some index set S of size s to obtain

$$\|-\mathbf{v}_S - \Delta(\mathsf{A}(-\mathbf{v}_S))\|_2 \leq 0, \qquad \text{i.e.,} \quad -\mathbf{v}_S = \Delta(\mathsf{A}(-\mathbf{v}_S)) = \Delta(\mathsf{A}\mathbf{v}_{\overline{S}}),$$
$$\|\mathbf{v}_{\overline{S}} - \Delta(\mathsf{A}\mathbf{v}_{\overline{S}})\|_2 \leq \frac{C}{\sqrt{s}}\sigma_s(\mathbf{v}_{\overline{S}})_1, \quad \text{i.e.,} \quad \|\mathbf{v}\|_2 \leq \frac{C}{\sqrt{s}}\sigma_s(\mathbf{v}_{\overline{S}})_1 \leq \frac{C}{\sqrt{s}}\|\mathbf{v}\|_1.$$

Given that the latter inequality holds for all $\mathbf{v} \in \ker \mathsf{A}$, which is a subspace of ℓ_2^N of codimension at most m, we deduce that $d^m(B_1^N, \ell_2^N) \leq C/\sqrt{s}$. But since $d^m(B_1^N, \ell_2^N) \geq c \min\{1, \sqrt{\ln(eN/m)/m}\}$, we conclude that $m \geq cs \ln(eN/m)$ (the case $s \leq C$ being set aside). This looks like, but is not exactly the same as, the desired condition $m \geq cs \ln(eN/s)$. It is not hard to see that these two conditions are actually equivalent up to changing the constant c, as justified in [39, Lemma C.6].

2.3 Consequence of the Upper Estimate

Briefly stated, *the upper estimate provides a pair* (A, Δ) *which is stable of order* $s \asymp m/ln(eN/m)$. Indeed, in view of $d^m(B_1^N, \ell_2^N) \le C\sqrt{\ln(eN/m)/m}$, if we set $s \approx m/(8C^2 \ln(eN/m))$, then there is a matrix $\mathsf{A} \in \mathbb{R}^{m \times N}$ (whose null space is an optimal subspace L^m in the definition of the Gelfand width) such that

$$\|\mathbf{v}\|_2 \le \frac{1}{\sqrt{8s}} \|\mathbf{v}\|_1 \qquad \text{for all } \mathbf{v} \in \ker \mathsf{A}.$$

The recovery map is defined, for any $\boldsymbol{y} \in \mathbb{R}^m$, by

$$\Delta(\boldsymbol{y}) = \underset{\boldsymbol{z} \in \boldsymbol{R}^N}{\operatorname{argmin}}\, \sigma_s(\boldsymbol{z})_1 \qquad \text{subject to } \mathsf{A}\boldsymbol{z} = \boldsymbol{y}. \tag{4}$$

For a vector $\boldsymbol{x} \in \mathbb{R}^N$, let us write $\widetilde{\boldsymbol{x}} := \Delta(\mathsf{A}\boldsymbol{x})$ for short and let us consider index sets S and $\widetilde{S}$ of s largest absolute entries of $\boldsymbol{x}$ and $\widetilde{\boldsymbol{x}}$, respectively. By the definition of $\widetilde{\boldsymbol{x}}$, we have

$$\|\widetilde{\boldsymbol{x}} - \widetilde{\boldsymbol{x}}_{\widetilde{S}}\|_1 \le \|\boldsymbol{x} - \boldsymbol{x}_S\|_1 = \sigma_s(\boldsymbol{x})_1.$$

In particular, there holds

$$\|\widetilde{\boldsymbol{x}} - \widetilde{\boldsymbol{x}}_{S \cup \widetilde{S}}\|_1 \le \|\widetilde{\boldsymbol{x}} - \widetilde{\boldsymbol{x}}_{\widetilde{S}}\|_1 \le \sigma_s(\boldsymbol{x})_1, \quad \text{as well as } \|\boldsymbol{x} - \boldsymbol{x}_{S \cup \widetilde{S}}\|_1 \le \|\boldsymbol{x} - \boldsymbol{x}_S\|_1 = \sigma_s(\boldsymbol{x})_1.$$

Moreover, the fact that $(\boldsymbol{x} - \widetilde{\boldsymbol{x}})_{S \cup \widetilde{S}}$ is $2s$-sparse implies that

$$\|(\boldsymbol{x} - \widetilde{\boldsymbol{x}})_{S \cup \widetilde{S}}\|_1 \le \sqrt{2s}\, \|(\boldsymbol{x} - \widetilde{\boldsymbol{x}})_{S \cup \widetilde{S}}\|_2 \le \sqrt{2s}\, \|\boldsymbol{x} - \widetilde{\boldsymbol{x}}\|_2.$$

Therefore, observing that $\boldsymbol{x} - \widetilde{\boldsymbol{x}} \in \ker \mathsf{A}$, we derive

$$\begin{aligned} \|\boldsymbol{x} - \widetilde{\boldsymbol{x}}\|_2 &\le \frac{1}{\sqrt{8s}} \|\boldsymbol{x} - \widetilde{\boldsymbol{x}}\|_1 \le \frac{1}{\sqrt{8s}} \left(\|(\boldsymbol{x} - \widetilde{\boldsymbol{x}})_{S \cup \widetilde{S}}\|_1 + \|\boldsymbol{x} - \boldsymbol{x}_{S \cup \widetilde{S}}\|_1 + \|\widetilde{\boldsymbol{x}} - \widetilde{\boldsymbol{x}}_{S \cup \widetilde{S}}\|_1 \right) \\ &\le \frac{1}{\sqrt{8s}} \left(\sqrt{2s}\, \|\boldsymbol{x} - \widetilde{\boldsymbol{x}}\|_2 + 2\sigma_s(\boldsymbol{x})_1 \right) = \frac{1}{2} \|\boldsymbol{x} - \widetilde{\boldsymbol{x}}\|_2 + \frac{1}{\sqrt{2s}} \sigma_s(\boldsymbol{x})_1. \end{aligned} \tag{5}$$

After rearrangement, we obtain the desired inequality (3) with constant $C = \sqrt{2}$.

2.4 Discussion

Arguably, the story of compressive sensing could very well end here... After all, the standard compressive sensing problem has been solved in its stable version (3). Of course, the optimization program (4) is impractical, but it can be replaced by the computationally friendly recovery map Δ_{BP} arising from the ℓ_1-minimization (16). Indeed, it is known (though not really well known) that if a pair (A, Δ) is stable of order s, then the pair $(\mathsf{A}, \Delta_{\mathrm{BP}})$ is stable of order s/C, see [39, Exercise 11.5]. In fact, we will present at the beginning of Sect. 6 a short argument showing that the upper estimate implies that the pair $(\mathsf{A}, \Delta_{\mathrm{BP}})$ is stable of order $s \asymp m/\ln(eN/m)$. So why not end the story here, really? Well, for starter, we have to take these difficult Gelfand width estimates for granted. In contrast, the compressive sensing theory can now dissociate itself from Gelfand widths entirely, as we will see in Sect. 6.3. Besides, the theory created some tools that justify the Gelfand width estimates differently, as exposed in Sects. 6.1 and 6.2. The breakthrough in Kashin's original argument for the upper bound was the use of probabilisitic techniques. The compressive sensing justification of the upper estimate also relies on such techniques, but it has the appealing feature of decoupling the probabilistic ingredients from the deterministic ingredients. This feature is explained in the next section, where we focus on a property of the matrix A that implies stability of many pairs (A, Δ) and we prove that this property is satisfied with high probability by Gaussian random matrices. Whether one counts the story of compressive sensing by beginning from Gelfand widths or by building an independent theory yielding Gelfand width estimates as corollaries is probably a matter of personal taste. It is the latter alternative that has my preference, as it incidentally provided me with some intuition not only on widths but also on other aspects of Banach space geometry such as Kashin's decomposition theorem (see [39, Sect. 10.3]).

3 The Restricted Isometry Property

The deterministic property mentioned a few lines above is the so-called restricted isometry property (RIP) introduced in [18].

Definition 1 A matrix $\mathsf{A} \in \mathbb{C}^{m \times N}$ satisfies the restricted isometry property of order s with constant $\delta \in (0, 1)$ if

$$(1-\delta)\|z\|_2^2 \le \|\mathsf{A}z\|_2^2 \le (1+\delta)\|z\|_2^2 \qquad \text{for all } s\text{-sparse vectors } z \in \mathbb{C}^N. \tag{6}$$

The smallest such constant δ is called the restricted isometry constant of order s. It is denoted by δ_s and is also expressed as

$$\delta_s = \max_{\mathrm{card}(S)=s} \|\mathsf{A}_S^*\mathsf{A}_S - \mathsf{I}\|_{2\to 2},$$

where A_S represents the column submatrix of A indexed by S. Our first task is to verify that, in the optimal regime of parameters s, m, and N, RIP is really a nonvoid property. This is achieved via probabilistic techniques. Next, we present a simple algorithm which, coupled to a matrix with RIP, solves the standard compressive sensing problem in an expeditious fashion.

3.1 RIP Matrices Do Exist

For a matrix A with real entries, it is straightforward to see that the complex RIP (6) holds if and only if its real counterpart involving only real-valued vectors z holds. It is this real version that is established for Gaussian matrices in the theorem below. Its proof is included because the argument for the concentration inequality is simpler in the Gaussian case than in the more general case treated in [39, Sect. 9.1].

Theorem 1 *Let* $\mathsf{A} \in \mathbb{R}^{m\times N}$ *be a random matrix populated by independent Gaussian variables with mean zero and variance* $1/m$. *If* $m \geq C\delta^{-2}s\ln(eN/s)$, *then, with probability at least* $1 - 2\exp(-c\delta^2 m)$, *the matrix* A *satisfies the restricted isometry property of order* s *with constant* δ.

Proof The classical proof presented here consists of three steps.
Concentration inequality: For a fixed vector $\boldsymbol{x} \in \mathbb{R}^N$, and for any $i \in [\![1:m]\!]$, since a linear combination of Gaussian random variables is still Gaussian, we have

$$(\mathsf{A}\boldsymbol{x})_i = \sum_{j=1}^{N} A_{i,j}x_j = \frac{\|\boldsymbol{x}\|_2}{\sqrt{m}} g_i,$$

where g_i denotes a standard Gaussian random variable (i.e., it has mean zero and variance one). Then, given $t \in (0,1)$, since $\|\mathsf{A}\boldsymbol{x}\|_2^2 = \sum_{i=1}^m (\mathsf{A}\boldsymbol{x})_i^2$, we have

$$\begin{aligned}\mathbb{P}\left(\|\mathsf{A}\boldsymbol{x}\|_2^2 > (1+t)\|\boldsymbol{x}\|_2^2\right) &= \mathbb{P}\left(\sum_{i=1}^{m} g_i^2 \geq m(1+t)\right)\\ &= \mathbb{P}\left(\exp\left(u\sum_{i=1}^{m} g_i^2\right) > \exp(um(1+t))\right),\end{aligned}$$

where the specific value of $u > 0$ will be chosen later. Using Markov's inequality (see, e.g., [39, Theorem 7.3]), the independence of the random variables $g_1, \ldots, g_m$, as well as the expression of moment-generating function $\theta \mapsto \mathbb{E}(\exp(\theta g^2))$ of a squared standard Gaussian random variable (see, e.g., [39, Lemma 7.6]), we derive

$$\begin{aligned}\mathbb{P}\left(\|\mathsf{A}\boldsymbol{x}\|_2^2 > (1+t)\|\boldsymbol{x}\|_2^2\right) &\leq \frac{\mathbb{E}\left(\exp\left(u\sum_{i=1}^m g_i^2\right)\right)}{\exp(um(1+t))} = \frac{\mathbb{E}\left(\prod_{i=1}^m \exp\left(ug_i^2\right)\right)}{\prod_{i=1}^m \exp(u(1+t))}\\ &= \prod_{i=1}^{m} \frac{\mathbb{E}\left(\exp(ug_i^2)\right)}{\exp(u(1+t))} = \left(\frac{1/\sqrt{1-2u}}{\exp(u(1+t))}\right)^m.\end{aligned}$$

We note that imposing $u < 1/4$ allows us to write

$$\frac{1}{\sqrt{1-2u}} = \left(1 + \frac{2u}{1-2u}\right)^{1/2} \leq (1+2u(1+4u))^{1/2} \leq \exp(2u(1+4u))^{1/2}$$
$$= \exp(u+4u^2).$$

It follows that

$$\mathbb{P}\left(\|\mathsf{A}\boldsymbol{x}\|_2^2 > (1+t)\|\boldsymbol{x}\|_2^2\right) \leq \left(\exp(4u^2 - ut)\right)^m = \exp\left(\frac{t^2}{16} - \frac{t^2}{8}\right)^m = \exp\left(-\frac{mt^2}{16}\right),$$

where we have made the optimal choice $u = t/8 < 1/4$. Likewise, we can show that

$$\mathbb{P}\left(\|\mathsf{A}\boldsymbol{x}\|_2^2 < (1-t)\|\boldsymbol{x}\|_2^2\right) \leq \exp\left(-\frac{mt^2}{16}\right).$$

All in all, for a fixed $\boldsymbol{x} \in \mathbb{R}^N$, we have obtained the concentration inequality

$$\mathbb{P}\left(\left|\|\mathsf{A}\boldsymbol{x}\|_2^2 - \|\boldsymbol{x}\|_2^2\right| > t\|\boldsymbol{x}\|_2^2\right) \leq 2\exp\left(-\frac{mt^2}{16}\right).$$

Covering argument: We rely on the fact that the unit sphere of $\mathbb{R}^s$ can be covered with $n \leq (1+2/\rho)^s$ balls of radius ρ (see, e.g., [39, Proposition C.3]). Fixing an index set S of size s and identifying the space $\Sigma_S := \{\boldsymbol{z} \in \mathbb{R}^N : \mathrm{supp}(\boldsymbol{z}) \subseteq S\}$ with $\mathbb{R}^s$, this means that there are ℓ_2-normalized vectors $\mathbf{u}_1, \ldots, \mathbf{u}_n \in \Sigma_S$, $n \leq (1+2/\rho)^s$, such that any ℓ_2-normalized $\boldsymbol{x} \in \Sigma_S$ satisfies $\|\boldsymbol{x} - \mathbf{u}_k\|_2 \leq \rho$ for some $k \in [\![1:n]\!]$. Given $t \in (0,1)$, the concentration inequality reads, for any $k \in [\![1:n]\!]$,

$$\mathbb{P}\left(|\langle(\mathsf{A}_S^*\mathsf{A}_S - \mathsf{I})\mathbf{u}_k, \mathbf{u}_k\rangle| > t\right) = \mathbb{P}\left(\left|\|\mathsf{A}\mathbf{u}_k\|_2^2 - \|\mathbf{u}_k\|_2^2\right| > t\right) \leq 2\exp\left(-\frac{mt^2}{16}\right).$$

Therefore, writing $\mathsf{B} := \mathsf{A}_S^*\mathsf{A}_S - \mathsf{I}$ for short, by the union bound

$$\mathbb{P}\left(|\langle\mathsf{B}\mathbf{u}_k, \mathbf{u}_k\rangle| > t \text{ for some } k \in [\![1:n]\!]\right) \leq n \times 2\exp\left(-\frac{mt^2}{16}\right)$$
$$\leq 2\left(1+\frac{2}{\rho}\right)^s \exp\left(-\frac{mt^2}{16}\right). \quad (7)$$

Let us place ourselves in the likely situation, where $|\langle\mathsf{B}\mathbf{u}_k, \mathbf{u}_k\rangle| \leq t$ for all $k \in [\![1:n]\!]$. Then, for $\boldsymbol{x} \in \Sigma_S$ with $\|\boldsymbol{x}\|_2 = 1$, we consider $k \in [\![1:n]\!]$ such that $\|\boldsymbol{x} - \mathbf{u}_k\|_2 \leq \rho$. We obtain, thanks to the self-adjointness of B,

$$|\langle\mathsf{B}\boldsymbol{x}, \boldsymbol{x}\rangle| = |\langle\mathsf{B}\mathbf{u}_k, \mathbf{u}_k\rangle + \langle\mathsf{B}(\boldsymbol{x}+\mathbf{u}_k), \boldsymbol{x}-\mathbf{u}_k\rangle| \leq |\langle\mathsf{B}\mathbf{u}_k, \mathbf{u}_k\rangle| + |\langle\mathsf{B}(\boldsymbol{x}+\mathbf{u}_k), \boldsymbol{x}-\mathbf{u}_k\rangle|$$
$$\leq t + \|\mathsf{B}\|_{2\to2}\|\boldsymbol{x}+\mathbf{u}_k\|_2\|\boldsymbol{x}-\mathbf{u}_k\|_2 \leq t + 2\rho\|\mathsf{B}\|_{2\to2}.$$

Taking the supremum over $\boldsymbol{x}$ yields

$$\|\mathsf{B}\|_{2\to 2} \le t + 2\rho \|\mathsf{B}\|_{2\to 2}.$$

With the choice $\rho = 1/4$ and $t = \delta/2$, a rearrangement gives

$$\|\mathsf{B}\|_{2\to 2} \le \frac{t}{1-2\rho} = \delta.$$

This occurs with failure probability at most (7), i.e.,

$$\begin{aligned}\mathbb{P}\left(\|\mathsf{A}_S^*\mathsf{A}_S - \mathsf{I}\|_{2\to 2} > \delta\right) &\le 2\left(1 + \frac{2}{1/4}\right)^s \exp\left(-\frac{m(\delta/2)^2}{16}\right)\\ &= 2\exp\left(\ln(9)s - \frac{m\delta^2}{64}\right).\end{aligned}$$

Union bound: It remains to unfix the index set S. This leads to

$$\begin{aligned}\mathbb{P}(\delta_s > \delta) &= \mathbb{P}\left(\|\mathsf{A}_S^*\mathsf{A}_S - \mathsf{I}\|_{2\to 2} > \delta \text{ for some } S \subseteq [\![1:N]\!] \text{ with } \mathrm{card}(S) = s\right)\\ &\le \binom{N}{s} \times 2\exp\left(\ln(9)s - \frac{m\delta^2}{64}\right) \le 2\exp\left(s\ln\left(\frac{eN}{s}\right) + \ln(9)s - \frac{m\delta^2}{64}\right)\\ &\le 2\exp\left(\ln(9e)s\ln\left(\frac{eN}{s}\right) - \frac{m\delta^2}{64}\right),\end{aligned}$$

where we have used the inequality $\binom{N}{s} \le (eN/s)^s$ (see, e.g., [39, Lemma C.5]). We conclude that

$$\mathbb{P}(\delta_s > \delta) \le 2\exp\left(-\frac{m\delta^2}{128}\right),$$

provided that $\ln(9e)s\ln(eN/s) \le m\delta^2/128$, i.e., that $m \ge C\delta^{-2}s\ln(eN/s)$ with $C = 128\ln(9e)$, which is exactly our starting assumption. □

3.2 Recovery for the Impatient: Iterative Hard Thresholding

The RIP guarantees that various sparse recovery procedures are stable and robust. Typically, a condition of the type $\delta_{\kappa s} < \delta_*$ for some small integer $\kappa = 1, 2, 3, \ldots$ and some threshold $\delta_* \in (0, 1)$ allows one to prove a robustness estimate of the type

$$\|\boldsymbol{x} - \Delta(\mathsf{A}\boldsymbol{x} + \mathbf{e})\|_2 \le D\|\mathbf{e}\|_2$$

valid for all s-sparse $\boldsymbol{x} \in \mathbb{C}^N$ and all vectors $\mathbf{e} \in \mathbb{C}^m$. We will study recovery maps Δ associated with orthogonal matching pursuit and with ℓ_1-minimizations in the next two sections. But the sparse recovery procedure with the easiest justification turns out to be iterative hard thresholding (IHT), so we present the argument here. The algorithm consists in mimicking classical iterative methods for solving the square system $\mathsf{A}^*\mathsf{A}z = \mathsf{A}^* y$ and 'sparsifying' each iteration. Precisely, starting from $x^0 = \mathbf{0}$, we construct a sequence $(x^n)_{n\geq 0}$ according to

$$x^{n+1} = H_s(x^n + \mathsf{A}^*(y - \mathsf{A}x^n)), \qquad n \geq 0.$$

The operator H_s is the hard thresholding operator that keeps s largest absolute entries of a vector and sends the other ones to zero. The condition $\delta_{3s} < 1/2$ below is not optimal, as a little more effort reveals that $\delta_{3s} < 1/\sqrt{3}$ is also sufficient, see [39, Theorem 6.18].

Theorem 2 *Suppose that a matrix* $\mathsf{A} \in \mathbb{C}^{m\times N}$ *has a restricted isometry constant* $\delta_{3s} < 1/2$. *Then, for all s-sparse* $x \in \mathbb{C}^N$ *and all* $\mathbf{e}\in\mathbb{C}^m$, *the output* $x^\infty := \lim_{n\to\infty} x^n$ *of IHT run on* $\mathbf{y} = \mathsf{A}x + \mathbf{e}$ *satisfies*

$$\|x - x^\infty\|_2 \leq \frac{d}{1 - 2\delta_{3s}}\|\mathbf{e}\|_2.$$

Proof It is enough to prove that there are constants $0 < \rho < 1$ and $d > 0$ such that, for any $n \geq 0$,

$$\|x - x^{n+1}\|_2 \leq \rho\|x - x^n\|_2 + d\|\mathbf{e}\|_2.$$

To derive this inequality, we first remark that x^{n+1} is a better s-term approximation to $x^n + \mathsf{A}^*(y - \mathsf{A}x^n)$ than x is, so that

$$\|x^n + \mathsf{A}^*(y - \mathsf{A}x^n) - x^{n+1}\|_2^2 \leq \|x^n + \mathsf{A}^*(y - \mathsf{A}x^n) - x\|_2^2.$$

Next, introducing x in the left-hand side and expanding the square gives

$$\begin{aligned}\|x - x^{n+1}\|_2^2 &\leq -2\langle x - x^{n+1}, x^n + \mathsf{A}^*(y - \mathsf{A}x^n) - x\rangle \\ &= -2\langle x - x^{n+1}, (\mathsf{A}^*\mathsf{A} - \mathsf{I})(x - x^n) + \mathsf{A}^*\mathbf{e}\rangle \\ &= -2\left(\langle x - x^{n+1}, (\mathsf{A}_T^*\mathsf{A}_T - \mathsf{I})(x - x^n)\rangle + \langle \mathsf{A}(x - x^{n+1}), \mathbf{e}\rangle\right),\end{aligned}$$

where the index set $T := \mathrm{supp}(x) \cup \mathrm{supp}(x^n) \cup \mathrm{supp}(x^{n+1})$ has size at most $3s$. It then follows that

$$\begin{aligned}\|x - x^{n+1}\|_2^2 &\leq 2\|x - x^{n+1}\|_2\|\mathsf{A}_T^*\mathsf{A}_T - \mathsf{I}\|_{2\to 2}\|x - x^n\|_2 + 2\|\mathsf{A}(x - x^{n+1})\|_2\|\mathbf{e}\|_2 \\ &\leq 2\delta_{3s}\|x - x^{n+1}\|_2\|x - x^n\|_2 + 2\sqrt{1+\delta_{3s}}\|x - x^{n+1}\|_2\|\mathbf{e}\|_2.\end{aligned}$$

After simplifying by $\|x - x^{n+1}\|_2$, we obtain the desired result with $\rho := 2\delta_{3s} < 1$ and $d := 2\sqrt{1+\delta_{3s}} < \sqrt{6}$. □

4 Orthogonal Matching Pursuit

The IHT procedure presented above has a critical deficiency, in that an estimation of the sparsity level is needed for the algorithm to run. This is not the case for the recovery procedures put forward in this section and the next.[4] We start by reviewing the orthogonal matching pursuit (OMP) procedure.

4.1 The Algorithm

Orthogonal matching pursuit prevails as a popular algorithm in approximation theory, specifically in the subfield of sparse approximation, where it is better known as orthogonal greedy algorithm (see the book [64] for a comprehensive treatment). There is a subtle difference in perspective, though, in that compressive sensing aims at approximating a vector $\boldsymbol{x} \in \mathbb{C}^N$ by some $\widetilde{\boldsymbol{x}} \in \mathbb{C}^N$ computed from $\boldsymbol{y} = \mathsf{A}\boldsymbol{x} \in \mathbb{C}^m$ only, while the goal in sparse approximation is to approximate a vector $\boldsymbol{y} \in \mathbb{C}^m$, not necessarily of the form $\boldsymbol{y} = \mathsf{A}\boldsymbol{x}$, by some $\mathsf{A}\widetilde{\boldsymbol{x}} \in \mathbb{C}^m$ where $\widetilde{\boldsymbol{x}} \in \mathbb{C}^N$ is sparse. Under RIP, these tasks are somewhat equivalent since $\|\mathsf{A}(\boldsymbol{x} - \widetilde{\boldsymbol{x}})\|_2 \approx \|\boldsymbol{x} - \widetilde{\boldsymbol{x}}\|_2$ when $\boldsymbol{x}$ and $\widetilde{\boldsymbol{x}}$ are sparse. Regardless of the intended usage, the OMP algorithm does not change—it is an iterative algorithm that consists in building a candidate support by adding new indices one by one and producing candidate vectors with the candidate support that best fit the measurements. More formally, starting from the index set $S^0 = \emptyset$ and the vector $\boldsymbol{x}^0 = \boldsymbol{0}$, a sequence $(S^n, \boldsymbol{x}^n)_{n \geq 0}$ is constructed according to

$$S^{n+1} := S^n \cup \{j^{n+1}\}, \qquad \text{where } j^{n+1} := \underset{j \in [\![1:N]\!]}{\operatorname{argmin}} |(\mathsf{A}^*(\boldsymbol{y} - \mathsf{A}\boldsymbol{x}^n))_j|, \tag{OMP$_1$}$$

$$\boldsymbol{x}^{n+1} := \underset{\boldsymbol{z} \in \mathbb{C}^N}{\operatorname{argmin}} \|\boldsymbol{y} - \mathsf{A}\boldsymbol{z}\|_2 \text{ subject to } \operatorname{supp}(\boldsymbol{z}) \subseteq S^{n+1}, \tag{OMP$_2$}$$

until a stopping criterion is met. The rule for choosing the new index j^{n+1} is, in retrospect, justified by (10) when the columns $\boldsymbol{a}_1, \ldots, \boldsymbol{a}_N$ of A are ℓ_2-normalized, which we assume for the rest of this subsection. As a matter of fact, if one wishes to be as greedy as one can be and decrease the norm of the residual as much as possible at every iteration, the new index j^{n+1} should maximize the quantity $|(\mathsf{A}^*(\boldsymbol{y} - \mathsf{A}\boldsymbol{x}^n))_j|/\operatorname{dist}(\boldsymbol{a}_j, \operatorname{span}\{\boldsymbol{a}_{j^1}, \ldots, \boldsymbol{a}_{j^n}\})$ instead of $|(\mathsf{A}^*(\boldsymbol{y} - \mathsf{A}\boldsymbol{x}^n))_j|$. Under RIP, the former is approximately equal to the latter, though. See [30] for details.

[4]Arguably, orthogonal matching pursuit could require an estimation of the sparsity level if an estimation of the magnitude of the measurement error is not available.

4.2 RIP Analysis

If one wants to recover s-sparse vectors via OMP, it seems quite natural to stop the algorithm after s iterations (although this would require a prior estimation of s). In fact, when $m \asymp s\ln(N)$, it was established in [67] that, given a fixed s-sparse vector $\boldsymbol{x} \in \mathbb{C}^N$, one has $\mathbb{P}(\boldsymbol{x}^s = \boldsymbol{x}) \geq 1 - 1/N$. But a stronger result of the type $\mathbb{P}(\boldsymbol{x}^s = \boldsymbol{x}$ for all s-sparse $\boldsymbol{x} \in \mathbb{C}^N) \geq 1 - o(N)$ is not possible, see [27, Sect. 7]. However, such a result becomes possible if the number of iterations is allowed to exceed s. Of course, this number of iterations should remain small, ideally at most proportional to s, as is the case for, e.g., Hard Thresholding Pursuit [9]. This objective is achievable under RIP. Here is a precise statement, taking robustness into account. In this statement, one could take $\kappa = 12$ and $\delta_* = 1/6$, but other choices could be made, too.

Theorem 3 *Suppose that a matrix* $\mathsf{A} \in \mathbb{C}^{m\times N}$ *with* ℓ_2*-normalized columns has restricted isometry constant* $\delta_{(1+\kappa)s} < \delta_*$. *Then, for all* s*-sparse* $\boldsymbol{x} \in \mathbb{C}^N$ *and all* $\mathbf{e} \in \mathbb{C}^m$, *the output of the OMP algorithm run on* $\boldsymbol{y} = \mathsf{A}\boldsymbol{x} + \mathbf{e}$ *satisfies*

$$\|\boldsymbol{y} - \mathsf{A}\boldsymbol{x}^{\kappa s}\|_2 \leq D\|\mathbf{e}\|_2. \tag{8}$$

The original proof of this result is rather convoluted, see [69]. It was slightly simplified in [30] (see also, [39, Theorem 6.25]), and later much simplified in [24]. Even so, the arguments feel artificial and a natural proof is still coveted. There is an easy explanation, presented below, for the recovery of sparse vectors that are 'flat' on their supports. Numerical evidence suggests that this is in fact the worst case, see Fig. 1. The best case seems to be the one of the sparse vectors whose nonincreasing rearrangements decay geometrically, because each new index j^{n+1} can be shown

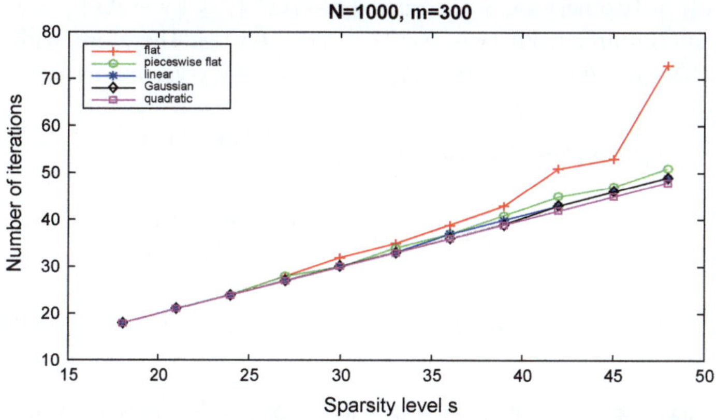

Fig. 1 Number of iterations OMP needs to recover s-sparse vectors, maximized over 200 tries, when the sparse vectors are as follows: flat ($x_j = 1$), piecewise flat ($x_j \in \{1, 2\}$), linear ($x_j = j$), Gaussian ($x_j \sim \mathcal{N}(0, 1)$), and quadratic ($x_j = j^2$)

to belong to the true support. Intuitively, one feels that there should be a simple reasoning bridging these two extreme cases.

Proposition 1 *Given* $\gamma \geq 1$, *let* $\boldsymbol{x} \in \mathbb{C}^N$ *be an* s*-sparse vector which satisfies* $\max\{|x_j|, j \in \mathrm{supp}(\boldsymbol{x})\} \leq \gamma \min\{|x_j|, j \in \mathrm{supp}(\boldsymbol{x})\}$. *With* $t := \lceil 12\gamma^4 s \rceil$, *if a matrix* $\mathsf{A} \in \mathbb{C}^{m \times N}$ *with* ℓ_2*-normalized columns has restricted isometry constant* $\delta_{s+t} \leq 1/5$, *then for all* $\mathbf{e} \in \mathbb{C}^m$, *the output of the OMP algorithm run on* $\boldsymbol{y} = \mathsf{A}\boldsymbol{x} + \mathbf{e}$ *satisfies*

$$\|\boldsymbol{y} - \mathsf{A}\boldsymbol{x}^t\|_2 \leq 4\|\mathbf{e}\|_2.$$

Proof We recall the following observations (see [39, Lemmas 3.4 and 3.3]):

$$\left(\mathsf{A}^*(\boldsymbol{y} - \mathsf{A}\boldsymbol{x}^n)\right)_{S^n} = \mathbf{0}, \tag{9}$$

$$\|\boldsymbol{y} - \mathsf{A}\boldsymbol{x}^{n+1}\|_2^2 \leq \|\boldsymbol{y} - \mathsf{A}\boldsymbol{x}^n\|_2^2 - |(\mathsf{A}^*(\boldsymbol{y} - \mathsf{A}\boldsymbol{x}^n))_{j^{n+1}}|^2. \tag{10}$$

With S denoting the support of $\boldsymbol{x}$, we now remark that

$$\begin{aligned}
\|\boldsymbol{x}_{S \setminus S^n}\|_1 |(\mathsf{A}^*(\boldsymbol{y} - \mathsf{A}\boldsymbol{x}^n))_{j^{n+1}}| &\geq \sum_{j \in S \setminus S^n} x_j (\mathsf{A}^*(\boldsymbol{y} - \mathsf{A}\boldsymbol{x}^n))_j \\
&= \sum_{j \in S \cup S^n} (\boldsymbol{x} - \boldsymbol{x}^n)_j (\mathsf{A}^*(\boldsymbol{y} - \mathsf{A}\boldsymbol{x}^n))_j = \langle \boldsymbol{x} - \boldsymbol{x}^n, \mathsf{A}^*(\boldsymbol{y} - \mathsf{A}\boldsymbol{x}^n) \rangle \\
&= \langle \mathsf{A}(\boldsymbol{x} - \boldsymbol{x}^n), \boldsymbol{y} - \mathsf{A}\boldsymbol{x}^n \rangle = \|\mathsf{A}(\boldsymbol{x} - \boldsymbol{x}^n)\|_2^2 + \langle \mathsf{A}(\boldsymbol{x} - \boldsymbol{x}^n), \mathbf{e} \rangle \\
&\geq \|\mathsf{A}(\boldsymbol{x} - \boldsymbol{x}^n)\|_2^2 - \|\mathsf{A}(\boldsymbol{x} - \boldsymbol{x}^n)\|_2 \|\mathbf{e}\|_2.
\end{aligned}$$

For $n < t$, we may assume that $\|\mathbf{e}\|_2 \leq (1 - c)\|\mathsf{A}(\boldsymbol{x} - \boldsymbol{x}^n)\|_2$, where $c := \sqrt{5}/4$, otherwise $\|\boldsymbol{y} - \mathsf{A}\boldsymbol{x}^n\|_2 \leq \|\mathsf{A}(\boldsymbol{x} - \boldsymbol{x}^n)\|_2 + \|\mathbf{e}\|_2 \leq (1/(1 - c) + 1)\|\mathbf{e}\|_2 \leq 4\|\mathbf{e}\|_2$, so the result is acquired from $\|\boldsymbol{y} - \mathsf{A}\boldsymbol{x}^t\|_2 \leq \|\boldsymbol{y} - \mathsf{A}\boldsymbol{x}^n\|_2$. We may also assume that $\|\boldsymbol{x}_{S \setminus S^n}\|_1 > 0$, otherwise $S \subseteq S^n$ and $\|\boldsymbol{y} - \mathsf{A}\boldsymbol{x}^n\|_2 \leq \|\boldsymbol{y} - \mathsf{A}\boldsymbol{x}\|_2 = \|\mathbf{e}\|_2$, so the result is again acquired from $\|\boldsymbol{y} - \mathsf{A}\boldsymbol{x}^t\|_2 \leq \|\boldsymbol{y} - \mathsf{A}\boldsymbol{x}^n\|_2$. Therefore, with $\delta := \delta_{s+t}$, we obtain $|(\mathsf{A}^*(\boldsymbol{y} - \mathsf{A}\boldsymbol{x}^n))_{j^{n+1}}| \geq c\|\mathsf{A}(\boldsymbol{x} - \boldsymbol{x}^n)\|_2^2 / \|\boldsymbol{x}_{S \setminus S^n}\|_1$, and in turn

$$\begin{aligned}
\|\boldsymbol{y} - \mathsf{A}\boldsymbol{x}^{n+1}\|_2^2 &\leq \|\boldsymbol{y} - \mathsf{A}\boldsymbol{x}^n\|_2^2 - \frac{c^2 \|\mathsf{A}(\boldsymbol{x} - \boldsymbol{x}^n)\|_2^4}{\|\boldsymbol{x}_{S \setminus S^n}\|_1^2} \\
&\leq \|\boldsymbol{y} - \mathsf{A}\boldsymbol{x}^n\|_2^2 - \frac{c^2 (1 - \delta)^2 \|\boldsymbol{x} - \boldsymbol{x}^n\|_2^4}{\|\boldsymbol{x}_{S \setminus S^n}\|_1^2}.
\end{aligned}$$

With $\alpha := \min\{|x_j|, j \in S\}$ and $\beta := \max\{|x_j|, j \in S\}$, we now notice that

$$\|\boldsymbol{x} - \boldsymbol{x}^n\|_2^2 \geq \|\boldsymbol{x}_{S \setminus S^n}\|_2^2 \geq \alpha^2 \mathrm{card}(S \setminus S^n) = \frac{\beta^2}{\gamma^2} \mathrm{card}(S \setminus S^n),$$

$$\|\boldsymbol{x}_{S \setminus S^n}\|_1 \leq \beta \mathrm{card}(S \setminus S^n).$$

These inequalities allow us to derive that

$$\|y - Ax^{n+1}\|_2^2 \leq \|y - Ax^n\|_2^2 - \frac{c^2(1-\delta^2)}{\gamma^4}\beta^2 \leq \|y - Ax^n\|_2^2 - \frac{c^2(1-\delta^2)}{\gamma^4}\frac{\|x\|_2^2}{s}.$$

By immediate induction, we arrive at

$$\begin{aligned}\|y - Ax^t\|_2^2 &\leq \|y\|_2^2 - t\frac{c^2(1-\delta^2)}{\gamma^4}\frac{\|x\|_2^2}{s} \leq 2\|Ax\|_2^2 + 2\|\mathbf{e}\|_2^2 - \frac{t}{s}\frac{c^2(1-\delta^2)}{\gamma^4}\|x\|_2^2 \\ &\leq \left(2(1+\delta) - \frac{t}{s}\frac{c^2(1-\delta)^2}{\gamma^4}\right)\|x\|_2^2 + 2\|\mathbf{e}\|_2^2 \\ &= \left(\frac{12}{5} - \frac{t}{s}\frac{1}{5\gamma^4}\right)\|x\|_2^2 + 2\|\mathbf{e}\|_2^2 \leq 2\|\mathbf{e}\|_2^2,\end{aligned}$$

where the last two steps followed from the choices $\delta = 1/5$ and $t = \lceil 12\gamma^4 s \rceil$. □

4.3 *Stability via the Sort-and-Split Technique*

The bound (8) takes care of measurement error, but what about sparsity defect? As mentioned in Sect. 1, this can be incorporated into the measurement error. Below is an argument that leads to stability and robustness estimates. For $x \in \mathbb{C}^N$, not necessarily s-sparse, we consider an index set T of $t = 2s$ largest absolute entries of x and we write $y = Ax + \mathbf{e}$ as $y = Ax_T + \mathbf{e}'$ with $\mathbf{e}' := Ax_{\overline{T}} + \mathbf{e}$. Then, provided $\delta_{2(1+\kappa)s} < \delta_*$, we can apply Theorem 3 to obtain

$$\|y - Ax^{2\kappa s}\|_2 \leq D\|\mathbf{e}'\|_2 \leq D\|Ax_{\overline{T}}\|_2 + D\|\mathbf{e}\|_2.$$

To get a bound for $\|x - x^{2\kappa s}\|_2$ instead—a 'compressive sensing' estimate rather than a 'sparse approximation' estimate, so to speak—we start by noticing that

$$\|x - x^{2\kappa s}\|_2 \leq \|x_T - x^{2\kappa s}\|_2 + \|x_{\overline{T}}\|_2 \leq \|x_T - x^{2\kappa s}\|_2 + \frac{1}{\sqrt{s}}\sigma_s(x)_1, \tag{11}$$

where we have applied Stechkin bound for the ℓ_q-norm of the tail of a vector in terms of its ℓ_p-norm, $p < q$ (see [39, Proposition 2.3]). Then, using the RIP, we have

$$\begin{aligned}\|x_T - x^{2\kappa s}\|_2 &\leq \frac{1}{\sqrt{1-\delta_*}}\|A(x_T - x^{2\kappa s})\|_2 = \frac{1}{\sqrt{1-\delta_*}}\|y - Ax^{2\kappa s} - (Ax_{\overline{T}} + \mathbf{e})\|_2 \\ &\leq \frac{1}{\sqrt{1-\delta_*}}\left(\|y - Ax^{2\kappa s}\|_2 + \|Ax_{\overline{T}}\|_2 + \|\mathbf{e}\|_2\right) \\ &\leq C\|Ax_{\overline{T}}\|_2 + D\|\mathbf{e}\|_2. \end{aligned} \tag{12}$$

It would be nice to try and bound $\|\mathrm{A}\boldsymbol{x}_{\overline{T}}\|_2$ by an absolute constant times $\|\boldsymbol{x}_{\overline{T}}\|_2$, but this is doomed (in the terminology of [23], it would imply ℓ_2-instance optimality, which is only possible for $m \asymp N$). Instead, the key is to use a technique that has now become routine in compressive sensing—I call it the sort-and-split technique in this survey. We decompose $[\![1:N]\!]$ into index sets $S_0, S_1, S_2, \ldots$ of size s in such a way that S_0 contains s largest absolute entries of $\boldsymbol{x}$, S_1 contains s next largest absolute entries of $\boldsymbol{x}$, etc. We invoke the RIP again to write

$$\|\mathrm{A}\boldsymbol{x}_{\overline{T}}\|_2 = \left\|\mathrm{A}\left(\sum_{k\geq 2}\boldsymbol{x}_{S_k}\right)\right\|_2 \leq \sum_{k\geq 2}\|\mathrm{A}\boldsymbol{x}_{S_k}\|_2 \leq \sqrt{1+\delta_*}\sum_{k\geq 2}\|\boldsymbol{x}_{S_k}\|_2.$$

Now, since any absolute entry of $\boldsymbol{x}_{S_k}$ is bounded by any absolute entry of $\boldsymbol{x}_{S_{k-1}}$, we can easily check that, for each $k \geq 2$,

$$\|\boldsymbol{x}_{S_k}\|_2 \leq \frac{1}{\sqrt{s}}\|\boldsymbol{x}_{S_{k-1}}\|_1.$$

This yields

$$\|\mathrm{A}\boldsymbol{x}_{\overline{T}}\|_2 \leq \frac{\sqrt{1+\delta_*}}{\sqrt{s}}\sum_{k\geq 2}\|\boldsymbol{x}_{S_{k-1}}\|_1 \leq \frac{\sqrt{1+\delta_*}}{\sqrt{s}}\|\boldsymbol{x}_{\overline{S_0}}\|_1 = \frac{\sqrt{1+\delta_*}}{\sqrt{s}}\sigma_s(\boldsymbol{x})_1. \quad (13)$$

Putting (11)–(13) together gives

$$\|\boldsymbol{x} - \boldsymbol{x}^{2\kappa s}\|_2 \leq \frac{C}{\sqrt{s}}\sigma_s(\boldsymbol{x})_1 + D\|\mathbf{e}\|_2. \quad (14)$$

This is the stability and robustness estimate incorporating (3) that we were after. Note that the argument remains valid for other algorithms, e.g., for IHT by starting from the result of Theorem 2.

We finally point out that a stability and robustness estimate in the 'weaker' sense of footnote 3 can also be obtained via the sort-and-split technique. Indeed, instead of (11), we could also have written

$$\|\boldsymbol{x} - \boldsymbol{x}^{2\kappa s}\|_1 \leq \|\boldsymbol{x}_T - \boldsymbol{x}^{2\kappa s}\|_1 + \|\boldsymbol{x}_{\overline{T}}\|_1 \leq \sqrt{2(1+\kappa)s}\,\|\boldsymbol{x}_T - \boldsymbol{x}^{2\kappa s}\|_2 + \sigma_s(\boldsymbol{x})_1.$$

The latter term is at most the right-hand side of (11) multiplied by $\sqrt{2(1+\kappa)s}$. Therefore, calling upon (12) and (13) again, we arrive at the following recovery estimate measuring error in ℓ_1 rather than ℓ_2:

$$\|\boldsymbol{x} - \boldsymbol{x}^{2\kappa s}\|_1 \leq C\sigma_s(\boldsymbol{x})_1 + D\sqrt{s}\|\mathbf{e}\|_2. \quad (15)$$

5 Basis Pursuits

In this section, we consider both the equality-constrained ℓ_1-minimization

$$\underset{z \in \mathbb{K}^N}{\text{minimize}} \, \|z\|_1 \qquad \text{subject to } \mathsf{A}z = y, \tag{16}$$

as well as the quadratically constrained ℓ_1-minimization

$$\underset{z \in \mathbb{K}^N}{\text{minimize}} \, \|z\|_1 \qquad \text{subject to } \|\mathsf{A}z - y\|_2 \leq \eta. \tag{17}$$

The first program is relevant when the measurements are perfectly accurate, while the second program seems more relevant when there is some measurement error in $y = \mathsf{A}x + \mathbf{e}$, but a bound $\|\mathbf{e}\|_2 \leq \eta$ on the magnitude of the measurement error is available. There are several intuitive explanations for why ℓ_1-minimization does promote sparsity. One of them is the fact that a minimizer of (16) or (17) is automatically m-sparse, provided it is unique (see [39, Theorem 3.1 and Exercise 3.3]). Beware here of a subtle difference between the real and complex settings—this statement, true when $\mathbb{K} = \mathbb{R}$, becomes false when $\mathbb{K} = \mathbb{C}$ (see [39, Exercise 3.2]). As another subtle difference between the two settings, the optimization problem (16) can be transformed into a linear program when $\mathbb{K} = \mathbb{R}$, but is only into a second-order cone program when $\mathbb{K} = \mathbb{C}$ (see [39, pp. 63–64]). These considerations set aside, for exact sparse recovery from uncorrupted measurements, the equivalence between the following two assertions is easy to verify (approximation theorists can view it as resulting from characterizations of best approximation in ℓ_1-norm):

(a) every s-sparse $x \in \mathbb{K}^N$ is the unique minimizer of (16) with $y = \mathsf{A}x$,
(b) for all $\mathbf{v} \in \ker \mathsf{A} \setminus \{\mathbf{0}\}$, for all index sets S of size s,

$$\|\mathbf{v}_S\|_1 < \|\mathbf{v}_{\overline{S}}\|_1.$$

The latter property is called the null space property (NSP). The name emphasizes that the success of sparse recovery via ℓ_1-minimization depends only on the null space of the measurement matrix, not on the matrix itself, so, e.g., one can safely rescale, reshuffle, or add measurements (see [39, Remark 4.6]). Note that, given a matrix $\mathsf{A} \in \mathbb{R}^{m \times N}$, (b) could be interpreted in two different ways, whether $\ker \mathsf{A}$ is considered a real vector space or a complex vector space. Remarkably, these two versions of the NSP are equivalent (see [34] or [39, Theorem 4.7]).

5.1 Robust Null Space Property

The recovery procedure considered in this subsection corresponds to (17) and is called quadratically constrained basis pursuit—as the reader has guessed already, basis pursuit is synonymous to ℓ_1-minimization. We will use a specific notation for the associated recovery map, namely

$$\Delta_{\mathrm{BP}}^{\eta}(\mathbf{y}) := \operatorname{argmin} \|\mathbf{z}\|_1 \qquad \text{subject to } \|\mathsf{A}\mathbf{z} - \mathbf{y}\|_2 \leq \eta.$$

We target stability and robustness estimates akin to (14), with η replacing $\|\mathbf{e}\|_2$, i.e.,

$$\|\mathbf{x} - \Delta_{\mathrm{BP}}^{\eta}(\mathsf{A}\mathbf{x} + \mathbf{e})\|_2 \leq \frac{C}{\sqrt{s}}\sigma_s(\mathbf{x})_1 + D\eta \tag{18}$$

valid for all $\mathbf{x} \in \mathbb{C}^N$ and all $\mathbf{e} \in \mathbb{C}^m$ with $\|\mathbf{e}\|_2 \leq \eta$. Assuming that (18) holds, if we make the particular choice $\mathbf{x} = \mathbf{v} \in \mathbb{C}^N$, $\mathbf{e} = -\mathsf{A}\mathbf{v}$, and $\eta = \|\mathsf{A}\mathbf{v}\|_2$, we have $\Delta_{\mathrm{BP}}^{\eta}(\mathsf{A}\mathbf{x} + \mathbf{e}) = \mathbf{0}$, hence

$$\|\mathbf{v}\|_2 \leq \frac{C}{\sqrt{s}}\sigma_s(\mathbf{v})_1 + D\|\mathsf{A}\mathbf{v}\|_2.$$

It follows in particular that, for any $\mathbf{v} \in \mathbb{C}^N$ and any index set S of size s,

$$\|\mathbf{v}_S\|_2 \leq \frac{C}{\sqrt{s}}\|\mathbf{v}_{\overline{S}}\|_1 + D\|\mathsf{A}\mathbf{v}\|_2.$$

This is called the robust null space property (RNSP) of order s with constants $C, D > 0$. We have just pointed out that it is a necessary condition for the stability and robustness estimate (18) to hold. As it turns out, the RNSP is also a sufficient condition, with the proviso that $C < 1$ (this proviso makes sense in view of the NSP mentioned in (b)). Here is a more general result (see [39, Theorem 4.25]). To see why (19) below is more general than (18), just apply (19) to $\mathbf{z} := \Delta_{\mathrm{BP}}^{\eta}(\mathsf{A}\mathbf{x} + \mathbf{e})$, so that $\|\mathbf{z}\|_1 \leq \|\mathbf{x}\|_1$ and $\|\mathsf{A}(\mathbf{z} - \mathbf{x})\|_2 \leq \|\mathbf{y} - \mathsf{A}\mathbf{z}\|_2 + \|\mathbf{y} - \mathsf{A}\mathbf{x}\|_2 \leq 2\eta$.

Theorem 4 *If a matrix $\mathsf{A} \in \mathbb{C}^{m\times N}$ satisfies the RNSP of order s with constants $0 < \rho < 1$ and $\tau > 0$, then for any $\mathbf{x}, \mathbf{z} \in \mathbb{C}^N$,*

$$\|\mathbf{x} - \mathbf{z}\|_2 \leq \frac{C}{\sqrt{s}}\left(\|\mathbf{z}\|_1 - \|\mathbf{x}\|_1 + 2\sigma_s(\mathbf{x})_1\right) + D\|\mathsf{A}(\mathbf{z} - \mathbf{x})\|_2, \tag{19}$$

with constants $C, D > 0$ depending only on ρ and τ.

We now point out that the RNSP can be deduced from the RIP. The easiest way to do so would be to reproduce the argument proposed in [39, Theorem 6.9 and Exercise 6.12], which invokes the condition $\delta_{2s} < 1/3$. We provide a less stringent condition below. For several years, researchers improved sufficient conditions of the type $\delta_{2s} < \delta_*$ by raising the threshold δ_*. As mentioned in [26], it is impossible to take $\delta_* > 1/\sqrt{2}$. It has recently been proved in [14] that one can take $\delta_* = 1/\sqrt{2}$. We paraphrase the argument to fit the RNSP framework.

Theorem 5 *If a matrix* $\mathsf{A} \in \mathbb{C}^{m \times N}$ *has restricted isometry constant*

$$\delta_{2s} < \frac{1}{\sqrt{2}},$$

then it satisfies the RNSP with constants $0 < \rho < 1$ *and* $\tau > 0$ *depending only on* δ_{2s}.

Proof Given a vector $\mathbf{v} \in \mathbb{C}^N$ and an index set S of size s, our goal is to find constants $0 < \rho < 1$ and $\tau > 0$ such that

$$\|\mathbf{v}_S\|_2 \le \frac{\rho}{\sqrt{s}} \|\mathbf{v}_{\overline{S}}\|_1 + \tau \|\mathsf{A}\mathbf{v}\|_2.$$

Note that we may assume that $\mathbf{v} \in \mathbb{R}^N$. Indeed, we can multiply by a diagonal matrix D whose nonzero entries have modulus one and replace $\mathbf{v} \in \mathbb{C}^N$ by $\mathsf{D}\mathbf{v} \in \mathbb{R}^N$. At the same time, we would replace A by $\mathsf{A}\mathsf{D}^{-1}$, which does not change the restricted isometry constants. We now partition the complement $\overline{S}$ of S as $\overline{S} = S' \cup S''$, where

$$S' := \left\{ j \in \overline{S} : |v_j| > \frac{\|\mathbf{v}_{\overline{S}}\|_1}{s} \right\}, \qquad S'' := \left\{ j \in \overline{S} : |v_j| \le \frac{\|\mathbf{v}_{\overline{S}}\|_1}{s} \right\}.$$

With $s' := \mathrm{card}(S')$, we derive that $s' \le s$ from $\|\mathbf{v}_{S'}\|_1 \ge s' \|\mathbf{v}_{\overline{S}}\|_1 / s \ge s' \|\mathbf{v}_{S'}\|_1 / s$. We further remark that

$$\|\mathbf{v}_{S''}\|_1 = \|\mathbf{v}_{\overline{S}}\|_1 - \|\mathbf{v}_{S'}\|_1 \le (s - s') \frac{\|\mathbf{v}_{\overline{S}}\|_1}{s},$$

while we also have

$$\|\mathbf{v}_{S''}\|_\infty \le \frac{\|\mathbf{v}_{\overline{S}}\|_1}{s}.$$

Thus, the vector $\mathbf{v}_{S''}$ belongs to a scaled version of the polytope $(s - s') B_1^{s''} \cap B_\infty^{s''}$, $s'' := \mathrm{card}(S'')$. A key to the argument of [14] was to recognize and prove that this polytope can be represented as the convex hull of $(s - s')$-sparse vectors. In fact, this result is similar to [54, Lemma 5.2]. It allows us to write $\mathbf{v}_{S''}$ as a convex combination

$$\mathbf{v}_{S''} = \sum_{i=1}^{n} \lambda_i \mathbf{u}^i,$$

of $(s - s')$-sparse vectors $\mathbf{u}^1, \ldots, \mathbf{u}^n$ supported on S'' that satisfy $\|\mathbf{u}^i\|_\infty \le \|\mathbf{v}_{\overline{S}}\|_1 / s$ for all $i \in [\![1 : n]\!]$. We now observe that

$$\sum_{i=1}^{n} \lambda_i \left[\|\mathsf{A}((1+\delta)(\mathbf{v}_S + \mathbf{v}_{S'}) + \delta \mathbf{u}^i)\|_2^2 - \|\mathsf{A}((1-\delta)(\mathbf{v}_S + \mathbf{v}_{S'}) - \delta \mathbf{u}^i)\|_2^2 \right]$$

$$= \sum_{i=1}^{n} \lambda_i \left[\langle \mathsf{A}(2(\mathbf{v}_S + \mathbf{v}_{S'})), \mathsf{A}(2\delta(\mathbf{v}_S + \mathbf{v}_{S'}) + 2\delta \mathbf{u}^i) \rangle \right]$$

$$= 4\delta \langle \mathsf{A}(\mathbf{v}_s + \mathbf{v}_{S'}), \mathsf{A}(\mathbf{v}_S + \mathbf{v}_{S'} + \sum_{i=1}^{n} \lambda_i \mathbf{u}^i) \rangle$$

$$= 4\delta \langle \mathsf{A}(\mathbf{v}_S + \mathbf{v}_{S'}), \mathsf{A}\mathbf{v} \rangle. \tag{20}$$

Because the sparsity of the vectors involved in the left-hand side of (20) is at most $s + s' + (s - s') = 2s$, with $\delta := \delta_{2s}$, each summand in this left-hand side is bounded from below by

$$\begin{aligned}
&(1-\delta)\|(1+\delta)(\mathbf{v}_S + \mathbf{v}_{S'}) + \delta \mathbf{u}^i\|_2^2 - (1+\delta)\|(1-\delta)(\mathbf{v}_S + \mathbf{v}_{S'}) - \delta \mathbf{u}^i\|_2^2 \\
&= (1-\delta)\left((1+\delta)^2 \|\mathbf{v}_S + \mathbf{v}_{S'}\|_2^2 + \delta^2 \|\mathbf{u}^i\|_2^2\right) - (1+\delta)\left((1-\delta)^2 \|\mathbf{v}_S + \mathbf{v}_{S'}\|_2^2 + \delta^2 \|\mathbf{u}^i\|_2^2\right) \\
&= (1-\delta^2) 2\delta \|\mathbf{v}_S + \mathbf{v}_{S'}\|_2^2 - 2\delta^3 \|\mathbf{u}^i\|_2^2.
\end{aligned}$$

Using the fact that $\|\mathbf{u}^i\|_2^2 \leq (s - s')\|\mathbf{u}^i\|_\infty^2 \leq (s - s')\|\mathbf{v}_{\overline{S}}\|_1^2/s^2 \leq \|\mathbf{v}_{\overline{S}}\|_1^2/s$, multiplying by λ_i and summing over i shows that the left-hand side of (20) is bounded from below by

$$(1-\delta^2) 2\delta \|\mathbf{v}_S + \mathbf{v}_{S'}\|_2^2 - 2\delta^3 \frac{\|\mathbf{v}_{\overline{S}}\|_1^2}{s}.$$

On the other hand, the right-hand side of (20) is bounded from above by

$$4\delta \|\mathsf{A}(\mathbf{v}_S + \mathbf{v}_{S'})\|_2 \|\mathsf{A}\mathbf{v}\|_2 \leq 4\delta \sqrt{1+\delta} \|\mathbf{v}_S + \mathbf{v}_{S'}\|_2 \|\mathsf{A}\mathbf{v}\|_2.$$

Combining the two bounds gives

$$(1-\delta^2) \|\mathbf{v}_S + \mathbf{v}_{S'}\|_2^2 - \delta^2 \frac{\|\mathbf{v}_{\overline{S}}\|_1^2}{s} \leq 2\sqrt{1+\delta} \|\mathbf{v}_S + \mathbf{v}_{S'}\|_2 \|\mathsf{A}\mathbf{v}\|_2,$$

which is equivalent to

$$\left(\sqrt{1-\delta^2} \|\mathbf{v}_S + \mathbf{v}_{S'}\|_2 - \frac{1}{\sqrt{1-\delta}} \|\mathsf{A}\mathbf{v}\|_2 \right)^2 \leq \delta^2 \frac{\|\mathbf{v}_{\overline{S}}\|_1^2}{s} + \frac{1}{1-\delta} \|\mathsf{A}\mathbf{v}\|_2^2.$$

As a result, we obtain

$$\sqrt{1-\delta^2}\|\mathbf{v}_S + \mathbf{v}_{S'}\|_2 \leq \left(\delta^2 \frac{\|\mathbf{v}_{\overline{S}}\|_1^2}{s} + \frac{1}{1-\delta}\|\mathsf{A}\mathbf{v}\|_2^2\right)^{1/2} + \frac{1}{\sqrt{1-\delta}}\|\mathsf{A}\mathbf{v}\|_2$$
$$\leq \delta \frac{\|\mathbf{v}_{\overline{S}}\|_1}{\sqrt{s}} + \frac{2}{\sqrt{1-\delta}}\|\mathsf{A}\mathbf{v}\|_2.$$

It remains to use the fact that $\|\mathbf{v}_S\|_2 \leq \|\mathbf{v}_S + \mathbf{v}_{S'}\|_2$ to arrive at

$$\|\mathbf{v}_S\|_2 \leq \frac{\delta}{\sqrt{1-\delta^2}} \frac{\|\mathbf{v}_{\overline{S}}\|_1}{\sqrt{s}} + \frac{2}{(1-\delta)\sqrt{1+\delta}}\|\mathsf{A}\mathbf{v}\|_2,$$

which is the desired inequality with $\rho := \delta/\sqrt{1-\delta^2} < 1$ when $\delta < 1/\sqrt{2}$ and with $\tau := 2/((1-\delta)\sqrt{1+\delta}) < 5.25$. □

From Theorems 4 and 5, we now know that, under the condition $\delta_{2s} < 1/\sqrt{2}$, the stability and robustness estimate (18) holds. Similarly to OMP, we can also give a stability and robustness estimate with recovery error measured in ℓ_1 instead of ℓ_2. One way to achieve this is based on the following lemma.

Lemma 1 *If $\boldsymbol{x} \in \mathbb{C}^N$ is s-sparse and if $\boldsymbol{x}^\sharp = \Delta_{\mathrm{BP}}(\mathsf{A}\boldsymbol{x})$ or $\boldsymbol{x}^\sharp = \Delta_{\mathrm{BP}}^{\eta}(\mathsf{A}\boldsymbol{x} + \mathbf{e})$ with $\|\mathbf{e}\|_2 \leq \eta$, then $\boldsymbol{x} - \boldsymbol{x}^\sharp$ is effectively $4s$-sparse, meaning that*

$$\|\boldsymbol{x} - \boldsymbol{x}^\sharp\|_1 \leq \sqrt{4s}\|\boldsymbol{x} - \boldsymbol{x}^\sharp\|_2.$$

More generally, for any vector $\boldsymbol{x} \in \mathbb{C}^N$ and any index set $S \subseteq [\![1:N]\!]$, if $\boldsymbol{x}^\sharp$ is a minimizer of $\|\boldsymbol{z}\|_1$ subject to some constraint met by $\boldsymbol{x}$, then

$$\|(\boldsymbol{x} - \boldsymbol{x}^\sharp)_{\overline{S}}\|_1 \leq \|(\boldsymbol{x} - \boldsymbol{x}^\sharp)_S\|_1 + 2\|\boldsymbol{x}_{\overline{S}}\|_1.$$

Proof To establish the second statement, we combine the fact that $\|\boldsymbol{x}^\sharp\|_1 \leq \|\boldsymbol{x}\|_1$ with the observations that

$$\|\boldsymbol{x}^\sharp\|_1 = \|\boldsymbol{x}^\sharp_S\|_1 + \|\boldsymbol{x}^\sharp_{\overline{S}}\|_1 \geq \|\boldsymbol{x}^\sharp_S\|_1 + \|(\boldsymbol{x} - \boldsymbol{x}^\sharp)_{\overline{S}}\|_1 - \|\boldsymbol{x}_{\overline{S}}\|_1,$$
$$\|\boldsymbol{x}\|_1 = \|\boldsymbol{x}_S\|_1 + \|\boldsymbol{x}_{\overline{S}}\|_1 \leq \|\boldsymbol{x}^\sharp_S\|_1 + \|(\boldsymbol{x} - \boldsymbol{x}^\sharp)_S\|_1 + \|\boldsymbol{x}_{\overline{S}}\|_1.$$

To prove the first statement, we write the second one as $\|(\boldsymbol{x} - \boldsymbol{x}^\sharp)_{\overline{S}}\|_1 \leq \|(\boldsymbol{x} - \boldsymbol{x}^\sharp)_S\|_1$ when $\boldsymbol{x}$ is s-sparse and S is the support of $\boldsymbol{x}$. We then derive that

$$\|\boldsymbol{x} - \boldsymbol{x}^\sharp\|_1 \leq 2\|(\boldsymbol{x} - \boldsymbol{x}^\sharp)_S\|_1 \leq 2\sqrt{s}\|(\boldsymbol{x} - \boldsymbol{x}^\sharp)_S\|_2 \leq \sqrt{4s}\|\boldsymbol{x} - \boldsymbol{x}^\sharp\|_2,$$

which is the desired result. □

We sum up the considerations about quadratically constrained basis pursuit presented so far with the following statement about the recovery error in ℓ_p for $p = 1$ and $p = 2$. It is not hard to extend it to ℓ_p for any $p \in [1, 2]$.

Theorem 6 *If a matrix $\mathsf{A} \in \mathbb{C}^{m\times N}$ has restricted isometry constant $\delta_{2s} < 1/\sqrt{2}$, then, for any $\boldsymbol{x} \in \mathbb{C}^N$ and any $\mathbf{e} \in \mathbb{C}^m$ with $\|\mathbf{e}\|_2 \le \eta$, a minimizer of the quadratically constrained ℓ_1-minimization* (17) *approximates $\boldsymbol{x}$ with error*

$$\|\boldsymbol{x} - \Delta_{\mathrm{BP}}^{\eta}(\mathsf{A}\boldsymbol{x} + \mathbf{e})\|_1 \le C\sigma_s(\boldsymbol{x})_1 + D\sqrt{s}\eta, \tag{21}$$

$$\|\boldsymbol{x} - \Delta_{\mathrm{BP}}^{\eta}(\mathsf{A}\boldsymbol{x} + \mathbf{e})\|_2 \le \frac{C}{\sqrt{s}}\sigma_s(\boldsymbol{x})_1 + D\eta. \tag{22}$$

5.2 *Quotient Property*

Although the result of Theorem 6 is quite satisfying, there is a minor issue one wishes to overcome, namely an estimation of the magnitude η of the measurement error is needed for (17) to be executed (in parallel, recall that typically iterative algorithms require an estimation of the sparsity level s to run). Could it happen that recovery via equality-constrained ℓ_1-minimization provides stability and robustness even in the presence of measurement error? The answer is affirmative, as we shall see. Precisely, using the notation

$$\Delta_{\mathrm{BP}}(\boldsymbol{y}) := \operatorname{argmin} \|\boldsymbol{z}\|_1 \qquad \text{subject to } \mathsf{A}\boldsymbol{z} = \boldsymbol{y}, \tag{23}$$

we target an estimate of the type (22), i.e.,

$$\|\boldsymbol{x} - \Delta_{\mathrm{BP}}(\mathsf{A}\boldsymbol{x} + \mathbf{e})\|_2 \le \frac{C}{\sqrt{s}}\sigma_s(\boldsymbol{x})_1 + D\|\mathbf{e}\|_2 \tag{24}$$

valid for all $\boldsymbol{x} \in \mathbb{C}^N$ and all $\mathbf{e} \in \mathbb{C}^m$ in the optimal regime of parameters $s \asymp s_*$, where

$$s_* := \frac{m}{\ln(eN/m)}.$$

Note that (24) holds for all $s \lesssim s_*$ as soon as it holds for $s \asymp s_*$. We also target, for $s \asymp s_*$, an estimate similar to (21) which measures recovery error in ℓ_1-norm, precisely

$$\|\boldsymbol{x} - \Delta_{\mathrm{BP}}(\mathsf{A}\boldsymbol{x} + \mathbf{e})\|_1 \le C\sigma_{s_*}(\boldsymbol{x})_1 + D\sqrt{s_*}\|\mathbf{e}\|_2. \tag{25}$$

We can see that the RNSP is still a necessary condition to guarantee (24)—as before, we make the choice $\boldsymbol{x} = \mathbf{v} \in \mathbb{C}^N$ and $\mathbf{e} = -\mathsf{A}\mathbf{v}$. Moreover, if we make the choice $\boldsymbol{x} = \mathbf{0}$ in (25), we obtain that, for any $\mathbf{e} \in \mathbb{C}^m$,

$$\|\Delta_{\mathrm{BP}}(\mathbf{e})\|_1 \le D\sqrt{s_*}\|\mathbf{e}\|_2.$$

From the definition of $\Delta_{\mathrm{BP}}(\mathbf{e})$, this is equivalent to the following statement:

$$\text{For every } \mathbf{e} \in \mathbb{C}^m, \text{ there exists } \mathbf{u} \in \mathbb{C}^N \text{ with } \mathsf{A}\mathbf{u} = \mathbf{e} \text{ and } \|\mathbf{u}\|_1 \leq D\sqrt{s_*}\|\mathbf{e}\|_2.$$

This is called the quotient property (QP). We finish this subsection by proving that the RNSP and the QP are not only necessary conditions for the estimates (24) and (25) to hold, but they are also sufficient conditions. The argument given below simplifies the ones from [39, Sect. 11.2]. Of course, we should establish that matrices with the RNSP and QP actually do exist. Propitiously, Gaussian matrices with $N \geq 2m$ do exhibit the QP with high probability, as proved in [39, Theorem 11.19].

Theorem 7 *If a matrix $\mathsf{A} \in \mathbb{C}^{m\times N}$ satisfies the QP and the RNSP of order cs_*, $s_* := m/\ln(eN/m)$, then for any $\boldsymbol{x} \in \mathbb{C}^N$ and any $\mathbf{e} \in \mathbb{C}^m$, a minimizer of the equality-constrained ℓ_1-minimization* (16) *applied to $\mathbf{y} = \mathsf{A}\boldsymbol{x} + \mathbf{e}$ approximates $\boldsymbol{x}$ with error*

$$\|\boldsymbol{x} - \Delta_{\mathrm{BP}}(\mathsf{A}\boldsymbol{x} + \mathbf{e})\|_1 \leq C\sigma_{cs_*}(\boldsymbol{x})_1 + D\sqrt{s_*}\|\mathbf{e}\|_2,$$
$$\|\boldsymbol{x} - \Delta_{\mathrm{BP}}(\mathsf{A}\boldsymbol{x} + \mathbf{e})\|_2 \leq \frac{C}{\sqrt{s_*}}\sigma_{cs_*}(\boldsymbol{x})_1 + D\|\mathbf{e}\|_2.$$

Proof By the quotient property, there exists $\mathbf{u} \in \mathbb{C}^N$ such that

$$\mathsf{A}\mathbf{u} = \mathbf{e} \quad \text{and} \quad \|\mathbf{u}\|_1 \leq D\sqrt{s_*}\|\mathbf{e}\|_2.$$

We notice that $\boldsymbol{x} + \mathbf{u}$ satisfies $\mathsf{A}(\boldsymbol{x} + \mathbf{u}) = \mathsf{A}\boldsymbol{x} + \mathbf{e} = \mathbf{y}$. Therefore, by definition of $\boldsymbol{x}^\sharp := \Delta_{\mathrm{BP}}(\mathsf{A}\boldsymbol{x} + \mathbf{e})$, we have

$$\|\boldsymbol{x}^\sharp\|_1 \leq \|\boldsymbol{x} + \mathbf{u}\|_1 \leq \|\boldsymbol{x}\|_1 + \|\mathbf{u}\|_1 \leq \|\boldsymbol{x}\|_1 + D\sqrt{s_*}\|\mathbf{e}\|_2.$$

We also notice that $\mathsf{A}(\boldsymbol{x}^\sharp - \boldsymbol{x}) = \mathbf{y} - \mathsf{A}\boldsymbol{x} = \mathbf{e}$. By Theorem 4, the RNSP implies

$$\begin{aligned}\|\boldsymbol{x} - \boldsymbol{x}^\sharp\|_2 &\leq \frac{C}{\sqrt{cs_*}}\left(\|\boldsymbol{x}^\sharp\|_1 - \|\boldsymbol{x}\|_1 + 2\sigma_{cs_*}(\boldsymbol{x})_1\right) + D\|\mathsf{A}(\boldsymbol{x}^\sharp - \boldsymbol{x})\|_2 \\ &\leq \frac{C}{\sqrt{s_*}}\left(D\sqrt{s_*}\|\mathbf{e}\|_2 + 2\sigma_{cs_*}(\boldsymbol{x})_1\right) + D\|\mathbf{e}\|_2 \leq \frac{C}{\sqrt{s_*}}\sigma_{cs_*}(\boldsymbol{x})_1 + D\|\mathbf{e}\|_2.\end{aligned}$$

This is the result for $p = 2$. To obtain the result for $p = 1$, we make use of Lemma 1 applied to $\boldsymbol{x} + \mathbf{u}$ and to an index set S of cs_* largest absolute entries of $\boldsymbol{x}$. We obtain

$$\begin{aligned}\|\boldsymbol{x} - \boldsymbol{x}^\sharp\|_1 &\leq \|\boldsymbol{x} + \mathbf{u} - \boldsymbol{x}^\sharp\|_1 + \|\mathbf{u}\|_1 \leq 2\|(\boldsymbol{x} + \mathbf{u} - \boldsymbol{x}^\sharp)_S\|_1 + 2\|\boldsymbol{x}_{\overline{S}}\|_1 + \|\mathbf{u}\|_1 \\ &\leq 2\|(\boldsymbol{x} - \boldsymbol{x}^\sharp)_S\|_1 + 2\sigma_{cs_*}(\boldsymbol{x})_1 + 3\|\mathbf{u}\|_1 \\ &\leq 2\sqrt{cs_*}\|\boldsymbol{x} - \boldsymbol{x}^\sharp\|_2 + 2\sigma_{cs_*}(\boldsymbol{x})_1 + 3D\sqrt{s_*}\|\mathbf{e}\|_2,\end{aligned}$$

so that the bound for $p = 1$ follows from the bound for $p = 2$ derived above. □

5.3 Regularizations

We have not discussed algorithms to perform the ℓ_1-minimizations (16) and (17). There are efficient algorithms[5] to perform this task exactly, but it is also popular, especially in statistics, to solve the related unconstrained optimization problem

$$\underset{z}{\text{minimize}} \;\; \|z\|_1 + \lambda \|\mathsf{A}z - y\|_2^2. \tag{26}$$

Intuitively, choosing a large parameter λ imposes $\|\mathsf{A}z - y\|_2$ to be small, so a solution of (26) is potentially also a solution of (17). To be a little more precise, (17) and (26) are equivalent in the sense that a minimizer of (17) is also a minimizer of (26) for some choice of parameter λ, and conversely a minimizer of (26) is also a minimizer of (17) for some choice of parameter η (see, e.g., [39, Proposition 3.2] for details). The catch is that the parameter in one program cannot be chosen beforehand as it depends on the minimizer of the other program. Problem (17) is also equivalent (with the same meaning of equivalence) to the optimization problem

$$\underset{z}{\text{minimize}} \;\; \|\mathsf{A}z - y\|_2 \qquad \text{subject to } \|z\|_1 \le \tau. \tag{27}$$

Problem (27) is called LASSO and Problem (26) is called basis pursuit denoising, but there is often some confusion in the terminology between (17), (26), and (27).

Instead of (26), I personally prefer a regularized problem where the ℓ_1-norm is squared, namely

$$\underset{z}{\text{minimize}} \;\; \|z\|_1^2 + \lambda \|\mathsf{A}z - y\|_2^2. \tag{28}$$

Apart from the seemingly more natural homogeneity in z, the reason for my preference comes from the fact that (28) can be transformed into a nonnegative least squares problem (see [33] for a theoretical analysis[6]). For a quick explanation, consider the situation where the sparse vector $\boldsymbol{x} \in \mathbb{R}^N$ to recover is nonnegative (it is a problem in metagenomics, where $\boldsymbol{x}$ is a vector of bacterial concentrations—that motivated this scenario—see [47, 48] for the end results). In such a case, we readily see that the optimization problem

$$\underset{z}{\text{minimize}} \;\; \|z\|_1^2 + \lambda \|\mathsf{A}z - y\|_2^2 \qquad \text{subject to } z \ge \mathbf{0}$$

can be recast as the nonnegative least squares problem

$$\underset{z}{\text{minimize}} \;\; \left\| \begin{bmatrix} \mathbf{1} \\ \hline \sqrt{\lambda}\mathsf{A} \end{bmatrix} z - \begin{bmatrix} 0 \\ \hline \sqrt{\lambda} y \end{bmatrix} \right\|_2^2 \qquad \text{subject to } z \ge \mathbf{0},$$

---

[5]For instance, ℓ_1-MAGIC, NESTA, and YALL1 are freely available online.

[6]See also the MATLAB reproducible for a numerical illustration.

where $\mathbf{1}$ is the row vector of all ones. A classical algorithm for nonnegative least squares, due to Lawson and Hanson, dates back to 1973 with the first edition of [50], see Chap. 23. It turns out to be very well suited for sparse recovery because it is an 'active set' method. In order to give some details, let us change notation and consider the problem

$$\underset{z}{\text{minimize}}\ \|\widetilde{\mathsf{A}}z - \widetilde{y}\|_2^2 \qquad \text{subject to } z \geq \mathbf{0}.$$

Lawson–Hanson algorithm proceeds iteratively in a manner reminiscent of OMP. In fact, the similarity is striking by writing the algorithm as

$$S^{n+1} := S^n \cup \{j^{n+1}\}, \qquad \text{where } j^{n+1} := \underset{j\in[\![1:N]\!]}{\text{argmin}}\, (\widetilde{\mathsf{A}}^*(\widetilde{y} - \widetilde{\mathsf{A}}x^n))_j \tag{LH$_1$}$$

$$x^{n+1} := \underset{z\in\mathbf{R}^N}{\text{argmin}}\, \|\widetilde{y} - \widetilde{\mathsf{A}}z\|_2 \text{ subject to } \text{supp}(z) \subseteq S^{n+1} \tag{LH$_2$}$$

& enter an inner loop to adjust x^{n+1} until all its entries are nonnegative.

We see that (LH$_2$) is the same as (OMP$_2$), save for the presence of the inner loop (which in practice is rarely entered), and that (LH$_1$) is the same as (OMP$_1$), save for the absence of the absolute value (which guarantees the creation of one nonnegative entry, in view of $x^{n+1}_{j^{n+1}} = (\widetilde{\mathsf{A}}^*(\widetilde{y} - \widetilde{\mathsf{A}}x^n))_{j^{n+1}}/\text{dist}(\widetilde{a}_{j^{n+1}}, \text{span}\{\widetilde{a}_{j^1}, \ldots, \widetilde{a}_{j^n}\})$, see [30] for details).

6 Gelfand Widths Estimates: By-product of Compressive Sensing

It is time to come back to the connection between Gelfand widths and compressive sensing, now that we are more informed about the compressive sensing theory from Sects. 3–5. The intent is to turn the story around by showing that the compressive sensing theory is self-sufficient and that it provides Gelfand width estimates as side results. Before doing so, I promised in Sect. 2.4 to replace the impractical recovery map (4) by the recovery map (23) while maintaining the message of Sect. 2.3, namely that the upper estimate for the Gelfand width provides a measurement matrix A such that the pair $(\mathsf{A}, \Delta_{\text{BP}})$ is stable of order $s \asymp m/\ln(eN/m)$. This can be easily seen by replacing $\widetilde{x}$ by $x^\sharp = \Delta_{\text{BP}}(\mathsf{A}x)$ until (5), where we would call upon Lemma 1 to derive

$$\begin{aligned}\|x - x^\sharp\|_2 &\leq \frac{1}{\sqrt{8s}}\|x - x^\sharp\|_1 = \frac{1}{\sqrt{8s}}\left(\|(x - x^\sharp)_{\overline{S}}\|_1 + \|(x - x^\sharp)_S\|_1\right) \\ &\leq \frac{1}{\sqrt{8s}}\left(2\|(x - x^\sharp)_S\|_1 + 2\sigma_s(x)_1\right) \leq \frac{1}{\sqrt{2}}\|x - x^\sharp\|_2 + \frac{1}{\sqrt{2s}}\sigma_s(x)_1,\end{aligned}$$

so a rearrangement provides the desired inequality (3) with constant $C = \sqrt{2} + 1$.

After this digression, let us come back to our main goal of proving the upper and lower estimates for the Gelfand width of the ℓ_1-ball B_1^N relative to ℓ_2^N using compressive sensing tools only and highlighting that Gelfand widths could be left out of the compressive sensing theory if so desired. We return to the real setting for the rest of this section.

6.1 Derivation of the Upper Estimate

Bounding $d^m(B_1^N, \ell_2^N)$ from above requires the RIP and the sort-and-split technique. Theorem 1 guarantees that we can find a matrix $\mathsf{A} \in \mathbb{R}^{m \times N}$ with restricted isometry constant $\delta_{2s} < 4/5$, say, for $s \asymp m/ln(eN/m)$. Since the space $L^m := \ker \mathsf{A}$ has codimension at most m, the definition of the Gelfand width yields

$$d^m(B_1^N, \ell_2^N) \le \sup_{\mathbf{v} \in \ker \mathsf{A}} \frac{\|\mathbf{v}\|_2}{\|\mathbf{v}\|_1}.$$

Then, given $\mathbf{v} \in \ker \mathsf{A}$, we notice that $\mathsf{A}(\mathbf{v}_S) = -\mathsf{A}(\mathbf{v}_{\overline{S}})$ with S denoting an index set of s largest absolute entries of $\mathbf{v}$. Thanks to the RIP and the sort-and-split technique, we obtain (as in (13))

$$\|\mathbf{v}_S\|_2 \le \frac{1}{\sqrt{1-\delta}} \|\mathsf{A}\mathbf{v}_S\|_2 = \frac{1}{\sqrt{1-\delta}} \|\mathsf{A}\mathbf{v}_{\overline{S}}\|_2 \le \frac{\sqrt{1+\delta}}{\sqrt{1-\delta}} \frac{\|\mathbf{v}\|_1}{\sqrt{s}}.$$

Moreover, Stechkin bound gives $\|\mathbf{v}_{\overline{S}}\|_2 \le \|\mathbf{v}\|_1/\sqrt{s}$. It follows that

$$\|\mathbf{v}\|_2 \le \|\mathbf{v}_S\|_2 + \|\mathbf{v}_{\overline{S}}\|_2 \le \left(\frac{\sqrt{1+\delta}}{\sqrt{1-\delta}} + 1 \right) \frac{\|\mathbf{v}\|_1}{\sqrt{s}} \le 4 \frac{\|\mathbf{v}\|_1}{\sqrt{s}}.$$

Taking $s \asymp m/ln(eN/m)$ into account, we arrive at

$$d^m(B_1^N, \ell_2^N) \le C \sqrt{\frac{\ln(eN/m)}{m}}.$$

Since $d^m(B_1^N, \ell_2^N) \le 1$ (the ℓ_2-ball contains the ℓ_1-ball), the upper bound is proved.

6.2 Derivation of the Lower Estimate

Section 2.2 can give the impression that the lower estimate for $d^m(B_1^N, \ell_2^N)$ is essential in the compressive sensing theory, as it enables to prove that the number of measurements $m \asymp s \ln(eN/s)$ is optimal. The next subsection shows that this dependence on the lower estimate can be lifted, but for now we want to stress that the compressive sensing theory provides tools, and perhaps more importantly intuition, to retrieve this lower estimate. Basically, the two insights:

(i) small $d^m(B_1^N, \ell_2^N)$ yields exact s-sparse recovery via ℓ_1-minimization for 'large' s,

(ii) exact s-sparse recovery via ℓ_1-minimization is only possible for 'moderate' s,

inform us that the Gelfand width cannot be too small, i.e., they establish a lower bound. A formal argument can be found in [38], where the ℓ_1-ball B_1^N is replaced by the ℓ_p-balls B_p^N for $0 < p \leq 1$. Here, we settle for an informal justification of the two insights (i) and (ii).

For (i), we pick an optimal subspace L^m in the definition of the Gelfand width and a matrix $\mathsf{A} \in \mathbb{R}^{m\times N}$ with $\ker \mathsf{A} = L^m$. For any $\mathbf{v} \in \ker \mathsf{A}$, we have $\|\mathbf{v}\|_2 \leq \omega \|\mathbf{v}\|_1$, where we wrote $\omega := d^m(B_1^N, \ell_2^N)$ for short. Then, for any index set S of size s,

$$\|\mathbf{v}_S\|_1 \leq \sqrt{s}\|\mathbf{v}_S\|_2 \leq \sqrt{s}\|\mathbf{v}\|_2 \leq \sqrt{s}\omega\|\mathbf{v}\|_1 = \sqrt{s}\omega\|\mathbf{v}_S\|_1 + \sqrt{s}\omega\|\mathbf{v}_{\overline{S}}\|_1.$$

Choosing s such that $\sqrt{s}\omega = 1/2$, this reduces to the NSP (b), which is equivalent to exact s-sparse recovery as stated in (a).

For (ii), the key ingredient is a combinatorial lemma that appears independently in many places (we have seen it in [13, 41, 53, 55, 56]) and we refer to [39, Lemma 10.12] for a proof.

Lemma 2 *One can find a large number*

$$n \geq \left(\frac{N}{4s}\right)^{s/2}$$

of subsets $S_1, \ldots, S_n$ of $[\![1 : N]\!]$ with size s that are almost disjoint in the sense that

$$\mathrm{card}(S_i \cap S_j) < \frac{s}{2} \quad \textit{whenever } i \neq j.$$

Now, consider the vectors $\boldsymbol{x}^1, \ldots, \boldsymbol{x}^n \in \mathbb{R}^N$ supported on $S_1, \ldots, S_n$, respectively, that equal $1/s$ on their supports. We have $\|\boldsymbol{x}^i\|_1 = 1$ for all $i \in [\![1 : n]\!]$, and thanks to Lemma 2, $\|\boldsymbol{x}^i - \boldsymbol{x}^j\|_1 \geq 1$ for all distinct $i, j \in [\![1 : n]\!]$. Since the $\boldsymbol{x}^i$ and $\boldsymbol{x}^i - \boldsymbol{x}^j$ are sparse vectors, they are ℓ_1-minimizers, so that $\|[\boldsymbol{x}^i]\| = 1$ and $\|[\boldsymbol{x}^i] - [\boldsymbol{x}^j]\| \geq 1$ in the quotient space $\ell_1^N / \ker \mathsf{A}$ equipped with the norm $\|[z]\| := \inf\{\|z'\|_1 : \mathsf{A}z' = \mathsf{A}z\}$. This supplies a set of $n \geq (N/4s)^{s/2}$ points on the unit sphere of the m-dimensional space $\ell_1^N / \ker \mathsf{A}$ which are all separated by a distance at least 1. Since such a set cannot be too large (see, e.g., [39, Proposition C.3]), this imposes the desired restriction on the size of s.

6.3 Distantiation from the Gelfand Width

Sections 2 and 6 described a close connection between the Gelfand width of ℓ_1-balls and compressive sensing and how one topic can be seen as a ramification of the other and vice versa. As an epilogue, I want to highlight that the detour through Gelfand widths is no longer necessary to build the compressive sensing theory (the shortcut presented below was discovered after studying Gelfand widths, though). Specifically, given $p \in [1,2]$, let us show that a pair (A, Δ) providing the stability estimate

$$\|\boldsymbol{x} - \Delta(\mathsf{A}\boldsymbol{x})\|_p \leq \frac{C}{s^{1-1/p}} \sigma_s(\boldsymbol{x})_1$$

valid for all $\boldsymbol{x} \in \mathbb{C}^N$ can only exist if $m \geq C s \ln(eN/s)$. Note that, for $p = 1$, the result could not be established using Gelfand widths. We still consider the s-sparse vectors $\boldsymbol{x}^1, \ldots, \boldsymbol{x}^n$ built from the index sets $S_1, \ldots, S_n$ of Lemma 2. With $\rho := 1/(4(C+1))$, we consider the subsets of $\mathbb{R}^m$ defined by

$$E_i := \mathsf{A}\left(\boldsymbol{x}^i + \rho\left(B_1^N \cap \frac{1}{s^{1-1/p}} B_p^N\right)\right), \qquad i \in [\![1:n]\!].$$

We claim that $E_1, \ldots, E_n$ are all disjoint. Indeed, if there was an element in $E_i \cap E_j$, $i \neq j$, say $\boldsymbol{y} = \mathsf{A}(\boldsymbol{x}^i + \rho z) = \mathsf{A}(\boldsymbol{x}^j + \rho z')$ with $z, z' \in B_1^N \cap (1/s^{1-1/p}) B_p^N$, then

$$\begin{aligned}
\|\boldsymbol{x}^i - \boldsymbol{x}^j\|_p &\leq \|\boldsymbol{x}^i + \rho z - \Delta(\boldsymbol{y})\|_p + \|\boldsymbol{x}^j + \rho z' - \Delta(\boldsymbol{y})\|_p + \|\rho z - \rho z'\|_p \\
&\leq \frac{C}{s^{1-1/p}} \sigma_s(\boldsymbol{x}^i + \rho z)_1 + \frac{C}{s^{1-1/p}} \sigma_s(\boldsymbol{x}^j + \rho z')_1 + \rho \|z - z'\|_p \\
&\leq \frac{C}{s^{1-1/p}} \rho \left(\|z\|_1 + \|z'\|_1\right) + \rho\left(\|z\|_p + \|z'\|_p\right) \leq \frac{C}{s^{1-1/p}} 2\rho + \rho \frac{2}{s^{1-1/p}} \\
&\leq \frac{2(C+1)\rho}{s^{1-1/p}} = \frac{1}{2 s^{1-1/p}}.
\end{aligned}$$

This contradicts the fact that $\|\boldsymbol{x}^i - \boldsymbol{x}^j\|_p \geq (1/s)\mathrm{card}(S_i \Delta S_j)^{1/p} \geq 1/s^{1-1/p}$. We now also claim that, because all the $\boldsymbol{x}^i$ belong to $B_1^N \cap (1/s^{1-1/p}) B_p^N$, all the sets E_i are contained in the ball $(1+\rho)\mathsf{A}\left(B_1^N \cap (1/s^{1-1/p}) B_p^N\right)$. From these two claims, we derive that $\sum_{i=1}^n \mathrm{Vol}(\mathrm{E_i}) = \mathrm{Vol}\left(\cup_{i=1}^n E_i\right) \leq \mathrm{Vol}\left((1+\rho)\mathsf{A}\left(B_1^N \cap (1/s^{1-1/p}) B_p^N\right)\right)$. With $\mathcal{V} := \mathrm{Vol}\left(\mathsf{A}\left(B_1^N \cap (1/s^{1-1/p}) B_p^N\right)\right)$ and $r := \mathrm{rank}(\mathsf{A}) \leq m$, this inequality reads $n\rho^r \mathcal{V} \leq (1+\rho)^r \mathcal{V}$. This implies that

$$\left(\frac{N}{4s}\right)^{s/2} \leq n \leq \left(1 + \frac{1}{\rho}\right)^r \leq \left(1 + \frac{1}{\rho}\right)^m, \quad \text{hence} \quad \frac{s}{2} \ln\left(\frac{N}{4s}\right) \leq m \ln\left(1 + \frac{1}{\rho}\right).$$

We have obtained $m \geq cs\ln(N/4s)$, which is equivalent, up to changing the constant, to the desired inequality $m \geq cs\ln(eN/s)$, see [39, Lemma C.6].

7 Other Matrices Suitable for Compressive Sensing

The measurement matrices put forward so far have been realizations of Gaussian matrices. We now discuss other types of matrices that can be used in compressive sensing. Again, they will be realizations of some random matrices.

7.1 *Subgaussian Matrices*

A mean-zero random variable ξ is called subgaussian if its tails are dominated by the tails of a Gaussian random variable, which is equivalent to the existence of constants $\alpha, \beta > 0$ such that

$$\mathbb{P}(|\xi| > t) \le \alpha \exp(-\beta t^2) \qquad \text{for all } t > 0.$$

Another equivalent condition involving the moment-generating function of ξ reads

$$\mathbb{E}\left(\exp(\theta\xi)\right) \le \exp(\gamma\theta^2) \qquad \text{for all } \theta \in \mathbb{R}.$$

Gaussian random variables are subgaussian, of course. Other important examples of subgaussian random variables include random variables uniformly distributed on an interval $[-c, c]$ and Rademacher random variables taking values -1 and $+1$ with probability $1/2$ each. By a subgaussian random matrix, we mean a matrix $\mathsf{A} \in \mathbb{R}^{m\times N}$ populated by independent (but not necessarily identically distributed) subgaussian random variables with a common parameter γ and normalized to have variance $1/m$. Such matrices are suitable compressive sensing matrices because they possess, with overwhelming probability, the RIP in the optimal regime $m \asymp s\ln(eN/s)$. The argument is almost identical to the proof of Theorem 1: More work is required to establish the concentration inequality, but the covering argument and the union bound are similar. See [39, Sect. 9.1] for details.

7.2 *Subexponential Matrices*

A mean-zero random variable is called subexponential if there exist constants $\alpha, \beta > 0$ such that

$$\mathbb{P}(|\xi| > t) \le \alpha \exp(-\beta t) \qquad \text{for all } t > 0.$$

For instance, Laplace random variables are subexponential. By a subexponential random matrix, we mean a matrix $\mathsf{A} \in \mathbb{R}^{m\times N}$ populated by independent (but not necessarily identically distributed) subexponential random variables with common parameters α, β and normalized to have variance $1/m$. Such matrices do not have the RIP in the optimal regime, since $m \geq C s \ln(eN/s)^2$ is required, as shown in [1]. But these matrices still satisfy the RNSP when $m \asymp s \ln(eN/s)$, as shown in [36]; hence, they allow for stable and robust sparse recovery via quadratically constrained basis pursuit. In fact, they also allow for stable and robust sparse recovery via other algorithms, as shown in [35]. This can be seen as a consequence of the modified RIP

$$(1-\delta) /\!\!/ z /\!\!/ \leq \|\mathsf{A}z\|_1 \leq (1+\delta) /\!\!/ z /\!\!/ \qquad \text{for all } s\text{-sparse vectors } z \in \mathbb{R}^N,$$

which features the ℓ_1-norm as the inner norm. The outer norm, whose precise expression is somewhat irrelevant, depends a priori on the probability distributions of the entries, but it satisfies, for some absolute constants $c, C > 0$,

$$c\sqrt{m}\|z\|_2 \leq /\!\!/ z /\!\!/ \leq C\sqrt{m}\|z\|_2 \qquad \text{for all vectors } z \in \mathbb{C}^N.$$

In particular, the result applies to Weibull matrices with parameters $r \in [1, \infty]$, whose entries are independent symmetric random variables obeying

$$\mathbb{P}(|\xi| > t) = \exp\left(-\left(\sqrt{\Gamma(1+2/r)m}\, t\right)^r\right).$$

If we further make the restriction $r \in [1, 2]$, then, with overwhelming probability, these matrices also satisfy the QP, as shown in [31]. Therefore, they will allow for stable and robust sparse recovery via equality-constrained basis pursuit even in the presence of measurement error.

We close this subsection by mentioning that the NSP can also be obtained, with high probability, for a wider random matrix class than subexponential matrices. Indeed, [51] establishes the NSP under rather weak conditions on the moments of the entries of the random matrices.

7.3 *Fourier Submatrices and Affiliates*

Consider the matrix $\mathsf{A} \in \mathbb{C}^{m \times N}$ with entries

$$A_{k,\ell} = \exp(\mathrm{i}2\pi \ell t_k), \qquad k \in [\![1:m]\!], \quad \ell \in [\![1:N]\!],$$

where the points $t_1, \ldots, t_m \in [0, 1]$ are independently chosen uniformly at random from the grid $\{0, 1/N, \ldots, (N-1)/N\}$ or from the interval $[0, 1]$. These two cases are of practical importance: The first case corresponds to randomly selecting rows of the discrete $N \times N$ Fourier matrix, so it provides fast matrix-vector multiplications exploited in the recovery algorithms; in the second case, samples of a trigonometric polynomial $f(t) = \sum_{\ell=0}^{N-1} x_\ell \exp(\mathrm{i}2\pi \ell t)$ at random points $t_1, \ldots, t_m \in [0, 1]$ produce a vector $\mathsf{A}x \in \mathbb{C}^m$, from where f can be recovered provided the coefficient vector $\boldsymbol{x} \in \mathbb{C}^N$ is s-sparse and $m \asymp s \ln^4(N)$—this drastically improves on $m \asymp N$ predicted from the Shannon–Nyquist theorem traditionally invoked in signal processing. It can indeed be proved that both types of random matrices possess the RIP of order s when $m \asymp s \ln^4(N)$. This remains true for the random selection of rows from Hadamard matrices, and more generally for random sampling matrices associated with bounded orthonormal systems (BOS). These matrices have entries $\Phi_\ell(t_k)$, $k \in [\![1 : m]\!]$, $\ell \in [\![1 : N]\!]$, where the functions $\Phi_1, \ldots, \Phi_N$ are uniformly bounded on a domain $\mathscr{D}$ by some absolute constant K and form an orthonormal system for the inner product associated with a probability measure ν on $\mathscr{D}$, and where the points $t_1, \ldots, t_m \in \mathscr{D}$ are independently chosen at random according to ν. We refer to [39, Chap. 12] for details. The power of the logarithm factor in the number of required measurements is improvable, with $m \asymp s \ln^2(s) \ln(N)$ seemingly the best result so far (see [22]). It is conceivable that the logarithmic factors may be further reduced, but not below $\ln(N)$ (see [39, Sect. 12.2]), thus not down to $\ln(eN/s)$ as is the case for subgaussian matrices.

7.4 Adjacency Matrices of Lossless Expanders

Expander graphs, and their bipartite counterpart lossless expanders, are very popular objects in computer science. They have made their appearance in compressive sensing via [6] and constitute now an important part of the theory, too. In short, an (s, d, θ)-lossless expander is a bipartite graph with left vertices indexed by $[\![1 : N]\!]$ and right vertices indexed by $[\![1 : m]\!]$ such that each left vertex is connected to d right vertices and that

$$\operatorname{card}(R(S)) \geq (1 - \theta)\, d \operatorname{card}(S)$$

for every set S of left vertices of size at most s, with $R(S)$ denoting the set of right vertices connected to S. Using probabilistic arguments, it can be shown that (s, d, θ)-lossless expanders exist with $d \asymp \theta^{-1} \ln(eN/s)$ and $m \asymp \theta^{-2} s \ln(eN/s)$. The adjacency matrix of such a graph is the matrix $\mathsf{A} \in \{0, 1\}^{m \times N}$ defined for $i \in [\![1 : m]\!]$ and $j \in [\![1 : N]\!]$ by

$$A_{i,j} = \begin{cases} 1 & \text{if there is an edge from } j \text{ to } i, \\ 0 & \text{otherwise.} \end{cases}$$

Although these matrices (properly normalized) do not satisfy the RIP, they still allow for stable and robust sparse recovery (in the weak sense of footnote 3) via basis pursuit or via an adapted iterative thresholding algorithm. All the details can be found in [39, Chap. 13].

8 Nonstandard Compressive Sensing

The name compressive sensing originates from the possibility of simultaneously acquiring (sensing) and compressing high-dimensional vectors $\boldsymbol{x} \in \mathbb{C}^N$ by taking only few linear measurements $y_1 = \langle \boldsymbol{a}_1, \boldsymbol{x} \rangle, \ldots, y_m = \langle \boldsymbol{a}_m, \boldsymbol{x} \rangle$ (with the understanding, of course, that efficient decompression procedures are also provided). The sparsity of the vectors of interest is what made this specific task feasible. But any task that exploits some structure of high-dimensional objects of interest to simultaneously acquire and compress them by means of few observations (while providing efficient decompression procedures) is also to be put under the umbrella of compressive sensing. We present below several extensions of the standard compressive sensing problem that fall in this category.

8.1 Low-Rank Recovery

The standard compressive sensing problem can sometimes be referred to as sparse recovery. The low-rank recovery problem is very similar: s-sparse vectors $\boldsymbol{x} \in \mathbb{C}^N$ are replaced by rank-r matrices $\mathbf{X} \in \mathbb{C}^{n_1 \times n_2}$—for simplicity of exposition, we assume here that $n_1 = n_2 =: n$. The matrices are still acquired by means of m linear measurements organized in the vector $\boldsymbol{y} = \mathscr{A}(\mathbf{X}) \in \mathbb{C}^m$, where $\mathscr{A} : \mathbb{C}^{n \times n} \to \mathbb{C}^m$ is a linear map. The minimal number of measurements for stable and robust rank-r recovery is $m \asymp rn$—there is no logarithmic factor. The theoretical explanation follows a route almost identical to the sparse recovery route, as outlined below. A popular optimization-based recovery procedure consists in solving

$$\underset{\mathbf{Z} \in \mathbb{C}^{n \times n}}{\text{minimize}} \; \|\mathbf{Z}\|_* \qquad \text{subject to } \|\mathscr{A}(\mathbf{Z}) - \boldsymbol{y}\|_2 \leq \eta,$$

where the nuclear norm is the sum of the singular values $\sigma_1(\mathbf{Z}) \geq \cdots \geq \sigma_n(\mathbf{Z}) \geq 0$, i.e.,

$$\|\mathbf{Z}\|_* = \sum_{k=1}^{n} \sigma_k(\mathbf{Z}). \tag{29}$$

An analogy with basis pursuit is offered by the facts that a matrix has rank at most r if and only if its vector of singular values is r-sparse and that the nuclear norm is just the ℓ_1-norm of the vector of singular values. The problem (29) can be solved by semidefinite programming, thanks to the observation that

$$\|\mathbf{Z}\|_* = \inf \left\{ \operatorname{tr}(\mathbf{M}) + \operatorname{tr}(\mathbf{N}), \; \begin{bmatrix} \mathbf{M} & \mathbf{Z} \\ \mathbf{Z}^* & \mathbf{N} \end{bmatrix} \succeq \mathbf{0} \right\}. \tag{30}$$

There is a null space property which is equivalent to the success of exact rank-r recovery via nuclear norm minimization (see [39, Theorem 4.40]) and a robust version that guarantees stability and robustness of the recovery (see [39, Exercise 4.20]). With constants $0 < \rho < 1$ and $\tau > 0$, it reads:

$$\text{for all } \mathbf{M} \in \mathbb{C}^{n \times n}, \qquad \left[\sum_{k=1}^{r} \sigma_k(\mathbf{M})^2\right]^{1/2} \leq \frac{\rho}{\sqrt{r}} \sum_{k=r+1}^{n} \sigma_k(\mathbf{M}) + \tau \|\mathscr{A}(\mathbf{M})\|_2. \quad (31)$$

This property is implied by a matrix version of the restricted isometry property, which reads

$$(1-\delta)\|\mathbf{Z}\|_F^2 \leq \|\mathscr{A}(\mathbf{Z})\|_2^2 \leq (1+\delta)\|\mathbf{Z}\|_F^2 \qquad \text{for all } \mathbf{Z} \in \mathbb{C}^{n \times n} \text{ with rank } (\mathbf{Z}) \leq r.$$

As in the vector case, we write δ_r for the smallest such constant δ. Given a linear map $\mathscr{A} : \mathbf{M} \in \mathbb{C}^{n \times n} \mapsto \left(\sum_{k,\ell=1}^{n} \mathscr{A}_{i,k,\ell} M_{k,\ell}\right)_{i=1}^{m} \in \mathbb{C}^m$ where the $\mathscr{A}_{i,k,\ell}$ are independent Gaussian random variables with mean zero and variance $1/m$, it can be shown that $\delta_r < \delta_*$ with overwhelming probability provided that $m \geq C\delta_*^{-2} rn$ (see [39, Exercise 9.12]). The logarithmic factor disappears because a union bound is unnecessary to estimate the covering number of the set $\{\mathbf{Z} \in \mathbb{C}^{n \times n} : \|\mathbf{Z}\|_F = 1, \operatorname{rank}(\mathbf{Z}) \leq r\}$. Finally, seeing why the restricted isometry condition $\delta_{2r} < 1/\sqrt{2}$ (or any sufficient condition $\delta_{\kappa r} < \delta_*$ from the vector case) implies the matrix version of the RNSP requires almost no work. Indeed, given $\mathbf{M} \in \mathbb{C}^{n \times n}$, we consider the singular value decomposition $\mathbf{M} = \mathbf{U}\operatorname{diag}[\boldsymbol{\sigma}(\mathbf{M})]\mathsf{V}^*$, and we define a matrix $\mathsf{A} \in \mathbb{C}^{m \times n}$ via its action

$$z \in \mathbb{C}^n \mapsto \mathscr{A}(\mathbf{U}\operatorname{diag}[z]\mathsf{V}^*) \in \mathbb{C}^m.$$

This matrix obeys the restricted isometry condition, because, for every $2r$-sparse $z \in \mathbb{C}^n$, the matrix $\mathbf{U}\operatorname{diag}[z]\mathsf{V}^*$ has rank at most $2r$; hence

$$\begin{aligned} \left|\|\mathsf{A}(z)\|_2^2 - \|z\|_2^2\right| &= \left|\|\mathscr{A}(\mathbf{U}\operatorname{diag}(z)\mathsf{V}^*)\|_2^2 - \|\mathbf{U}\operatorname{diag}[z]\mathsf{V}^*\|_F^2\right| \leq \delta_{2r}\|\mathbf{U}\operatorname{diag}[z]\mathsf{V}^*\|_F^2 \\ &< \delta_*\|z\|_2^2. \end{aligned}$$

Therefore, the matrix A possesses the RNSP of order r. In particular, writing it for the vector $\boldsymbol{\sigma}(\mathbf{M})$ and the index set $S = \{1, \ldots, r\}$ reduces to (31).

8.2 *Low-Rank Matrix Completion*

Low-rank matrix completion is a particular instance of low-rank recovery where the linear measurements are restricted in nature. Precisely, what is observed consists of few entries of the matrix. It can be proved that $m \asymp rn$ entries observed at random suffice to efficiently recover, with high probability, 'most' rank-r matrices. Note that the recovery of all rank-r matrices is impossible, as the zero matrix and the rank-1 matrix with only one nonzero entry cannot be distinguished from a typical set of observations. For details, see [17] where theoretical results were first derived,

[42, 61] where such results were improved and simplified, [16] where robustness to observation errors was taken into account, and [7] where uniform recovery results were obtained.

8.3 Sparse Phaseless Recovery

The sparse phaseless recovery scenario is very close to the plain sparse recovery scenario, except that the linear measurements $\langle \boldsymbol{a}_i, \boldsymbol{x} \rangle$ lose their phase, so only their magnitude is available. In other words, sparse vectors $\boldsymbol{x} \in \mathbb{C}^N$ are acquired through the nonlinear measurements $y_i = |\langle \boldsymbol{a}_i, \boldsymbol{x} \rangle|^2$ and one wishes to recover them (up to a phase factor). The nonsparse version of the problem has been around for a long time. The article [21] was the first one to propose a recovery strategy based on convex optimization. The following 'lifting trick' is a central part of the argument:

$$y_i = |\langle \boldsymbol{a}_i, \boldsymbol{x} \rangle|^2 = \langle \boldsymbol{x}, \boldsymbol{a}_i \rangle \langle \boldsymbol{a}_i, \boldsymbol{x} \rangle = \boldsymbol{a}_i^* \boldsymbol{x} \boldsymbol{x}^* \boldsymbol{a}_i = \operatorname{tr}(\boldsymbol{a}_i^* \boldsymbol{x} \boldsymbol{x}^* \boldsymbol{a}_i) = \operatorname{tr}(\boldsymbol{a}_i \boldsymbol{a}_i^* \boldsymbol{x} \boldsymbol{x}^*) = \operatorname{tr}(\boldsymbol{a}_i \boldsymbol{a}_i^* \mathbf{X}),$$

where $\mathbf{X}$ is the rank-one positive semidefinite matrix defined by $\mathbf{X} := \boldsymbol{x}\boldsymbol{x}^*$. Thus, the recovery problem reduces to

$$\underset{\mathbf{Z} \in \mathbb{C}^{n \times n}}{\text{minimize}} \ \operatorname{rank}(\mathbf{Z}) \qquad \text{subject to} \quad \operatorname{tr}(\boldsymbol{a}_i \boldsymbol{a}_i^* \mathbf{Z}) = y_i, \quad i \in [\![1 : m]\!],$$
$$\text{and to} \quad \mathbf{Z} \succeq \mathbf{0}.$$

Since this optimization problem is not computationally tractable, it is relaxed to a convex optimization problem by replacing the rank by the nuclear norm. The expression (30) allows one to treat the resulting optimization problem, namely

$$\underset{\mathbf{Z} \in \mathbb{C}^{n \times n}}{\text{minimize}} \ \|\mathbf{Z}\|_* \qquad \text{subject to} \quad \operatorname{tr}(\boldsymbol{a}_i \boldsymbol{a}_i^* \mathbf{Z}) = y_i, \quad i \in [\![1 : m]\!], \quad \text{and to} \quad \mathbf{Z} \succeq \mathbf{0} \tag{32}$$

as a semidefinite program. If a solution to (32) happens to be of rank one, call it $\mathbf{X} \in \mathbb{C}^{n \times n}$, then a solution to the original phaseless recovery problem is derived as $\boldsymbol{x} = (\|\mathbf{X}\|_F^{1/2} / \|\mathbf{X}\mathbf{u}\|_2)\mathbf{X}\mathbf{u}$ for an arbitrary $\mathbf{u} \in \mathbb{C}^n$. It was shown in [21] that taking $m \asymp N \ln(N)$ Gaussian measurements guarantees that the solution to (32) is indeed of rank one, and the estimate for the number of measurements was improved to $m \asymp N$ in [15]. The number of measurements can be further reduced if the vectors to be recovered are s-sparse: [52] originally showed (in the real setting) that the order $m \asymp s^2 \ln(N)$ is achievable via convex programming, but that $m \asymp s \ln(N)$ is not achievable for a class of naive semidefinite relaxations, and later [45] showed that there is a choice of $m \asymp s \ln(eN/s)$ phaseless measurements allowing for s-sparse recovery.

8.4 One-Bit Compressive Sensing

The one-bit compressive sensing problem is in some sense complementary to the sparse phaseless recovery problem: Instead of keeping only the magnitudes of linear measurements, these magnitudes are lost and only the sign information is kept—we are now working in the real setting. The problem, originally proposed in [10], was born out of the practical consideration that measurements really have to be quantized. Although sophisticated quantization scheme such as $\Sigma\Delta$ quantization should be used in practice (see [43] for a $\Sigma\Delta$ analysis of compressive sensing), the one-bit compressive sensing problem concentrates on the extreme scenario where only some binary information about the measurements $\langle \boldsymbol{a}_i, \boldsymbol{x} \rangle$ made on sparse vectors $\boldsymbol{x} \in \mathbb{R}^N$ is retained, i.e., only $y_1 = \mathrm{sgn}\langle \boldsymbol{a}_1, \boldsymbol{x} \rangle, \ldots, y_m = \mathrm{sgn}\langle \boldsymbol{a}_m, \boldsymbol{x} \rangle$ are now available. Of course, exact s-sparse recovery is now impossible. Instead, the question is phrased as: How few measurements allow us to recover the direction of s-sparse vectors up to accuracy ε? Note that we cannot recover more than the direction. The theory was mostly developed in [59, 60]. We propose below some swift shortcuts through the theory (to keep things simple, stability and robustness issues are left out), relying solely on a modified version of the restricted isometry property with ℓ_1-norm as the inner norm. Precisely, we consider a matrix $\mathsf{A} \in \mathbb{R}^{m \times N}$ such that

$$(1-\delta)\|\boldsymbol{z}\|_2 \leq \|\mathsf{A}\boldsymbol{z}\|_1 \leq (1+\delta)\|\boldsymbol{z}\|_2 \qquad \text{for all } s\text{-sparse vectors } \boldsymbol{z} \in \mathbb{R}^N, \tag{33}$$

in which case we write that A satisfies $\mathrm{RIP}_1(s,\delta)$. For Gaussian matrices populated by independent random variables with mean zero and standard deviation $\sqrt{\pi/2}/m$, $\mathrm{RIP}_1(s,\delta)$ occurs with high probability provided $m \asymp \delta^{-\mu} s \ln(eN/s)$. This can be shown in the same way as Theorem 1, with the caveat that the power of δ^{-1} will have a nonoptimal value $\mu > 2$. The value $\mu = 2$ can be achieved by other means, which in fact allow to prove this modified RIP not only for genuinely s-sparse vectors but also for effectively s-sparse vectors (see [8, 62]). We write $\mathrm{RIP}_1^{\mathrm{eff}}(s,\delta)$ for this extension of the modified RIP. Everything is now in place to present our simplified argument on sparse recovery from one-bit measurements via a thresholding-based strategy and via an optimization-based strategy.

Recovery via hard thresholding: The following observation is the crucial part in the proof of the next theorem.

Lemma 3 *If* A *satisfies* $\mathrm{RIP}_1(s,\delta)$*, then for any* ℓ_2*-normalized vector* $\boldsymbol{x} \in \mathbb{R}^N$ *supported on an index set* S *of size* s*,*

$$\left\| (\mathsf{A}^*\mathrm{sgn}(\mathsf{A}\boldsymbol{x}))_S - \boldsymbol{x} \right\|_2^2 \leq 5\delta.$$

Proof Expanding the square gives

$$\|(\mathsf{A}^*\mathrm{sgn}(\mathsf{A}\boldsymbol{x}))_S - \boldsymbol{x}\|_2^2 = \|(\mathsf{A}^*\mathrm{sgn}(\mathsf{A}\boldsymbol{x}))_S\|_2^2 - 2\langle (\mathsf{A}^*\mathrm{sgn}(\mathsf{A}\boldsymbol{x}))_S, \boldsymbol{x} \rangle + \|\boldsymbol{x}\|_2^2. \tag{34}$$

The first term on the right-hand side of (34) satisfies

$$\begin{aligned}\|(\mathsf{A}^*\mathrm{sgn}(\mathsf{A}\boldsymbol{x}))_S\|_2^2 &= \langle \mathsf{A}^*\mathrm{sgn}(\mathsf{A}\boldsymbol{x}), (\mathsf{A}^*\mathrm{sgn}(\mathsf{A}\boldsymbol{x}))_S\rangle = \langle \mathrm{sgn}(\mathsf{A}\boldsymbol{x}), \mathsf{A}\left((\mathsf{A}^*\mathrm{sgn}(\mathsf{A}\boldsymbol{x}))_S\right)\rangle \\ &\leq \left\|\mathsf{A}\left((\mathsf{A}^*\mathrm{sgn}(\mathsf{A}\boldsymbol{x}))_S\right)\right\|_1 \leq (1+\delta)\|(\mathsf{A}^*\mathrm{sgn}(\mathsf{A}\boldsymbol{x}))_S\|_2,\end{aligned}$$

so that, after simplification,

$$\|(\mathsf{A}^*\mathrm{sgn}(\mathsf{A}\boldsymbol{x}))_S\|_2 \leq (1+\delta).$$

For the second term on the right-hand side of (34), we notice that

$$\begin{aligned}\langle (\mathsf{A}^*\mathrm{sgn}(\mathsf{A}\boldsymbol{x}))_S, \boldsymbol{x}\rangle &= \langle \mathsf{A}^*\mathrm{sgn}(\mathsf{A}\boldsymbol{x}), \boldsymbol{x}\rangle = \langle \mathrm{sgn}(\mathsf{A}\boldsymbol{x}), \mathsf{A}\boldsymbol{x}\rangle = \|\mathsf{A}\boldsymbol{x}\|_1 \\ &\geq (1-\delta)\|\boldsymbol{x}\|_2 = (1-\delta).\end{aligned}$$

The third term on the right-hand side of (34) is simply $\|\boldsymbol{x}\|_2^2 = 1$. Altogether, we arrive at

$$\|(\mathsf{A}^*\mathrm{sgn}(\mathsf{A}\boldsymbol{x})_S - \boldsymbol{x}\|_2^2 \leq (1+\delta)^2 - 2(1-\delta) + 1 = 4\delta + \delta^2 \leq 5\delta,$$

which is the announced result. □

We now consider the simple (noniterative) hard thresholding procedure given by

$$\Delta_{\mathrm{1B-HT}}(\boldsymbol{y}) := H_s(\mathsf{A}^*\boldsymbol{y}). \tag{35}$$

Theorem 8 *If* A *satisfies* $\mathit{RIP}_1(2s,\delta)$*, then every* ℓ_2*-normalized and* s*-sparse vector* $\boldsymbol{x} \in \mathbb{R}^N$ *observed via* $\boldsymbol{y} = \mathrm{sgn}(\mathsf{A}\boldsymbol{x}) \in \{\pm 1\}^m$ *is approximated by the output of the hard thresholding procedure* (35) *with error*

$$\|\boldsymbol{x} - \Delta_{\mathrm{1B-HT}}(\boldsymbol{y})\|_2 \leq C\sqrt{\delta}.$$

Proof Let $S := \mathrm{supp}(\boldsymbol{x})$ and $T := \mathrm{supp}(\boldsymbol{x}^{\mathrm{HT}})$. We notice that $\boldsymbol{x}^{\mathrm{HT}}$ is also the best s-term approximation to $(\mathsf{A}^*\boldsymbol{y})_{S\cup T}$, so that

$$\begin{aligned}\|\boldsymbol{x} - \boldsymbol{x}^{\mathrm{HT}}\|_2 &\leq \|(\mathsf{A}^*\boldsymbol{y})_{S\cup T} - \boldsymbol{x}^{\mathrm{HT}}\|_2 + \|(\mathsf{A}^*\boldsymbol{y})_{S\cup T} - \boldsymbol{x}\|_2 \leq 2\|(\mathsf{A}^*\boldsymbol{y})_{S\cup T} - \boldsymbol{x}\|_2 \\ &\leq 2\sqrt{5\delta},\end{aligned}$$

where we made use of Lemma 3. This establishes our result. □

Recovery via linear programming: We consider here the optimization problem

$$\Delta_{\mathrm{1B-LP}}(\boldsymbol{y}) = \mathrm{argmin}\ \|\boldsymbol{z}\|_1 \quad \text{subject to } \mathrm{sgn}(\mathsf{A}\boldsymbol{z}) = \boldsymbol{y} \ \text{ and } \ \|\mathsf{A}\boldsymbol{z}\|_1 = 1. \tag{36}$$

By writing the constraints as $y_i \langle \boldsymbol{a}_i, \boldsymbol{z} \rangle \geq 0$ for $i \in [\![1:m]\!]$ and $\sum_{i=1}^m y_i \langle \boldsymbol{a}_i, \boldsymbol{z} \rangle = 1$, we see that this optimization problem can be recast as a linear program. We need the following intermediate result before justifying the recovery guarantee.

Lemma 4 *If* A *satisfies RIP*$_1(9s, 1/5)$ *and if* $\boldsymbol{x} \in \mathbb{R}^N$ *is* s*-sparse, then any convex combination of* $\boldsymbol{x}$ *and* $\Delta_{1\mathrm{B-LP}}(\mathrm{sgn}(\mathsf{A}\boldsymbol{x}))$ *is effectively* $9s$*-sparse. The same conclusion holds under RIP*$_1^{\mathrm{eff}}(9s, 1/5)$ *if* $\boldsymbol{x}$ *is effectively* s*-sparse.*

Proof Let us use the notation $t = 9s$, $\delta = 1/5$, and $\boldsymbol{x}^\sharp = \Delta_{1\mathrm{B-LP}}(\mathrm{sgn}(\mathsf{A}\boldsymbol{x}))$. By the defining property of $\boldsymbol{x}^\sharp$, we have

$$\|\boldsymbol{x}^\sharp\|_1 \leq \left\| \frac{\boldsymbol{x}}{\|\mathsf{A}\boldsymbol{x}\|_1} \right\|_1 \leq \frac{\sqrt{s}\|\boldsymbol{x}\|_2}{\|\mathsf{A}\boldsymbol{x}\|_1} \leq \frac{\sqrt{s}}{1-\delta} = \frac{5\sqrt{s}}{4}.$$

If $\widehat{\boldsymbol{x}} = (1-\lambda)\boldsymbol{x} + \lambda \boldsymbol{x}^\sharp$ denotes a convex combination of $\boldsymbol{x}$ and $\boldsymbol{x}^\sharp$, then we obtain

$$\|\widehat{\boldsymbol{x}}\|_1 \leq (1-\lambda)\|\boldsymbol{x}\|_1 + \lambda\|\boldsymbol{x}^\sharp\|_1 \leq (1-\lambda)\sqrt{s} + \lambda\frac{5\sqrt{s}}{4} = \left(1 + \frac{\lambda}{4}\right)\sqrt{s}.$$

Notice also that $\mathrm{sgn}(\mathsf{A}\widehat{\boldsymbol{x}}) = \mathrm{sgn}(\mathsf{A}\boldsymbol{x}^\sharp) = \mathrm{sgn}(\mathsf{A}\boldsymbol{x})$, so that

$$\|\mathsf{A}\widehat{\boldsymbol{x}}\|_1 = (1-\lambda)\|\mathsf{A}\boldsymbol{x}\|_1 + \lambda\|\mathsf{A}\boldsymbol{x}^\sharp\|_1 \geq (1-\lambda)(1-\delta) + \lambda = \frac{4}{5}\left(1 + \frac{\lambda}{4}\right).$$

Next, we use the sort-and-split technique and consider an index set T_0 of t largest absolute entries of $\widehat{\boldsymbol{x}}$, an index set T_1 of t next largest absolute entries of $\widehat{\boldsymbol{x}}$, an index set T_2 of t next largest absolute entries of $\widehat{\boldsymbol{x}}$, etc. Thus,

$$\begin{aligned}
\frac{4}{5}\left(1 + \frac{\lambda}{4}\right) &\leq \|\mathsf{A}\widehat{\boldsymbol{x}}\|_1 \leq \sum_{k\geq 0} \|\mathsf{A}\widehat{\boldsymbol{x}}_{T_k}\|_1 \leq (1+\delta)\left(\|\widehat{\boldsymbol{x}}_{T_0}\|_2 + \sum_{k\geq 1} \|\widehat{\boldsymbol{x}}_{T_k}\|_2 \right) \\
&\leq (1+\delta)\left(\|\widehat{\boldsymbol{x}}\|_2 + \frac{1}{\sqrt{t}}\|\widehat{\boldsymbol{x}}\|_1 \right) \leq (1+\delta)\left(\|\widehat{\boldsymbol{x}}\|_2 + \sqrt{\frac{s}{t}}\left(1 + \frac{\lambda}{4}\right)\right) \\
&= \frac{6}{5}\left(\|\widehat{\boldsymbol{x}}\|_2 + \frac{1}{3}\left(1 + \frac{\lambda}{4}\right)\right).
\end{aligned}$$

This implies that $\|\widehat{\boldsymbol{x}}\|_2 \geq (1 + \lambda/4)/3$. In turn, we derive

$$\frac{\|\widehat{\boldsymbol{x}}\|_1}{\|\widehat{\boldsymbol{x}}\|_2} \leq \frac{(1+\lambda/4)\sqrt{s}}{(1+\lambda/4)/3} = \sqrt{9s}.$$

This means that $\widehat{\boldsymbol{x}}$ is effectively $9s$-sparse, as announced. □

We can now state and prove the recovery result for the linear program.

Theorem 9 *If* A *satisfies* $RIP_1^{\rm eff}(9s,\delta)$ *with* $\delta \le 1/5$*, then every* ℓ_2*-normalized and effectively* s*-sparse vector* $\boldsymbol{x} \in \mathbb{R}^N$ *observed via* $\boldsymbol{y} = \operatorname{sgn}(\mathsf{A}\boldsymbol{x}) \in \{\pm 1\}^m$ *is approximated by the output of the linear programming procedure* (36) *with error*

$$\|\boldsymbol{x} - \Delta_{\rm 1B-LP}(\boldsymbol{y})\|_2 \le C\sqrt{\delta}.$$

Proof We still use the notation $\boldsymbol{x}^\sharp$ for $\Delta_{\rm 1B-LP}(\boldsymbol{y})$. By Lemma 4, the vector $(\boldsymbol{x}+\boldsymbol{x}^\sharp)/2$ is effectively $9s$-sparse, and using the fact that $\operatorname{sgn}(\mathsf{A}\boldsymbol{x}^\sharp) = \operatorname{sgn}(\mathsf{A}\boldsymbol{x})$, we have

$$\begin{aligned}\left\|\frac{\boldsymbol{x}+\boldsymbol{x}^\sharp}{2}\right\|_2 &\ge \frac{1}{1+\delta}\left\|\mathsf{A}\left(\frac{\boldsymbol{x}+\boldsymbol{x}^\sharp}{2}\right)\right\|_1 = \frac{1}{1+\delta}\frac{\|\mathsf{A}\boldsymbol{x}\|_1 + \|\mathsf{A}\boldsymbol{x}^\sharp\|_1}{2} \ge \frac{1}{1+\delta}\frac{(1-\delta)+1}{2}\\ &= \frac{1-\delta/2}{1+\delta}.\end{aligned}$$

Lemma 4 also implies that $\boldsymbol{x}^\sharp$ is effectively $9s$-sparse, so we can write

$$\begin{aligned}\left\|\frac{\boldsymbol{x}-\boldsymbol{x}^\sharp}{2}\right\|_2^2 &= \frac{1+\|\boldsymbol{x}^\sharp\|_2^2}{2} - \left\|\frac{\boldsymbol{x}+\boldsymbol{x}^\sharp}{2}\right\|_2^2 \le \frac{1+1/(1-\delta)^2}{2} - \frac{(1-\delta/2)^2}{(1+\delta)^2}\\ &\le C\delta.\end{aligned}$$

This immediately implies the announced result. □

Recovery of more than the direction: To sum up the previous results, it is possible to recover the direction of s-sparse vectors $\boldsymbol{x} \in \mathbb{R}^N$ acquired via $\boldsymbol{y} = \operatorname{sgn}(\mathsf{A}\boldsymbol{x}) \in \{\pm 1\}^m$ with accuracy ε provided $m \ge C\varepsilon^4 s \ln(eN/s)$. Stated differently, there are recovery maps $\Delta : \{\pm 1\}^m \to \mathbb{R}^N$ such that

$$\|\boldsymbol{x} - \Delta(\operatorname{sgn}(\mathsf{A}\boldsymbol{x}))\|_2 \le C\left[\frac{m}{s\ln(eN/s)}\right]^{1/4} \quad \text{whenever } \|\boldsymbol{x}\|_0 \le s,\ \|\boldsymbol{x}\|_2 = 1.$$

It is also possible to estimate the magnitude of $\boldsymbol{x}$ if one allows thresholds $\tau_1, \dots, \tau_m$ in the binary measurements $y_1 = \operatorname{sgn}(\langle \boldsymbol{a}_i, \boldsymbol{x}\rangle - \tau_i)$, $i \in [\![1:m]\!]$, see [4, 5, 49]. In fact, if these thresholds can be chosen adaptively, we proved in [4] that one can even achieve the bound

$$\|\boldsymbol{x} - \Delta(\operatorname{sgn}(\mathsf{A}\boldsymbol{x} - \boldsymbol{\tau}))\|_2 \le C\exp\left(-c\left[\frac{m}{s\ln(eN/s)}\right]\right) \quad \text{whenever } \|\boldsymbol{x}\|_0 \le s,\ \|\boldsymbol{x}\|_2 \le 1.$$

9 Open Questions

We close this survey by describing some compressive sensing problems that remain currently unresolved. The first one, concerning Gelfand widths, sits at the boundary of the topic, but it is included in view of the central part played by Gelfand widths in this survey. The other ones are bona fide compressive sensing problems.

Gelfand width of the ℓ_1-ball in ℓ_∞: We have seen how two-sided estimates for the Gelfand width $d^m(B_1^N, \ell_2^N)$ can be obtained from methods of compressive sensing. The same methods also yield two-sided estimates for the Gelfand width $d^m(B_1^N, \ell_q^N)$ for any $q \in [1, 2]$. But for $q > 2$, known lower and upper estimates do not quite match. In particular, for $q = \infty$, it is only known that

$$c \max\left\{\sqrt{\frac{1}{m}}, \min\left\{1, \frac{\ln(eN/m)}{m}\right\}\right\} \leq d^m(B_1^N, \ell_\infty^N) \leq C \min\left\{1, \sqrt{\frac{\ln(eN/m)}{m}}\right\},$$

see [38, 68] for details. Finding the correct order for $d^m(B_1^N, \ell_\infty^N)$ seems challenging.

The quest for deterministic compressive sensing matrices: We have seen that measurement matrices $\mathsf{A} \in \mathbb{C}^{m \times N}$ allowing for stable and robust s-sparse recovery in the optimal regime $m \asymp s \ln(eN/s)$ do exist. In fact, many such matrices should exist, as the argument consists in proving that random matrices satisfy, with high probability, some property that guarantee sparse recovery, e.g., the RIP. To date, however, the derandomization problem of exhibiting concrete examples of matrices allowing for stable and robust sparse recovery remains open—arguably, it is the most significant open problem in the field. Note that selecting a realization of a random matrix and verifying that it has the RIP, say, is not a viable strategy, since this task is NP-hard (see [3, 66]). Using the concept of coherence, it is easy (see, e.g., [39, Chapter 5]) to deterministically produce matrices that allow for stable and robust s-sparse recovery with a number m of rows of the order of s^2. Reducing this number of rows to $m \asymp s^\beta$ with $\beta < 2$ would already be a breakthrough. In fact, such a breakthrough was achieved in [12]: An explicit choice of $m \asymp s^{2-\varepsilon}$ rows from the discrete Fourier matrix was shown to yield a matrix with the RIP of order s, but $\varepsilon > 0$ was ever so small that this theoretical feat is of no practical interest. Finally, it has to be noted that, should one settle for a weaker notion of stability and robustness (in the spirit of footnote 3), deterministic measurement matrices in the near-optimal regime $m \asymp s^\alpha$ are supposedly available for any $\alpha > 1$, since [44] claims an explicit construction of (s, d, θ)-lossless expanders with this number of right vertices.

Universality of phase transition for random matrices: In [28], the authors look at the standard compressive sensing problem from a different perspective. Defining $\delta := m/N \in [0, 1]$ (not to be confused with a restricted isometry constant) and $\rho := s/m \in [0, 1]$, they investigate the phase diagram for the function

$$P(\delta, \rho) := \lim_{N \to \infty} \mathbb{P}(\boldsymbol{x} \text{ is exactly recovered from } \boldsymbol{y} = \mathsf{A}\boldsymbol{x} \text{ by } \ell_1\text{-minimization}),$$

where $\boldsymbol{x} \in \mathbb{R}^N$ is an arbitrary (but fixed) vector and where $\mathsf{A} \in \mathbb{R}^{m \times N}$ is a Gaussian random matrix. Exploiting sophisticated results about random polytopes, they could show that $P^{-1}(\{0\})$ and $P^{-1}(\{1\})$ partition the square $[0, 1]^2$. In other words, when

N is large, ℓ_1-recovery from random measurements either fails or succeeds with high probability depending on the values of the ratios s/m and m/N. They could also give an implicit expression of the curve separating the regions $P^{-1}(\{0\})$ and $P^{-1}(\{1\})$. Their original arguments have been simplified and extended in [2] by relying on tools from integral geometry. The authors also carried out extensive numerical experiments for non-Gaussian measurements. These experiments, reported in [29], suggest that the phase transition phenomenon is ubiquitous across a wide variety of random matrices, as the observed phase diagrams were in many cases identical to the one derived theoretically in [28]. An explanation for this universality is currently lacking.

Logarithmic factors in the RIP for BOS: We have seen in Sect. 7.3 that random sampling matrices associated with bounded orthonormal systems possess the RIP of order s provided $m \asymp s \ln^3(N)$. Reducing the power of the logarithm factor further seems to require considerable efforts. It constitutes a noteworthy open problem, not so much because of practical implications (as the hidden constants could dwarf the logarithm factor) but because of theoretical implications. Indeed, achieving $m \asymp s \ln(N)$ would solve a standing conjecture about the Λ_1-problem studied by Bourgain [11] and Talagrand [63]. See [39, Sect. 12.7] for details about this connection.

Dictionary sparsity: We have started this journey by identifying from the onset signals with their coefficient vectors in a given basis, and these vectors were assumed to be sparse. But it is often more realistic to assume that the signals have a sparse representation not in a basis but rather in a dictionary (i.e., an overcomplete system). In order to give a few details, suppose to simplify that $\mathsf{D} \in \mathbb{C}^{n \times N}$ is a tight frame, in the sense that $\mathsf{D}\mathsf{D}^* = \mathsf{I}$. Saying that a signal $\boldsymbol{f} \in \mathbb{C}^n$ is sparse with respect to the dictionary D may have two meanings:

- it is synthesis-sparse, i.e., $\boldsymbol{f} = \mathsf{D}\boldsymbol{x}$ for some sparse $\boldsymbol{x} \in \mathbb{C}^N$,
- it is analysis-sparse, i.e., $\mathsf{D}^*\boldsymbol{f} \in \mathbb{C}^N$ is sparse.

Synthesis sparsity is a more realistic assumption to make, and there are theoretical algorithms that allow for stable and robust s-synthesis-sparse recovery from a number $m \asymp s \ln(eN/s)$ of random measurements, see, e.g., [25]. The term 'theoretical' algorithms was used because these algorithms involve finding best or near-best approximations from the set of s-synthesis-sparse vectors, which is sometimes a difficult task in itself (see [65]). In contrast, the algorithms proposed for analysis-sparse recovery are really practical. The optimization-based strategy of [20] and the thresholding-based iterative strategy of [32] both yield the same stability and robustness estimate featuring the error $\sigma_s(\mathsf{D}^*\boldsymbol{f})_1$ of best s-term approximation to $\mathsf{D}^*\boldsymbol{f}$ in ℓ_1-norm. However, it is not clear in all situations that this is a small quantity. Certainly, if the columns of D are in general position, then $\sigma_s(\mathsf{D}^*\boldsymbol{f})_1$ cannot equal zero for $s \leq N - n$ unless $\boldsymbol{f} = \{\mathbf{0}\}$, otherwise $(\mathsf{D}^*\boldsymbol{f})_j = 0$ for $N - s$ indices j means that $\boldsymbol{f} \in \mathbb{R}^n \setminus \{\mathbf{0}\}$ is orthogonal to $N - s \geq n$ linearly independent vectors, which is impossible. To sum up, the two notions of dictionary sparsity are both undermined by specific drawbacks for certain dictionaries. It is desirable to determine which dictionaries can circumvent the pitfalls of synthesis-sparsity recovery and of analysis-sparsity recovery simultaneously.

Structure other than sparsity, combination of structures: We have argued that compressive sensing does not reduce to the standard problem (3)—it concerns more generally the recovery of structured high-dimensional objects acquired from fewer measurements than conventionally thought necessary, as exemplified in Sect. 8.1, say. There are other structured objects to be considered besides sparse vectors and low-rank matrices, e.g., the recovery of low-rank tensors is currently an active research niche. The recovery of objects possessing several structures simultaneously leads to further interesting inquiries. For fairly general structures, the authors of [57] showed that, roughly speaking, as far as the asymptotic number of Gaussian measurements is concerned and when optimization-based recovery methods are used, there is no gain in exploiting the combination of structures instead of exploiting each one of them separately. Such an observation was corroborated in [37], where a similar conclusion was reached for a quite restrictive combination of structures—sparsity and disjointedness—but for arbitrary measurement and recovery schemes. However, this does not seem to be a general phenomenon. As a case in point, it can be shown that recovering matrices that are both low-rank and sparse are possible from fewer measurements than the numbers needed for the recovery of low-rank matrices alone and for the recovery of sparse matrices alone. The recovery map is not optimization-based—this would contradict [57]—but it is a thresholding-based iterative algorithm, with the caveat that it involves the impractical task of finding best or near-best approximations from the set of simultaneously low-rank and sparse matrices. But besides solving this minor issue, a challenging program consists in understanding exactly when the combination of structures turns out to be beneficial. Such a program may need to be carried out on a case-to-case basis.

Acknowledgements I thank the organizers of the International Conference on Approximation Theory for running this important series of triennial meetings. It was a plenary address by Ron DeVore at the 2007 meeting that drove me into the subject of compressive sensing. His talk was entitled 'A Taste of Compressed sensing' and my title is clearly a reference to his. Furthermore, I acknowledge support from the NSF under the grant DMS-1622134. Finally, I am also indebted to the AIM SQuaRE program for funding and hosting a collaboration on one-bit compressive sensing.

References

1. R. Adamczak, A. Litvak, A. Pajor, N. Tomczak-Jaegermann, Restricted isometry property of matrices with independent columns and neighborly polytopes by random sampling. Constr. Approx. **34**, 61–88 (2011)
2. D. Amelunxen, M. Lotz, M. McCoy, J. Tropp, Living on the edge: Phase transitions in convex programs with random data. Information and Inference. iau005 (2014)
3. A. Bandeira, E. Dobriban, D. Mixon, W. Sawin, Certifying the restricted isometry property is hard. IEEE Trans. Inform. Theory **59**, 3448–3450 (2013)
4. R. Baraniuk, S. Foucart, D. Needell, Y. Plan, M. Wootters, Exponential decay of reconstruction error from binary measurements of sparse signals. IEEE Trans. Inform. Theory **63**(6), 3368–3385 (2017)
5. R. Baraniuk, S. Foucart, D. Needell, Y. Plan, M. Wootters, One-bit compressive sensing of dictionary-sparse signals (Information and Influence)

6. R. Berinde, A. Gilbert, P. Indyk, H. Karloff, M. Strauss, Combining geometry and combinatorics: a unified approach to sparse signal recovery, in *Proceedings of 46th Annual Allerton Conference on Communication, Control, and Computing* (2008), pp. 798–805
7. S. Bhojanapalli, P. Jain, Universal matrix completion, in *Proceedings of the 31st International Conference on Machine Learning (ICML)* (MIT Press, 2014)
8. D. Bilyk, M.T. Lacey, Random tessellations, restricted isometric embeddings, and one bit sensing (2015), arXiv:1512.06697
9. J.-L. Bouchot, S. Foucart, P. Hitczenko, Hard thresholding pursuit algorithms: number of iterations. Appl. Comput. Harmon. Anal. **41**, 412–435 (2016)
10. P. Boufounos, R. Baraniuk, 1-bit compressive sensing, in *Proceedings of the 42nd Annual Conference on Information Sciences and Systems (CISS)* (IEEE, 2008), pp. 16–21
11. J. Bourgain, Bounded orthogonal systems and the $\Lambda(p)$-set problem. Acta Math. **162**, 227–245 (1989)
12. J. Bourgain, S. Dilworth, K. Ford, S. Konyagin, D. Kutzarova, Explicit constructions of RIP matrices and related problems. Duke Math. J. **159**, 145–185 (2011)
13. H. Buhrman, P. Miltersen, J. Radhakrishnan, S. Venkatesh, Are bitvectors optimal? in *Proceedings of the 32nd Annual ACM Symposium on Theory of Computing (STOC)* (ACM, 2000), pp. 449–458
14. T. Cai, A. Zhang, Sparse representation of a polytope and recovery of sparse signals and low-rank matrices. IEEE Trans. Inform. Theory **60**, 122–132 (2014)
15. E. Candès, X. Li, Solving quadratic equations via PhaseLift when there are about as many equations as unknowns. Found. Comput. Math. **14**, 1017–1026 (2014)
16. E. Candès, Y. Plan, Matrix completion with noise. Proc. IEEE **98**, 925–936 (2010)
17. E. Candès, B. Recht, Exact matrix completion via convex optimization. Found. Comput. Math. **9**, 717–772 (2009)
18. E. Candès, T. Tao, Decoding by linear programming. IEEE Trans. Inform. Theory **51**, 4203–4215 (2005)
19. E. Candès, J. Romberg, T. Tao, Robust uncertainty principles: exact signal reconstruction from highly incomplete frequency information. IEEE Trans. Inform. Theory **52**, 489–509 (2006)
20. E. Candès, Y. Eldar, D. Needell, P. Randall, Compressed sensing with coherent and redundant dictionaries. Appl. Comput. Harmon. Anal. **31**, 59–73 (2011)
21. E. Candès, T. Strohmer, V. Voroninski, Phaselift: exact and stable signal recovery from magnitude measurements via convex programming. Commun. Pure Appl. Math. **66**, 1241–1274 (2013)
22. A. Chkifa, N. Dexter, H. Tran, C. Webster, Polynomial approximation via compressed sensing of high-dimensional functions on lower sets (Preprint)
23. A. Cohen, W. Dahmen, R. DeVore, Compressed sensing and best k-term approximation. J. Amer. Math. Soc. **22**, 211–231 (2009)
24. A. Cohen, W. Dahmen, R. DeVore, Orthogonal matching pursuit under the restricted isometry property. Constr. Approx. **45**, 113–127 (2017)
25. M. Davenport, D. Needell, M. Wakin, Signal space CoSaMP for sparse recovery with redundant dictionaries. IEEE Trans. Inform. Theory **59**, 6820–6829 (2013)
26. M. Davies, R. Gribonval, Restricted isometry constants where ℓ^p sparse recovery can fail for $0 < p \leq 1$. IEEE Trans. Inform. Theory **55**, 2203–2214 (2009)
27. D. Donoho, For most large underdetermined systems of linear equations the minimal ℓ^1 solution is also the sparsest solution. Commun. Pure Appl. Math. **59**, 797–829 (2006)
28. D. Donoho, J. Tanner, Counting faces of randomly projected polytopes when the projection radically lowers dimension. J. Am. Math. Soc. **22**, 1–53 (2009)
29. D. Donoho, J. Tanner, Observed universality of phase transitions in high-dimensional geometry, with implications for modern data analysis and signal processing. Philos. Trans. R. Soc. A: Math. Phys. Eng. Sci. **367**, 4273–4293 (2009)
30. S. Foucart, Stability and Robustness of Weak Orthogonal Matching Pursuits, in *Recent Advances in Harmonic Analysis and Applications*, ed. by D. Bilyk, L. De Carli, A. Petukhov, A.M. Stokolos, B.D. Wick (Springer, New York, 2012), pp. 395–405

31. S. Foucart, Stability and robustness of ℓ_1-minimizations with Weibull matrices and redundant dictionaries. Linear Algebra Appl. **441**, 4–21 (2014)
32. S. Foucart, Dictionary-sparse recovery via thresholding-based algorithms. J. Fourier Anal. Appl. **22**, 6–19 (2016)
33. S. Foucart, D. Koslicki, Sparse recovery by means of nonnegative least squares. IEEE Signal Proces. Lett. **21**, 498–502 (2014)
34. S. Foucart, R. Gribonval, Real vs. complex null space properties for sparse vector recovery. C. R. Math. Acad. Sci. Paris **348**, 863–865 (2010)
35. S. Foucart, G. Lecué, An IHT algorithm for sparse recovery from subexponential measurements (Preprint)
36. S. Foucart, M.-J. Lai, Sparse recovery with pre-Gaussian random matrices. Studia Math. **200**, 91–102 (2010)
37. S. Foucart, M. Minner, T. Needham, Sparse disjointed recovery from noninflating measurements. Appl. Comput. Harmon. Anal. **39**, 558–567 (2015)
38. S. Foucart, A. Pajor, H. Rauhut, T. Ullrich, The Gelfand widths of ℓ_p-balls for $0<p \leq 1$. J. Compl. **26**, 629–640 (2010)
39. S. Foucart, H. Rauhut, *A Mathematical Introduction to Compressive Sensing* (Birkhäuser, Boston, 2013)
40. A. Garnaev, E. Gluskin, On widths of the Euclidean ball. Sov. Math. Dokl. **30**, 200–204 (1984)
41. R. Graham, N. Sloane, Lower bounds for constant weight codes. IEEE Trans. Inform. Theory **26**, 37–43 (1980)
42. D. Gross, Recovering low-rank matrices from few coefficients in any basis. IEEE Trans. Inform. Theory **57**, 1548–1566 (2011)
43. C. Güntürk, M. Lammers, A. Powell, R. Saab, Ö. Yilmaz, Sigma-Delta quantization for compressed sensing, in *Proceedings of the 44th Annual Conference on Information Sciences and Systems (CISS)* (IEEE, 2010)
44. V. Guruswani, C. Umans, S. Vadhan: Unbalanced expanders and randomness extractors from Parvaresh-Vardy codes, in *IEEE Conference on Computational Complexity* (2007), pp. 237–246
45. M. Iwen, A. Viswanathan, Y. Wang, Robust sparse phase retrieval made easy. Appl. Comput. Harmon. Ana. **42**, 135–142 (2017)
46. B. Kashin, Diameters of some finite-dimensional sets and classes of smooth functions. Math. USSR, Izv **11**, 317–333 (1977)
47. D. Koslicki, S. Foucart, G. Rosen, Quikr: a method for rapid reconstruction of bacterial communities via compressive sensing. Bioinformatics **29**(17), 2096–2102 (2013). btt336
48. D. Koslicki, S. Foucart, G. Rosen, WGSQuikr: fast whole-genome shotgun metagenomic classification. PloS one **9**, e91784 (2014)
49. K. Knudson, R. Saab, R. Ward, One-bit compressive sensing with norm estimation. IEEE Trans. Inform. Theory **62**, 2748–2758 (2016)
50. C. Lawson, R. Hanson, *Solving Least Squares Problems* (SIAM, Philadelphia, 1995)
51. G. Lecué, S. Mendelson, Sparse recovery under weak moment assumptions. J. Eur. Math. Soc. **19**, 881–904 (2017)
52. X. Li, V. Voroninski, Sparse signal recovery from quadratic measurements via convex programming. SIAM J. Math. Anal. **45**, 3019–3033 (2013)
53. N. Linial, I. Novik, How neighborly can a centrally symmetric polytope be? Discrete. Comput. Geom. **36**, 273–281 (2006)
54. G. Lorentz, M. von Golitschek, Y. Makovoz, *Constructive Approximation: Advanced Problems* (Springer, Berlin, 1996)
55. S. Mendelson, A. Pajor, M. Rudelson, The geometry of random $\{-1, 1\}$-polytopes. Discrete. Comput. Geom. **34**, 365–379 (2005)
56. N. Noam, W. Avi, Hardness vs randomness. J. Comput. Syst. Sci. **49**, 149–167 (1994)
57. S. Oymak, A. Jalali, M. Fazel, Y. Eldar, B. Hassibi, Simultaneously structured models with application to sparse and low-rank matrices. IEEE Trans. Inform. Theory **61**, 2886–2908 (2015)
58. A. Pinkus, *n-Widths in Approximation Theory* (Springer, Berlin, 1985)

59. Y. Plan, R. Vershynin, One-bit compressed sensing by linear programming. Commun. Pure Appl. Math. **66**, 1275–1297 (2013)
60. Y. Plan, R. Vershynin, Robust 1-bit compressed sensing and sparse logistic regression: a convex programming approach. IEEE Trans. Inform. Theory **59**, 482–494 (2013)
61. B. Recht, A simpler approach to matrix completion. J. Mach. Learn. Res. **12**, 3413–3430 (2011)
62. G. Schechtman, Two observations regarding embedding subsets of Euclidean spaces in normed spaces. Adv. Math. **200**, 125–135 (2006)
63. M. Talagrand, Selecting a proportion of characters. Israel J. Math. **108**, 173–191 (1998)
64. V. Temlyakov, *Greedy Approximation* (Cambridge University Press, Cambridge, 2011)
65. A. Tillmann, R. Gribonval, M. Pfetsch, Projection onto the cosparse set is NP-hard, in *Proceedings of the 2014 Conference on Acoustics, Speech and Signal Processing (ICASSP)* (IEEE, 2014)
66. A. Tillmann, M. Pfetsch, The computational complexity of the restricted isometry property, the nullspace property, and related concepts in compressed sensing. IEEE Trans. Inform. Theory **60**, 1248–1259 (2014)
67. J. Tropp, A. Gilbert, Signal recovery from random measurements via orthogonal matching pursuit. IEEE Trans. Inform. Theory **53**, 4655–4666 (2007)
68. J. Vybíral, Widths of embeddings in function spaces. J. Complex. **24**, 545–570 (2008)
69. T. Zhang, Sparse recovery with orthogonal matching pursuit under RIP. IEEE Trans. Inform. Theory **57**, 6215–6221 (2011)

Computing with Functions on Domains with Arbitrary Shapes

Daan Huybrechs and Roel Matthysen

Abstract We describe an approximation scheme and an implementation technique that enables numerical computations with functions defined on domains with an arbitrary shape. The scheme is spectrally accurate for smooth functions. The main advantage of the technique is that, unlike most spectral approximation schemes in higher dimensions, it is not limited to domains with tensor-product structure. The scheme itself is a discrete least squares approximation in a redundant set (a frame) that originates from a basis on a bounding box. The implementation technique consists of representing a domain by its characteristic function, i.e., the function that indicates whether or not a point belongs to the set. We show in a separate paper that the least squares approximation with N degrees of freedom can be solved in $\mathcal{O}(N^2 \log^2 N)$ operations for any domain that has non-trivial volume. The computational cost improves to $\mathcal{O}(N \log^2 N)$ operations for domains that do have tensor-product structure. The scheme applies to domains even with fractal shapes, such as the Mandelbrot set, since such domains are defined precisely by their characteristic function.

Keywords Fourier series · Complex domains · Function approximation

1 Introduction

1.1 Motivation

Approximation schemes of a single variable usually extend to multivariate schemes only when the domains under consideration have tensor-product structure, the simplest domains being squares and rectangles. Functions on more general domains

D. Huybrechs (✉) · R. Matthysen
KU Leuven, Leuven, Belgium
e-mail: daan.huybrechs@kuleuven.be

R. Matthysen
e-mail: roel.matthysen@kuleuven.be

G.E. Fasshauer and L.L. Schumaker (eds.), *Approximation Theory XV: San Antonio 2016*, Springer Proceedings in Mathematics & Statistics 201,
DOI 10.1007/978-3-319-59912-0_5

can be approximated using mappings of tensor-product domains to more general domains, or by tiling a general domain with quadrilaterals. These approaches are powerful, and they are fundamental in scientific computing, though usually only low-order approximations are aimed for in that setting. Moreover, the complexity of the mapping or of the tiling clearly depends on the complexity of the geometry of the domain under consideration. This affects the performance of the scheme, even when the function to be approximated is perfectly smooth and, from an approximation point of view, 'well-behaved'.

2 The Approximation Scheme

Let the domain of interest be a compact subset B of $\mathbb{R}^n$. Let f be a target function of n variables in $\mathscr{L}^2_B$, the space of functions that are square integrable over B. We represent f through a basis $G = \{\phi_i\}$ for $\mathscr{L}^2_C$, where C is a domain that contains B in its interior and for which a basis can easily be found. Further on, we will use a bounding box and define a classical tensor-product basis on that box. In general, when a basis for $\mathscr{L}^2_C$ on C is restricted to a subdomain B, the resulting set is a frame for $\mathscr{L}^2_B$ in the sense of Duffin and Schaeffer [5].

Assuming some linear ordering of the elements of G, we denote a truncated frame by $G_N = \{\phi_i\}_{i=1}^N$ and its span by $\mathscr{G}_N = \operatorname{span} G_N$. We define the best approximation to f in this space using the associated norm over B,

$$f_N = \arg\min_{g \in \mathscr{G}_N} \|f - g\|_B. \tag{1}$$

This results in increasingly accurate approximations in the truncated set G_N, as N increases.

In order to arrive at a simple linear system, we substitute the $\mathscr{L}_2$ norm over B by a discrete norm over a point set P_M, consisting of $M \geq N$ points in B,

$$f_N = \arg\min_{g \in \mathscr{G}_N} \sum_{\mathbf{x} \in P_M} |f(\mathbf{x}) - g(\mathbf{x})|^2. \tag{2}$$

Here, we assume that besides being in $\mathscr{L}^2_B$, f is also continuous on B. With a similar linear ordering of the point set P_M, this leads to a least squares system

$$A\alpha = b, \qquad A_{ij} = \phi_j(\mathbf{x}_i), \qquad b_i = f(\mathbf{x}_i). \tag{3}$$

The matrix A has dimensions $M \times N$, and further on we typically choose $M = 2N$. We stress that both the norm in (1) and the points in (2) are confined to B. This implies that we make no assumption about the existence of f outside of the domain B, i.e., we require no information from f on the extension region $C \backslash B$.

There are very few practical restrictions on G_N, besides completeness in $\mathscr{L}^2_B$. A disadvantage compared to using a basis tailored to the domain B is that the conditioning of A in Eq. (3) can be arbitrarily bad. In fact, the expansion $f_N = \sum_{i=1}^N \alpha_i \phi_i$ might not even be unique and A can be singular.

Both the ill-conditioning and the potential lack of uniqueness correspond to the typical redundancy of a frame compared to a basis. From an approximation point of view, both of these effects are relatively benign. The numerical stability of this type of least squares approximation was studied in [2] for the case where B is an interval and G is a Fourier basis on a larger interval. This analysis is generalized to numerical approximations in more general frames in [1]. The discrete least squares approximation with oversampling ($M > N$) leads, somewhat surprisingly, to a stable approximation scheme for sufficiently large N, regardless of the shape of the domain $B \subset C$.

A compelling practical advantage of the present approach is that constructing bases for arbitrary domains is hard in general and requires at least some a priori domain knowledge. In contrast, our scheme can be based on any known classical basis that spans $\mathscr{L}^2$ on a bounding box C encompassing B. It requires only the generation of a suitable point set P_M, which we discuss further on.

Because of ill-conditioning of A, iterative solvers tend not to perform very well in these applications. Direct solvers, and in particular a truncated singular value decomposition which allows for some regularization of the solution, seem appropriate. However, they come at an $\mathscr{O}(N^3)$ cost.

2.1 The Fourier Extension Scheme

For certain choices of bases and point sets, the approximation scheme allows for very efficient solutions to Eq. (3). When the approximation space is that of periodic functions on C, and the point set is the intersection of an equispaced n-dimensional grid with B, the scheme is known as *Fourier Extension*. An illustration showing the equispaced grid restricted to a domain is shown in Fig. 1.

The least squares matrix A in this case can be separated into a well-conditioned part, and a lower-rank part that captures the ill-conditioning [8–10]. By solving the lower-rank part first, and the well-conditioned part quickly, the total complexity becomes $\mathscr{O}(N^{3-2/n} \log^2(N))$. This is linear up to a logarithmic term for $n = 1$ and can provide a substantial speedup in higher dimensions for sufficiently large N.

The fast algorithms of [9, 10] can also be employed when using Chebyshev polynomials on the bounding box and an associated tensor-product grid of Chebyshev points, restricted to the domain B.

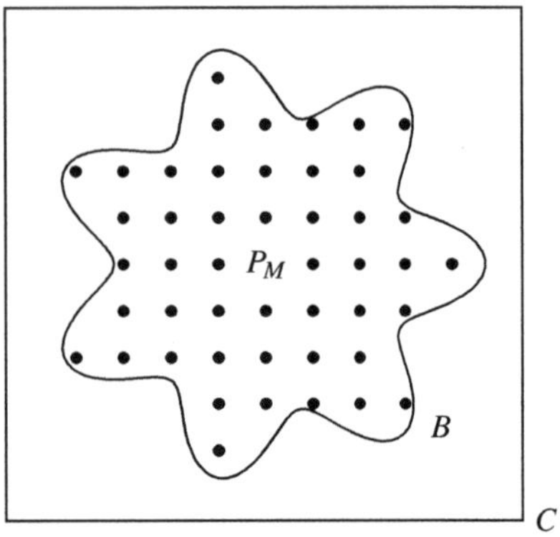

Fig. 1 The bounding box C encompassing the equidistant samples P_M

2.2 *Computing with a Spectral Basis*

As demonstrated by the Chebfun package [4], expansions in a spectral basis such as those resulting from the Fourier Extension allow you to compute with functions by manipulating the expansion coefficients. For example:

- Scalar multiplication and summation of approximations are performed in coefficient space.
- Pointwise multiplication of two Fourier series can be accomplished through convolution of coefficients.
- Constant coefficient differential operators are easy to apply in coefficient space. In the Fourier case, these operators are diagonal.
- Rootfinding and finding extrema can be found by applying existing methods for the bounding box and restricting the results to the domain of interest.

Limitations are that due to the nature of the Fourier basis singularities are difficult to deal with and restrict convergence rates. This may be ameliorated by adding suitable singularity functions to the frame, resulting in a slightly larger frame with better approximation properties [1]. Another limitation is that the expansion on the bounding box is naturally defined on the bounding box and not just on the domain. It is simple to compute the definite integral over the bounding box, but not straightforward to do so over the embedded domain.

The third bullet above (diagonalization of constant coefficient differential operators) makes Fourier Extensions an appealing scheme for boundary value problems on complicated domains. The use of implicit or explicit extensions to treat complex geometries arises already in several schemes in scientific computing, such as embedded domain methods, immersed or fictitious boundary methods, and volume-penalty methods [3, 7, 11]. The difference compared to our setting is that with oversampled Fourier Extensions, we obtain spectral accuracy with an efficient algorithm. Our current efforts in this direction are related to meshless methods, in particular to Kansa's method often used with radial basis functions [6].

3 Domain Representation

3.1 The Characteristic Function

The *characteristic function* χ, or *indicator function*, of a domain $B \subset \mathbb{R}^n$ is a function on $\mathbb{R}^n$ that has value 1 for points that belong to B and the value 0 for points that do not, i.e.,

$$\chi(x) := \begin{cases} 1, & x \in B, \\ 0, & x \notin B. \end{cases} \tag{4}$$

It is convenient in implementations to associate Boolean values with $\chi(x)$, so that it evaluates to true or false, rather than the numeric values 1 and 0.

Representing a domain by its characteristic function has a number of consequences. Two **advantages** are:

- The function is unique and well defined for any domain, be it open or closed, connected or disconnected, punctured, empty, a discrete set, finite or infinite, a fractal, …
- As we will see later on, the characteristic function is often easy to implement. For example, with $x = [x_1, x_2]$ in two dimensions, the half-open domain bounded by the parabola $x_2 = x_1^2$ and the straight line $x_2 = x_1$ has characteristic function

$$\chi(x) = (x_2 > x_1^2) \,\&\, (x_2 \leq x_1). \tag{5}$$

There is no need even to find the intersection points of both curves, as far as implementing the characteristic function is concerned.

Consequently, it is easy and very cheap to find the characteristic function of the domain that is bounded by, say, the level curves of a given function, even if the resulting domain is disconnected and contains many holes. This operation does not even require any numerical computation, as will be demonstrated later on.

Two **disadvantages** are:

- The characteristic function does not explicitly convey information about the boundary of the domain. This would be difficult for fractal domains, but it would be convenient to have for simpler domains, and essential to have for boundary value problems.
- The least squares approximation scheme requires point evaluations inside the domain. Though the characteristic function is well defined for domains that have no volume in $\mathbb{R}^n$, such as a line in $\mathbb{R}^2$ or a surface in $\mathbb{R}^3$, the concept is not suited for approximating functions on such domains.

3.2 Generating Points

The least squares approximation scheme requires M point evaluations of the given function f inside the domain B. Thus, one needs a way to find M points that belong to B.

It is convenient at this stage too to have at hand a bounding box C, or the knowledge of any other region C that is easily sampled for which $B \subset C$. Then, points inside B can be generated by sampling Q points y_j of C and checking whether $\chi_B(y_j)$ is true. This results in a set of M_Q points with $M_Q \leq Q$:

$$\{x_j\}_{j=1}^{M_Q} := \{y_j \mid \chi_B(y_j) = 1, j = 1, \ldots, Q\}$$

Only those points are retained, and the procedure is repeated with denser samplings, corresponding to increasing values of Q, until $M_Q \geq M$ points are retained.

For domains with nonzero volume in $\mathbb{R}^n$, it is guaranteed that M points will eventually be found if the sampling of C becomes uniformly denser. Though in principle any randomly chosen set of points $\{x_j\}$ is sufficient, for efficiency reasons it is better to choose a structured set. In particular, in combination with the Fourier Extension scheme, we use a bounding box C and an equidistant grid on C. The main advantage is that Fourier series approximations can be evaluated efficiently on that grid with the FFT. In several examples further on, the characteristic function of a domain itself is defined in terms of a Fourier series, and in that case too the characteristic function too can be evaluated efficiently on a structured grid using the FFT (Fig. 2).

From the point of view of the approximation problem, it may be better to have more points clustered toward the boundary of the domain. However, even for multivariate polynomial interpolation it is a very difficult problem to determine the best points on a general domain. Furthermore, since we make very few assumptions regarding our knowledge of the boundary, choosing more points near the boundary requires algorithmic work. Instead, we oversample.

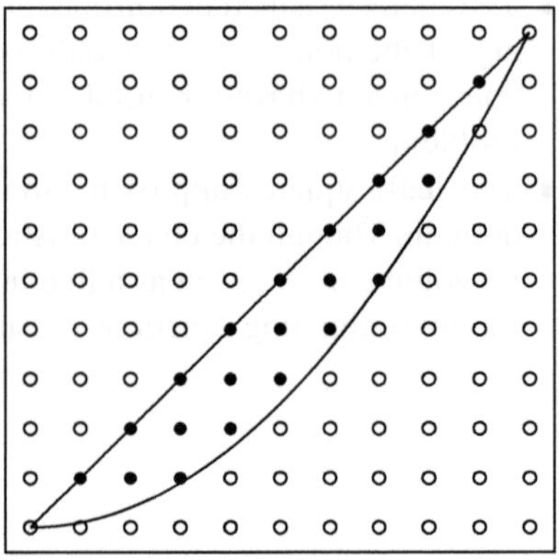

Fig. 2 The characteristic function (5) evaluated in Q points inside the bounding box C. The points are a subset of a structured equispaced grid on C

3.3 Implementation

In an object-oriented approach, a domain would be an object. Disregarding that property, the elements that make up a domain include:

- a bounding box,
- a routine to evaluate the characteristic function at a single point x,
- an optimized routine to evaluate the characteristic function on a regular grid of the bounding box.

The latter routine will be called the *grid evaluation* routine. It is not an essential part of the implementation, but it leads to much increased efficiency in particular when using the Fourier Extension approximation scheme. The goal is not merely to exploit the speed gained from vectorization, but to lower the computational complexity compared to calling the single evaluation routines many times.

For points on the boundary, the characteristic function can be true or false, corresponding to closed and open domains. This makes a difference in practice only in special circumstances, since in general the points that are sampled are unlikely to coincide with the boundary of the domain. In general, it is very difficult to distinguish between open and closed domains with the proposed techniques.

We have implemented the scheme in Julia,[1] with a user interface that is modeled after the software package Chebfun [4]. In our implementation, the domain is implemented as an object with three properties, corresponding to the three elements above.

4 Computing with Domains

4.1 Set Operations

Basic set operations have rather obvious ramifications for the characteristic function. The union, difference, and intersection of two domains give rise to logical relationships between the characteristic functions involved. Assume the domains B_i, $i = 1, 2, 3$, have characteristic functions χ_i. Then we have

$$\begin{aligned}
B_3 &= B_1 \cup B_2 \Rightarrow \chi_3(x) = \chi_1(x) \textbf{ or } \chi_2(x) \\
B_3 &= B_1 \cap B_2 \Rightarrow \chi_3(x) = \chi_1(x) \textbf{ and } \chi_2(x) \\
B_3 &= B_1 \setminus B_2 \Rightarrow \chi_3(x) = \chi_1(x) \textbf{ andnot } \chi_2(x) \\
B_3 &= (B_1 \cup B_2) \setminus (B_1 \cap B_2) \Rightarrow \chi_3(x) = \chi_1(x) \textbf{ xor } \chi_2(x).
\end{aligned}$$

[1]The code is publicly available in the online GitHub repository of the FrameFuns package, `http://github.com/daanhb/FrameFun.jl`.

These operations are easily implemented by defining χ_3 in terms of the supplied definitions of χ_1 and χ_2. Similarly, the grid evaluation routine of B_3 can be defined in terms of the grid evaluation routines in B_1 and B_2. This makes sure that a potentially fast implementation of this procedure for B_1 and B_2 leads to a fast implementation of this procedure also for B_3.

In Julia, this enables the following operations

```
» B3 = B1 & B2
» B3 = B1 | B2
» B3 = B1 \ B2
» B3 = xor(B1,B2)
```

by overloading the logical operators for domain objects.

4.2 Arithmetic Operations

Domains can be translated and scaled by adding a vector and by multiplying by a scalar, respectively. We have

$$\forall c \in \mathbb{R}^n : \quad B_2 = B_1 + c \Rightarrow \chi_2(x) = \chi_1(x - c)$$
$$\forall a \in \mathbb{R} : \quad B_2 = a * B_1 \Rightarrow \chi_2(x) = \chi_1(x/a).$$

It should be noted that while translation of a domain is independent of the location of the origin, scaling a domain like above does depend on the location of the origin. A circle centered around the origin would simply increase in size by a factor of a, but a circle centered at the point [1; 0] would also move a factor a to the right.

Arithmetic operations are also easily implemented, by defining χ_2 in terms of the supplied definition of χ_1 and similarly for the grid evaluation routines.

In Julia, we may write

```
» B2 = B1 + [1;0]
» B3 = 2*B1
```

Combined with the above, a moon-shaped domain can be defined in terms of a circle C with radius 1 by the statement

```
» moon = C \ (C + [1/2; 0])
```

Similarly, if C is centered around the origin, a domain with a hole is obtained by

```
» annulus = 2*C \ C
```

4.3 Implicitly Defined or Derived Domains

Finding the level curves of a function, say the set of points where $f(x) = 3$, requires algorithmic work and can become arbitrarily complicated depending on the complexity of the given function f. However, it is very easy to define the characteristic function of a domain that is bounded by this level curve. Say a function f is defined on B, and the domain C is the open domain where $f(x) > 3$. Then the characteristic function χ_C of C is given explicitly by

$$\chi_C(x) = \begin{cases} f(x) > 3, & \forall x \in B, \\ 0, & \text{otherwise.} \end{cases}$$

The implementation of the characteristic function is defined in terms of the inequality $f(x) > 3$, which is a Boolean expression for each x. The grid evaluation routine of C may be implemented in terms of the grid evaluation routine of f. Thus, if f can be evaluated efficiently via FFT for example, then the same holds for the characteristic function of the domain C.

In Julia, we may now write

```
>> C = f > 3
>> C = f > g
>> C = cos(f .^ 2) - 3 < sqrt(pi)
```

where both f and g are existing functions. In the second statement, the domain C is in addition restricted to the intersection of the domains of f and g, such that it makes sense to compare f and g.

Interestingly, from the point of view of implementation, it is irrelevant whether or not the resulting domains are connected or not. The shape of the resulting domain can be truly arbitrary and does not effect the computational cost of this new characteristic function. Of course, the geometry of the domain does play a role in the approximation problem to be solved, though even there its influence remains fairly minor.

4.4 Deciding on the Equivalence of Domains

When given two characteristic functions χ_1 and χ_2, the problem of deciding whether they represent the same domain is a difficult one and requires careful consideration. It is of course not possible to check for each and every point $x \in \mathbb{R}^2$ whether $\chi_1(x)$ equals $\chi_2(x)$. Two possible ways to treat this problem are as follows.

(1) Verify equivalence up to a certain resolution

The characteristic functions χ_1 and χ_2 are sampled on an equidistant grid with a certain specified resolution and covering both domains. Their equivalence at this resolution level is determined by their equivalence at the grid points.

(2) Construct a global table of domains with unique identifiers

The domain object can be extended with a unique identifier. Domains are compared by comparing their identifiers. Each operation that results in a new domain also results in a new identifier, which is kept in a global table. A function g that is computed from a function f inherits the domain of f, along with its identifier.

The first approach is costly and does not always give the right mathematical answer, in the sense that it may conclude equivalence for two domains that are not equivalent. It will never conclude inequivalence for equivalent domains. However, the approach applies to all domains and will always converge to the correct answer when increasing the resolution level.

The second approach is fast, but comes at a cost of having to construct a global table. This adds overhead and memory costs. The approach also does not always give the right mathematical answer, as two domains may be constructed in similar ways but independently of each other. Their identifiers will be different, though the domains may be the same. Avoiding this situation requires care from the user of the software.

5 Examples

5.1 *Characteristic Function*

For some domains, the characteristic function is simply the most convenient description. The Mandelbrot set is an example, defined by

$$\chi(x) = \left(\limsup_{n\to\infty} |z_{n+1}| \leq 2\right),$$
$$z_{n+1} = z_n^2 + x_1 + \mathrm{i}x_2, \qquad z_0 = 0.$$

An approximation of

$$F_m(x) = \cos(20x_1 + \mathrm{i}x_2) - 5x_1x_2 \tag{6}$$

is shown in Fig. 3a. It was obtained using an equispaced grid on $[-2, 2] \times [-1.5, 1.5]$. Using the Fourier Extension technique, convergence up to a tolerance of 10^{-12} was achieved for 32×32 basis functions (Fig. 3b).

5.2 *Domain Arithmetic*

As an example of computing with domains, Fig. 4 shows an approximation on a ring, obtained by the Julia commands

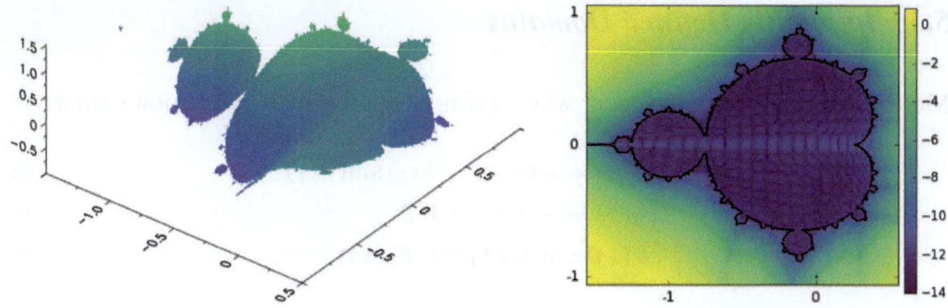

(a) Approximation (b) Error

Fig. 3 An approximation of F_m (Eq. (6)) on the Mandelbrot set. The *right* figure shows $\log_{10}(|f_N - F_m|)$. The approximation error is very small precisely on the Mandelbrot set. In the extension region, the functions f_N and F_m are both defined and they can be evaluated and compared, but they bear no resemblance. In particular, f_N is periodic on the box, while F_m is not

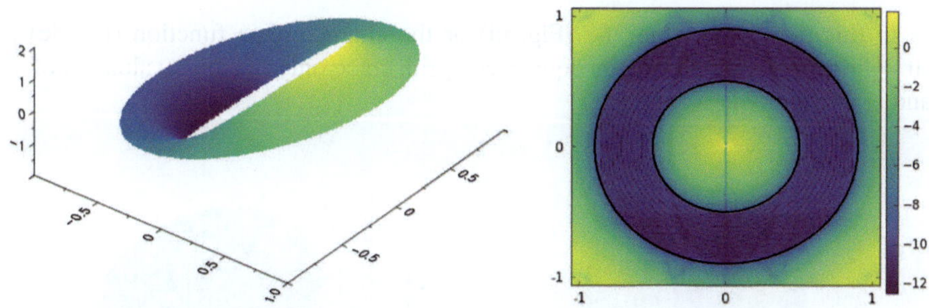

(a) Approximation (b) Error

Fig. 4 An approximation of F_r (Eq. (7)) on a ring-shaped domain. The *right* figure shows $\log_{10}(|f_N - F_r|)$

```
» B3 = Disk(0.9) \ Disk(0.5)
```

Of special note here is that the target function

$$F_r(x) = \frac{x_1}{x_1^2 + x_2^2} \tag{7}$$

has a singularity at (0, 0), enclosed in the domain. However, this has no influence on the approximation, as the exterior of the domain is never sampled. As Fig. 4b shows the approximation converges up to a tolerance of 10^{-10} for 32×32 basis functions.

5.3 Implicitly Defined Domains

More convoluted domains occur when trying to approximate a function such as

$$F_1(x) = \sin(5x_1 - 3x_2)\sin(7x_1) \tag{8}$$

$$F_2(x) = -0.5x_1 + 0.2 \tag{9}$$

$$F(x) = \max(F_1(x), F_2(x)) \tag{10}$$

on a disk. Given f_1 and f_2, approximations of F_1 and F_2 on the full disk (Fig. 5a, b), f_N is simply

$$f_N(x) = \begin{cases} f_1(x), & f_1(x) \geq f_2(x) \\ f_2(x), & f_1(x) < f_2(x) \end{cases}.$$

In this case, evaluating f_N (Fig. 5d) or the characteristic function (Fig. 5c) is straightforward, and fast on an equispaced grid, since only one full evaluation of f_1 and f_2 is required.

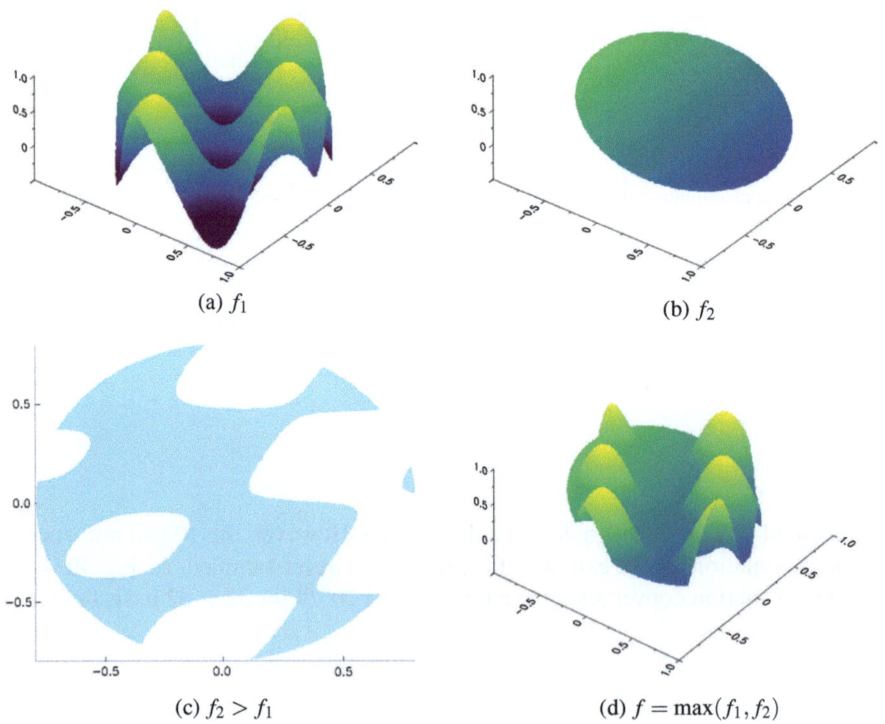

(a) f_1 (b) f_2 (c) $f_2 > f_1$ (d) $f = \max(f_1, f_2)$

Fig. 5 A piecewise approximation of F (Eq. (10)) and the implicit domain $f_2 > f_1$

Acknowledgements The authors are supported by FWO Flanders Projects G.A004.14 and G.0641.11

References

1. B. Adcock, D. Huybrechs, Frames and numerical approximation, Technical report TW-674, KU Leuven, December 2016
2. B. Adcock, D. Huybrechs, J. Martín-Vaquero, On the numerical stability of Fourier extensions. Found. Comput. Math. **14**, 635–687 (2014)
3. D. Boffi, N. Cavallini, L. Gastaldi, The finite element immersed boundary method with distributed Lagrange multiplier. SIAM J. Numer. Anal. **53**(6), 2584–2604 (2015)
4. T.A. Driscoll, N. Hale, L.N. Trefethen, *Chebfun Guide* (Pafnuty Publications, Oxford, 2014)
5. R.J. Duffin, A.C. Schaeffer, A class of nonharmonic Fourier series. Trans. Am. Math. Soc. **72**(2), 341–366 (1952)
6. E.J. Kansa, Multiquadrics - a scattered data approximation scheme with applications to computational fluid-dynamics II: solutions to parabolic, hyperbolic and elliptic partial differential equations. Comput. Math. Appl. **19**, 147–161 (1990)
7. F. Kasolis, E. Wadbro, M. Berggren, Analysis of fictitious domain approximations of hard scatterers. SIAM J. Numer. Anal. **2015**(5), 2347–2362 (2015)
8. M. Lyon, A fast algorithm for Fourier continuation. SIAM J. Sci. Comput. **33**(6), 3241–3260 (2011)
9. R. Matthysen, D. Huybrechs, Fast algorithms for the computation of Fourier extensions of arbitrary length. SIAM J. Sci. Comput. **36**(2), 828–845 (2015)
10. R. Matthysen, D. Huybrechs, Multi-dimensional Fourier frame approximations through collocation, In preparation, (2017)
11. D. Shirokoff, J. C. Nave, A Sharp-interface active penalty method for the incompressible Navier–Stokes equations. J. Sci. Comput. **62**(1), 53–77 (2015)

A Polygonal Spline Method for General Second-Order Elliptic Equations and Its Applications

Ming-Jun Lai and James Lanterman

Abstract We explain how to use polygonal splines to numerically solve second-order elliptic partial differential equations. The convergence of the polygonal spline method will be studied. Also, we will use this approach to numerically study the solution of some mixed parabolic and hyperbolic partial differential equations. Comparison with standard bivariate spline method will be given to demonstrate that our polygonal splines have some better numerical performance.

Keywords Polygonal splines · Generalized barycentric coordinates · Numerical solution of PDEs

1 Introduction

Traditionally, people use triangulations to numerically solve partial differential equations (PDEs). A new trend is to use more general polygonal meshes. That is, we can be more versatile and efficient than the standard finite element method when numerically solving PDEs. See [2–4] for a so-called virtual element method to solve PDEs based on arbitrary polygonal meshes. In [17], the researchers use C^0 generalized barycentric coordinate (GBC) elements to solve the standard Poisson equations in 2D and 3D over arbitrary convex polygonal partitions. In [20], the researchers construct C^0 quadratic finite elements based on GBC to solve PDEs. In [18], these researchers use the weak Galerkin method based on rectangular partitions to solve second-order elliptic PDEs. In [12], the researchers constructed a class of continuous polygonal

This research is partially supported by Simons Collaboration Grant 280646 and the National Science Foundation under the Grant #DMS 1521537.

M.-J. Lai (✉) · J. Lanterman
Department of Mathematics, University of Georgia, Athens, GA 30602, USA
e-mail: mjlai@math.uga.edu

J. Lanterman
e-mail: jay@math.uga.edu

G.E. Fasshauer and L.L. Schumaker (eds.), *Approximation Theory XV: San Antonio 2016*, Springer Proceedings in Mathematics & Statistics 201,
DOI 10.1007/978-3-319-59912-0_6

finite elements of arbitrary order d which allows for reproduction of polynomials of total degree d. These elements were implemented for numerical solution of Poisson equations. In this paper, we shall explore how to use these polygonal elements to solve general second-order elliptic PDEs and some mixed parabolic and hyperbolic partial differential equations.

A model PDE problem considered in this paper can be described as follows. Let Ω be a bounded open polyhedral domain in $\mathbb{R}^2$, and let $\Gamma = \partial\Omega$ be the boundary of Ω. We consider the following general second-order PDE:

$$\begin{cases} \mathscr{L}(u) = f, & \mathbf{x} \in \Omega \\ u = g, & \mathbf{x} \in \Gamma \end{cases} \tag{1}$$

where $\mathbf{x} = (x_1, x_2) \in \mathbb{R}^2$ and $\mathscr{L}$ is a partial differential operator in the following form:

$$\mathscr{L}(u) := -\sum_{i,j=1}^{2} \frac{\partial}{\partial x_j}\left(A_{ij}\frac{\partial}{\partial x_i}u\right) + \sum_{k=1}^{2} B_k \frac{\partial}{\partial x_k}u + Cu, \tag{2}$$

with $A_{ij} \in L_\infty(\Omega)$, $B_k \in L_\infty(\Omega)$, $C \in L_\infty(\Omega)$, f is a function in $L_2(\Omega)$, and $g \in L^\infty(\partial\Omega)$.

When the matrix $A = [A_{ij}]_{1\le i,j\le 2}$ is symmetric and positive definite over Ω, the PDE in (1) is said to be elliptic. A typical PDE of this type can be given by defining the operator $\mathscr{L}$ with the following weight functions: Let

$$\begin{bmatrix} A_{11} & A_{12} \\ A_{21} & A_{22} \end{bmatrix} = \begin{bmatrix} \varepsilon + x & xy \\ xy & \varepsilon + y \end{bmatrix}, \tag{3}$$

with $\varepsilon > 0$, $\mathbf{B} = (B_1, B_2) = (0, 0)$, and $C = \exp(-x^2 - y^2)$. Then, the corresponding PDE is elliptic in the first quadrant. Given the conditions listed above for A_{ij}, B_k, C, f, and g, we know that this type of PDE has a unique solution. See Theorem 3 in a later section.

There is a standard approach to use methods for solution of second-order elliptic PDE to study hyperbolic equations, transport equations, and mixed parabolic and hyperbolic equations. Indeed, consider a singularly perturbed elliptic PDE:

$$-\varepsilon\Delta u + (2 - y^2)D_x u + (2 - x)D_y u + (1 + (1 + x)(1 + y)^2)u = f, \quad (x, y) \in \Omega \tag{4}$$

where $\Omega = (0, 1) \times (0, 1)$,with $u|_{\partial\Omega} = g$ where f and g are any appropriate functions. When $\varepsilon = 0$, this is a hyperbolic test problem considered in [5, 14, 15]. One can numerically solve (4) for $\varepsilon > 0$ very small to approximate the solution of the hyperbolic problem with $\varepsilon = 0$.

For another example, the following is a singularly perturbed advection-diffusion problem:

$$-\varepsilon\Delta u + D_x u + D_y u = f, \quad (x, y) \in \Omega = (0, 1) \times (0, 1), \tag{5}$$

with $u|_{\partial\Omega} = g$ where f and g are appropriate functions. This example was studied in [15].

For another example, the following problem is parabolic for $y > 0$ and hyperbolic for $y \leq 0$:

$$\begin{aligned} -\varepsilon D_{yy}u + D_x u + c_1 u &= 0, \quad (x, y) \in (-1, 1) \times (0, 1), \\ D_x u + c_2 u &= 0, \quad (x, y) \in (-1, 1) \times (-1, 0], \end{aligned} \tag{6}$$

with $u|_{\partial\Omega} = g$, for any constants $c_1 > 0$ and $c_2 > 0$. It was also studied in [15]. We can use the following general elliptic PDE to study the above problem by considering

$$\begin{aligned} -\eta D_{xx}u - \varepsilon D_{yy}u + D_x u + c_1 u &= f_1, \quad (x, y) \in (-1, 1) \times (0, 1), \\ -\eta \Delta u + D_x u + c_2 u &= f_2, \quad (x, y) \in (-1, 1) \times (-1, 0], \end{aligned} \tag{7}$$

with $u|_{\partial\Omega} = g$ and $\eta > 0$ where f_1, f_2 and g are appropriate functions. We can approximate the solution to (62) by letting $\eta > 0$ go to zero and use spline functions which are not necessarily continuous at $y = 0$.

These examples show the usefulness of a numerical solution to the model problem (1) in this paper. On the other hand, it is known that not all second-order elliptic PDE has a unique solution because of the Fredholm alternative theorem (cf. e.g., [9]). Although there are many sufficient conditions to ensure the existence and uniqueness of the solution (1), it is interesting to know when such a PDE can be numerically solved and admits a numerical solution. These are our motivations to study (1). In the next section, we provide a standard sufficient condition to ensure the existence and uniqueness of the weak solution of (1) based on the Lax–Milgram theorem. The PDEs listed above may not satisfy that sufficient condition. However, we are still able to find their numerical solutions by using our polygonal splines.

This second-order elliptic PDE in (1) has been studied by many other methods before. For example, in [1], the researchers used bivariate spline method to numerically solve (1). See [21] for other numerical solutions of (1) using other spline methods. In particular, given a triangulation Δ of a domain Ω, let

$$S_d^r(\triangle) = \{s \in C^r(\Omega) : s|_T \in \mathscr{P}_d, \forall T \in \triangle\} \tag{8}$$

be the spline space of smoothness r and degree $d \geq r$. In general, we need to use $d \geq 3r + 2$ to have the spline space $S_d^r(\triangle)$ to be nonempty (cf. [16]). The researchers use $S_d^{-1}(\triangle)$ (i.e., discontinuous splines of degree d over $\triangle$) and add the smoothness conditions $H\mathbf{c} = 0$ as side constraints to approximate the following weak solution:

$$\sum_{i,j=1}^{2} \int_\Omega A_{ij} \frac{\partial}{\partial x_i} u \frac{\partial}{\partial x_j} v + \sum_{k=1}^{2} \int_\Omega [B_k \frac{\partial}{\partial x_k} u] v + \int_\Omega C u v = \int_\Omega f v \tag{9}$$

for all test function $v \in S_d^{-1}(\triangle)$. For another example, the researchers in [18] use the weak Galerkin method to solve (1). In this paper, we shall use the polygonal splines invented in [12] to solve (9). Polygonal splines have been shown to be more efficient than polynomial spline functions for numerical solution of the Poisson equation (cf. [12]). We shall use the polygonal splines to solve the general second-order elliptic PDE and compare the numerical solutions with the solutions by using bivariate spline functions. We continue to demonstrate that the polygonal splines numerical solutions are generally better than the ones obtained with bivariate splines of the same degrees.

This paper is organized as follows. We shall review polygonal splines in Sect. 2. Then, we review the ellipticity concept to explain when the PDE in (1) has a solution in Sect. 3. Implementation and convergence of polygonal spline solution will be explained in Sect. 4. These are generalizations of the standard results for Poisson equation: the Céa lemma and the Aubin–Nitsche technique for the optimal convergence rate in L_2 norm over a convex domain. Finally, we present many numerical results using polygonal splines to solve the PDEs given above.

2 Preliminary on Polygonal Splines

Let us begin with generalized barycentric coordinates (GBCs). There are many ways to define barycentric coordinates in a polygon with n sides, $n \geq 3$; see [10]. We will restrict our attention to convex polygons in this paper.

Let $P_n = \langle \mathbf{v}_1, \ldots, \mathbf{v}_n \rangle$ be a convex polygon. Any functions ϕ_i, $i = 1, \ldots, n$, will be called generalized barycentric coordinates (GBCs) if, for all $\mathbf{x} \in P_n$, $\phi_i(\mathbf{x}) \geq 0$ and

$$\sum_{i=1}^{n} \phi_i(\mathbf{x}) = 1, \qquad \text{and} \qquad \sum_{i=1}^{n} \phi_i(\mathbf{x})\mathbf{v}_i = \mathbf{x}. \tag{10}$$

When P_n is a triangle, the coordinates ϕ_1, ϕ_2, ϕ_3 are the usual barycentric coordinates. For $n > 3$, the ϕ_i are not uniquely determined by (10), but they share the basic property that they are piecewise linear on the boundary of P_n:

$$\begin{aligned} \phi_i(\mathbf{v}_j) &= \delta_{ij}, \text{ and} \\ \phi_i((1-\mu)\mathbf{v}_j + \mu\mathbf{v}_{j+1}) &= (1-\mu)\phi_i(\mathbf{v}_j) + \mu\phi_i(\mathbf{v}_{j+1}) \text{ for } \mu \in [0, 1]. \end{aligned} \tag{11}$$

Example 1 (*Wachspress (rational) coordinates*) For general, $n \geq 3$, let $\mathbf{n}_i \in \mathbb{R}^2$ be the outward unit normal to the edge $e_i = [\mathbf{v}_i, \mathbf{v}_{i+1}]$, $i = 1, \ldots, n$, and for any $\mathbf{x} \in P_n$, let $h_i(\mathbf{x})$ be the perpendicular distance of $\mathbf{x}$ to the edge e_i, so that

$$h_i(\mathbf{x}) = (\mathbf{v}_i - \mathbf{x}) \cdot \mathbf{n}_i = (\mathbf{v}_{i+1} - \mathbf{x}) \cdot \mathbf{n}_i. \tag{12}$$

Let

$$w_i(\mathbf{x}) = d_i \prod_{\substack{j=1,\dots,n \\ j\neq i-1,i}} h_j(\mathbf{x}), \text{ and } W = \sum_{j=1}^{n} w_j \tag{13}$$

where d_i is the cross product

$$d_i = \mathbf{n}_{i-1} \times \mathbf{n}_i = \begin{vmatrix} n_{i-1}^x & n_i^x \\ n_{i-1}^y & n_i^y \end{vmatrix},$$

and $\mathbf{n}_j = (n_j^x, n_j^y)$ is the normal of edge $[\mathbf{v}_i, \mathbf{v}_{i+1}]$. Then, the functions $\phi_i = w_i/W$, $i = 1, \dots, n$ are GBCs, which are rational functions of degree $(n-2, n-3)$. See [10] for several other representations of these coordinates.

For a convex polygon P_n with $n \geq 3$ sides, let $\phi_1, \dots, \phi_n$ be a set of GBCs. For any $d \geq 0$, and any multi-index $\mathbf{j} = (j_1, \dots, j_n) \in \mathbb{N}_0^n$ with $|\mathbf{j}| := j_1 + \cdots + j_n = d$, let

$$B_{\mathbf{j}}^d(\mathbf{x}) = \frac{d!}{j_1! \cdots j_n!} \phi_1^{j_1}(\mathbf{x}) \cdots \phi_n^{j_n}(\mathbf{x}), \qquad \mathbf{x} \in P_n, \tag{14}$$

which we will call a Bernstein–Bézier function. Note that for $n > 3$, $B_{\mathbf{j}}^d$ is not necessarily a polynomial. For any n, define $\Phi_d(P_n)$ as the linear space of functions of the form

$$s(\mathbf{x}) = \sum_{|\mathbf{j}|=d} c_{\mathbf{j}} B_{\mathbf{j}}^d(\mathbf{x}), \qquad \mathbf{x} \in P_n, \quad c_{\mathbf{j}} \in \mathbb{R}. \tag{15}$$

The following properties are known:

1 $\Pi_d \subset \Phi_d(P_n)$ where Π_d is the space of polynomials of degree $\leq d$.
2 Due to (11), the function $s(\mathbf{x})$ in (15) is a univariate polynomial of degree $\leq d$ on each edge of the polygon.
3 When $n \geq 4$, the functions $B_{\mathbf{j}}^d \in \Phi_d(P_n)$ are not linearly independent.

Based on the polynomial blossom property, when $d = 2$ we can construct a basis for a subspace which still contains the space of quadratic polynomials Π_2:

Theorem 1 *Where $\lambda_{i,j}$, $i = 1, ...n$, $j \in \{-1, 0, 1\}$ is the traditional barycentric coordinate associated with $\mathbf{v}_{i+j}$ in the triangle $\langle v_{i-1}, v_i, v_{i+1}\rangle$, let*

$$F_i = \phi_i \lambda_{i,0}, \qquad F_{i,1} = \phi_i \lambda_{i,1} + \phi_{i+1} \lambda_{i+1,-1}, \qquad i = 1, \dots, n,$$

and

$$\Psi_2(P_n) = span\{F_i, F_{i,1}, i = 1, \dots, n\}. \tag{16}$$

Then, $\Pi_2 \subset \Psi_2(P_n) \subset \Phi_2(P_n)$.

By specializing to specifically Wachspress coordinates, we can do the same for $d \geq 3$:

Theorem 2 *For $d \geq 3$ and with Wachspress coordinates ϕ_i, let*

$$F_i = \phi_i \lambda_{i,0}^{d-1}, \qquad i = 1, ..., n,$$

$$F_{i,k} = \binom{d-1}{k} \phi_i \lambda_{i,1}^{k} \lambda_{i,0}^{d-1-k} + \binom{d-1}{k-1} \phi_{i+1} \lambda_{i+1,0}^{k-1} \lambda_{i+1,-1}^{d-k},$$

$$i = 1, ..., n, \ k = 1, ..., d-1,$$

and

$$\Psi_d(P_n) := span\{F_i, \ i = 1, ..., n\} \oplus span\{F_{i,k}, \ i = 1, ..., n, \ k = 1, ..., d-1\} \oplus \frac{b}{\sum_{j=1}^{n} w_j} \Pi_{d-3} \tag{17}$$

where b is the bubble function given by

$$b := \prod_{k=1}^{n} h_k.$$

Then, $\Pi_d \subset \Psi_d(P_n) \subset \Phi_d(P_n)$.

We refer the reader to [12] for proofs of Theorems 1 and 2.

Given a domain Ω, let $\Delta = \{P_n\}$ be a partition of Ω into convex polygons, we can use the basis elements of $\Psi_d(P_n)$ over each polygon P_n to define spline functions. For any polygons $P_i, P_j \in \Delta$, we assume that the intersection $P_i \cap P_j$ is either empty, a common edge, or a common vertex. Then, define

$$\mathcal{S}_d(\Delta) := \{s \in C(\Omega), \quad s|_{P_n} \in \Psi_d(P_n), \forall P_n \in \Delta\}. \tag{18}$$

This is the spline space we will use to numerically solve PDEs of the form (1). In particular, our numerical trials were performed when $d = 2$ and $d = 3$. In [12], it was shown that one can use the part of $\Psi_d(P_n)$ associated with the boundary of P_n, that is, span$\{F_i, \ i = 1, ..., n\} \oplus$ span$\{F_{i,k}, \ i = 1, ..., n, \ k = 1, ..., d-1\}$ to form an interpolatory basis $\{L_j, \ j = 1, ..., dn\}$:

$$\begin{aligned} L_{di-k} &= \sum_{s=1}^{d-1} r_{ks} F_{i,s}, \quad i = 1, ..., n, \ k = 1, ..., d-1, \\ L_{di} &= F_i - \sum_{k=1}^{d-1} (1 - k/d)^d (L_{di-k} + L_{d(i-2)+k}), \quad i = 1, ..., n \end{aligned} \tag{19}$$

where $R = (r_{ks})_{k,s=1}^{d}$ is the inverse of a matrix built from coefficients of Lagrange interpolation of univariate Bernstein polynomials of degree d. The functions L_{di} interpolate at vertices of P_n, while the functions L_{di-k} interpolate at points on edges of P_n. When $d > 3$, there is additional work to be done, as there is still another component of the space $\Psi_d(P_n)$, namely $\frac{b}{W}\Pi_{d-3}$. Since we focus on degree 3 or less in this paper, the interested reader should refer to [12] for more information on this case.

3 Existence, Uniqueness, Stability, and Convergence of Solutions

We will review some sufficient conditions such that the elliptic PDE in (1) admits a unique weak solution with zero boundary condition, i.e., $g = 0$ on $\partial\Omega$. This PDE has been studied for a long time. We refer to [6, 7] for detail. For convenience, we include some basic theory to explain our numerical solution.

Of course, it would be beneficial to find necessary conditions as well, but these can be hard to pinpoint. In particular, it must be required that the associated homogeneous PDEs, where $f = g = 0$, have the unique solution $u = 0$; otherwise, when given a solution u^* of the PDE above, we would be able to build a distinct solution using $u^* + Ku$ for any constant K.

The weak formulation of this PDE is given by (9) for all $v \in H_0^1(\Omega)$. To do so, we shall use the following norm and semi-norm on $H^1(\Omega)$ for convenience: $\|u\|_{2,\Omega} = \|u\|_{L^2(\Omega)}$, $|u|_{1,2,\Omega} = \|\nabla u\|_{L^2(\Omega)}$, and $|u|_{2,2,\Omega} = |u|_{H^2(\Omega)}$. Similarly, $|u|_{d+1,2,\Omega} = |u|_{H^{d+2}(\Omega)}$.

Define by $a(u, v)$ the bilinear form in the left-hand side of the equation in (9). To find the weak solution in $H_0^1(\Omega)$, we must show that $a(u, v)$ is bounded above and coercive in order to use the Lax–Milgram theorem.

Definition 1 We say the PDE in (1) is uniform elliptic if the coefficient matrix $[A_{ij}]_{1\le i,j,\le 2}$ is symmetric and positive definite with smallest eigenvalue $\alpha > 0$ over Ω for a positive number α called the ellipticity of the PDE.

Theorem 3 *Suppose that the second-order PDE in (1) is uniformly elliptic with ellipticity $\alpha > 0$. Let $\beta := \|\mathbf{B}\|_{\infty,\Omega} < \infty$ and $C \ge \gamma > 0$. Suppose that there exists a positive constant c such that*

$$\alpha > \frac{\beta}{2c} \text{ and } \gamma \ge \frac{c\beta}{2}. \tag{20}$$

Then, the PDE (1) has a unique weak solution u in $H_0^1(\Omega)$ satisfying the weak formulation (9) for $v \in H_0^1(\Omega)$.

Proof A proof can be found in many standard finite element textbooks, e.g., [6, 7].

When B_1 is a function of y only and B_2 is a function of x only, we note that for all $u \in H_0^1(\Omega)$,

$$\int_\Omega B_1\left(\frac{\partial}{\partial x_1}u\right)u\,dxdy = -\int_\Omega B_1\left(\frac{\partial}{\partial x_1}u\right)u\,dxdy \tag{21}$$

by using integration by parts and zero boundary condition. Thus, $\int_\Omega B_1(\frac{\partial}{\partial x_1}u)udxdy = 0$. Similarly, $\int_\Omega B_2(\frac{\partial}{\partial x_2}u)udxdy = 0$. Hence, the terms involving first-order derivatives in $a(u, u)$ are zero and

$$\begin{aligned} a(u,u) =& \int_\Omega \left[\sum_{i,j=1}^{2} A_{ij} \frac{\partial}{\partial x_i} u \frac{\partial}{\partial x_j} u + Cu^2 \right] dxdy \\ \geq& \alpha |u|_{1,2,\Omega}^2 + \gamma \|u\|_{2,\Omega}^2 \end{aligned} \tag{22}$$

which implies that $a(u, u)$ is coercive. Thus, we have established the following:

Corollary 1 *Suppose that the second-order PDE in (1) is uniform elliptic with ellipticity $\alpha > 0$. Suppose that B_1 is a function of y only and B_2 is a function of x only. If $C \geq 0$, then the PDE (1) has a unique weak solution u in $H_0^1(\Omega)$ satisfying the weak formulation (9) for $v \in H_0^1(\Omega)$.*

By applying Theorem 3 and Corollary 1, we can establish the following result:

Corollary 2 *Suppose that the second-order PDE in (1) is uniform elliptic with ellipticity $\alpha > 0$. Suppose that $B_1(x, y) = \hat{B}_1(x, y) + B_1'(y)$ and $B_2(x, y) = \hat{B}_2(x, y) + B_2'(x)$ where $B_1'(y)$ is a function of y only and B_2' is a function of x only. Let $\hat{\beta} := \max\{\|\hat{B}_1\|_{\infty,\Omega}, \|\hat{B}_2\|_{\infty,\Omega}\} < \infty$ and $C \geq \gamma > 0$. Suppose that there exists a positive constant c such that*

$$\alpha > \frac{\hat{\beta}}{2c} \text{ and } \gamma \geq \frac{c\hat{\beta}}{2}. \tag{23}$$

Then, the PDE (1) has a unique weak solution u in $H_0^1(\Omega)$ satisfying the weak formulation (9) for $v \in H_0^1(\Omega)$.

In particular, when $B_1 = B_2 \equiv 0$, the PDE in (1) has a unique weak solution according to Theorem 3 and Corollary 2. In fact, we can establish the existence, uniqueness, and stability of the solution of (1) without using Lax–Milgram theorem. Indeed, in this case, it is easy to see that the weak form $a(u, v) = \langle f, v\rangle$ is the Euler–Langrange equation of the following minimization:

$$\min_{\substack{u\in H^1(\Omega)\\ u|_{\partial\Omega}=g}} J_f(u) \tag{24}$$

where $J_f(u) = \frac{1}{2}a(u, u) - \langle f, u\rangle$. To approximate the exact solution $u \in H^1(\Omega)$ with $u|_{\partial\Omega} = 0$, we can instead find the minimum among $u \in H_0^1(\Omega)$. To numerically solve the PDE, we can instead search for u in $\mathscr{S}_d := H_0^1(\Omega) \cap S_d(\Delta)$ where $S_d(\Delta)$ is the space of polygonal splines of order d which are defined with respect to a polygonal partition Δ of Ω as explained in the previous section. In the following analysis, we will consider this minimization for $u \in \mathscr{S}_d$.

Using a standard convex analysis, one can show

Theorem 4 *Suppose that* $[A_{ij}]_{1\le i,j\le 2}$ *is symmetric and positive definite. Suppose that* $B_1 = B_2 \equiv 0$. *If* $C \ge \gamma \ge 0$, *then* J_f *is strongly convex with convexity coefficient* μ *which is independent of* f; *therefore,* J_f *has a unique minimizer* u_f. *Hence, there exists a unique weak solution* u_f *satisfying (9).*

We can further derive the following result regarding the stability of the minimizer of J_f with respect to the source function f:

Theorem 5 *Suppose that the PDE in (1) satisfies the uniform ellipticity conditions in the hypotheses of Theorem 3. For two functions* f *and* g, *denote the minimizer of* J_f *by* u_f *and the minimizer of* J_g *by* u_g. *Then* $||u_f - u_g||_{L^2(\Omega)} \le \mu^{-1}||f - g||_{L^2(\Omega)}$.

Proof Since J_f and J_g are both μ-strongly convex, we can say that

$$J_f(u_g) - J_f(u_f) \ge \langle \nabla J_f(u_f), u_g - u_f\rangle + \frac{\mu}{2}||u_f - u_g||^2_{L^2(\Omega)} = \frac{\mu}{2}||u_f - u_g||^2_{L^2(\Omega)}$$

and

$$J_g(u_f) - J_g(u_g) \ge \langle \nabla J_g(u_g), u_f - u_g\rangle + \frac{\mu}{2}||u_f - u_g||^2_{L^2(\Omega)} = \frac{\mu}{2}||u_f - u_g||^2_{L^2(\Omega)}$$

where the last equalities in each equation come from the fact that u_f and u_g minimizing J_f and J_g, respectively, implies that both $\nabla J_f(u_f)$ and $\nabla J_g(u_g)$ are 0. If we sum two equations above, then after some simplification we see that

$$\langle f - g, u_f - u_g\rangle \ge \mu||u_f - u_g||^2_{L^2(\Omega)}.$$

Using Cauchy–Schwarz on the left side of this inequality gives us

$$||f - g||_{L^2(\Omega)}||u_f - u_g||_{L^2(\Omega)} \ge \mu||u_f - u_g||^2_{L^2(\Omega)}.$$

A simple division by $\mu||u_f - u_g||_{L^2(\Omega)}$ yields the desired result.

Finally, we discuss convergence of the numerical solutions. The discussion is divided into two parts. The first part shows the approximation power of $\mathscr{S}_d(\Delta)$. The second part is to apply the approximation property to establish the convergence of polygonal splines to the weak solution.

Proving the approximation power of this space is more complicated than in the cases of finite elements and splines over triangulations due to difficulties in bounding the gradients of the Wachspress coordinates. Fortunately, it has been shown in [10, 11] that

$$\sup_{\mathbf{x}\in P}\sum_{j=1}^{n}||\nabla\phi_j(\mathbf{x})||_2 \le \frac{4}{h_*} \tag{25}$$

where h^* is the shortest perpendicular distance from any vertex of P to a nonincident edge of P. To control this quantity, we will have to assume that Δ satisfies

$$0 < \alpha_1 < \theta_{P,i} < \alpha_2 < \pi, \quad i = 1, \cdots, n(P), \forall P \in \Delta \tag{26}$$

for two given positive constants α_1 and α_2 where $\theta_{P,i}$ is the interior angle of P at its ith vertex, and $n(P)$ stands for the number of sides of P. We shall assume that there exists a positive integer n_0 such that $n(P) \leq n_0$ for all $P \in \Delta$. For each $P \in \Delta$, let $|P|$ be the diameter of P, that is, the diameter of the smallest circle containing P and ρ_P be the radius of the largest circle contained in P. We use $\kappa_P = \frac{|P|}{\rho_P}$ to be the shape parameter of P (cf. [16]) or the chunkiness of P (cf. [7]). For convenience, we let $|\Delta|$ be the largest of $|P|$ for all $P \in \Delta$ instead of the longest of the lengths of edges of Δ which is commonly used for triangulations. Where $e(P)$ is the length of the shortest edge of P, let $e(\Delta) = \min\{e(P), P \in \Delta\}$. We will assume that the global shape parameter γ_Δ to satisfy

$$\gamma_\Delta = \frac{|\Delta|}{e(\Delta)} \leq \gamma < \infty \tag{27}$$

in the following study for a given $\gamma > 0$.

As $\mathscr{S}_d(\Delta)$ is a space of continuous functions over Ω, we cannot simply apply the Bramble–Hilbert lemma to establish the approximation property of $\mathscr{S}_d(\Delta)$. Instead, we follow the ideas in [16]. For simplicity, let us focus ourselves to the case $d = 2$; the case $d \geq 3$ can be done similarly.

First, we prove the following:

Lemma 1 *Let P be an n-gon in Δ. Let L_j be one of the functions listed in (19) which is supported on P. Then*

$$||L_j||_{2,P} \leq C_{n,\alpha_2,\gamma}|P| \tag{28}$$

and

$$|L_j|_{1,2,P} \leq C_{n,\alpha_1,\alpha_2,\gamma} \tag{29}$$

for two positive constants $C_{n,\alpha_2,\gamma}$ and $C_{n,\alpha_1,\alpha_2,\gamma}$.

Proof Since the functions L_j are built from linear combinations of the functions F_k and $F_{k,1}$ given in Theorem 1, we have for some constant C_n which depends only on n

$$||L_j||_{2,P} \leq C_n \max_{k=1,\ldots,n} \{||F_k||_{2,P}, ||F_{k,1}||_{2,P}\}.$$

Thus, we really need to bound $||F_k||_{2,P}$ and $||F_{k,1}||_{2,P}$. By the definition of F_k,

$$||F_k||_{2,P} \leq ||\phi_i \lambda_{i,0}||_{2,P} \leq ||\lambda_{i,0}||_{\infty,P}||\phi_i||_{2,P} \leq |P|||\lambda_{i,0}||_{\infty,P}.$$

To estimate $||\lambda_{i,0}||_{\infty,P}$, let $h_{\perp,i}$ be the perpendicular distance from $\mathbf{v}_i$ to the line connecting $\mathbf{v}_{i-1}$ to $\mathbf{v}_{i+1}$, and denote by m_i the point on this line which is a distance $h_{\perp,i}$ from $\mathbf{v}_i$. Then since $\lambda_{i,0}$ is a linear function, we have $||\nabla\lambda_{i,0}||_2 = \frac{1}{h_{\perp,i}}$.

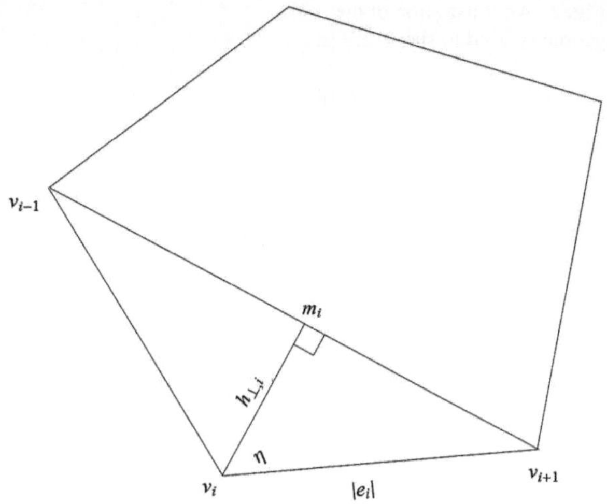

Fig. 1 An illustration to clarify the geometry used to show (30)

Let $|e_{i-1}|$ be the length of the edge between $\mathbf{v}_{i-1}$ and $\mathbf{v}_i$ and similar for $|e_i|$. Without loss of generality, suppose that $|e_i| \leq |e_{i-1}|$. If we draw the triangle $\tau = \langle \mathbf{v}_i, \mathbf{v}_{i+1}, m_i \rangle$ (see Fig. 1), we can see that $h_{\perp,i} = |e_i| \cos \eta$ where η is the interior angle of τ at $\mathbf{v}_i$. Since $|e_i| \leq |e_{i-1}|$, we have that $\eta < \frac{1}{2}\theta_{P,i}$. By (26), $0 < \frac{1}{2}\theta_{P,i} < \frac{\pi}{2}$, so

$$\cos \eta \geq \cos(\frac{1}{2}\theta_{P,i}) = \sqrt{\frac{1 + \cos \theta_{P,i}}{2}} \geq \sqrt{\frac{1 + \cos \alpha_2}{2}} = C_{\alpha_2} \tag{30}$$

for a constant C_{α_2} which depends on α_2.

Therefore, we have that

$$h_{\perp,i} \geq C_{\alpha_2} |e_i| \geq C_{\alpha_2} e(P). \tag{31}$$

Now, $||\lambda_{i,0}||_{\infty,P}$ will be achieved by $\lambda_{i,0}$ at the point in P which has the largest perpendicular distance from the line connecting $\mathbf{v}_{i-1}$ and $\mathbf{v}_{i+1}$, which is of course a distance at most $|P|$ from this line. Since $\lambda_{i,0}$ is linear, we have that

$$||\lambda_{i,0}||_{\infty,P} \leq \frac{|P|}{h_{\perp,i}} \leq \frac{1}{C_{\alpha_2}} \frac{|P|}{e(P)} \leq C_{\alpha_2} \gamma. \tag{32}$$

Therefore, we have $||F_k||_{2,P} \leq C_{n(P),\alpha_2,\gamma} |P|$. A similar argument shows that

$$||F_{k,1}||_{2,P} \leq \frac{2}{C_2} \frac{|P|^2}{e(P)} \leq C_{n,\alpha_2,\gamma} |P|,$$

which completes the proof of (28).

Fig. 2 An illustration of the geometry used to show (33)

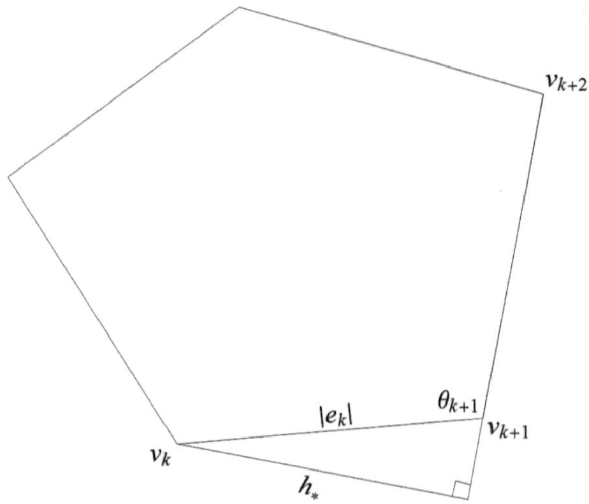

To prove (29), we will follow a similar strategy. As

$$|L_j|_{1,2,P} \leq C_n \max_{k=1,\ldots,n} \{|F_k|_{1,2,P}, |F_{k,1}|_{1,2,P}\},$$

we need to bound $|F_k|_{1,2,P}$ and $|F_{k,1}|_{1,2,P}$.

$$\begin{aligned}
|F_k|_{1,2,P}^2 &= \int_P (\lambda_{i,0} D_x \phi_i + \phi_i D_x \lambda_{i,0})^2 + (\lambda_{i,0} D_y \phi_i + \phi_i D_y \lambda_{i,0})^2 \, dx \, dy \\
&\leq 2 \sup_{\mathbf{x} \in P} ||\nabla \phi_i(\mathbf{x})||_2^2 \int_P \lambda_{i,0}^2 \, dx \, dy + 2 \sup_{\mathbf{x} \in P} ||\nabla \lambda_{i,0}(\mathbf{x})||_2^2 \int_P \phi_i^2 \, dx \, dy \\
&\leq 2 \sup_{\mathbf{x} \in P} ||\nabla \phi_i(\mathbf{x})||_2^2 (||\lambda_{i,0}||_{\infty,P}^2) |P|^2 + \frac{2}{h_{\perp,i}^2} |P|^2 \\
&\leq C_{n(P)} \sup_{\mathbf{x} \in P} ||\nabla \phi_i(\mathbf{x})||_2^2 \left(\frac{|P|}{h_{\perp,i}}\right)^2 |P|^2 + C_{n(P)} \left(\frac{|P|}{h_{\perp,i}}\right)^2 \\
&\leq C_{n(P)} \frac{16|P|^2}{h_*^2} \left(\frac{|P|}{h_{\perp,i}}\right)^2 + C_{n(P)} \left(\frac{|P|}{h_{\perp,i}}\right)^2 = C_{n(P),\alpha_2} \left(\frac{|P|}{e(P)}\right)^2 (1 + \frac{16|P|^2}{h_*^2})
\end{aligned}$$

by using (25) and (31).

Now, we will show that h_* is comparable to $e(P)$. In particular, since P is convex, h_* ought to be realized by a line drawn from a vertex of P, say v_k, to an edge to an edge which is a graph-distance of 2 from v_k, say the edge between v_{k+1} and v_{k+2}. If we draw in this line (see Fig. 2), a right triangle is formed which shows that

$$h_* = \sin \theta_{k+1} |e_k| \geq \min_{\mathbf{x}} \{\sin \alpha_1, \sin \alpha_2\} e(P). \tag{33}$$

Therefore, we can say

$$\sup_{\mathbf{x}\in P} ||\nabla\phi_i(\mathbf{x})||_2^2 \leq \frac{16}{h_*^2} \leq \frac{C_{\alpha_1,\alpha_2}}{e(P)^2}. \tag{34}$$

Combining (34) and our above analysis shows that

$$|F_k|_{1,2,P}^2 \leq C_{n(P)} \tfrac{|P|^2}{e(P)^2}(1 + 16C_{\alpha_1,\alpha_2} \tfrac{|P|^2}{e(P)^2})$$
$$\Rightarrow |F_k|_{1,2,P} \leq C_{\alpha_1,\alpha_2} \tfrac{|P|}{e(P)}(1 + \tfrac{|P|}{e(P)})C_{\alpha_1,\alpha_2,\gamma}.$$

A similar argument will show that $|F_{k,1}|_{1,2,P} \leq C_{\alpha_1,\alpha_2,\gamma}$, so we have that $|L_j|_{1,2,P} \leq C_{n,\alpha_1,\alpha_2,\gamma}$ as desired in (29).

We are nearly ready to establish the approximation power of $\mathscr{S}_d(\Delta)$. Our main result is the following theorem:

Theorem 6 *Suppose that Δ satisfies four assumptions: $\gamma_\Delta \leq \gamma$, $0 < \alpha_1 \leq \theta_{P,i} \leq \alpha_2 < \pi$, $\kappa_P \leq \kappa < \infty$ and $n(\Delta) \leq n_0$. Then for any $u \in H^{d+1}(\Omega)$, there exists a polygonal spline $Q(u) \in \mathscr{S}_d(\Delta)$ such that*

$$\|u - Q(u)\|_{2,\Omega} \leq C_{n_0,\alpha_1,\alpha_2,\kappa,\gamma}|\Delta|^{d+1}|u|_{d+1,2,\Omega} \tag{35}$$

and

$$|u - Q(u)|_{1,2,\Omega} \leq C_{n_0,\alpha_1,\alpha_2,\kappa,\gamma}|\Delta|^{d}|u|_{d+1,2,\Omega} \tag{36}$$

for constant $C(n_0, \alpha_1, \alpha_2, \kappa, \gamma)$ which is independent of u and $|\Delta|$, but may depend on the domain Ω if Ω is nonconvex.

Similarly, for any $u \in H^{d+1}(\Omega) \cap H_0^1(\Omega)$, there exists a polygonal spline $Q_0(u) \in \mathscr{S}_d(\Delta) \cap H_0^1(\Omega)$ such that

$$\|u - Q_0(u)\|_{2,\Omega} \leq C_{n_0,\alpha_1,\alpha_2,\kappa,\gamma}|\Delta|^{d+1}|u|_{d+1,2,\Omega} \tag{37}$$

and

$$|u - Q_0(u)|_{1,2,\Omega} \leq C_{n_0,\alpha_1,\alpha_2,\kappa,\gamma}|\Delta|^{d}|u|_{d+1,2,\Omega} \tag{38}$$

for another constant $C(n_0, \alpha_1, \alpha_2, \kappa, \gamma)$ which is independent of u and $|\Delta|$, but may depend on the domain Ω if Ω is nonconvex.

We will require bit more discussion, along with another lemma, to prove this theorem. For convenience, we consider the case $d = 2$ first. Let us recall locally supported basis functions in $\mathscr{S}_d(\Delta)$ from [12]. For each vertex $\mathbf{v} \in \Delta$, let $L_v \in \mathscr{S}_2(\Delta)$ be the spline supported on the collection $\Omega_\mathbf{v}$ of polygons which has $\mathbf{v}$ as one of its vertices satisfying $L_v(\mathbf{v}) = 1$ and zero for other vertices and zero for all midpoints of edges. For each edge $e \in \Delta$, let $L_e \in \mathscr{S}_2(\Delta)$ be the spline supported on the collection Ω_e of polygons which share the edge e satisfying $L_e(\mathbf{w}_e) = 1$ and zero for all vertices and zero for all midpoints of other edges where $\mathbf{w}_e$ is the midpoint of e. By the definition of these locally supported polygonal splines, we have that

$$\mathscr{S}_2(\Delta) = \text{span}\{L_v, \forall \mathbf{v} \in \Delta\} \oplus \text{span}\{L_e, \forall e \in \Delta\}.$$

We will construct quasi-interpolatory splines $Q(u) \in \mathscr{S}_2(\Delta)$ and $Q_0(u) \in \mathscr{S}_2(\Delta) \cap H_0^1(\Omega)$.

To this end, we first extend any $u \in H^3(\Omega)$ to a function in $H^3(\mathbb{R}^2)$ with the property $\|u\|_{H^3(\mathbb{R}^2)} \le E\|u\|_{H^3(\Omega)}$ with a positive constant E dependent only on Ω (cf. [22]) and call it u again for convenience.

For each $\mathbf{v}$, let $\Omega_\mathbf{v}$ be the collection of all polygons sharing the vertex $\mathbf{v}$. Let $B_\mathbf{v}$ be largest disk contained in $\Omega_\mathbf{v}$ if $\mathbf{v}$ is an interior vertex. If $\mathbf{v}$ is a boundary vertex, we let $B_\mathbf{v}$ be the largest disk contained in the convex hull $co(\Omega_\mathbf{v})$ of $\Omega_\mathbf{v}$. Let $F_\mathbf{v}(u)$ is the averaged Taylor polynomial of degree 2 associated with u based on the disk $B_\mathbf{v}$ (cf. [16]). Define by

$$c_\mathbf{v}(u) = F_\mathbf{v}(u)|_\mathbf{v}. \tag{39}$$

Let $T_\mathbf{v} \in \Omega_\mathbf{v}$ be a triangle with a vertex $\mathbf{v}$. We simply use the polynomial property $\|p\|_{\infty,T} \le \frac{K_1}{|T|}\|p\|_{2,T}$ for any triangle T along with the property that $\|F_\mathbf{v}(u)\|_{2,co(\Omega_\mathbf{v})} \le K_2\|u\|_{2,co(\Omega_\mathbf{v})}$ (cf. [16]) to have

$$|c_\mathbf{v}(u)| \le \|F_\mathbf{v}(u)\|_{\infty,T} \le \frac{K_1}{|T_\mathbf{v}|}\|F_\mathbf{v}(u)\|_{2,T} \le \frac{K_1K_2}{|T_\mathbf{v}|}\|u\|_{2,co(\Omega_\mathbf{v})} \tag{40}$$

for two constants K_1 and K_2 independent of u and $T_\mathbf{v}$. Similarly, for $e \in \Delta$, let Ω_e be the union of two polygons in Δ if e is an interior edge. Let B_e be a largest disk contained in Ω_e. If e is a boundary edge, we can choose a disk B_e contained in the polygon with edge e. Then we let $F_e(u)$ be the averaged Taylor polynomial of degree d based on B_e. Choose c_e to be the value at $F_e(u)$ evaluated at the midpoint $\mathbf{w}_e$ of e. Choose a good triangle T_e containing $\mathbf{w}_e$. Then, $c_e(u)$ will satisfy a similar property in (40). Our quasi-interpolatory spline is defined by

$$Q(u) = \sum_{\mathbf{v}\in\Delta} c_\mathbf{v}(u)L_v + \sum_{e\in\Delta} c_e(u)L_e. \tag{41}$$

Similarly, for $u \in H_0^1(\Omega)$, we let

$$Q_0(u) = \sum_{\substack{\mathbf{v}\in\Delta\\ \mathbf{v}\notin\partial\Omega}} c_\mathbf{v}(u)L_v + \sum_{\substack{e\in\Delta\\ e\notin\partial\Omega}} c_e(u)L_e. \tag{42}$$

Thus, $Q_0(u) \in \mathscr{S}_d(\Delta) \cap H_0^1(\Omega)$. Let us show that $Q(u)$ and $Q_0(u)$ are bounded operators on $L^2(\Omega)$. That is,

Lemma 2 *For any $u \in L^2(\Omega)$, we have*

$$\|Q(u)\|_{2,\Omega} \le K_3\|u\|_{2,\Omega} \tag{43}$$

for a positive constant K_3 independent of u. In addition, for nonnegative integers α, β with $\alpha + \beta = 1$,

$$\|D_x^\alpha D_y^\beta Q(u)\|_{2,\Omega} \le \frac{K_4}{e(\Delta)} \|u\|_{2,\Omega} \tag{44}$$

for another positive constant K_4 independent of u. The same estimates hold for Q_0.

Proof For each polygon $P \in \Delta$, denote by Ω_P the union of polygons which share an edge or a vertex of P. Note that $L_{\mathbf{v}}|_P$ is just L_j for some j and so is $L_e|_P$. Then, we use Lemma 1 to have

$$\begin{aligned}
\|Q(u)\|_{2,P} &= \left[\int_P |\sum_{\mathbf{v}\in P} c_{\mathbf{v}}(u)L_v + \sum_{e\in P} c_e(u)L_e|^2 dx\, dy\right]^{1/2} \\
&\le \sum_{\mathbf{v}\in P} |c_{\mathbf{v}}(u)| \left(\int_P |L_v|^2 dx\, dy\right)^{1/2} + \sum_{e\in P} |c_e(u)| \left(\int_P |L_e|^2 dx\, dy\right)^{1/2} \\
&\le \sum_{\mathbf{v}\in P} \frac{K_1 K_2}{|T_{\mathbf{v}}|} \|u\|_{2,\Omega_{\mathbf{v}}} ||L_v||_{2,P} + \sum_{e\in P} \frac{K_1 K_2}{|T_e|} \|u\|_{2,\Omega_e} ||L_e||_{2,P} \\
&\le C_{n(P),\alpha_2,\gamma} \frac{|P|}{\min\limits_{\mathbf{v},e\in P} \{|T_{\mathbf{v}}|, |T_e|\}} ||u||_{2,\Omega_P} \\
&\le C_{n_0,\alpha_2,\gamma} \frac{|\Delta|}{e(\Delta)} ||u||_{2,\Omega_P} \le C_{n_0,\alpha_2,\gamma} ||u||_{2,\Omega_P}
\end{aligned}$$

for a constant $C_{n_0,\alpha_2,\gamma}$ as we are always able to choose $|T_{\mathbf{v}}| \ge e(P)$ and $|T_e| \ge e(P)$. Hence,

$$\begin{aligned}
\|Q(u)\|_{2,\Omega}^2 &= \sum_{P\in\Delta} \|Q(u)\|_{2,P}^2 \le C_{n_0,\alpha_2,\gamma}^2 \sum_{P\in\Delta} ||u||_{2,\Omega_P}^2 \\
&\le C_{n_0,\alpha_1,\alpha_2,\gamma}^2 \sum_{P\in\Delta} \|u\|_{2,P}^2 = C_{n_0,\alpha_1,\alpha_2,\gamma}^2 \|u\|_{2,\Omega}^2
\end{aligned} \tag{45}$$

where we have used the fact that $\sum_{P\in\Delta} ||u||_{2,\Omega_P}^2 \le C(n_0, \alpha_1) \sum_{P\in\Delta} ||u||_{2,p}^2$ for a positive constant $C(n_0, \alpha_1)$ since for each polygon $q \in \Delta$, $q \in \Omega_P$ for at most $n_0 2\pi/\alpha_1$ polygons $P \in \Delta$.

Similarly, for nonnegative integers α and β such that $\alpha + \beta = 1$, we have

$$\|D_x^\alpha D_y^\beta Q(u)\|_{2,P} = \left[\int_P |\sum_{\mathbf{v}\in P} c_{\mathbf{v}}(u) D_x^\alpha D_y^\beta L_{\mathbf{v}} + \sum_{e\in P} c_e(u) D_x^\alpha D_y^\beta L_e|^2 dxdy\right]^{1/2}$$
$$\leq \sum_{\mathbf{v}\in P} |c_{\mathbf{v}}(u)| \left(\int_P |D_x^\alpha D_y^\beta L_{\mathbf{v}}|^2\right)^{1/2} + \sum_{e\in P} |c_e(u)| \left(\int_P |D_x^\alpha D_y^\beta L_e|^2 dxdy\right)^{1/2}$$
$$\leq \sum_{\mathbf{v}\in P} \frac{K_1 K_2}{|T_{\mathbf{v}}|} \|u\|_{2,\Omega_{\mathbf{v}}} |L_{\mathbf{v}}|_{1,2,\Omega_{\mathbf{v}}} + \sum_{e\in P} \frac{K_1 K_2}{|T_e|} \|u\|_{2,\Omega_e} |L_e|_{1,2,\Omega_e}$$
$$\leq C_{n_0,\alpha_1,\alpha_2,\gamma} \frac{||u||_{2,\Omega_P}}{e(\Delta)}.$$

Hence, we have

$$\|D_x^\alpha D_y^\beta Q(u)\|_{2,\Omega}^2 = \sum_{P\in\Delta} \|D_x^\alpha D_y^\beta Q(u)\|_{2,P}^2$$
$$\leq C_{n_0,\alpha_2,\gamma}^2 \sum_{P\in\Delta} ||u||_{2,\Omega_P}^2 \frac{1}{e(\Delta)^2} = \frac{C_{n_0,\alpha_1,\alpha_2,\gamma}^2}{e(\Delta)^2} \|u\|_{2,\Omega}^2. \tag{46}$$

By taking the square root both sides, we finish the proof of (44).
Similarly, we have the same estimates for Q_0.

Now, we are ready to prove the main result:

Proof (*of Theorem* 6) For simplicity, let us consider the approximation in $L^2(\Omega)$ first. It is easy to see

$$\begin{aligned} \|u - Q(u)\|_{L^2(\Omega)}^2 &= \sum_{P\in\Delta} \|u - Q(u)\|_{L^2(P)}^2 \\ &\leq 2 \sum_{P\in\Delta} \|u - F_{P,2}(u)\|_{L^2(P)}^2 + \|F_{P,2}(u) - Q(u)\|_{L^2(P)}^2 \end{aligned} \tag{47}$$

where $F_{P,2}(u)$ is the averaged Taylor polynomial of degree 2 associated with u based on the largest disk B_P inside P. We know from [16] that

$$||u - F_{P,2}(u)||_{2,P} \leq C_{\kappa_P} |P|^3 |u|_{3,2,P}. \tag{48}$$

For $\mathbf{v} \in P$, $F_{\mathbf{v}}(F_{P,2}(u)) = F_{P,2}(u)$ and for $e \in P$, $F_e(F_{P,2}(u)) = F_{P,2}(u)$. we have $Q(F_{P,2}(u)) = F_{P,2}(u)$, and hence, by Lemma 2,

$$\|F_{P,2}(u) - Q(u)\|_{2,P} = \|Q(F_{P,2}(u) - u)\|_{2,P}$$
$$\leq K_3 \|u - F_{P,2}(u)\|_{2,\Omega_P} \leq K_3 C_{\kappa_P} |\Omega_P|^3 |u|_{3,2,\Omega_P}.$$

Combining this with (47) and (48), we have the following:

$$\|u - Q(u)\|^2_{2,\Omega)} \le 2\sum_{P\in\Delta} ||u - F_{P,2}(u)||^2_{2,P} + \|F_{P,2}(u) - Q(u)\|^2_{2,P}$$
$$\le \sum_{P\in\Delta} C^2_{\kappa_P}|P|^6|u|^2_{3,2,P} + K^2C^2_{\kappa_P}|\Omega_P|^6|u|^2_{3,2,\Omega_P}$$
$$\le K_3^2(1+C_\kappa^2)\sum_{P\in\Delta}|\Omega_P|^6|u|^2_{3,2,\Omega} \le K_3^2(1+C_\kappa^2)|\Delta|^6|u|^2_{3,2,\Omega}$$

where we have noted that the number of polygons containing each vertex is bounded by $2\pi/\alpha_1$, and hence, the number of polygons $p \in \Delta$ such that $p \subset \Omega_P$ is bounded by $n_0 2\pi/\alpha_1$ and that $|\Omega_P| \le 3|\Delta|$. Therefore,

$$||u - Q(u)||_{2,\Omega} \le C_\kappa|\Delta|^3|u|_{3,2,\Omega}.$$

Now, we consider $|u - Q(u)|_{1,2,\Omega}$. Recall that the averaged Taylor polynomial has the property that $D_x^\alpha D_y^\beta F_{P,d}(u) = F_{P,d-\alpha-\beta}(D_x^\alpha D_y^\beta u)$, so we use Lemma 2, i.e., (44) to have

$$|u - Q(u)|^2_{1,2,\Omega} = \sum_{\alpha+\beta=1} ||D_x^\alpha D_y^\beta(u - Q(u))||^2_{2,\Omega}$$
$$= \sum_{P\in\Delta}\sum_{\alpha+\beta=1} ||D_x^\alpha D_y^\beta(u - Q(u))||^2_{2,P}$$
$$\le 2\sum_{P\in\Delta}\sum_{\alpha+\beta=1} ||D_x^\alpha D_y^\beta u - F_{P,1}(D_x^\alpha D_y^\beta u)||^2_{2,P} + ||D_x^\alpha D_y^\beta(F_{P,1}(u) - Q(u))||^2_{2,P}$$
$$= 2\sum_{P\in\Delta}\sum_{\alpha+\beta=1} ||D_x^\alpha D_y^\beta u - F_{P,1}(D_x^\alpha D_y^\beta u)||^2_{2,P} + \frac{K_4^2}{e(\Delta)^2}\|F_{P,2}(u) - u\|^2_{2,\Omega_P}$$
$$\le 2\sum_{P\in\Delta}\sum_{\alpha+\beta=1} C_{\kappa_P}|P|^4|D_x^\alpha D_y^\beta u|^2_{2,2,\Omega_P} + \frac{K_4^2}{e(\Delta)^2}C_{\kappa_P}|P|^6|u|^2_{3,\Omega_P}$$

which completes the proof of (36).

To prove (37), we have

$$\|u - Q_0(u)\|^2_{2,\Omega} \le 2\|u - Q(u)\|^2_{2,\Omega} + 2\|Q(u) - Q_0(u)\|^2_{2,\Omega}. \tag{49}$$

We use (35) for the first term on the right-hand side. Let us focus on the second term. Note from (41)

$$Q(u) - Q_0(u) = \sum_{\mathbf{v}\in\partial\Omega} c_\mathbf{v}(u)L_\mathbf{v} + \sum_{e\in\partial\Omega} c_e(u)L_e.$$

From (39), we have $c_\mathbf{v}(u) = F_\mathbf{v}(u)|_\mathbf{v} = (F_\mathbf{v}(u) - u)|_\mathbf{v}$ since $u = 0$ on $\partial\Omega$. Thus, $|c_\mathbf{v}(u)| \le \|F_\mathbf{v}(u) - u\|_{\infty,co(\Omega_\mathbf{v})}$. Using the estimate on page 9 of [16] and letting $q \to \infty$, we have

$$\|u - F_{\mathbf{v}}(u)\|_{\infty,co(\Omega_{\mathbf{v}})} \leq \frac{K}{2}(1+\kappa_{co(\Omega_{\mathbf{v}})}) \sum_{\alpha+\beta=3} \frac{3!}{\alpha!\beta!} \times$$

$$\| \int_{\langle (x,y), B_{\mathbf{v}} \rangle} |D_\nu^\alpha D_\mu^\beta u(\nu,\mu)| [(x-\mu)^2 + (y-\nu)^2]^{1/2} d\nu d\mu \|_{\infty,co(\Omega_{\mathbf{v}})}$$

$$\leq K_8 |\Omega_{\mathbf{v}}| \sum_{\alpha+\beta=3} \int_{\Omega_{\mathbf{v}}} |D_\nu^\alpha D_\mu^\beta u(\nu,\mu)| d\nu d\mu \leq K_8 |\Omega_{\mathbf{v}}|^2 |u|_{3,2,co(\Omega_{\mathbf{v}})}$$

where we have used the Cauchy–Schwarz inequality in the last inequality and $K_8 > 0$ is a constant independent of u and $|\Delta|$. We can use a similar estimate for $c_e(u)$. Hence, due to the boundedness of $L_{\mathbf{v}}$ and L_e, we conclude

$$\begin{aligned}
\|Q(u) - Q_0(u)\|_{2,\Omega}^2 &\leq 2 \sum_{\substack{P\in\Delta \\ P\cap\partial\Omega\neq\emptyset}} \| \sum_{\mathbf{v}\in P\cap\partial\Omega} c_{\mathbf{v}}(u) L_{\mathbf{v}} \|_{2,P}^2 + \| \sum_{e\in P\cap\partial\Omega} c_e(u) L_e \|_{2,P}^2 \\
&\leq 2 \sum_{\mathbf{v}\in\partial\Omega} K_8^2 |\Omega_{\mathbf{v}}|^4 |u|_{3,2,co(\Omega_{\mathbf{v}})}^2 \|L_{\mathbf{v}}\|_{2,co(\Omega_{\mathbf{v}})}^2 \\
&\quad + 2 \sum_{e\in\partial\Omega} K_8^2 |\Omega_e|^4 |u|_{3,2,\Omega_e}^2 \|L_e\|_{2,\Omega_e}^2 \\
&\leq K_9 |\Delta|^6 |u|_{3,2,\Omega}^2
\end{aligned}$$

by using Lemma 1 where $K_9 > 0$ is a constant independent of u. This finishes the estimate for the second term in (49) and hence, the estimate in (37) follows.

The estimate in (38) is derived similarly.

Theorem 7 *Suppose that the PDE in (1) satisfies the assumptions in Theorem 3 and Δ satisfies (27) and (26). Suppose that the weak solution u of the PDE in (1) is in $H^{d+1}(\Omega)$. Let $u_S \in \mathcal{S}_d$ be the weak solution satisfying $a(u_S, v) = \langle f, v \rangle$ for all $v \in \mathcal{S}_d$. Then*

$$|u - u_S|_{1,2,\Omega} \leq K |u|_{d+1,2,\Omega} |\Delta|^d \tag{50}$$

where $|\Delta|$ is the length of the longest edge in Δ, and $K = K(\Omega, \Delta, A, \mathbf{B}, C)$ is a positive constant depending only on the domain Ω, the partition Δ, the largest eigenvalue Λ of A and $\|C\|_{\infty,\Omega}$.

Proof We must prove some preliminary results in order to prove the results in this theorem. First, notice that in the proof of Theorem 3, we actually have

$$a(v, v) \geq \mu |v|_{1,2,\Omega}^2 \tag{51}$$

where $\mu = \alpha - \frac{c\beta}{2}$ for $c > 0$ such that $\gamma - \frac{\beta}{2c} \geq 0$. In addition, we can show that $a(u, v)$ is bounded. Indeed,

$$
\begin{aligned}
a(u,v) &= \int_\Omega \sum_{i,j=1}^{2} A_{ij}\frac{\partial u}{\partial x_i}\frac{\partial v}{\partial x_j} + \int_\Omega \sum_{k=1}^{2} B_k \frac{\partial u}{\partial x_k} v + \int_\Omega Cuv \\
&\le \Lambda ||\nabla u||_{L^2}||\nabla v||_{L^2} + \beta ||\nabla u||_{L^2}||v||_{L^2} + \|C\|_\infty ||u||_{L^2}||v||_{L^2} \\
&\le M_1\left(||\nabla u||_{L^2}||\nabla v||_{L^2} + ||\nabla u||_{L^2}||v||_{L^2} + ||u||_{L^2}||v||_{L^2}\right) \\
&\le M_1\left(|u|_{1,2,\Omega}|v|_{1,2,\Omega} + |u|_{1,2,\Omega}(K_1|v|_{1,2,\Omega}) + (K_1|u|_{1,2,\Omega})(K_1|v|_{1,2,\Omega})\right) \\
&\le 3\max\{M_1, M_1K_1, M_1K_1^2\}|u|_{1,2,\Omega}|v|_{1,2,\Omega}
\end{aligned}
$$

where $\Lambda > 0$ is the largest eigenvalue of $[A_{ij}]_{1\le i,j\le 2}$, $M_1 = \max\{\Lambda, \beta, \|C\|_{\infty,\Omega}\}$, and K_1 is the constant given by Poincaré's inequality, which depends only on Ω. That is,

$$
a(u,v) \le M|u|_{1,2,\Omega}|v|_{1,2,\Omega}. \tag{52}
$$

for another positive constant M. By definition of weak solution, we know that for all $v \in H_0^1(\Omega)$, $a(u,v) = \langle f, v\rangle$, and for all $v \in \mathcal{S}_d$, $a(u_S, v) = \langle f, v\rangle$. Since $\mathcal{S}_d \subset H_0^1(\Omega)$, we can say that for all $v \in \mathcal{S}_d$,

$$
a(u - u_S, v) = 0, \quad \forall v \in \mathcal{S}_d. \tag{53}
$$

Now, define $u_{best} := \arg\min_{s\in\mathcal{S}_d} |u - s|_{1,2,\Omega}$. Then, we have

$$
\begin{aligned}
\mu|u_{best} - u_S|^2_{1,2,\Omega} &\le a(u_{best} - u_S, u_{best} - u_S) \\
&= a(u_{best} - u, u_{best} - u_S) \\
&\le M|u_{best} - u|_{1,2,\Omega}\,|u_{best} - u_S|_{1,2,\Omega} \\
\Rightarrow \mu|u_{best} - u_S|_{1,2,\Omega} &\le M|u_{best} - u|_{1,2,\Omega} \\
\mu|u - u_S|_{1,2,\Omega} &\le \mu|u - u_{best}|_{1,2,\Omega} + \mu|u_{best} - u_S|_{1,2,\Omega} \\
&\le \mu|u - u_{best}|_{1,2,\Omega} + M|u_{best} - u|_{1,2,\Omega} \\
\Rightarrow \qquad |u - u_S|_{1,2,\Omega} &\le \frac{\mu + M}{\mu}|u - u_{best}|_{1,2,\Omega} \le \frac{\mu + M}{\mu} C|u|_{d+1,2,\Omega}|\Delta|^d.
\end{aligned}
$$

These complete the proof.

We next explain the convergence in L_2 norm. When Ω is a convex domain, the convergence rate $||u - u_{best}||_{L_2(\Omega)}$ should be optimal based on a generalization of the well-known Aubin–Nitsche technique (cf. [8]) for Poisson equation. That is, we have

Theorem 8 *Suppose that the conditions of Theorem 7 are satisfied, and further suppose that the underlying Lipschitz domain Ω is convex. Let u_S be the weak solution of (1). Then for $d \ge 1$,*

$$
\|u - u_S\|_{L_2(\Omega)} \le C|\Delta|^{d+1}|u|_{d+1,2,\Omega} \tag{54}
$$

for a positive constant C independent of u.

Proof For $u - u_S \in L_2(\Omega)$, we can find the weak solution $w \in H_0^1(\Omega)$ satisfying

$$a(v, w) = \langle u - u_S, v\rangle, \quad \forall v \in H_0^1(\Omega). \tag{55}$$

Indeed, let $\hat{a}(u, v) = a(v, u)$ be a new bilinear form. By using the same proof of Theorem 3, we can show $\hat{a}(u, v)$ is a bounded bilinear form and $\hat{a}(u, u)$ is coercive since $\hat{a}(u, u) = a(u, u)$. By the Lax–Milgram theorem, there exists a weak solution w satisfying (55). It is known that $w \in H^2(\Omega)$ when Ω is convex (cf. [13]) and satisfies $|w|_{2,2,\Omega} \leq C\|u - u_S\|_{L^2(\Omega)}$ for a positive constant $C > 0$ independent of u and u_S.

Thus, we use (53) with an appropriate $v \in \mathscr{S}_d$,

$$\begin{aligned}
\|u - u_S\|^2_{L_2(\Omega)} &= \langle u - u_S, u - u_S\rangle = a(u - u_S, w) \\
&= a(u - S, w - v) \leq M|u - u_S|_{1,2,\Omega}\ |w - v|_{1,2,\Omega} \\
&\leq MC|\Delta|^d|u|_{d+1,2,\Omega}C|\Delta||w|_{2,2,\Omega} \\
&\leq MC|\Delta|^{d+1}|u|_{d+1,2,\Omega}C\|u - u_S\|_{L_2(\Omega)},
\end{aligned}$$

for positive constants C which are different in different lines. It now follows that

$$\|u - u_S\|_{L_2(\Omega)} \leq C|\Delta|^{d+1}|u|_{d+1,2,\Omega}$$

for another positive constant C. This completes the proof.

4 Numerical Results

4.1 Description of Our Implementation

In this section, we explain our implementation to numerically solve general second-order elliptic PDEs. It is an adaptation of the method detailed in [12] which uses polygonal splines to solve Poisson equations, which is in turn based on the method in [1], which details how to solve both linear and nonlinear PDEs with multivariate splines.

Our goal will be to solve for a vector of coefficients $\mathbf{u}$. We can begin in the same place as in [12], first constructing a matrix H to determine continuity conditions. If two polygons share an edge, then there are $d + 1$ pairs of elements supported on that edge between the two polygons, and each pair share values on the edge. Hence we force their respective coefficients to match, resulting in a linear system summarized by $H\mathbf{u} = 0$. We can similarly represent our boundary conditions by a linear system $B\mathbf{u} = \mathbf{g}$.

An important difference arises from here: We will need to form a different "stiffness" matrix than in the simpler Poisson case. In particular, in $\mathbb{R}^D$, using degree d polygonal splines, the new left-hand side of the weak form of the problem can be simplified to the following:

$$\sum_{P_n\in\Delta}\sum_{k=1}^{dn} u_k\left[\sum_{i,j=1}^{D}\int_{P_n} A_{ij}\frac{\partial v}{\partial x_i}\frac{\partial L_k}{\partial x_j}+\int_{P_n} cvL_k\right]$$

where we have expressed $u \approx u_S = \sum_{k=1}^{dn} u_k L_k$ for some coefficients u_k where L_k is an ordering of the interpolatory basis of $\mathscr{S}_d(\Delta)$ (which, when restricted to the domain P_n, is simply $\Psi_d(P_n)$). Similarly, write $f \approx f_S = \sum_{k=1}^{dn} f_k L_k$ and notice that the right-hand side of the weak form will be equal to $\sum_{P_n\in\Delta}\sum_{k=1}^{dn} f_k \int_{P_n} vL_k$ for any $v \in \mathscr{S}_d(\Delta)\cap H_0^1(\Omega)$. Hence, it must be true for $v = L_m$ for $m = 1, 2, ..., dn$. We can thus construct the following matrices:

$$M = [M_{P_n}]_{P_n\in\Delta} \text{ where } M_{P_n} = \left(M_{P_n,p,q}\right)_{p,q=1}^{D}, \text{ and } M_{P_n,p,q} = \int_{P_n} L_p L_q\ ;$$

$$\mathscr{K} = [\mathscr{K}_{P_n}]_{P_n\in\Delta} \text{ where } \mathscr{K}_{P_n} = \sum_{i,j=1}^{D}\mathscr{K}_{P_n}^{i,j} \text{ and}$$

$$\mathscr{K}_{P_n}^{i,j} = \left(\mathscr{K}_{P_n,p,q}^{i,j}\right)_{p,q=1}^{D} \text{ where } \mathscr{K}_{P_n,p,q}^{i,j} = \int_{P_n} A_{ij}\frac{\partial L_p}{\partial x_i}\frac{\partial L_q}{\partial x_j}\ ;$$

$$\mathscr{M}_{P_n} = \left(\mathscr{M}_{P_n,p,q}\right)_{p,q=1}^{D} \text{ where } \mathscr{M}_{P_n,p,q} = \int_{P_n} CL_pL_q\ ;$$

$$K = [\mathscr{K}_{P_n}]_{P_n\in\Delta} + [\mathscr{M}_{P_n}]_{P_n\in\Delta};\quad \mathbf{u} = (u_k)_{k=1}^{D}\ ;\quad \mathbf{f} = (f_k)_{k=1}^{D}\ ;$$

where the integrals are computed by first decomposing each polygon into quadrilaterals, and then using the tensor product of the Gauss quadrature formula of high order, say order 5 × 5, on each quadrilateral.

Then, notice that we can rewrite our weak formulation as

$$K\mathbf{u} = M\mathbf{f}.$$

Our minimization in (24) can be recast in terms of polygonal splines as

$$\min_{\substack{\mathbf{u}\\ H\mathbf{u}=0,\,B\mathbf{u}=\mathbf{g}}} \frac{1}{2}\mathbf{u}^T K\mathbf{u} - \mathbf{f}^T M\mathbf{u}$$

which is a constrained minimization problem which can be solved using the well-known iterative method described in [1]. We have implemented the computational scheme in MATLAB and experimented with many second-order elliptic PDEs. Some numerical results will be shown in the next section. Some of these results also involve first derivatives; these are implemented as another stiffness matrix added to K: First, we define

$$\mathcal{J}_{P_n} = \sum_{k=1}^{2} \mathcal{J}_{P_n}^k$$

where $\mathcal{J}_{P_n}^k = \left(\mathcal{J}_{P_n,p,q}^k\right)_{p,q=1}^D$ and $\mathcal{J}_{P_n,p,q}^k = \int_{P_n} B_k L_p \frac{\partial L_q}{\partial x_k}$ where B_k is the appropriate coefficient function. Then, we instead use $K = [\mathcal{K}_{P_n}]_{P_n \in \Delta} + [\mathcal{J}_{P_n}]_{P_n \in \Delta} + [\mathcal{M}_{P_n}]_{P_n \in \Delta}$.

It is worth mentioning that other finite element methods accommodate continuity conditions directly rather than solving a linear system, which saves computational power and time. However, our approach is designed conveniently to implement more complex continuity conditions; see Example 7.

4.2 Numerical Results of Elliptic PDEs

This section is divided into two parts. In the first part, we demonstrate the power of polygonal splines as a tool for numerically solving some general second-order elliptic PDEs. In the second part, we show the potential to use these numerical solutions to approximate the solutions of parabolic equations and hyperbolic equations. In all the following examples, we denote by u_S the spline solution, and by u the exact solution. To approximate the L^2 error, we report the root-mean-squared (RMS) error $E_{RMS} = ||u - u_S||_{RMS}$ of the spline solution based on 1001×1001 equally spaced points over Ω. Since $\nabla(u - u_S) = \left(\frac{\partial}{\partial x}(u - u_S), \frac{\partial}{\partial y}(u - u_S)\right)$, we report the RMS error $\nabla E_{RMS} = ||\nabla(u - u_S)||_{RMS}$, which is the average of the RMS error of $\frac{\partial}{\partial x}(u - u_S)$ and $\frac{\partial}{\partial y}(u - u_S)$. We also report the mesh size (i.e., the longest edge length) of the partition at each iteration and the computed rate of convergence in reference to the mesh size—in light of Theorems 7 and 8, we expect the rates to be 4 for degree 3 in the L^2 norm, 3 for degree 2 in the L^2 norm and degree 3 in the H^1 norm, and 2 for degree 2 in the H^1 norm.

Let us begin with numerical solutions of some standard second-order elliptic PDEs.

Example 2 We return to Example 3 on the unit square $\Omega = (0, 1) \times (0, 1)$ to demonstrate convergence of the method. We will set $\varepsilon = 10^{-5}$ and choose f and g so that

$$u(x, y) = \frac{(1 + x)^2}{4} \sin(2\pi xy) \tag{56}$$

is the exact solution. We use the following polygonal partition shown in Fig. 3.

We employ our polygonal spline method to solve (1) with exact solution in (56). Our numerical results are shown below in Tables 1 and 2.

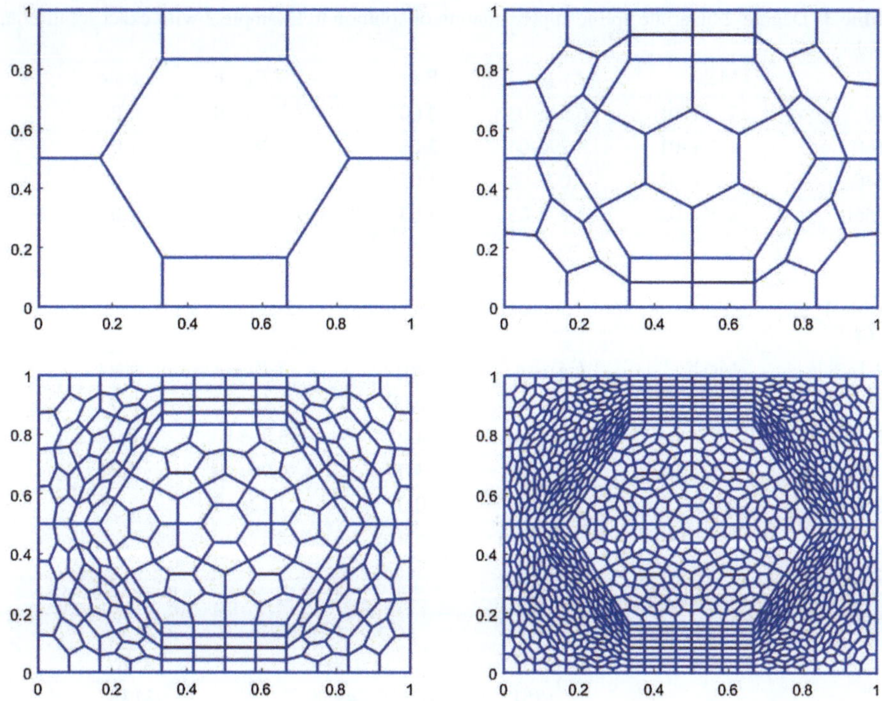

Fig. 3 A partition of the unit square and a few refinements

Table 1 Degree-2 polygonal spline approximation of solution to Example 2 with exact solution in (56)

# P	Mesh	E_{RMS}	Rate	∇E_{RMS}	Rate
39	2.50e-01	5.47e-03	0.00	1.34e-01	0.00
219	1.25e-01	4.16e-04	3.72	2.60e-02	2.36
1251	6.25e-02	3.68e-05	3.50	5.17e-03	2.33
7251	3.13e-02	3.29e-06	3.48	1.03e-03	2.33

Table 2 Degree-3 polygonal spline approximation of solution to Example 2 with exact solution in (56)

# P	Mesh	E_{RMS}	Rate	∇E_{RMS}	Rate
39	2.50e-01	9.01e-04	0.00	1.61e-02	0.00
219	1.25e-01	2.74e-05	5.04	1.41e-03	3.52
1251	6.25e-02	1.25e-06	4.45	1.31e-04	3.43
7251	3.13e-02	6.92e-08	4.18	1.22e-05	3.43

Table 3 Degree-2 bivariate spline approximation of solution to Example 2 with exact solution in (56)

# T	Mesh	E_{RMS}	Rate	∇E_{RMS}	Rate
40	3.54e-01	6.94e-03	0.00	2.09e-01	0.00
160	1.77e-01	8.29e-04	3.06	5.38e-02	1.96
640	8.84e-02	1.00e-04	3.05	1.34e-02	2.00
2560	4.42e-02	1.22e-05	3.03	3.32e-03	2.01

Table 4 Degree-3 bivariate spline approximation of solution to Example 2 with exact solution in (56)

# T	Mesh	E_{RMS}	Rate	∇E_{RMS}	Rate
40	3.54e-01	7.16e-04	0.00	3.15e-02	0.00
160	1.77e-01	4.43e-05	4.01	3.96e-03	2.99
640	8.84e-02	5.33e-06	3.06	4.96e-04	3.00
2560	4.42e-02	5.52e-06	–0.05	7.22e-05	2.78

Table 5 Polygonal splines' degrees of freedom

# P	DoF ($d = 2$)	DoF ($d = 3$)
39	179	313
219	886	1657
1251	4958	9313
7251	28654	53857

The numerical results in Tables 1 and 2 show that the polygonal spline method works very well. We shall compare with the solution using degree-2 and degree-3 bivariate splines triangulation of the same domain. We chose a grid-based initial triangulation with close to the same number of elements as our initial polygonal partition.

From Tables 1, 2, 3, and 4, we can see that polygonal splines can produce a more accurate solution on polygonal partitions containing a similar number of polygons as a triangulation of the same domain.

It is worth noting the difference in degrees of freedom in this example. In particular, the polygonal splines have significantly more degrees of freedom than each iteration of triangular spline. However, this does not seem representative in general—for example, one could imagine removing edges from a triangulation to form a polygonal partition of the same domain while reducing the number of degrees of freedom. Regardless, there is no doubt that our polygonal spline methods are more numerically taxing than traditional bivariate spline methods. At each iteration, we have the following numbers of degrees of freedom (Tables 5 and 6).

As we will use the same partitions for each example in this paper, the reader can refer back to these tables.

Table 6 Bivariate splines' degrees of freedom

# T	DoF ($d = 2$)	DoF ($d = 3$)
40	97	205
160	353	769
640	1345	2977
2560	5249	11713

Table 7 Degree-2 polygonal spline approximation of solution to Example 3 with exact solution (58)

# P	Mesh	E_{RMS}	Rate	∇E_{RMS}	Rate
39	2.50e-01	1.30e-04	0.00	3.82e-03	0.00
219	1.25e-01	1.08e-05	3.59	7.47e-04	2.35
1251	6.25e-02	1.09e-06	3.30	1.54e-04	2.28
7251	3.13e-02	1.57e-07	2.80	3.40e-05	2.18

Example 3 Here is another example of an elliptic second-order PDE: Let

$$\begin{bmatrix} A_{11} & A_{12} \\ A_{21} & A_{22} \end{bmatrix} = \begin{bmatrix} 1+\varepsilon & 1 \\ 1 & 1+\varepsilon \end{bmatrix}$$

for some $\varepsilon > 0$, and let $C = 1$, and solve the PDE given by

$$-\sum_{i,j=1}^{2} \frac{\partial}{\partial x_j}\left(A_{ij}\frac{\partial u}{\partial x_i}\right) + \frac{\partial u}{\partial x_1} + \frac{\partial u}{\partial x_2} + Cu = f \text{ in } \Omega; \qquad u = g \text{ on } \partial\Omega. \tag{57}$$

To test our method, we shall choose f and g so that

$$u = (1 + x^2 + y^2)^{-1} \tag{58}$$

is the exact solution.

According to Corollary 1, this elliptic PDE has a unique weak solution. Although this PDE technically does not fit with our computational scheme in the previous section because the PDE involves the first-order derivatives, our minimization scheme (24) with additional first-order derivative terms still produces good results. In fact, we can even use $\varepsilon = 0$, which makes this PDE nonelliptic, and still produce good solutions. We use the same partition as in Example 2 to solve this PDE. Tables 7 and 8 show the results of our minimization, using the nonelliptic condition $\varepsilon = 0$.

Similarly, the minimization (24) with first-order derivatives based on bivariate splines can also produce good numerical results. For comparison, Tables 9 and 10 tabulate the results of the same computation using bivariate splines of degree 2 and degree 3 over grid-based right triangulations of the same domain.

Table 8 Degree-3 polygonal spline approximation of solution to Example 3 with exact solution (58)

# P	Mesh	E_{RMS}	Rate	∇E_{RMS}	Rate
39	2.50e-01	7.64e-06	0.00	2.41e-04	0.00
219	1.25e-01	3.70e-07	4.37	2.41e-05	3.32
1251	6.25e-02	1.99e-08	4.21	2.36e-06	3.35
7251	3.13e-02	1.25e-09	4.00	2.41e-07	3.29

Table 9 Degree-2 bivariate spline approximation of solution to Example 3 with exact solution in (58)

# T	Mesh	E_{RMS}	Rate	∇E_{RMS}	Rate
40	3.54e-01	4.80e-04	0.00	1.14e-02	0.00
160	1.77e-01	5.23e-05	3.20	2.70e-03	2.07
640	8.84e-02	6.21e-06	3.07	7.14e-04	1.92
2560	4.42e-02	8.53e-07	2.86	2.29e-04	1.64

Table 10 Degree-3 bivariate spline approximation of solution to Example 3 with exact solution in (58)

# T	Mesh	E_{RMS}	Rate	∇E_{RMS}	Rate
40	3.54e-01	2.43e-05	0.00	9.87e-04	0.00
160	1.77e-01	1.81e-06	3.75	1.44e-04	2.77
640	8.84e-02	1.29e-07	3.81	2.01e-05	2.84
2560	4.42e-02	9.69e-09	3.74	2.97e-06	2.76

4.3 Numerical Solutions of Parabolic and Hyperbolic PDEs

Example 4 We return again to Example (3) on the unit square $\Omega = (0, 1) \times (0, 1)$, but this time with $\varepsilon = 0$. We will choose f and g so that

$$u(x, y) = \frac{(1 + x)^2}{4} \sin(2\pi xy) \tag{59}$$

is the exact solution. Notice that, in this case, the PDE is not elliptic. However, our method still approximates the true solution quite well. We will show the convergence of our approximations for decreasing values of ε.

For comparison, let us show the results of the same PDE using bivariate splines over a triangulation of the same domain instead (Tables 11, 12, 13, 14, 15, 16, 17, 18, 19, 20, 21, and 22).

The numerical results in the tables above show that the polygonal spline method is efficient in approximating the solutions of nonelliptic PDEs.

Table 11 Degree-2 polygonal spline approximation of solution to Example 4 with exact solution in (59) and $\varepsilon = 10^{-3}$

# P	Mesh	E_{RMS}	Rate	∇E_{RMS}	Rate
39	2.50e-01	5.27e-03	0.00	1.33e-01	0.00
219	1.25e-01	6.41e-04	3.04	2.61e-02	2.35
1251	6.25e-02	5.54e-04	0.21	6.14e-03	2.09
7251	3.13e-02	5.56e-04	–0.00	3.71e-03	0.73

Table 12 Degree-3 polygonal spline approximation of solution to Example 4 with exact solution in (59) and $\varepsilon = 10^{-3}$

# P	Mesh	E_{RMS}	Rate	∇E_{RMS}	Rate
39	2.50e-01	1.39e-03	0.00	1.71e-02	0.00
219	1.25e-01	5.71e-04	1.29	3.70e-03	2.21
1251	6.25e-02	5.56e-04	0.04	3.55e-03	0.06
7251	3.13e-02	5.56e-04	–0.00	3.65e-03	–0.04

Table 13 Degree-2 polygonal spline approximation of solution to Example 4 with exact solution in (59) and $\varepsilon = 10^{-5}$

# P	Mesh	E_{RMS}	Rate	∇E_{RMS}	Rate
39	2.50e-01	5.47e-03	0.00	1.34e-01	0.00
219	1.25e-01	4.15e-04	3.72	2.60e-02	2.36
1251	6.25e-02	3.69e-05	3.49	5.17e-03	2.33
7251	3.13e-02	6.45e-06	2.52	1.03e-03	2.33

Table 14 Degree-3 polygonal spline approximation of solution to Example 4 with exact solution in (59) and $\varepsilon = 10^{-5}$

# P	Mesh	E_{RMS}	Rate	∇E_{RMS}	Rate
39	2.50e-01	9.06e-04	0.00	1.61e-02	0.00
219	1.25e-01	3.11e-05	4.86	1.41e-03	3.52
1251	6.25e-02	6.10e-06	2.35	1.37e-04	3.37
7251	3.13e-02	5.62e-06	0.12	4.29e-05	1.67

Table 15 Degree-2 polygonal spline approximation of solution to Example 4 with exact solution in (59) and $\varepsilon = 10^{-10}$

# P	Mesh	E_{RMS}	Rate	∇E_{RMS}	Rate
39	2.50e-01	5.47e-03	0.00	1.34e-01	0.00
219	1.25e-01	4.16e-04	3.72	2.60e-02	2.36
1251	6.25e-02	3.68e-05	3.50	5.17e-03	2.33
7251	3.13e-02	3.29e-06	3.48	1.03e-03	2.33

Table 16 Degree-3 polygonal spline approximation of solution to Example 4 with exact solution in (59) and $\varepsilon = 10^{-10}$

# P	Mesh	E_{RMS}	Rate	∇E_{RMS}	Rate
39	2.50e-01	9.01e-04	0.00	1.61e-02	0.00
219	1.25e-01	2.74e-05	5.04	1.41e-03	3.52
1251	6.25e-02	1.25e-06	4.45	1.31e-04	3.43
7251	3.13e-02	6.92e-08	4.18	1.22e-05	3.43

Table 17 Degree-2 bivariate spline approximation of solution to Example 4 with exact solution in (59) and $\varepsilon = 10^{-3}$

# T	Mesh	E_{RMS}	Rate	∇E_{RMS}	Rate
40	3.54e-01	6.91e-03	0.00	2.09e-01	0.00
160	1.77e-01	9.63e-04	2.84	5.38e-02	1.96
640	8.84e-02	5.55e-04	0.80	1.37e-02	1.97
2560	4.42e-02	5.53e-04	0.00	4.68e-03	1.55

Table 18 Degree-3 bivariate spline approximation of solution to Example 4 with exact solution in (59) and $\varepsilon = 10^{-3}$

# T	Mesh	E_{RMS}	Rate	∇E_{RMS}	Rate
40	3.54e-01	7.40e-04	0.00	3.14e-02	0.00
160	1.77e-01	5.37e-04	0.46	4.99e-03	2.66
640	8.84e-02	5.52e-04	–0.04	3.36e-03	0.57
2560	4.42e-02	5.55e-04	–0.01	3.60e-03	–0.10

Table 19 Degree-2 bivariate spline approximation of solution to Example 4 with exact solution in (59) and $\varepsilon = 10^{-5}$

# T	Mesh	E_{RMS}	Rate	∇E_{RMS}	Rate
40	3.54e-01	6.94e-03	0.00	2.09e-01	0.00
160	1.77e-01	8.29e-04	3.06	5.38e-02	1.96
640	8.84e-02	1.00e-04	3.05	1.34e-02	2.00
2560	4.42e-02	1.32e-05	2.92	3.32e-03	2.01

Table 20 Degree-3 bivariate spline approximation of solution to Example 4 with exact solution in (59) and $\varepsilon = 10^{-5}$

# T	Mesh	E_{RMS}	Rate	∇E_{RMS}	Rate
40	3.54e-01	7.16e-04	0.00	3.15e-02	0.00
160	1.77e-01	4.43e-05	4.01	3.96e-03	2.99
640	8.84e-02	5.33e-06	3.06	4.96e-04	3.00
2560	4.42e-02	5.52e-06	–0.05	7.22e-05	2.78

Table 21 Degree-2 bivariate spline approximation of solution to Example 4 with exact solution in (59) and $\varepsilon = 10^{-10}$

# T	Mesh	E_{RMS}	Rate	∇E_{RMS}	Rate
40	3.54e-01	6.94e-03	0.00	2.09e-01	0.00
160	1.77e-01	8.29e-04	3.06	5.38e-02	1.96
640	8.84e-02	1.00e-04	3.05	1.34e-02	2.00
2560	4.42e-02	1.22e-05	3.03	3.32e-03	2.01

Table 22 Degree-3 bivariate spline approximation of solution to Example 4 with exact solution in (59) and $\varepsilon = 10^{-10}$

# T	Mesh	E_{RMS}	Rate	∇E_{RMS}	Rate
40	3.54e-01	7.18e-04	0.00	3.15e-02	0.00
160	1.77e-01	4.59e-05	3.97	3.96e-03	2.99
640	8.84e-02	2.93e-06	3.97	4.95e-04	3.00
2560	4.42e-02	1.85e-07	3.98	6.19e-05	3.00

Table 23 Weak Galerkin approximation of solution to Example 5

# Poly's	Mesh	$\|\|u - u_{WG}\|\|_{L^2}$	Rate	$\|\|\nabla u - \nabla u_{WG}\|\|_{H^1}$	Rate
64	1.25e-01	1.46e-03	0.00	2.52e-02	0.00
256	6.25e-02	3.74e-04	1.96	1.23e-02	9.98e-01
1024	3.13e-02	9.47e-05	1.98	6.31e-03	9.98e-01
4096	1.56e-02	2.39e-05	1.99	3.16e-03	9.98e-01

Example 5 Let

$$\begin{bmatrix} A_{11} & A_{12} \\ A_{21} & A_{22} \end{bmatrix} = \begin{bmatrix} xy & 0 \\ 0 & xy \end{bmatrix}$$

and $C = 0$. Choose f and g so that

$$u = x(1 - x)y(1 - y) \tag{60}$$

is the exact solution. This was studied in [19]. As in Example 4, this is a "nearly elliptic" PDE, but with some degeneracy at the origin. We shall use a different partition of the unit square this time, simply using a uniform grid of squares, as was the case in the original paper [19]. The weak Galerkin method presented in this paper retrieved the results shown in Table 23.

We use our method with polygonal splines to solve the PDE above and find that our method can produce much better results.

Comparison of Tables 23, 24, and 25 shows that our polygonal spline method produces a much more accurate solution. These results call for some remarks. First,

Table 24 Degree-2 polygonal spline approximation of solution to Example 5

# P	Mesh	E_{RMS}	Rate	∇E_{RMS}	Rate
64	1.25e-01	1.83e-06	0.00	1.39e-04	0.00
256	6.25e-02	9.85e-08	4.22	1.60e-05	3.12
1024	3.13e-02	5.65e-09	4.12	1.91e-06	3.07
4096	1.56e-02	3.42e-10	4.05	2.33e-07	3.04

Table 25 Degree-3 polygonal spline approximation of solution to Example 5

# P	Mesh	E_{RMS}	Rate	∇E_{RMS}	Rate
64	2.50e-01	3.59e-12	0.00	5.63e-11	0.00
256	1.25e-01	1.40e-11	–1.96	2.07e-10	–1.88
1024	6.25e-02	2.34e-11	–0.74	2.71e-10	–0.39
4096	3.13e-02	4.47e-11	–0.93	5.61e-10	–1.05

Table 26 Degree-2 polygonal spline approximation of solution to Example 5 over nongrid partition

# P	Mesh	E_{RMS}	Rate	∇E_{RMS}	Rate
39	2.50e-01	3.66e-05	0.00	1.19e-03	0.00
219	1.25e-01	3.09e-06	3.57	2.29e-04	2.38
1251	6.25e-02	2.75e-07	3.49	4.59e-05	2.32

Table 27 Degree-3 polygonal spline approximation of solution to Example 5 over nongrid partition

# P	Mesh	E_{RMS}	Rate	∇E_{RMS}	Rate
39	2.50e-01	3.10e-06	0.00	9.17e-05	0.00
219	1.25e-01	1.22e-07	4.66	7.07e-06	3.70
1251	6.25e-02	4.59e-09	4.74	5.96e-07	3.57

it is worth pointing out that our MATLAB code can only achieve $1e-11$ accuracy. In Table 25, the rates of convergence become negative due to round-off errors. That is, polygonal splines of degree-3 converged to the solution *virtually instantly*. Some improved performance has been observed before when the solution is a polynomial, but we would not normally expect immediate retrieval of the solution of a degree-4 polynomial using only degree-3 splines. The degree-2 splines also appear to have an increased order of convergence $O(h^4)$. We are interested in why the performance has been increased here, and our investigation seems to show that the partition plays a role. If we run a few iterations to solve the same problem over the unit square based on the partition from Example 2, we retrieve the following results shown in Tables 26 and 27.

Table 28 Degree-2 polygonal spline approximation of solution to Example 6 with $\varepsilon = 10^{-3}$

# P	Mesh	E_{RMS}	Rate	∇E_{RMS}	Rate
39	2.50e-01	1.28e-03	0.00	5.58e-02	0.00
219	1.25e-01	4.22e-04	1.60	2.38e-02	1.23
1251	6.25e-02	4.04e-04	0.07	2.28e-02	0.06
7251	3.13e-02	3.99e-04	0.02	2.15e-02	0.09

Table 29 Degree-3 polygonal spline approximation of solution to Example 6 with $\varepsilon = 10^{-3}$

# P	Mesh	E_{RMS}	Rate	∇E_{RMS}	Rate
39	2.50e-01	4.62e-04	0.00	2.02e-02	0.00
219	1.25e-01	4.07e-04	0.18	2.24e-02	–0.15
1251	6.25e-02	4.00e-04	0.02	2.18e-02	0.04
7251	3.13e-02	3.99e-04	0.01	1.81e-02	0.26

Table 30 Degree-2 polygonal spline approximation of solution to Example 6 with $\varepsilon = 10^{-5}$

# P	Mesh	E_{RMS}	Rate	∇E_{RMS}	Rate
39	2.50e-01	1.83e-03	0.00	7.70e-02	0.00
219	1.25e-01	2.97e-04	2.62	3.03e-02	1.35
1251	6.25e-02	4.51e-05	2.72	1.25e-02	1.28
7251	3.13e-02	6.32e-06	2.84	3.67e-03	1.77

We can see that this time the numerical solutions are closer to the expected rate of convergence. Thus, the grid partition plays a role to the solution of this problem. We would like to invite the interested reader to investigate it further.

Example 6 Consider the following example:

$$-\varepsilon \Delta u + (2 - y^2) D_x u + (2 - x) D_y u + (1 + (1 + x)(1 + y)^2) u = f, (x, y) \in \Omega \tag{61}$$

with $\Omega = (0, 1)^2$, and $u|_{\partial\Omega} = g$. The function f is so chosen that the exact solution is

$$u(x, y) = 1 + \sin(\pi(1 + x)(1 + y)^2/8).$$

When $\varepsilon = 0$, this is a hyperbolic test problem considered in [5, 14, 15]. However, for positive values of ε, this is an elliptic PDE. We can well approximate a solution to the hyperbolic problem by using very small positive values of ε.

For comparison, here are the results of the same computation using bivariate splines over a triangulation of the same domain (Tables 28, 29, 30, 31, 32, 33, 34, 35, 36, 37, 38, 39, and 40).

Table 31 Degree-3 polygonal spline approximation of solution to Example 6 with $\varepsilon = 10^{-5}$

# P	Mesh	E_{RMS}	Rate	∇E_{RMS}	Rate
39	2.50e-01	3.71e-05	0.00	2.50e-03	0.00
219	1.25e-01	6.66e-06	2.48	8.58e-04	1.54
1251	6.25e-02	5.19e-06	0.36	1.42e-03	–0.72
7251	3.13e-02	4.36e-06	0.25	2.02e-03	–0.51

Table 32 Degree-2 polygonal spline approximation of solution to Example 6 with $\varepsilon = 10^{-10}$

# P	Mesh	E_{RMS}	Rate	∇E_{RMS}	Rate
39	2.50e-01	1.84e-03	0.00	7.73e-02	0.00
219	1.25e-01	3.05e-04	2.59	3.10e-02	1.32
1251	6.25e-02	5.26e-05	2.54	1.43e-02	1.12
7251	3.13e-02	8.46e-06	2.63	5.60e-03	1.35

Table 33 Degree-3 polygonal spline approximation of solution to Example 6 with $\varepsilon = 10^{-10}$

# P	Mesh	E_{RMS}	Rate	∇E_{RMS}	Rate
39	2.50e-01	3.49e-05	0.00	2.37e-03	0.00
219	1.25e-01	2.00e-06	4.13	3.46e-04	2.78
1251	6.25e-02	1.24e-07	4.01	5.26e-05	2.72
7251	3.13e-02	1.84e-08	2.75	1.88e-05	1.49

Table 34 Degree-2 bivariate spline approximation of solution to Example 6 with $\varepsilon = 10^{-3}$

# T	Mesh	E_{RMS}	Rate	∇E_{RMS}	Rate
40	3.54e-01	1.52e-04	0.00	4.07e-03	0.00
160	1.77e-01	4.58e-05	1.73	1.70e-03	1.26
640	8.84e-02	2.87e-05	0.67	1.13e-03	0.58
2560	4.42e-02	2.67e-05	0.10	1.13e-03	–0.00

Table 35 Degree-3 bivariate spline approximation of solution to Example 6 with $\varepsilon = 10^{-3}$

# T	Mesh	E_{RMS}	Rate	∇E_{RMS}	Rate
40	3.54e-01	3.08e-05	0.00	7.95e-04	0.00
160	1.77e-01	2.82e-05	0.13	1.03e-03	–0.37
640	8.84e-02	2.69e-05	0.07	1.13e-03	–0.13
2560	4.42e-02	2.66e-05	0.02	1.21e-03	–0.10

Table 36 Degree-2 bivariate spline approximation of solution to Example 6 with $\varepsilon = 10^{-5}$

# T	Mesh	E_{RMS}	Rate	∇E_{RMS}	Rate
40	3.54e-01	1.58e-04	0.00	4.64e-03	0.00
160	1.77e-01	3.93e-05	2.01	2.21e-03	1.07
640	8.84e-02	9.81e-06	2.00	1.08e-03	1.04
2560	4.42e-02	2.40e-06	2.03	5.04e-04	1.09

Table 37 Degree-3 bivariate spline approximation of solution to Example 6 with $\varepsilon = 10^{-5}$

# T	Mesh	E_{RMS}	Rate	∇E_{RMS}	Rate
40	3.54e-01	5.36e-06	0.00	2.65e-04	0.00
160	1.77e-01	6.10e-07	3.13	5.66e-05	2.23
640	8.84e-02	3.13e-07	0.96	3.86e-05	0.55
2560	4.42e-02	2.99e-07	0.07	6.72e-05	–0.80

Table 38 Degree-2 bivariate spline approximation of solution to Example 6 with $\varepsilon = 10^{-10}$

# T	Mesh	E_{RMS}	Rate	∇E_{RMS}	Rate
40	3.54e-01	1.58e-04	0.00	4.65e-03	0.00
160	1.77e-01	3.94e-05	2.01	2.22e-03	1.06
640	8.84e-02	9.90e-06	1.99	1.10e-03	1.01
2560	4.42e-02	2.46e-06	2.01	5.48e-04	1.01

Table 39 Degree-3 bivariate spline approximation of solution to Example 6 with $\varepsilon = 10^{-10}$

# T	Mesh	E_{RMS}	Rate	∇E_{RMS}	Rate
40	3.54e-01	5.43e-06	0.00	2.67e-04	0.00
160	1.77e-01	5.68e-07	3.26	5.57e-05	2.26
640	8.84e-02	6.93e-08	3.03	1.42e-05	1.97
2560	4.42e-02	8.55e-09	3.02	3.40e-06	2.06

We can see that the polygonal spline solutions approximate the exact solution very well. However, as in (59), we see that this PDE has a unique weak solution but does not satisfy the assumptions of Theorem 5. Nevertheless, our method works well as shown in Tables 36 and 37.

Example 7 For another example, the following problem is parabolic for $y > 0$ and hyperbolic for $y \leq 0$:

$$
\begin{aligned}
-\varepsilon D_{yy}u + D_x u + c_1 u = 0, &\quad (x, y) \in (-1, 1) \times (0, 1) \\
D_x u + c_2 u = 0, &\quad (x, y) \in (-1, 1) \times (-1, 0]
\end{aligned}
\tag{62}
$$

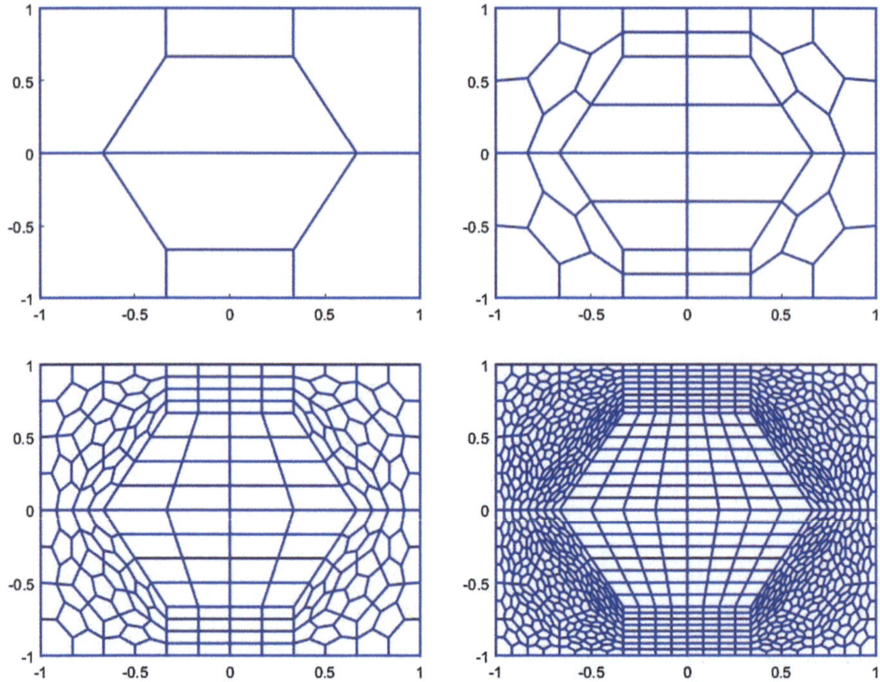

Fig. 4 A partition of $\Omega = [-1, 1]^2$ and a few refinements

with $u|_{\partial\Omega} = g$, for any constants $c_1 > 0$ and $c_2 > 0$. This PDE was studied in [15]. Note that the solution is discontinuous at $y = 0$. We can solve the following general elliptic PDE to estimate the solution to this problem:

$$\begin{aligned} -\eta D_{xx}u - \varepsilon D_{yy}u + D_x u + c_1 u &= 0, \quad (x, y) \in (-1, 1) \times (0, 1) \\ -\eta \Delta u + D_x u + c_2 u &= 0, \quad (x, y) \in (-1, 1) \times (-1, 0] \end{aligned} \tag{63}$$

with $u|_{\partial\Omega} = g$ and $\eta > 0$. We can approximate the solution to (62) by letting $\eta > 0$ go to zero and use spline functions which are not necessarily continuous at $y = 0$. Let the exact solution, $u(x, y)$, of (62) be the following piecewise function:

$$\begin{array}{ll} \sin(\pi(1 + y)/2)\exp(-(c_1 + \varepsilon\pi^2/4)(1 + x)), & -1 \le x \le 1, 0 \le y \le 1 \\ \sin(\pi(1 + y)/2)\exp(-c_2(1 + x)), & -1 \le x \le 1, -1 < y \le 0. \end{array} \tag{64}$$

We set $\varepsilon = 0.05$ and use a similar partition to the one from Example 2, scaled to cover the domain $\Omega = [-1, 1]^2$ and with an added edge to account for the discontinuity at $y = 0$. See Fig. 4

Table 40 Degree-2 polygonal spline approximation of solution to (62) with exact solution (64) when $\eta = 10^{-10}, c_1 = c_2 = 0.1$

# P	Mesh	E_{RMS}	Rate	∇E_{RMS}	Rate
40	6.67e-01	6.80e-03	0.00	2.13e-01	0.00
208	3.33e-01	2.45e-03	1.46	2.10e-01	0.02
1120	1.67e-01	1.15e-03	1.10	2.03e-01	0.05
6208	8.33e-02	4.98e-04	1.21	1.76e-01	0.20

Table 41 Degree-2 polygonal spline approximation of solution to (62) with exact solution (64) when $\eta = 10^{-10}, C_1 = 0.1, c_2 = c_1 + \varepsilon\pi^2/4$

# P	Mesh	E_{RMS}	Rate	∇E_{RMS}	Rate
40	6.67e-01	1.64e-03	0.00	2.82e-02	0.00
208	3.33e-01	2.61e-04	2.65	1.03e-02	1.45
1120	1.67e-01	3.86e-05	2.76	3.60e-03	1.52
6208	8.33e-02	5.68e-06	2.76	1.23e-03	1.55

Table 42 Degree-2 polygonal spline approximation of solution to (62) with exact solution (64) when $\eta = 10^{-10}, C_1 = 0.1, c_2 = c_1 + \varepsilon\pi^2/4$

# P	Mesh	E_{RMS}	Rate	∇E_{RMS}	Rate
40	6.67e-01	1.65e-03	0.00	2.62e-02	0.00
208	3.33e-01	2.48e-04	2.73	8.87e-03	1.56
1120	1.67e-01	3.80e-05	2.71	3.33e-03	1.42
6208	8.33e-02	5.65e-06	2.75	1.20e-03	1.47

Although this PDE does not technically fit our computational scheme due to the inclusion of first-order terms, we can still get a fairly good estimate of the true solution using degree-2 polygonal splines. Numerical results are shown in Table 40.

If we change the value of c_2 to $0.1 + \varepsilon\pi^2/4$, so that the solution is continuous, we retrieve the following results (without forcing continuity over the line $y = 0$). See Table 41.

Enforcing continuity over the line $y = 0$ leads to the results in Table 42.

From Tables 41 and 42, we can see that the computational results are very similar.

References

1. G. Awanou, M.-J. Lai, P. Wenston, The multivariate spline method for scattered data fitting and numerical solution of partial differential equations, in *Wavelets and Splines: Athens* (2006), pp. 24–74
2. L. Beirao da Veiga, F. Brezzi, A. Cangiani, G. Manzini, L. Marini, A. Russo, Basic principles of virtual element methods. Math. Models Methods Appl. Sci. **23** (2013)
3. L. Beirao da Veiga, K. Lipnikov, G. Manzini, Arbitrary-order nodal mimetic discretizations of elliptic problems on polygonal meshes. SIAM J. Numer. Anal. **49**, 1737–1760 (2011)
4. L. Beirao da Veiga, G. Manzini, A virtual element method with arbitrary regularity. IMA J. Numer. Anal. **34**, 759–781 (2014)
5. K. Bey, J. Oden, hp-version discontinuous galerkin methods for hyperbolic conservation laws. Comput. Methods Appl. Mech. Eng. **133**, 259–286 (1996)
6. D. Braess, *Finite Elements* (Cambridge University Press, Cambridge, 1997)
7. S. Brennet, L. Scott, *The Mathematical Theory of Finite Element Methods* (Springer, Berlin, 1994)
8. P. Ciarlet, *The Finite Element Method for Elliptic Problems* (North-Holland, 1978)
9. L. Evans, *Partial Differential Equations* (American Math. Society, 1998)
10. M. Floater, Generalized barycentric coordinates and applications. Acta Numerica **24**, 161–214 (2015)
11. M. Floater, A. Gillette, N. Sukumar, Gradient bounds for wachspress coordinates on polytopes. SIAM J. Numer. Anal. **52**, 515–532 (2014)
12. M. Floater, M.-J. Lai, Polygonal spline spaces and the numerical solution of the poisson equation. SIAM J. Numer. Anal. **54**, 797–824 (2016)
13. P. Grisvard, *Elliptic Problems in Nonsmooth Domains* (Pitman Advanced Pub. Program, 1985)
14. P. Houston, C. Schwab, E. Suli, Stabilized hp-finite element methods for first-order hyperbolic problems. SIAM J. Numer. Anal. **37**, 1618–1643 (2000)
15. P. Houston, C. Schwab, E. Suli, Discontinuous hp-finite element methods for advection-diffusion problems. SIAM J. Numer. Anal. **39**, 2133–2163 (2002)
16. M.-J. Lai , L. Schumaker, *Spline Functions Over Triangulations* (Cambridge University Press, Cambridge, 2007)
17. G. Manzini, A. Russo, N. Sukumar, New perspectives on polygonal and polyhedral finite element methods. Math. Models Methods Appl. Sci **24**, 1665–1699 (2014)
18. L. Mu, J. Wang, Y. Wang, X. Ye, A computational study of the weak galerkin method for second-order elliptic equations. Numer. Algorithms **63**, 753–777 (2013)
19. L. Mu, J. Wang, X. Ye, Weak galerkin finite element methods on polytopal meshes. Int. J. Numer. Anal. **12**, 31–53 (2015)
20. A. Rand, A. Gillette, C. Bajaj, Quadratic serendipity finite elements on polygons using generalized barycentric coordinates. Math. Comput. **83**, 2691–2716 (2014)
21. L. Schumaker, *Spline Functions: Computational Methods* (SIAM, 2015)
22. E. Stein, *Singular Integrals and Differentiability Properties of Functions* (Princeton University Press, 1970)

An Adaptive Triangulation Method for Bivariate Spline Solutions of PDEs

Ming-Jun Lai and Clayton Mersmann

Abstract We report numerical performance of our adaptive triangulation algorithms to improve numerical solutions of PDEs using bivariate spline functions. Our ultimate goal is to find a PDE-solution-dependent triangulation which improves both the accuracy and computational efficiency of the spline solution. We present little theory to guide our search for such a triangulation, but instead approach the problem numerically. Starting with some initial triangulation $\triangle$, we use the gradient values of the spline solution based on $\triangle$ to generate an updated triangulation $\triangle'$ and compute a new spline solution. We consider both refining and coarsening the initial triangulation $\triangle$ in order to make the spline solution more effective. As we add vertices to and remove vertices from the vertex set, we use a global retriangulation instead of local refinement techniques. We introduce a new concept of mesh efficiency to measure the effectiveness of a spline solution over a given triangulation. Extensive numerical experiments have been conducted and are summarized in this paper. In addition, we report a heuristic for generating an initial solution-dependent triangulation and show numerical evidence that this algorithm produces an initial triangulation which yields a better spline solution than one based on a uniform initial mesh.

Keywords Bivariate splines · Adaptive triangulation · PDEs

This research is partially supported by Simons collaboration grant 280646 and the National Science Foundation under the grant #DMS 1521537.

M.-J. Lai (✉) · C. Mersmann
Department of Mathematics, University of Georgia, Athens, GA 30602, USA
e-mail: mjlai@uga.edu

C. Mersmann
e-mail: mersmann@uga.edu

G.E. Fasshauer and L.L. Schumaker (eds.), *Approximation Theory XV: San Antonio 2016*, Springer Proceedings in Mathematics & Statistics 201,
DOI 10.1007/978-3-319-59912-0_7

1 Introduction

Besides the standard popular finite element method (cf. [9]), there are many other approaches available to solve boundary value problems of partial differential equations over an arbitrary polygonal domain. Among these are the bivariate spline method in [4, 17], the discontinuous Galerkin method (cf. [3]), the virtual element method (cf. [7]), the weak Galerkin method (cf. [22]), and etc. In this paper, we are interested in how we can solve PDEs more effectively when using the bivariate spline functions proposed in [4].

For PDEs whose solutions are difficult to approximate, uniform refinements lead to triangulations with huge numbers of triangles and unwieldy spline solutions. For these types of problems, it is necessary to use a PDE-solution-dependent triangulation to solve the PDE more effectively. Traditionally, researchers use adaptive triangulation methods to increase the accuracy of the solution. We refer to [2, 21] for classic adaptive methods for finite element solutions which are based on *a posteriori* error estimates. See [8, 11] for convergence analysis of adaptive triangulation methods based on linear and higher order finite element approximation. Also, [14] performs adaptive computations using splines over triangulations with hanging nodes, developed in [18], for numerical solution of PDEs. The goal of our research is to construct a PDE-solution-dependent triangulation over which the spline solutions exhibit better numerical performance. Our goals for improved performance can be stated as follows: (1) fixing the number of triangles in a triangulation, we seek a mesh which generates more accurate solutions; (2) fixing a level of accuracy, we seek suitable spline solutions over a smaller triangulation (i.e., fewer triangles); or (3) a suitable combination of (1) and (2).

In classic adaptive triangulation schemes, an error function is computed from an initial numerical solution to estimate the places in the domain where large errors may occur. The triangulation is then refined only in those areas. One common way to identify which parts of the domain should be refined is to compute the residual $e := |f - \mathscr{L}(u_s)|$ over each triangle where u_s is an initial approximate solution and $\mathscr{L}$ is the partial differential operator. In [6], error indicators based on energy and $||\cdot||_{1,p}$ norms of the residual are derived for FEM solutions; the main result of [5] establishes an estimation of those norms using jumps in the gradients of the numerical solution across element boundaries. Another approach, as described in [1, 23, 24], utilizes some post-processing of an initial finite element solution to produce a "recovered" gradient estimate; error indicators are then based on differences between these recovered gradients and the gradient computed from the initial solution.

The adaptive triangulation algorithms we present in this paper are not based on rigorous error estimation, but are intended for straightforward application to spline functions with smoothness $r \geq 0$. For C^r splines with $r > 0$, the jump of the gradient across any interior edge is zero, so we are unable to use those discontinuities to compute an error function. Instead, we utilize the gradient of the approximate solution at the centroids of triangles as a simple indicator of the magnitude of the change of the solution. The larger the magnitude of the gradient, the more the solution changes over the triangle. If the gradient is large, the centroid is added to the vertex set, and if the gradient is near zero, the vertices of the triangle are removed.

Our simple gradient indicator is easy to compute, and numerical results show that it works well when constructing approximate solutions of PDEs that change far more drastically in some parts of the domain than others. One difference of our approach from many traditional adaptive methods (but not all; cf [10]) is that we retriangulate globally by a Delaunay triangulation method instead of using local refinements. For any fixed vertex set, like the ones returned by our algorithms, the Delaunay triangulation is the one with the max–min angle property. This mitigates decrease of the smallest angle of $\triangle$ due to local refinements via bisection. The small angle problem can also be avoided by using H-triangulations, i.e., triangulations with hanging nodes [18], as implemented for *hp*-finite elements in [20] and for splines in S_d^0 in [14]. Our approach uses ordinary triangulations due to their ubiquity in spline applications and to avoid the added difficulty of computing C^r splines over H-triangulations for $r \geq 1$.

This paper is organized as follows. In Sect. 2, we review the bivariate spline functions used in [4]. In Sect. 3, we describe our algorithm and introduce a simple *mesh efficiency* metric (*me*) to help measure relative improvements in a triangulation's adaptedness to a given PDE. We then define the motivating problem and present the results of our study in the context of the Poisson equation. In Sect. 3.1, we adapt the original uniform mesh by adding the locations where large "gradient" values occur to the vertex set; this amounts to a kind of h-refinement. In Sect. 3.2, we present an algorithm that also removes vertices of $\triangle$ from the vertex set in areas where the solution is flat. In both sections, we present extensive numerical results to demonstrate that our approach offers significant improvement over a uniform refinement scheme.

In Sect. 3.3, we present an approach that uses information from the source function f to generate an initial triangulation. This mesh produces a better *initial* spline solution than one generated over a uniform mesh with the same number of triangles and might be considered as a kind of r-refinement. We determine an initial number of desired triangles and generate a nonuniform vertex set whose Delaunay triangulation meets that requirement. We then apply the algorithms from Sect. 3.2 to generate a mesh that is highly adapted to the given PDE.

Finally, we comment that our adaptive approach can be extended to deal with other partial differential equations, e.g., the biharmonic equation.

2 Spline Functions on Triangulations

Given a polygonal region Ω, a collection $\triangle := \{T_1, ..., T_n\}$ of triangles is an ordinary triangulation of Ω if $\Omega = \cup_{i=1}^n T_i$ and if any two triangles T_i, T_j intersect at most at a common vertex or a common edge. There are many different possible triangulations $\triangle_j$ on a given vertex set. In some sense, the best triangulation $\triangle$ is the one that maximizes the minimum angle $\theta(\triangle)$ of the set of triangles $\{T_i\} \in \triangle_j$; this is partly because error bounds for spline approximations often depend on the minimum angle of a triangulation, but also because computation of spline coefficients can become numerically unstable if $\theta(\triangle)$ is too small. When Ω is convex, this $max - min$ angle triangulation is equivalent to the Delaunay triangulation of the region (cf. [13]).

For a triangle $T_i \in \Omega$, $T_i = (v_1, v_2, v_3)$, we define the barycentric coordinates (b_1, b_2, b_3) of a point $(x_o, y_o) \in \Omega$. These coordinates are the solution to the following system of equations

$$\begin{aligned} b_1 + b_2 + b_3 &= 1 \\ b_1 v_{1,x} + b_2 v_{2,x} + b_3 v_{3,x} &= x_o \\ b_1 v_{1,y} + b_2 v_{2,y} + b_3 v_{3,y} &= y_o, \end{aligned}$$

and are nonnegative if $(x_o, y_o) \in T_i$. The barycentric coordinates are then used to define the Bernstein polynomials of degree d:

$$B^T_{i,j,k}(x, y) := \frac{d!}{i!j!k!} b_1^i b_2^j b_3^k, \qquad i + j + k = d,$$

which form a basis for the space $\mathscr{P}_d$ of polynomials of degree d. Therefore, we can represent all $p \in \mathscr{P}_d$ in B-form:

$$p = \sum_{i+j+k=d} c_{ijk} B^T_{ijk}$$

where the B-coefficients c_{ijk} are uniquely determined by p.

We define the spline space $S^0_d := \{s \in C^0(\Omega) : s|_{T_i} \in \mathscr{P}_d\}$ where T_i is a triangle in a triangulation $\triangle$ of Ω. We use this piecewise continuous polynomial space to define the space $S^r_d := C^r(\Omega) \cap S^0_d(\triangle)$. The C^r condition for splines can be guaranteed across edges of an ordinary triangulation $\triangle$ by enforcing linear conditions on the B-coefficients of neighboring triangles.

Computations involving splines written in B-form can be performed recursively using the de Casteljau's algorithm. In fact, these spline functions have numerically stable, closed-form formulas for differentiation, integration, inner products (cf. [13]), and triple products. Spline functions in $S^r_d(\Omega)$ on ordinary triangulations have optimal approximation power if $d \geq 3r + 2$. In the following, we exhibit the flexibility of spline solutions to the Poisson problem by solving over $S^r_d(\Omega)$ for various choices of d and r. For a more complete discussion on the properties of bivariate splines in B-form, see [13] or [17]. The implementations in this paper are based on those discussed in [4]. In particular, we do not work with minimal determining sets, but instead impose smoothness conditions as side constraints. We also do not enforce supersmoothness conditions around vertices, as PDE solutions themselves may not exhibit this smoothness phenomenon at those points in the domain.

3 Numerical Results for Our Adaptive Approach

3.1 Part 1: Adding Triangles

To efficiently solve PDEs which change far more drastically in some areas of the domain than others, we should add more vertices only as needed, instead of refining uniformly. To decide where we need to add more vertices, we use the gradient of the spline solution based on an initial triangulation. We introduce our first algorithm as follows:

Algorithm 1 Input a triangulation $\triangle$ with vertex set V and triangle list T, and initial approximate spline solution s. Then do the following steps:

- compute centroids $\mathbf{c}_t$ of all triangles t in T;
- compute values of $D_t(s) = |D_x s| + |D_y s|$ at the location $\mathbf{c}_t$ for all $t \in T$;
- sort the values of D_t by size;
- choose the $\mathbf{c}_t$ corresponding to the largest D_t values above a certain threshold to be added to V;
- use a triangulation method (Delaunay triangulation) to find a new triangulation of the updated vertex set;
- output V and T of the resulting triangulation.

To help measure triangulation improvement in any setting, we introduce a simple *mesh efficiency* metric, or *me*, for a spline solution s over a triangulation $\triangle$, defined by

$$me(s) := \#(\triangle) \times \text{RMSE}(s)$$

where $\#(\triangle)$ is the number of triangles in the underlying triangulation $\triangle$, and RMSE(s) is the root-mean-squared error of spline solution, computed on a grid of 1001×1001 points spread evenly over the domain.

This metric helps measure relative improvements in a triangulation's adaptedness to a given PDE; a smaller *me* value represents an improvement in the triangulation size, approximation error, or both. Therefore, comparing reduction in *me* values from one mesh to another gives us a clear idea of how the triangulation is improving. For comparing mesh 1 to mesh 2, we refer to the percent reduction in *me* values from one mesh to the other as

$$\text{Net Gain} := 1 - me_2/me_1.$$

Example 1 Our study is motivated by the Poisson problem with exact solution $u(x, y) = \tanh(40y - 80x^2) - \tanh(40x - 80y^2)$ on $[0, 1] \times [0, 1]$; its graph is shown in Fig. 1. We first use our bivariate spline method to compute numerical solutions in S_7^0 for this particular problem over uniform type-1 triangulations of varying fineness. The initial triangulations shown in Fig. 2 are examples of this type of triangulation; in the following, we simply use "uniform triangulation" to mean uniform type-1 triangulation.

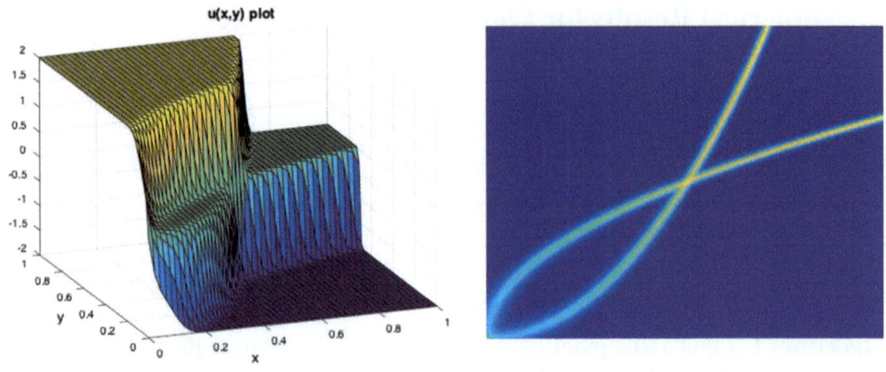

Fig. 1 A plot of the function $u(x, y) = \tanh(40y - 80x^2) - \tanh(40x - 80y^2)$ on $[0, 1] \times [0, 1]$ (*right*), and a color plot of its $|D_x u| + |D_y u|$ values (*left*). This function is hard to approximate based on a uniform triangulation

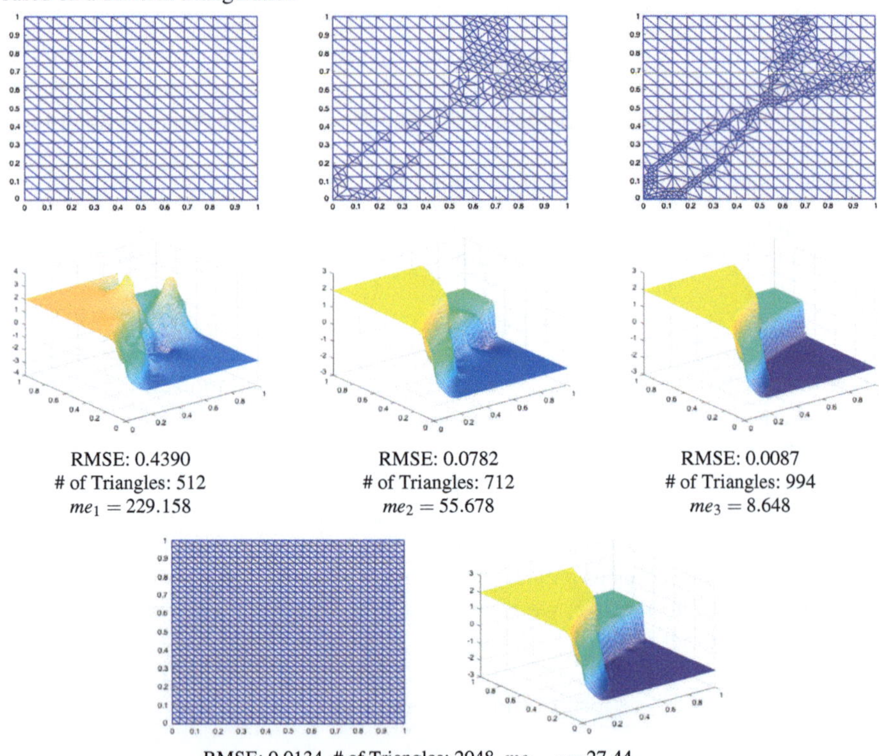

Fig. 2 Example 1: Here, we solve the Poisson problem on $[0, 1] \times [0, 1]$ with exact solution $u(x, y) = \tanh(40y - 80x^2) - \tanh(40x - 80y^2)$ over spline space S_7^0. A uniform triangulation (*top left*), a triangulation after one iteration of Algorithm 1 (*top middle*), and a triangulation after a second iteration of the algorithm (*top right*). In the *bottom row*, we include the triangulation and solution that result from a uniform refinement of the original mesh; our adapted triangulation generates a more accurate solution even though it uses less than half the number of triangles as the uniformly refined mesh

Table 1 RMSE results for 4 uniform meshes of varying fineness; Numerical results are based on spline spaces $S_7^0(\triangle)$

Number of triangles	RMSE	Number of spline coefficients
800	1.92e-1	28,800
1152	5.17e-2	41,472
1568	2.96e-2	56,488
2048	1.34e-2	73,728

From Table 1, we can see that one has to use quite a large uniform triangulation to get a reasonably accurate numerical solution. This makes it an ideal candidate for application of Algorithm 1.

Based on the implementation in [4], we present numerical data for spline solutions in the space S_7^0 to this particular Poisson problem. Here, we apply Algorithm 1 twice; the process and the resulting meshes are shown in Fig. 2. Comparison of the plots of the numerical solutions with the exact plot in Fig. 1 shows that solving the PDE over the adapted meshes produces a much more accurate *looking* approximate solution. We can also compare the numerical accuracy of the solution that results from solving over the adapted mesh shown in the top right of Fig. 2, which has 994 triangles, to the accuracy of a solution generated from a uniform mesh with many more triangles– say 2048. We look at the root-mean-squared error (RMSE) which we compute over 1001×1001 equally spaced points on $[0, 1] \times [0, 1]$. The solution from our adapted triangulation has RMSE 0.0087, while the RMSE from the (globally denser) uniform triangulation is 0.0134. This means that in this case, our mesh generates a solution that is 35% more accurate even while using a triangulation less than half the size of the uniform mesh. Using the *me* metric introduced in (Sect. 3.1) above, our adapted mesh represents a 75% $\textit{Net Gain} = 1 - me_3/me_1$ over the uniform triangulation.

More numerical results for this Poisson problem are presented in Table 2. We control the number of vertices added into the vertex set by our algorithm so that the resulting adapted triangulation has the same number of triangles as a different uniform mesh. Then, we can compare the accuracy of the solutions that result from the distinct triangulations and claim that any error reduction is due to the adaptedness of our mesh to the particular PDE.

The meshes in the "Adapted mesh" column were formed from the uniform meshes in the previous row. The appropriate number of vertices was added to that smaller uniform triangulation to generate a triangulation of exactly the same size as the uniform mesh in the next row. For example, the first triangulation was a uniform mesh T_u over vertex set V_u with 288 triangles. An initial approximate solution u_h was computed over T_u; then, $D_t(u_h)$ values were computed, and just the right number N of centroids $\mathbf{c}_t$ was added so that the new vertex set $V_1 := V \cup_{n=1}^{N} \mathbf{c}_n$ generated a triangulation T_1 with $|T_1| = 512$. We compute the percent reduction in RMSE from the uniform error e_u and the adapted error e_a in the same manner that *Net Gain* is calculated.

Table 2 Example 1: RMSE results for the Poisson problem on $[0, 1] \times [0, 1]$ with exact solution $u(x, y) = \tanh(40y - 80x^2) - \tanh(40x - 80y^2)$. Here, the approximate solutions are in $S_7^0(\triangle)$

Compared performance of uniform mesh and Algorithm 1 adapted mesh

Uniform mesh		Adapted mesh		Percent improvement		
Num Tri	RMSE	Num Tri	RMSE	Num Tri	RMSE reduction(%)	Net Gain(%)
288	6.01e-1	–	–	–	–	–
512	4.39e-1	512	3.29e-1	–	45.4	45.4
800	1.92e-1	800	4.70e-2	–	75.5	75.5
1152	5.18e-2	1152	1.44e-2	–	72.2	72.2
1568	2.96e-2	1568	3.31e-3	–	88.9	88.9
2048	1.34e-2	2048	8.56e-4	–	94.0	94.0

Table 3 Triangulation times for Example 1, in seconds; triangulation times make up a negligible component of the total computational time

Triangulation times for Example 1

Initial triangulation		After 1 iteration		After 2 iterations	
Num Tri	Tri Time	Num Tri	Tri Time	Num Tri	Tri Time
512	0.1953	800	0.1783	1152	0.1355
800	0.1467	1152	0.1259	1568	0.1735
1152	0.2140	1568	0.1716	2048	0.2205
1568	0.2637	2048	0.2206	2592	0.2924

We emphasize that we retriangulate globally with a Delaunay algorithm, assuring that the triangulation we use for the fixed vertex set returned by Algorithm 1 has the max–min angle property. But does this global retriangulation have a high computational cost? Does it add significantly to computation times? To address this potential concern, we report triangulation and retriangulation times for the meshes generated in Example 1. In Table 3, we see that the time spent retriangulating does not significantly impact computational times. In fact, because the initial triangulation requires a generation of a point grid across the domain, retriangulations of the vertex set passed by Algorithm 1 are often faster than the generation of the initial triangulation. The total time required to generate the table of data for Example 1 is about 5 min on a 2009 MacBook Pro with 4GB RAM; the total triangulation and retriangulation times for all the meshes reported in the table are less than 5 s. Since these triangulation times are negligible, we do not continue to report them in the remaining examples.

Although generating Delaunay tetrahedralizations of 3-dimensional vertex sets is a more complex undertaking, we expect that remeshing times will remain small relative to total computational time. On the same computer, we computed a Delaunay

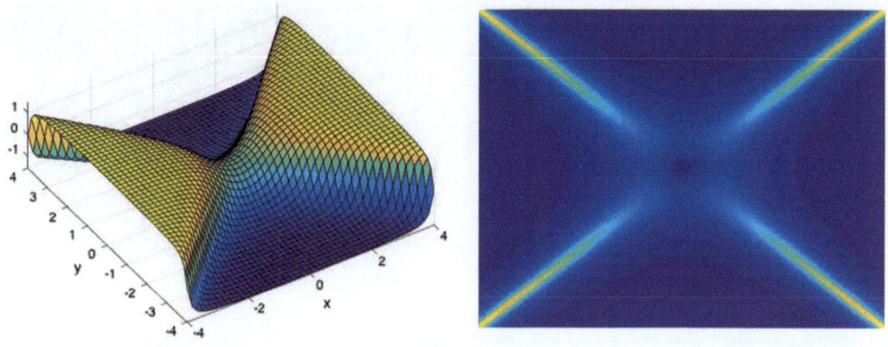

Fig. 3 A plot of the function $u(x, y) = \arctan(x^2 - y^2)$ on $[-4, 4] \times [-4, 4]$ (*left*), and a 2D color plot of its $|u_x| + |u_y|$ values (*right*)

tetrahedralization of 10,000 random vertices (x, y, z) in MATLAB; the meshing took just 0.3078 s. Thus, for reasonably sized vertex sets, we do not anticipate that retetrahedralization times will pose a difficulty in generalizing our approach to 3-dimensional elliptic problems.

Example 2 Here, we solve the Poisson problem with exact solution $u(x, y) = \arctan(x^2 - y^2)$. This function is also fairly difficult to approximate if one uses only a uniform triangulation. Its graph and a 2D color plot of its D_t values are shown in Fig. 3.

Here, the solution is studied over $\Omega = [-4, 4] \times [-4, 4]$ using spline space S_9^1. Again, we shall use the RMSE and number of triangles to measure the goodness of the performance of our spline method.

We show the algorithm's performance visually in Fig. 4. For comparison, a uniform refinement of the original mesh (the first one in the top row of Fig. 4) yields a triangulation with 800 triangles and produces a spline solution with a RMSE of 0.0030. The adaptive method achieves this level of accuracy while using a triangulation that less than half the size. Comparing these two results shows that the adapted mesh produces a *Net Gain* of 55%. As in the first example, we present more numerical results for this Poisson problem in Table 4 below.

3.2 *Part 2: Adding and Removing Triangles*

We now want to consider removing vertices from the triangulation as needed. This approach is worth consideration for reasons of portability and ease of use. If our spline solution contains more coefficients than necessary in parts of the domain where the function is not changing rapidly, then we would like to create a smaller spline solution by reducing the density of triangles in those regions. If we can do this

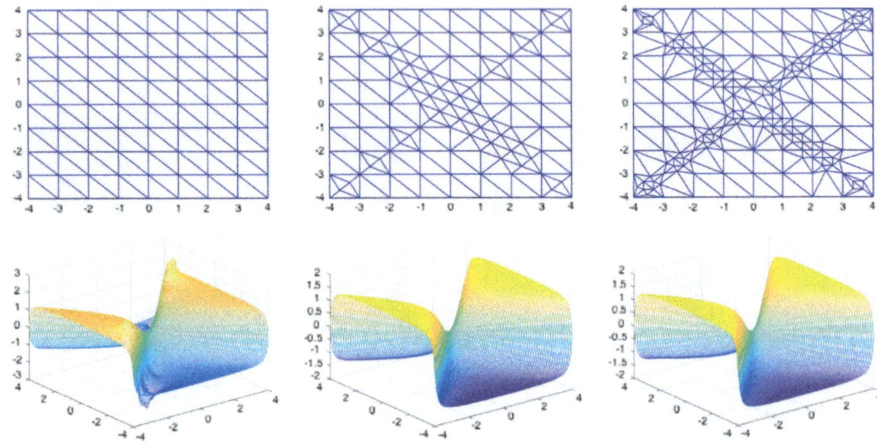

RMSE: 0.1631 | RMSE: 0.0102 | RMSE: 0.0029
# of Triangles: 128 | # of Triangles: 186 | # of Triangles: 364
$me_1 = 20.87$ | $me_2 = 1.897$ | $me_3 = 1.0556$

Fig. 4 Example 2: Here, we solve the Poisson problem on $[-4, 4] \times [-4, 4]$ with exact solution $u(x, y) = \arctan(x^2 - y^2)$ over spline space S_9^1. A uniform triangulation (*top left*), a triangulation produced after 1 iteration of Algorithm 1 (*top middle*), and the triangulation resulting from a second pass of the algorithm (*top right*). Spline solutions (*on the second row*) based on triangulations shown in the *top row*. The RMSEs, numbers of triangles, and *me* are explained in the *third row*

Table 4 Example 2: RMSE results for the Poisson problem on $[-4, 4] \times [-4, 4]$ with exact solution $u(x, y) = \arctan(x^2 - y^2)$. Here, the approximate solutions are in S_9^1

Compared performance of uniform mesh and Algorithm 1 adapted mesh						
Uniform mesh		Adapted mesh		Percent improvement		
Num Tri	RMSE	Num Tri	RMSE	Num Tri	RMSE reduction	Net Gain
288	2.71e-2	–	–	–	–	–
512	8.72e-3	512	6.57e-4	–	92.5%	92.5%
800	3.03e-3	800	4.56e-4	–	84.9%	84.9%
1152	1.13e-3	1152	7.66e-5	–	93.2%	93.2%
1568	4.72e-4	1568	2.17e-5	–	95.4%	95.4%
2048	2.08e-4	2048	1.33e-5	–	93.6%	93.6%

without affecting numerical accuracy, the resulting spline solution would be easier to transmit, store, and perform calculations with. To this end, our algorithm is presented as follows:

Algorithm 2 Input a triangulation Δ with vertex set V and triangle list T, and initial spline approximation s, and tolerance TOL. Then do the following steps:

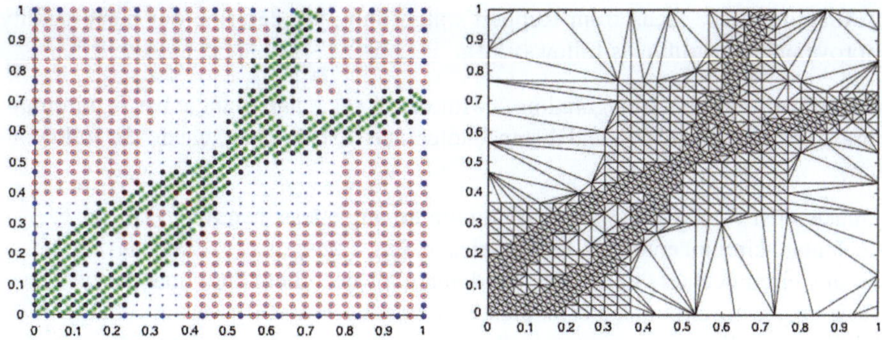

Fig. 5 Shown *left*, a color-coded vertex set demonstrating Algorithm 2 applied to the Poisson problem with exact solution $u(x, y) = \tanh(40y - 80x^2) - \tanh(40x - 80y^2)$, shown *left*. The dark vertices around the boundary form the list $\mathscr{P}$; the dark vertices in the interior of the domain form the list $\mathscr{L}$; the vertices forming the *ribbon shape* through the diagonal of the domain are the new vertices $\mathbf{c}_t$; the other *red-circled vertices* are to be removed (*left*). The resulting triangulation (*right*)

- compute centroids $\mathbf{c}_t$ of all triangles t in T;
- compute values of $D_t(s) = |D_x s| + |D_y s|$ at the location $\mathbf{c}_t$ for all $t \in T$;
- sort values of D_t by size;
- choose the $\mathbf{c}_t$ corresponding to the largest D_t values above a certain threshold to be added to the vertex set V;
- make a list $\mathscr{L}$ of the vertices of all triangles to which new points $\mathbf{c}_t$ were added;
- make a list $\mathscr{P}$ of a certain proportion of boundary vertices of V to be protected from removal;
- remove all vertices not in $\mathscr{L}$ or $\mathscr{P}$ with D_t value less than TOL;
- find a Delaunay triangulation nT of the new vertex set $\tilde{V}$, and output both.

This algorithm has great utility in applications where the solution is flat over large portions of the domain. We exhibit the process in Fig. 5.

Frequently, deleting vertices from the original mesh results in a new triangulation with a smaller minimum angle than that of the original one. Having highly acute triangles in a mesh is problematic for accurate numerical approximation in theory and in practice; constants in many error bounds depend on this minimum angle, and computing spline coefficients over especially skinny triangles leads to numerical instability.

To overcome the difficulty, we implement ideas from Ruppert's algorithm to improve the minimum angle characteristic of the mesh, without reintroducing too many vertices. Ruppert's algorithm works roughly by alternately splitting "encroached" edges and inserting circumcenters of skinny triangles into the vertex set. Given some reasonable assumptions about the domain to be triangulated, Ruppert's algorithm is guaranteed to converge to a mesh with minimum angle greater than 20°. See [19] for more details.

Adapting these ideas from Ruppert's algorithm, we describe our mesh quality improvement algorithm as follows:

Algorithm 3 Input a triangulation Δ with vertex set V and triangle list T, minimum angle tolerance $ANGTOL$, and distance tolerance $DTOL$. Then, we do the following steps:

- make a list T_s of all triangles with minimum angle less than $ANGTOL$;
- calculate circumcenters cr_t of triangles in T_s;
- if any circumcenters lie outside the domain, project them onto boundary;
- if a projected circumcenter is closer to existing boundary vertices than $DTOL$, delete from list;
- iterate: if any pair of circumcenters cr_i and cr_j are closer than $DTOL$, replace the pair with their midpoint;
- add cr_t to the vertex set V;
- retriangulate and output the new vertex set $\tilde{V}$ and the new triangle list $\tilde{T}$.

During our experiments, numerical accuracy suffered when computing over a triangulation with minimum angle below 4°. Therefore, we implemented Algorithm 3 with $ANGTOL = 4$. We choose one triangle from the original mesh and set $DTOL$ to be half than the minimum edge length of this triangle; because the initial triangulation is uniform, the functionality of $DTOL$ does not depend on which triangle we choose. In all cases tested with $ANGTOL \leq 4$, the output of Algorithm 3 was a triangulation with a greater minimum angle characteristic. The resulting meshes also often substantially improve numerical results, especially as applied in Sect. 3.3. The process is shown in Fig. 6.

Example 3 We return to the Poisson problem with exact solution $u(x, y) = \tanh(40y - 80x^2) - \tanh(40x - 80y^2)$ from Example 1 to show the effectiveness of adding *and* removing points from the vertex set. We again solve the problem in the spline space S_7^0. Because of the added variability in the size of the vertex set produced by Algorithms 2 and 3, we make good use of the *mesh efficiency* metric

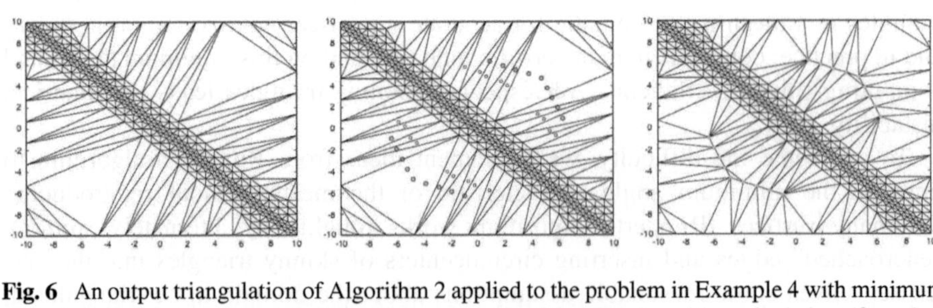

Fig. 6 An output triangulation of Algorithm 2 applied to the problem in Example 4 with minimum angle 2.5° (*left*) is input into Algorithm 3; the centroids of the triangles with angles below 4° marked by *small red circles* and the circumcenters of those triangles shown as *larger green dots* (*middle*); those *larger green dots* become vertices in the new triangulation (*right*) which has minimum angle 5.1°

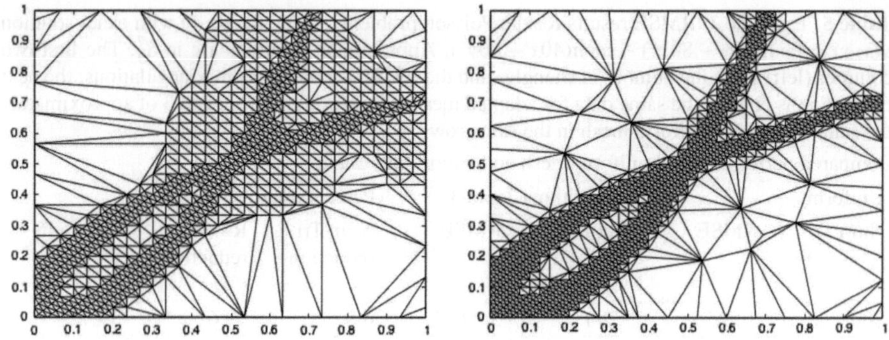

Fig. 7 A triangulation from Algorithm 2 with minimum angle above 4° (*left*); a different triangulation resulting from Algorithms 2 and 3, right, where the minimum angle was raised above 4° (*right*)

to gauge relative improvements in the adaptedness of our triangulation to the PDE. Figure 7 shows a few triangulations produced by the algorithms (from different initial uniform meshes) in this setting.

Table 5 compares numerical results for uniform meshes and adapted triangulations resulting from the algorithms. This table shows the number of triangles and RMSE for an initial uniform mesh Δ_u and initial solution $s_u \in S_7^0$, along with data for the solution computed over an adapted mesh of similar size produced by Algorithms 2 and 3. The *Net Gain* of the adapted mesh over the original uniform mesh is also shown.

While it is no surprise that these algorithms produce a better mesh for solving this PDE than the uniform approach, it is noteworthy that they also offer an improvement over Algorithm 1. Numerical results demonstrating this are shown in Table 6; here again, we made an effort to compare *me* data for solutions produced over similarly sized triangulations.

The percent improvement columns in Tables 5 and 6 are calculated in the same way as the *Net Gain* percentage described above. This number is the percentage of triangles that must be removed from Δ_u in order to produce a triangulation the same size as Δ_a.

Example 4 We present one more numerical example—the Poisson problem on $[-10, 10] \times [-10, 10]$ with exact solution $u = \dfrac{1}{1 + \exp^{-10(x+y)}}$. In this example, we solve for an approximate spline solution $s \in S_8^1$. Figure 8 shows a plot of the solution together with a 2D color plot of its derivative values. The triangulations shown in Fig. 6 resulted from applying Algorithms 2 and 3 to this problem. As in the previous example, we compare performance of the mesh generated by Algorithms 2 and 3 both to uniform meshes and to meshes of similar size from Algorithm 1. The results are shown in Tables 7 and 8, respectively.

Table 5 Example 3: RMSE results for the Poisson problem on [0, 1] × [0, 1] with exact solution $u(x, y) = \tanh(40y - 80x^2) - \tanh(40x - 80y^2)$. Approximate solutions are in S_0^7. The first two columns (left) show the number of triangles and the RMSE for the uniform triangulations; the next two columns contain the same data for adapted meshes from Algorithms 2 and 3 of approximately the same size as the uniform mesh in the same row

Compared performance of uniform mesh and Algorithms 2 and 3 meshes						
Uniform		Algorithms 2 and 3		Percent improvement		
Num Tri	RMSE	Num Tri	RMSE	Num Tri reduction (%)	RMSE reduction (%)	Net Gain (%)
800	1.92e-1	794	3.29e-2	0.8	79.6	79.7
1152	5.18e-2	1147	3.33e-3	0.4	93.6	93.6
1568	2.96e-2	1533	5.27e-4	2.2	98.2	98.2
2048	1.34e-2	1978	7.14e-5	3.4	99.5	99.5

Table 6 Example 3: RMSE results for the Poisson problem on [0, 1] × [0, 1] with exact solution $u(x, y) = \tanh(40y - 80x^2) - \tanh(40x - 80y^2)$. Approximate solutions are in S_7^0. The first two columns (left) triangulation data and RMSE for meshes and spline solutions resulting from Algorithm 1 where points are merely added to the vertex set; the next two columns show the same data for meshes and splines produced by Algorithms 2 and 3 where vertices may added to and removed from the original vertex set

Compared performance of Algorithm 1 meshes and Algorithms 2 and 3 Meshes						
Algorithm 1		Algorithms 2 and 3		Percent improvement		
Num Tri	RMSE	Num Tri	RMSE	Num Tri reduction (%)	RMSE reduction (%)	Net Gain (%)
800	4.70e-2	794	3.29e-2	0.8	30.0	30.5
1152	1.44e-2	1147	3.33e-3	0.4	76.8	76.9
1568	3.31e-3	1533	5.27e-4	2.2	84.1	84.1
2048	8.56e-4	1978	7.14e-5	3.4	91.7	91.9

3.3 *Part 3: A Good Initial Guess Triangulation*

We now present a heuristic method to generate a better initial triangulation for solving Poisson's equation with a nonzero source function. We produce a mesh that is denser in regions where the source function is large. The previous algorithms can then be applied to spline solutions produced over this initial mesh. The algorithm is presented as follows:

Algorithm 4 Input a source function f for Poisson's equation $-\Delta u = f$, and a parameter n related to desired mesh fineness. Then,

- with density specified by parameter n, generate 2 point grids and G_2 so that G_2 is twice as dense as G_1 and $G_1 \subset G_2$;

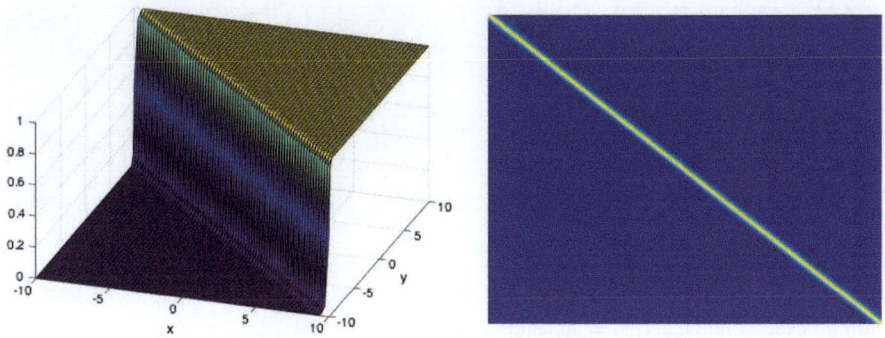

Fig. 8 A plot of the function $u = \dfrac{1}{1+\exp^{-10(x+y)}}$ on $[-10, 10] \times [-10, 10]$ (*left*), and a 2D color plot of its $|u_x| + |u_y|$ values (*right*)

Table 7 Example 4: RMSE results for the Poisson problem on $[-10, 10] \times [-10, 10]$ with exact solution $u(x, y) = \dfrac{1}{1+\exp^{-10(x+y)}}$. Approximate solutions are in S_8^1. The first two columns (left) show the number of triangles and the RMSE for the uniform triangulations; the next two columns contain the same data for adapted meshes from Algorithms 2 and 3 of approximately the same size as the uniform mesh in the same row

Compared performance of uniform meshes and Algorithm 2 & 3 adapted meshes

Uniform		Algorithm 2 and 3		% improvement		
Num Tri	RMSE	Num Tri	RMSE	Tri Red. (%)	RMSE Red. (%)	Net Gain (%)
392	2.06e-2	386	2.84e-3	1.5	86.2	86.4
512	7.26e-3	494	2.78e-4	3.5	96.2	96.3
648	7.75e-4	640	1.31e-4	1.2	83.1	83.3
800	3.55e-3	738	6.05e-5	7.8	98.3	98.4

Table 8 Example 4: RMSE results for the Poisson problem on $[-10, 10] \times [-10, 10]$ with exact solution $u(x, y) = \dfrac{1}{1+\exp^{-10(x+y)}}$. Approximate solutions are in S_8^1. The first two columns (left) triangulation data and RMSE for meshes and spline solutions resulting from Algorithm 1; the next two columns show the same data for meshes and splines produced by Algorithms 2 and 3

Compared performance of Algorithm 1 meshes and Algorithms 2 and 3 meshes

Algorithm 1		Algorithms 2 and 3		Percent improvement		
Num Tri	RMSE	Num Tri	RMSE	Num Tri reduction (%)	RMSE reduction (%)	Net Gain (%)
800	2.05e-3	590	1.32e-4	26.3	93.6	95.3
1080	8.70e-4	660	6.17e-5	38.9	92.9	95.7
1568	3.83e-4	1148	4.65e-6	26.8	98.8	99.1
2048	1.27e-4	1887	2.29e-6	7.8	98.2	98.3

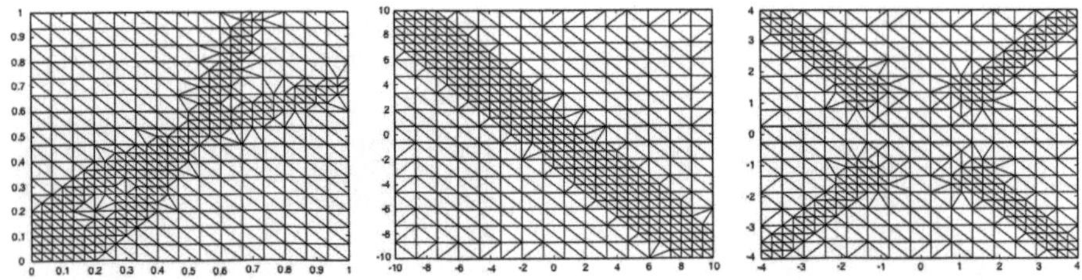

Fig. 9 Triangulations from Algorithm 4; a mesh for Poisson problem with exact solution $u(x, y) = \tanh(40y - 80x^2) - \tanh(40x - 80y^2)$ (*left*); mesh for exact solution $u = \frac{1}{1+\exp^{-10(x+y)}}$ (*middle*); mesh for exact solution $u(x, y) = \arctan(x^2 - y^2)$ (*right*)

- take a point $g \in G_2$, and let its horizontally and vertically adjacent neighbors be g_1, g_2, g_3, g_4;
- assign to g the value $L(g) = \frac{1}{5}(|f(g)| + |f(g_1)| + \ldots + |f(g_4)|)$;
- find the quartile of points Q of G_2 with the largest L values;
- build the vertex set $V := G_1 \bigcup Q$ and triangulate using Delaunay triangulation.

This algorithm places smaller triangles around areas where the values of the source function f are the greatest. The values of f obviously tell us something about changes in the solution function u, and while the correspondence between large values of f and large changes in u is not explicit, making the initial guess that they occur in the same regions of the domain results in improved numerical accuracy. Examples of triangulations produced by Algorithm 4 are shown in Fig. 9. We refer to these types of "initial guess" triangulations as *ig*-meshes in the following.

Example 5 In this example, we compare the numerical performance of spline solutions over the *ig*-meshes to those generated by uniform initial triangulations.

We begin with the Poisson problem with exact solution $u(x, y) = \tanh(40y - 80x^2) - \tanh(40x - 80y^2)$; Table 9 compares the mesh efficiencies of uniform meshes and *ig*-meshes from Algorithm 4 of similar size. For all the numerical results in this section, we use splines $s \in S^1_5(\triangle)$.

Table 9 shows that the *ig*-mesh offers improved numerical results in comparison with an initial uniform triangulation. Tables 10 and 11 show similar results for the other PDE solutions tested in Sects. 3.1 and 3.2. This finding has great potential utility, since computing the *ig*-mesh is far computationally cheaper than producing adapted meshes via Algorithms 1 and 2. Here, no initial solution is required; we simply use source function values to give an indication where the solution will be changing most rapidly. Of course, for some PDEs, the source functions are 0 or constant. In these settings, the algorithm will be useless.

The improvements from the *ig*-mesh seem to be maintained even when that triangulation is fed into Algorithms 2 and 3. A large *Net Gain* is observed when comparing against uniform meshes, but also when compared to the adapted triangulations produced by the algorithms in the previous sections. The tables below contain numerical data that supports this claim.

Table 9 Comparison of the RMSE for spline solutions produced over uniform meshes and *ig*-meshes of approximately the same size. This data is from the Poisson problem on $[0, 1] \times [0, 1]$ with exact solution $u(x, y) = \tanh(40y - 80x^2) - \tanh(40x - 80y^2)$. Solutions are in spline space S_5^1

Compared performance of uniform meshes and *ig*-meshes						
Uniform		*ig*-mesh		% improvement		
Num Tri	RMSE	Num Tri	RMSE	Tri Red.	RMSE Red. (%)	Net Gain (%)
800	1.44e-1	800	2.34e-2	–	83.7	83.7
1250	5.72e-2	1146	7.88e-3	8.3%	86.2	87.4
1800	2.36e-2	1706	3.08e-3	5.2%	87.0	87.6

Table 10 Comparison of the RMSE for spline solutions produced over uniform meshes and *ig*-meshes of approximately the same size. This data is from the Poisson problem on $[-10, 10] \times [-10, 10]$ with exact solution $u = \frac{1}{1+\exp^{-10(x+y)}}$. Solutions are in spline space S_5^1

Comparison of the performance of uniform meshes and *ig*-meshes						
Uniform		*ig*-mesh		% improvement		
Num Tri	RMSE	Num Tri	RMSE	Tri Red. (%)	RMSE Red. (%)	Net Gain (%)
1152	1.34e-2	1147	2.28e-3	0.4	83.0	83.0
1568	3.79e-3	1558	2.29e-3	0.6	39.6	40.0
2592	2.28e-3	2568	8.99e-4	0.9	60.6	60.9

Table 11 Comparison of the RMSE for spline solutions produced over uniform meshes and *ig*-meshes of approximately the same size. This data is from the Poisson problem on $[-4, 4] \times [-4, 4]$ with exact solution $u(x, y) = \arctan(x^2 - y^2)$. Solutions are in spline space S_5^1

Comparison of the performance of uniform meshes and *ig*-meshes						
Uniform		*ig*-mesh		% improvement		
Num Tri	RMSE	Num Tri	RMSE	Tri Red.	RMSE Red. (%)	Net Gain (%)
800	1.24e-2	800	2.04e-3	–	83.5	83.5
1250	5.17e-3	1150	7.04e-4	8.0%	86.4	87.5
1800	2.06e-3	1706	2.56e-4	5.2%	87.5	88.2

Example 6 Now, we compare the performance of the *ig*-mesh with the application of Algorithms 2 and 3 to the performance of a uniform mesh with the application of the algorithms. The results below show that starting with an *ig*-mesh offers improved numerical performance over starting with a uniform triangulation. We test all three of the Poisson problems mentioned in the previous sections and solve over spline space S_5^1. The data is presented in Tables 12, 13, and 14.

Table 12 RMSE comparison for spline solutions over triangulations from Algorithms 2 and 3 applied to uniform meshes and from Algorithms 2 and 3 applied to *ig*-meshes. This data is for the Poisson problem on $[0, 1] \times [0, 1]$ with exact solution $u(x, y) = \tanh(40y - 80x^2) - \tanh(40x - 80y^2)$. Solutions are in spline space S_5^1

Performance of Algorithms 2 and 3 applied to uniform and *ig*-meshes

Uniform + Alg		*ig*-mesh + Alg		% improvement		
Num Tri	RMSE	Num Tri	RMSE	Tri Red. (%)	RMSE Red. (%)	Net Gain (%)
1064	5.72e-2	892	5.30e-3	16.2	90.7	92.2
1385	1.39e-2	1213	2.49e-3	12.4	82.1	84.3
1496	6.89e-3	1860	1.22e-3	−24.3	82.4	78.1

Table 13 RMSE comparison for spline solutions over triangulations from Algorithms 2 and 3 applied to uniform meshes and from Algorithms 2 and 3 applied to *ig*-meshes. This data is for the Poisson problem on $[-10, 10] \times [-10, 10]$ with exact solution $u = \frac{1}{1+\exp^{-10(x+y)}}$. Solutions are in spline space S_5^1

Performance of Algorithms 2 and 3 applied to uniform and *ig*-meshes

Uniform + Alg		*ig*-mesh + Alg		% improvement		
Num Tri	RMSE	Num Tri	RMSE	Tri Red. (%)	RMSE Red. (%)	Net Gain (%)
574	3.37e-3	588	7.08e-4	−2.4	79.0	78.5
802	5.44e-4	800	3.73e-4	0.2	31.4	31.6
1068	5.09e-4	1236	5.32e-5	−15.7	89.6	87.9

Table 14 RMSE comparison for spline solutions over triangulations from Algorithms 2 and 3 applied to uniform meshes and from Algorithms 2 and 3 applied to *ig*-meshes. This data is for the Poisson problem on $[-4, 4] \times [-4, 4]$ with exact solution $u(x, y) = \arctan(x^2 - y^2)$. Solutions are in spline space S_5^1

Performance of Algorithms 2 and 3 applied to uniform and *ig*-meshes

Uniform + Alg		*ig*-mesh + Alg		% improvement		
Num Tri	RMSE	Num Tri	RMSE	Tri Red. (%)	RMSE Red.(%)	Net Gain (%)
910	3.23e-3	1027	5.29e-4	−12.9	83.6	81.5
1441	5.36e-4	1620	1.61e-4	−12.4	70.1	66.3
2036	3.41e-4	2474	4.92e-5	−21.5	85.6	82.5

4 Remarks and Conclusions

We have presented numerical results of our algorithms applied to three Poisson problems whose solutions are difficult to approximate accurately. For the examples shown in Sect. 3.1, we adjusted Algorithm 1 to terminate when the adapted mesh had the same number of triangles as a uniform mesh for easy comparison of performance; for the numerical experiments in Sects. 3.2 and 3.3, the tolerance which centroids $\mathbf{c}_t$ are added to the vertex set is determined within the algorithm. We computed the mean m and standard deviation σ of the middle 80% of the D_t values and added only the $\mathbf{c}_t$ with D_t above $m + 2\sigma$ to the vertex set. When applying Algorithm 2 to the ig-mesh, we added *all* boundary vertices of Q to the protected list $\mathscr{P}$. Although it can often improve accuracy to apply the algorithms multiple times, for ease and consistency in this paper, the numerical results in the data tables resulted from only one iteration of the given algorithm.

It is important to note that the adapted meshes produced by our algorithms are only effective when applied to functions that are especially difficult to approximate in a certain region of the domain. If, after an initial approximation, errors are somewhat uniformly distributed across the domain, there will be little gained by refining the

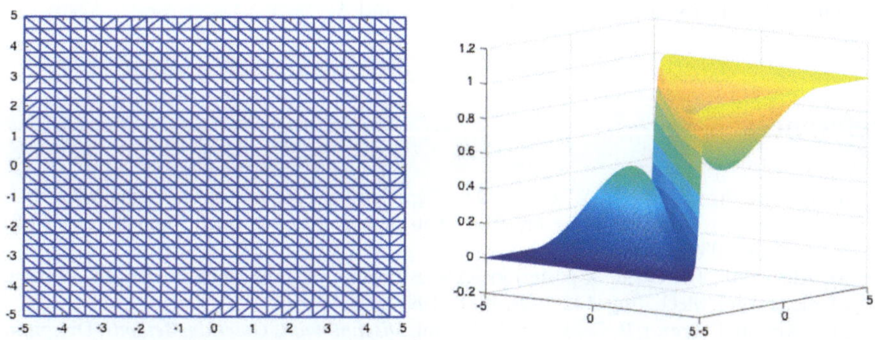

MAX ERR: 0.960, # of Triangles: 1250, $me_1 = 1200.25$

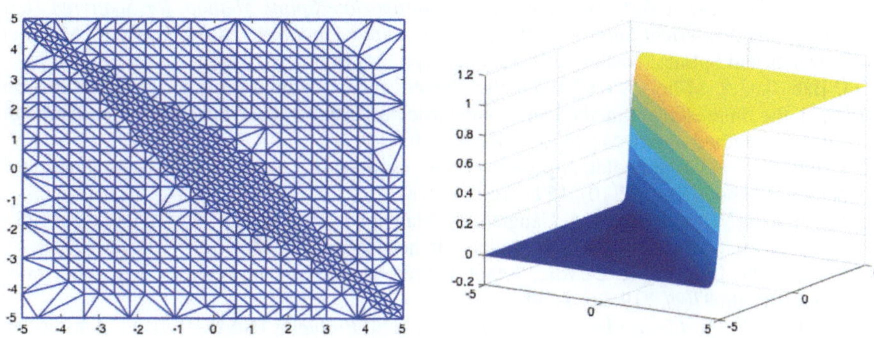

MAX ERR 0.036, # of Triangles: 1473, $me_2 = 53.028$.

Fig. 10 This image shows spline solutions in $S_5^1(\triangle)$ of the biharmonic equation over [0, 1] × [0, 1] with exact solution $u(x, y) = \tanh(40y - 80x^2) - \tanh(40x - 80y^2)$. A uniform triangulation (*top left*), a triangulation after one iteration of Algorithms 2 and 3 (*bottom left*). The solution images (*right*) are based on the triangulations shown in the corresponding row. Here, maximum errors are reported. The adapted mesh results in a *Net Gain* of 96.2%; a uniform refinement of does not reach the accuracy achieved by this adapted mesh until the number of triangles in the mesh is about 4500 mesh in some areas more than others. However, when solutions experience drastic change in certain areas, like the examples shown in this paper, or others like the steep wavefront problem in [15], our adapted meshes have great utility. Our approach is not only more efficient than uniform refinement, but it also offers a straightforward way to improve numerical performance when uniform refinement is unfeasible due to the size of the existing triangulation.

These ideas can be expanded to a wider scope; for example, we have also used this adaptive method when solving more complicated PDEs like the biharmonic equation,

$$\begin{cases} \Delta^2 u = f, & \text{on } \Omega, \\ u(x) = g(x), & \text{on } \partial\Omega, \\ \frac{\partial u}{\partial \mathbf{n}} = h, & \text{on } \partial\Omega. \end{cases}$$

Figure 10 shows an initial approximation and an improved solution generated over an adapted mesh resulting from Algorithms 2 and 3.

References

1. M. Ainsworth, J.Z. Zhu, A.W. Craig, O.C. Zienkiewicz, Analysis of the zienkiewicz zhua-posteriori error estimator in the finite element method. Int. J. Numer. Methods Eng **28**(9), 2161–2174 (1989)
2. M. Ainsworth, J.T. Oden, A posteriori error estimation in finite element analysis. Comput. Methods Appl. Mech. Eng. **142**(1–2), 1–88 (1997)
3. D.N. Arnold, F. Brezzi, B. Cockburn, D. Marini, *Discontinuous Galerkin Methods*, Discontinuous Galerkin Methods for Elliptic Problems (Springer, Berlin, 2000), pp. 89–101
4. G. Awanou, M.-J. Lai, P. Wenston, in *The Multivariate Spline Method, for Scattered Data Fitting and Numerical Solution of Partial Differential Equations*, ed. By G. Chen, M.J. Lai. Wavelets and Splines (Nashboro Press, 2006), pp. 24–74
5. I. Babuška, A. Miller, A feedback finite element method with a posteriori error estimation: Part i. the finite element method and some basic properties of the a posteriori error estimator. Comput. Methods Appl. Mech. Eng. **61**(1), 1–40 (1987)
6. I. Babuška, W.C. Rheinboldt, A-posteriori error estimates for the finite element method. Int. J. Numer. Methods Eng. **12**(10), 1597–1615 (1978)
7. L. Beirão da Veiga, F. Brezzi, A. Cangiani, G. Manzini, L.D. Marini, A. Russo, Basic principles of virtual element methods. Math. Models Methods Appl. Sci. **23**(1), 199–214 (2013)
8. P. Binev, W. Dahmen, R. DeVore, Adaptive finite element methods with convergence rates. Numer. Math. **97**(2), 219–268 (2004)
9. P.G. Ciarlet, *The Finite Element Method for Elliptic Problems* (North–Holland, Amsterdam, 1978)
10. P. Fernandes, P. Girdinio, M. Repetto, G. Secondo, Refinement strategies in adaptive meshing. IEEE Trans. Mag. **28**(2), 1739–1742 (1992)
11. F.D. Gaspoz, P. Morin, Approximation classes for adaptive higher order finite element approximation. Math. Comp. **83**(289), 2127–2160 (2014)
12. X. Hu, D. Han, M.-J. Lai, Bivariate splines of various degrees for numerical solution of PDE. SIAM J. Sci. Comput. **29**, 1338–1354 (2007)
13. M.-J. Lai, L.L. Schumaker, *Spline Functions on Triangulations* (Cambridge University Press, Cambridge, 2007)

14. S. Li, L.L. Schumaker, Adaptive Computation with Splines on Triangulations with Hanging Vertices, in *Approximation Theory XV: San Antonio 2016*, Springer Proc. in Math. and Stat. vol. 201, ed. by G.E. Fasshauer, L.L. Schumaker (Springer-Verlag, 2017), pp. 197–218
15. W.F. Mitchell, A collection of 2d elliptic problems for testing adaptive grid refinement algorithms. Appl. Math. Comput. **220**, 350–364 (2013)
16. J.-M. Mirebeau, A. Cohen, Greedy bisection generates optimally adapted triangulations. Math. Comput. **81**(278), 811–837 (2012)
17. L.L. Schumaker, *Spline Functions: Computational Methods* (SIAM Publication, Philadelphia, 2015)
18. L.L. Schumaker, L. Wang, Splines on triangulations with hanging vertices. Constr. Approx. **36**(3), 487–511 (2012)
19. J.R. Shewchuk, Delaunay refinement algorithms for triangular mesh generation. Comput. Geom. **22**(1–3), 21–74 (2002)
20. P. Solin, L. Dubcova, J. Cerveny, I. Dolezel, Adaptive hp-fem with arbitrary-level hanging nodes for maxwells equations. Adv. Appl. Math. Mech **2**(4), 518–532 (2010)
21. R. Verfürth, A posteriori error estimation and adaptive mesh-refinement techniques. J. Comput. Appl. Math. **50**(1–3), 67–83 (1994)
22. J. Wang, X. Ye, A weak Galerkin mixed finite element method for second-order elliptic problems. Math. Comp. **83**, 2101–2126 (2014)
23. O.C. Zienkiewicz, J.Z. Zhu, The super convergent patch recovery and a posteriori error estimates. part 1: the recovery technique. Int. J. Numer. Methods Eng. **33**(7), 1331–1364 (1992)
24. O.C. Zienkiewicz, J.Z. Zhu, The superconvergent patch recovery and a posteriori error estimates. part 2: errror estimates and adaptivity. Int. J. Numer. Methods Eng. **33**(7), 1365–1382 (1992)

Refinable Functions with PV Dilations

Wayne Lawton

Abstract A PV number is an algebraic integer α of degree $d \geq 2$ all of whose Galois conjugates other than itself have modulus less than 1. Erdös [8] proved that the Fourier transform $\widehat{\varphi}$, of a nonzero compactly supported scalar-valued function satisfying the refinement equation $\varphi(x) = \frac{|\alpha|}{2}\varphi(\alpha x) + \frac{|\alpha|}{2}\varphi(\alpha x - 1)$ with PV dilation α, does not vanish at infinity so by the Riemann–Lebesgue lemma φ is not integrable. Dai, Feng, and Wang [5] extended his result to scalar-valued solutions of $\varphi(x) = \sum_k a(k)\varphi(\alpha x - \tau(k))$ where $\tau(k)$ are integers and a has finite support and sums to $|\alpha|$. In ([22], Conjecture 4.2), we conjectured that their result holds under the weaker assumption that τ has values in the ring of polynomials in α with integer coefficients. This paper formulates a stronger conjecture and provides support for it based on a solenoidal representation of $\widehat{\varphi}$, and deep results of Erdös and Mahler [9]; Odoni [26] that gives lower bounds for the asymptotic density of integers represented by integral binary forms of degree > 2; degree $= 2$, respectively. We also construct an integrable vector-valued refinable function with PV dilation.

Keywords PV number · Refinable · Integral binary form · Real analytic · Lojasiewicz's structure theorem

1 Introduction

In this paper, $\mathbb{Z}, \mathbb{N} = \{1, 2, 3, ...\}, \mathbb{Q}, \mathbb{A}, \mathscr{O}, \mathbb{R}, \mathbb{C}$ are the integer, natural, rational, algebraic, algebraic integer, real, and complex numbers. For a ring R, $R[X]$; $R[X, X^{-1}]$ is the ring of polynomials; Laurent polynomials with coefficients in R in the indeterminant X. If $\alpha \in \mathbb{A}$, then $\mathbb{Q}[\alpha]$ equals the algebraic number field generated by α and we define $\mathscr{O}_\alpha = \mathscr{O} \bigcap \mathbb{Q}[\alpha]$, the degree function $d : \mathbb{A} \rightarrow \mathbb{N}$, and the trace and norm functions $T; N : \mathbb{A} \rightarrow \mathbb{Q}$. Their restrictions to $\mathscr{O}$ are integer valued. For

W. Lawton (✉)
School of Mathematics & Statistics, University of Western Australia,
Perth, Australia
e-mail: wlawton50@gmail.com

G.E. Fasshauer and L.L. Schumaker (eds.), *Approximation Theory XV: San Antonio 2016*, Springer Proceedings in Mathematics & Statistics 201,
DOI 10.1007/978-3-319-59912-0_8

$\alpha \in \mathbb{A}$, $P_\alpha(X) \in \mathbb{Q}[X]$ is its minimal degree monic polynomial, and $L(\alpha)$ is the least common multiple of the denominators of the coefficients of $P_\alpha(X)$. $\mathscr{O} \bigcap \mathbb{Q} = \mathbb{Z}$, $\alpha \in \mathbb{A} \Rightarrow L(\alpha)\alpha \in \mathscr{O}$, and $\alpha \in \mathscr{O} \Rightarrow P_\alpha(X) \in \mathbb{Z}[X]$. There exists $B(\alpha) \in \mathbb{N}$ with

$$\mathbb{Z}[\alpha] = \mathbb{Z} + \alpha\mathbb{Z} \cdots + \alpha^{d(\alpha)-1}\mathbb{Z} \subseteq \mathscr{O}_\alpha \subseteq \frac{1}{B(\alpha)}\mathbb{Z}[\alpha], \tag{1}$$

and hence, since $N(\alpha)\alpha^{-1} \in \mathscr{O}_\alpha$,

$$N(\alpha)B(\alpha)\alpha^{-1} \in \mathbb{Z}[\alpha]. \tag{2}$$

Example 1 If $\alpha \in \mathbb{A}$ and $P_\alpha(X) = X^3 - X^2 - 2X - 8 = 0$, Dedekind showed ([7], pp. 30–32, [25], p. 64) that $\{1, \alpha, \alpha(\alpha+1)/2\}$ is an integral basis for $\mathscr{O}_\alpha$. For this (1) holds with $B(\alpha) = 2$ and both inclusions are proper.

$\mathbb{T} = \mathbb{R}/\mathbb{Z}$; $\mathbb{T}_c = \{w \in \mathbb{C} : |w| = 1\}$ is the circle group represented additively; multiplicatively. For $x \in \mathbb{R}$, we define $||x|| = \min_{k\in\mathbb{Z}} |x-k| \in [0, \frac{1}{2}]$ and observe that $||x+y|| \leq ||x|| + ||y||$. Since $x + \mathbb{Z} = y + \mathbb{Z} \Rightarrow ||x|| = ||y||$, we can define $||\ ||$: $\mathbb{T} \rightarrow [0, \frac{1}{2}]$ by $||x + \mathbb{Z}|| = ||x||$. For $\alpha \in \mathbb{R} \setminus [-1, 1]$, define its Pisot set

$$\Lambda_\alpha = \{\, \lambda \in \mathbb{R} \setminus \{0\} \ : \ \lim_{j\rightarrow\infty} ||\lambda\alpha^j|| = 0 \,\}. \tag{3}$$

A Pisot–Vijayaraghavan (PV) number [3, 27] is $\alpha = \alpha_1 \in \mathscr{O}$ with $d(\alpha) \geq 2$ whose Galois conjugates $\alpha_2, ..., \alpha_d$ have moduli < 1. The Golden Mean $\frac{1+\sqrt{5}}{2} \approx 1.6180$ has Galois conjugate $\frac{1-\sqrt{5}}{2} \approx -0.6180$ so it is a PV number.

Theorem 1 (Pisot, Vijayaraghavan) *If $\alpha \in \mathbb{A}\setminus[-1, 1]$ has degree $d \geq 2$ and $\Lambda_\alpha \neq \phi$, then α is a PV number and*

$$\Lambda_\alpha = \{\, \alpha^m\mu \ : m \in \mathbb{Z},\ \mu \in \mathbb{Q}[\alpha] \setminus \{0\},\ T(\mu\alpha^j) \in \mathbb{Z},\ j = 0, ..., d-1 \,\}. \tag{4}$$

Furthermore, for $\lambda \in \Lambda_\alpha$, $||\lambda\alpha^j|| \rightarrow 0$ exponentially fast.

Proof Cassels ([3], Chapter VIII, Theorem 1) gives a simplified version, based on the properties of recursive sequences, of Pisot's proof in [27]. We relaxed the assumption that α is positive since α is a PV number iff $-\alpha$ is a PV number. The sequence $s(j) = T(\mu\alpha^j)$, $j \geq 0$ satisfies $s(j) = -c_{d-1}s(j-1) - \cdots - c_0 s(j-d),\ j \geq d$, where $P_\alpha(X) = X^d + c_{d-1}X^{d-1} + \cdots + c_0$. Then (4) implies that s has values in $\mathbb{Z}$. If $\lambda = \alpha^m\mu$ and $\mu = \mu_1, \mu_2, ..., \mu_d$ are the Galois conjugates of μ, then

$$||\lambda\alpha^j|| \leq |\mu\alpha^{j+m} - T(\mu\alpha^{j+m})| \leq \sum_{k=2}^{d} |\mu_k|\, |\alpha_k|^{j+m}, \quad j \geq -m, \tag{5}$$

converges to 0 exponentially fast as $j \rightarrow \infty$ since $|\alpha_k| < 1, k = 2, ..., d$.

This paper studies *refinable functions*, nonzero complex scalar- or vector-valued distributions φ satisfying a refinement equation

$$\varphi(x) = \sum_{k=1}^{\infty} a(k)\varphi(\alpha x - \tau(k)), \tag{6}$$

and whose Fourier transform $\widehat{\varphi}(y) = \int_{-\infty}^{\infty} \varphi(x)e^{-2\pi\, ixy}dx$ is continuous at $y = 0$ and $\widehat{\varphi}(0) \neq 0$. Here the *dilation* $\alpha \in \mathbb{R} \setminus [-1, 1]$, the coefficient sequence a, which is matrix valued for vector-valued refinable functions, decays exponentially fast, and τ takes values in $\mathbb{Z}[\alpha, \alpha^{-1}]$. Refinable functions constructed from integer $\alpha \geq 2$ and integer-valued τ include Daubechies' scaling functions used to construct orthonormal wavelet bases [6], basis functions constructed by Cavaretta, Dahmen, and Micchelli from stationary subdivision algorithms [4], and multiwavelets constructed from vector-valued refinable functions [15]. Equation (6) is equivalent to

$$\widehat{\varphi}(y) = \widehat{a}(y\alpha^{-1})\widehat{\varphi}(y\alpha^{-1}) \tag{7}$$

$$\widehat{a}(y) = |\alpha|^{-1} \sum_{k=1}^{\infty} a(k)e^{-2\pi i\tau(k)y}. \tag{8}$$

For scalar-valued φ, $\widehat{a}(0) = 1$, for vector-valued φ, $\widehat{a}(0)\,\widehat{\varphi}(0) = \widehat{\varphi}(0)$, and

$$\widehat{\varphi}(y\alpha^{J}) = \left(\prod_{j<J} \widehat{a}(y\alpha^{j})\right).\widehat{\varphi}(0), \quad J \in \mathbb{Z}, \tag{9}$$

where for matrices the factors move right with decreasing j.

Example 2 We call $\varphi = \chi_{[0,1]}$ the boxcar function. It satisfies (6) with $\alpha = 2$, $a(0) = a(1) = 1$, $\tau(0) = 0$, $\tau(1) = 1$. Then $\widehat{a}(y) = e^{\pi iy}\cos \pi y$ and

$$\widehat{\varphi}(y2^{J}) = e^{\pi iy2^{J}} \frac{\sin \pi y2^{J}}{\pi y2^{J}} = e^{\pi iy2^{J}} \prod_{j<J} \cos(\pi y2^{J}). \tag{10}$$

φ is integrable and $\widehat{\varphi}$ vanishes at infinity, despite the fact that $\widehat{\varphi}(\pi\lambda 2^{j})$ converges to 1 for every $\lambda \in \mathbb{Z}[2, \frac{1}{2}]$ (the set of dyadic rational numbers), because for every such λ there exists $j \in \mathbb{Z}$ such that $\cos(\pi\lambda 2^{j}) = 0$.

Example 3 $\alpha = 2$, $a(k) = \tau(k) = 2^{1-k}$, $k \in \mathbb{N}$ gives the *dyadic* function and $\widehat{a}(y) = \sum_{k\in\mathbb{N}} 2^{-k}\exp(-2\pi iy2^{1-k})$ is Bohr almost periodic [2] since

$$|\widehat{a}(y + 2^{L-1}) - \widehat{a}(y)| \leq 2^{-L}, \quad y \in \mathbb{R},\ L \in \mathbb{N}. \tag{11}$$

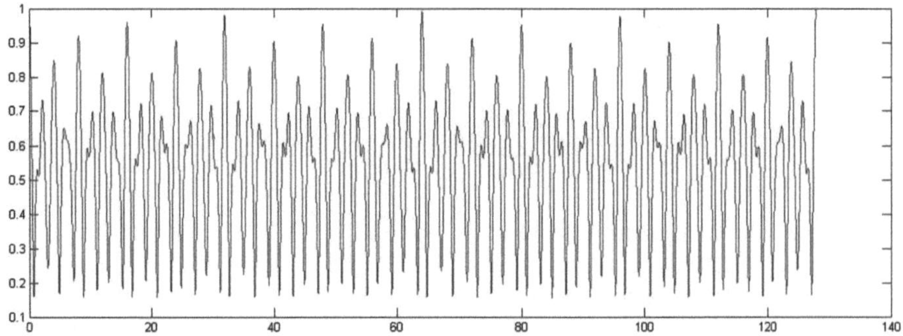

Fig. 1 Fourier transform of the dyadic refinement sequence

Furthermore, $\inf_{y\in\mathbb{R}} |\widehat{a}(y)| > 0$, and $\lim_{j\to\infty} \widehat{a}(\lambda 2^j) = 1$ for every $\lambda \in \mathbb{Z}[2, \frac{1}{2}]$. Fig. 1 shows the modulus of $\widehat{a}$ over $[0, 128]$. $\widehat{\varphi}$ does not vanish at infinity since $\lim_{j\to\infty} \widehat{\varphi}(2^j) \approx 0.2578 + 0.0692i$.

Example 4 If $\alpha = \frac{1+\sqrt{5}}{2}$ is the Golden Mean, then $\varphi = \begin{bmatrix} \chi_{[0,\alpha^{-1}]} \\ \chi_{[0,1]} \end{bmatrix}$ satisfies (6) with $a(1) = \begin{bmatrix} 0 & 1 \\ 0 & 1 \end{bmatrix}$, $a(2) = \begin{bmatrix} 0 & 0 \\ 1 & 0 \end{bmatrix}$, $\tau(1) = 0$, $\tau(2) = 1$.

$$\widehat{a}(y) = \alpha^{-1} \begin{bmatrix} 0 & 1 \\ e^{-2\pi i y} & 1 \end{bmatrix}, \quad \widehat{\varphi}(0) = \begin{bmatrix} \alpha^{-1} \\ 1 \end{bmatrix}. \tag{12}$$

$\widehat{a}(y)$ is never zero and satisfies $\lim_{j\to\infty} \widehat{a}(\lambda\alpha^j) = \widehat{a}(0)$ for all $\lambda \in \mathbb{Z}[\alpha, \alpha^{-1}]$, and

$$\widehat{\varphi}(y\alpha^J) = \begin{bmatrix} e^{-\pi i y\alpha^{J-1}} \frac{\sin(\pi y\alpha^{J-1})}{\pi y\alpha^J} \\ e^{-\pi i y\alpha^J} \frac{\sin(\pi y\alpha^J)}{\pi y\alpha^J} \end{bmatrix} = \left(\prod_{j<J} \alpha^{-1} \begin{bmatrix} 0 & 1 \\ e^{-2\pi i y\alpha^j} & 1 \end{bmatrix} \right) \widehat{\varphi}(0). \tag{13}$$

This example is related to the wavelets with PV dilations constructed by Gazeau, Patera, and Spiridinov [13, 14] and multiresolution analyses on quasicrystals [22]. For $\lambda \in \mathbb{Z}[\alpha, \alpha^{-1}]$, $\widehat{\varphi}(\lambda\alpha^j)$ is close to the eigenspace of $\widehat{a}(0)$ with eigenvalue $\alpha; -\frac{1}{\alpha}$ for $j \approx -\infty; \infty$. This fact makes ϕ integrable despite $\widehat{a}$ never vanishing, in contrast to scalar refinable functions with PV dilations discussed below.

Erdös [8] proved that if α is a PV number, then the refinable function satisfying

$$\varphi(x) = \frac{|\alpha|}{2}\varphi(\alpha x) + \frac{|\alpha|}{2}\varphi(\alpha x - 1) \tag{14}$$

(a Bernoulli convolution) is not integrable by showing that

$$\widehat{\varphi}(\alpha^J) = e^{-\pi i \frac{\alpha^J}{\alpha - 1}} \prod_{j<J} \cos(\pi \alpha^j) \tag{15}$$

fails to converge to 0 as $J \to \infty$. We give his proof. Since $||\alpha^j||$ converges to 0 exponentially fast as $|j| \to \infty$, $\cos \pi \alpha^j$ converges to ± 1 exponentially fast and hence $\varphi(\alpha^J)$ converges to 0 iff there exists $j \in \mathbb{Z}$ such that $\cos \pi \alpha^j = 0$, or equivalently if $\alpha^j \in \frac{1}{2} + \mathbb{Z}$. This is impossible because if $\alpha^j \in \mathbb{Q}$ then the Galois conjugates of α would be multiples of each other by roots of a cyclotomic polynomial and thus have identical moduli. Dai, Feng, and Wang [5] extended Erdös' result to scalar-valued refinable functions φ that satisfy (6) where a has finite support, α is a PV number, and τ is integer valued. We give their proof. Since τ is integer valued, $\widehat{a}$ in (8) has period 1. Since a has finite support, $\widehat{a}$ is real analytic so the set of its zeros in [0, 1) is a finite set F, so the set of zeros of $\widehat{a}$ equals $F + \mathbb{Z}$. If $\widehat{\varphi}$ vanished at infinity then for every $m \in \mathbb{N}$, $\lim_{J\to\infty} \widehat{\varphi}(m\alpha^J) = 0$ and since $||m\alpha^j||$ converges to 0 exponentially fast as $|j| \to \infty$, the same argument used to obtain Erdös' result implies that there exists $j_m \in \mathbb{Z}$ such that $m\alpha^{j_m} \in F + \mathbb{Z}$. Let $d = d(\alpha)$. Equation (1) implies that there exist unique $k_m \in \mathbb{Z}$ and $\beta_m \in \alpha\mathbb{Z} + \cdots + \alpha^{d-1}\mathbb{Z}$ with

$$m\alpha^{j_m} = mk_m + m\beta_m, \quad j_m \geq 0 \tag{16}$$

and (2) implies that

$$m\alpha^{j_m} = (N(\alpha)B(\alpha)))^{j_m} mk_m + (N(\alpha)B(\alpha)))^{j_m} m\beta_m, \quad j_m < 0. \tag{17}$$

Since $\alpha^{j_m} \notin \mathbb{Q}$, $\beta_m \neq 0$, choosing a prime p that does not divide $N(\alpha)B(\alpha)$ and letting $m = p^n$, $n \in \mathbb{N}$ produces a set of values of $(N(\alpha)B(\alpha)))^{j_m} m\beta_m$ that are infinite modulo $\mathbb{Z}$ (infinite modulo $\mathbb{Q}$ since $1, \alpha, ..., \alpha^{d-1}$ are rationally independent), contradicting the fact that F is finite.

This argument holds under the weaker assumption that a decays exponentially fast because then $\widehat{a}$ is a real-analytic function with period 1. The remainder of this paper provides support for the following extension of these results:

Conjecture 1 If ϕ is a scalar-valued refinable function satisfying (6), α is a PV number, a decays exponentially fast, and τ has values in $\mathbb{Z}[\alpha, \alpha^{-1}]$, then $\widehat{\varphi}$ does not vanish at infinity and hence φ is not integrable.

2 Solenoidal Representation

Let $\mathbb{T}^{\mathbb{Z}}$ be the group of functions $g : \mathbb{Z} \to \mathbb{T}$ under pointwise addition and the product topology. Tychonoff's theorem [31] implies that $\mathbb{T}^{\mathbb{Z}}$ is compact. Define the *shift automorphism* $\sigma : \mathbb{T}^{\mathbb{Z}} \to \mathbb{T}^{\mathbb{Z}}$ by $(\sigma g)(j) = g(j+1)$, $g \in \mathbb{T}^{\mathbb{Z}}$, $j \in \mathbb{Z}$, and homomorphisms $\rho_n : \mathbb{T}^{\mathbb{Z}} \rightarrow \mathbb{T}^n$ by $\rho_n(g) = [g(0), ..., g(n-1)]^T, \;\; g \in \mathbb{T}^{\mathbb{Z}}, \; n \in \mathbb{N}.$ Let α be a PV number of degree $d = d(\alpha)$. Its Galois conjugates $\alpha = \alpha_1, ...\alpha_d$ are the roots of its minimal polynomial $P_\alpha(X) = X^d + c_{d-1}X^{d-1} + \cdots + c_0 \in \mathbb{Z}[X]$. Define the Vandermonde V, Frobenius companion C, and Diagonal D, matrices

$$V = \begin{bmatrix} 1 & \dots & 1 \\ \alpha_1 & \dots & \alpha_d \\ \vdots & \dots & \vdots \\ \alpha_1^{d-1} & \dots & \alpha_d^{d-1} \end{bmatrix} C = \begin{bmatrix} 0 & 1 & \dots & 0 \\ \vdots & \ddots & \ddots & 0 \\ 0 & \dots & 0 & 1 \\ -c_0 & \dots & \dots & -c_{d-1} \end{bmatrix} D = \begin{bmatrix} \alpha_1 & \dots & 0 \\ \vdots & \ddots & \vdots \\ 0 & \dots & \alpha_d \end{bmatrix}$$

$\det V = \prod_{i<j}(\alpha_j - \alpha_i) \neq 0 \Rightarrow V$ is invertible, $CV = VD$, and $V^{-1}C = DV^{-1}$. Let $S_- = \{[0, s_2, ..., s_d]^T : s_j \in \mathbb{C}, \;\; \alpha_j = \overline{\alpha}_k \Rightarrow s_j = \overline{s}_k\}, \;\; S_+ = \{[s_1, 0, ..., 0]^T : s_1 \in \mathbb{R}\}$, $S = S_+ + S_-$. These are real subspaces of $\mathbb{C}^d$ with dimensions $d-1, 1, d$. Construct homomorphisms $\vartheta : S \rightarrow \mathbb{T}^{\mathbb{Z}}$ and $\theta : \mathbb{R} \rightarrow \mathbb{T}^{\mathbb{Z}}$,

$$\vartheta(s)(j) = \sum_{k=1}^{d} s_k \alpha_k^j + \mathbb{Z}, \;\; s \in S, \; j \in \mathbb{Z}, \tag{18}$$

$$\theta(y)(j) = y\alpha^j + \mathbb{Z}, \;\; y \in \mathbb{R}, \; j \in \mathbb{Z}, \tag{19}$$

and σ invariant subgroups $M_\pm = \vartheta(S_\pm)$, $M = M_+ + M_-$, and $G = \overline{M}$. Then $DS_\pm = S_\pm$, $S = V^{-1}R^d$, $M = \vartheta(S)$, $\rho_d(M) = \mathbb{T}^d$, $\rho_d(G) = \mathbb{T}^d$, and hence G is a compact, connected, abelian group with dimension $\geq d$. The restriction $\sigma : G \rightarrow G$ gives a dynamical system (G, σ) that is an *extension* [11] of the dynamical system $(\mathbb{T}^d, C)$ since $C \circ \rho_d = \rho_d \circ \sigma$. $0 \in G$ is an equilibrium point of σ and M_+; M_- is the unstable;stable manifold of 0. Define $K =$ kernel of $\rho_d : G \rightarrow \mathbb{T}^d$, $K_- = \bigcup_{j \in \mathbb{Z}} \sigma^j(K)$, $G_+ = M_+$, $G_- = M_- + K_-$, and $H = G_+ \cap G_-$. Then $G = M + K = G_+ + G_-$ and the orbits of points in H are *homoclinic* since $h \in H \Rightarrow \lim_{j \rightarrow \pm\infty} \sigma^j h = 0$.

Remark 1 Following Lind and Ward ([23], p. 411), we define a *solenoid* to be a compact, connected, dimension $n < \infty$ abelian group. Its dual group (see Sect. 4) is a discrete, torsion-free, rank n abelian group, or equivalently, a subgroup of $\mathbb{Q}^n$. The dynamical system (G, σ) is *expansive*, as defined by Lam [19], if there exists a neighborhood U of 0 such that $\bigcap_{j \in \mathbb{Z}} \sigma^j(U) = \{0\}$. In [20], we used *topological entropy* and Pontryagin duality to prove that every group that admits an expansive automorphism is a solenoid. Schmidt ([29], Chap. III) used algebraic methods to characterize more general expansive transformation groups.

Theorem 2 *The assumptions above imply that:*

1. *G is isomorphic to a group extension of $\mathbb{T}^d$ by K.*
2. *If $|c_0| = 1$, then $K = \{0\}$ and G is isomorphic to $\mathbb{T}^d$.*
3. *If $|c_0| > 1$, then K is homeomorphic to Cantor's set.*

4. *G is a d-dimensional solenoid and (G, σ) is expansive.*
5. *$\theta(\mathbb{R}) = \vartheta(S_+)$ is dense in G.*
6. *$\theta(\Lambda_\alpha) = H$.*

Proof (1) holds since $\rho_d : G \to \mathbb{T}^d$ is surjective. (2–3) hold since (18) $\Rightarrow$ every $g \in G_0$, hence every $g \in G$, satisfies

$$(P_\alpha(\sigma)g)(k) = c_0 g(k) + \cdots + c_{d-1} g(k+d-1) + g(k+d) = 0, \;\; k \in \mathbb{Z},$$

so $K = \{g \in \mathbb{T}^{\mathbb{Z}} : g(k) = 0, k \geq 0 \text{ and } c_0^{-k} g(k) = 0, k < 0\}$. (4) holds since $G = G_+ + G_-$. (5) holds since $(\sigma\theta(x))(j) = \theta(x)(j+1) = \theta(x\alpha) \Rightarrow \theta(\mathbb{R})$ is σ-invariant and $\rho_d(\theta(\mathbb{R})) = R[1, \alpha, ..., \alpha^{d-1}]^T + \mathbb{Z}^d$ is dense in $\mathbb{T}^d$ by the Kronecker–Weyl theorem [33], (Appendix, Theorem 2) since the entries of $[1, \alpha, ..., \alpha^{d-1}]^T$ are rationally independent. (6) holds since $\theta(s_1) \in H \Leftrightarrow \theta(s_1) \in M_- + K_-$ iff there exist $-[0, s_2, ..., s_d]^T \in S_-, m \in \mathbb{Z}$ with $\theta(s_1) + \vartheta([0, s_2, ..., s_d]^T) \in \sigma^m K \Leftrightarrow \rho_d(\sigma^{-m}\vartheta(s)) = 0$ where $s = [s_1, s_2, ..., s_d]^T$. Theorem 1 implies that this condition holds iff $VD^{-m}s = [T(\alpha^{-m}s_1), ..., T(\alpha^{-m+d-1}s_1)]^T \in \mathbb{Z}^d \Leftrightarrow s_1 \in \Lambda_\alpha$.

Assume that α is a PV number of degree $d = d(\alpha)$, $a : \mathbb{N} \to \mathbb{C}$ decays exponentially fast, $\sum_{k \in \mathbb{N}} a(k) = 1$, and $\tau : \mathbb{N} \to \mathbb{Z}[\alpha, \alpha^{-1}]$. For $k \in \mathbb{N}$ define $\tau_k : \mathbb{Z} \to \mathbb{Z}$ such that $\tau(k) = \sum_{j \in \mathbb{Z}} \tau_k(j)\alpha^j$. Each τ_k is finitely supported. Construct $A : G \to \mathbb{C}$

$$A(g) = |\alpha|^{-1} \sum_{k \in \mathbb{N}} a(k) \exp\left(-2\pi i \sum_{j \in \mathbb{Z}} \tau_k(j) g(k)\right), \quad g \in G. \tag{20}$$

Theorem 3 *These assumptions give the solenoidal representation $\widehat{a} = A \circ \theta$.*

Proof Follows from (8) and (19).

3 Zero Sets

This section develops relationships between and properties of the zero sets $\mathscr{S}(\widehat{\phi})$, $\mathscr{S}(\widehat{a})$, and $\mathscr{S}(A)$. Since a has exponential decay, A is a real-analytic function and $\mathscr{S}(A)$ is a real-analytic set. Theorem 3 implies $\theta(\mathscr{S}(\widehat{a})) \subset \mathscr{S}(A)$ and (7) implies

$$\mathscr{S}(\widehat{\varphi}) = \alpha \mathscr{S}(\widehat{a}) + \alpha \mathscr{S}(\widehat{\varphi}), \tag{21}$$

(+ of zero multiplicities) so the lower $\underline{d}$ and upper $\overline{d}$ asymptotic densities satisfy

$$\underline{d}(\mathscr{S}(\widehat{a})) = (|\alpha| - 1)\, \underline{d}(S(\widehat{\varphi})) \leq \overline{d}(\mathscr{S}(\widehat{a})) = (|\alpha| - 1)\, \overline{d}(S(\widehat{\varphi})). \tag{22}$$

Theorem 4 *If A is not zero, then $\overline{d}(\mathscr{S}(\widehat{a})) < \infty$.*

Proof $\mathscr{H}(G) = \{$closed subsets of $G\}$ with the Hausdorff topology is a compact space ([10], pp. 205, 255), and $\mathscr{T}(G) = \{\theta([0, 1]) + g : g \in G\}$ is a closed subset and hence compact. If $\overline{d}(\mathscr{S}(\widehat{a})) = \infty$ then there exist $g \in G$ and a sequence $x_n \in \mathbb{R}$, $n \in \mathbb{N}$ such that A has at least n zeros in $\theta([x_n, x_n + 1])$ and $\lim_{n\to\infty} \theta(x_n) = g$. Define $f(x) = |A(\theta(x) + g)|^2$, $x \in \mathbb{R}$. Then either f has an infinite number of zeros in the interval $[0, 1]$ or Rolle's theorem implies that it has a zero of infinite order in $[0, 1]$. Since A and hence f is real-analytic, $f = 0$. Since $\theta(\mathbb{R})$, and hence $\theta(\mathbb{R}) + g$, is dense in G, A is zero thus giving a contradicting.

For $m \in \mathbb{Z}$ and $\varepsilon = [\varepsilon_2, ..., \varepsilon_d]$ satisfying $\varepsilon_k > 0$ and $\alpha_j = \overline{\alpha}_k \Rightarrow \varepsilon_j = \varepsilon_k$ define

$$U(m, \varepsilon) = \sigma^m K + \{\, \vartheta(s) \,:\, s \in S_- \text{ and } |s_k| < \varepsilon_k,\ k = 2, ..., d \,\} \subset G_-. \tag{23}$$

Since $U(m, \varepsilon)$ decreases to $\{0\}$ as $m \to \infty$ and $\varepsilon_k \to 0$ we can, and will, choose m and ε so that A never vanishes on $U(m, \varepsilon)$. Also, $U(m, \varepsilon)$ is σ invariant since $\sigma U(m, \varepsilon) = U(m + 1, [\varepsilon_2|\alpha_2|, ..., \varepsilon_d|\alpha_d|]) \subset U(m, \varepsilon)$. Let a be the number of real α_k, $k = 2, ..., d$ and b be the number of complex conjugate pairs of α_k, $k = 2, ..., d$, and let $\gamma = |\det V|\, |c_0|^{-m}\, 2^a \pi^b\, \varepsilon_1 \cdots \varepsilon_d$. For $L > 0$ define

$$W(L) = \{\, V D^{-m}[y, s_2, ..., s_d]^T \,:\, y \in (-L, L), |s_k| < \varepsilon_k,\ k = 2, ..., d \,\},$$

$$Y(L) = \{\, y \in (-L, L) \,:\, \theta(y) \in U(m, \varepsilon) \,\}.$$

The sets $W(L)$, $L > 0$ are convex cylinders parallel to the vector $[1, \alpha, ..., \alpha^{d-1}]^T$ whose entries are rationally independent, therefore

$$\lim_{L\to\infty} \frac{1}{2L} \text{ card } W(L) \cap \mathbb{Z}^d = \frac{1}{2L} \text{vol } W(L) = \gamma. \tag{24}$$

Lemma 1 *The* $\mathbb{R}$*-linear function* $\xi : W(L) \cap \mathbb{Z}^d \to (-L, L)$ *defined by* $\xi(w) = [1, 0, ..., 0]\, D^m V^{-1} w$ *is a bijection onto* $Y(L)$*. Therefore*

$$\lim_{L\to\infty} \frac{1}{2L} \textit{ card } Y(L) = \gamma. \tag{25}$$

Proof Assume $w \in W(L) \cap \mathbb{Z}^d$. If $\xi(w) = 0$, then $w \in V D^{-m} S_-$ so $\lim_{\ell\to\infty} C^\ell w = 0$. Since $w \in \mathbb{Z}^d$, $w = 0$, so ξ is injective. Assume that $y \in (-L, L)$. Then $\theta(y) \in U(m, \varepsilon)$ iff there exist $-s_k$ that satisfy $|s_k| < \varepsilon_k$, $k = 2, ..., d$ and

$$\theta(y) \in \sigma^j K + \vartheta([0, -s_2, ..., -s_d]^T) \Leftrightarrow \sigma^{-m}\vartheta([y, s_2, ..., s_d]^T) \in K$$

$$\Leftrightarrow \rho_d(\sigma^{-m}\vartheta([y, s_2, ..., s_d]^T)) = 0 \Leftrightarrow V D^{-m}[y, s_2, ..., s_d]^T \in \mathbb{Z}^d.$$

Since $V D^{-m}[y, s_2, ..., s_d]^T \in W(L)$, the last inclusion holds iff $y \in \xi(W(L) \cap \mathbb{Z}^d)$. This shows that ξ maps $W(L) \cap \mathbb{Z}^d$ onto $Y(L)$. Then (24) $\Rightarrow$ (25).

Theorem 5 *If $\widehat{\varphi}$ vanishes at infinity, then $\underline{d}(\mathscr{S}(\widehat{\varphi})) \geq \gamma$.*

Proof (25) implies that it suffices to show that $\widehat{\varphi}$ vanishes on $Y(L)$ for every $L > 0$. Since $\widehat{\varphi}$ vanishes at infinity,

$$0 = \lim_{J\to\infty} \widehat{\varphi}(\alpha^J y) = \lim_{J\to\infty} \widehat{\varphi}(y) \prod_{j=1}^{J} \widehat{a}(\alpha^j y).$$

Since $y \in Y(L)$ and $j \geq 1$ imply $\theta(\alpha^j y) = \sigma^j \theta(y) \in U(m, \varepsilon)$, and since A never vanishes on $U(m, \varepsilon)$, (3) implies that $\widehat{a}(\alpha^j y) = A(\sigma^j(\theta(y)) \neq 0$. Since $\sigma^j(\theta(y))$ converges to 0 exponentially fast, $\prod_{j=1}^{\infty} \widehat{a}(\alpha^j y) \neq 0$, and hence $\widehat{\varphi}(y) = 0$.

Corollary 1 *If $\widehat{\varphi}$ vanishes at infinity, then $\underline{d}(\mathscr{S}(\widehat{a})) \geq (|\alpha| - 1)\,\gamma$. $\mathscr{S}(A)$ is a union of embedded manifolds in G and has dimension $d - 1$.*

Proof The first assertion follows from (22). Since $\mathscr{S}(A)$ is a real-analytic set, it is homeomorphic to a union of embedded manifolds by Lojasiewicz's structure theorem for real-analytic sets [17, 21, 24]. Since $\theta(\mathbb{R})$ is a uniformly distributed embedding, if the dimension of $\mathscr{S}(A)$ was less than $d - 1$ then $\underline{d}(\mathscr{S}(\widehat{a})) = 0$.

Theorem 6 *If α is a PV number of degree $d \geq 2$, then the set of norms $N(\Lambda_\alpha)$ is a set of rational numbers whose denominators have only a finite number of prime divisors and whose numerators are values of integral forms (homogeneous polynomials) of degree d in d integer variables. The number of these integer values having modulus $\leq L$ is asymptotically bounded below by $O(L^{2/d})$ for $d \geq 3$ and by $O(L/(\log L)^p)$ for some $p \in (0, 1)$ for $d = 2$.*

Proof Let $\alpha = \alpha_1, ...\alpha_d$ be the Galois conjugates of α. Theorem 1 implies that $\lambda \in \Lambda_\alpha$ iff there exist $m \in \mathbb{Z}$ and $\mu_k \in \mathbb{Q}[\alpha_k)$ such that $\lambda = \mu_1 \alpha^j$ where $[\mu_1, ..., \mu_d]^T \in V^{-1}\mathbb{Z}^d$. The elements of the kth row of V^{-1}, being the coefficients of the Lagrange interpolating polynomial $Q_k(X) = \frac{P_\alpha(X)}{(X-\alpha_k)P'(\alpha_k)}$, belong to the field $\mathbb{Q}[\alpha_k]$ and the elements in every column of V^{-1} are Galois conjugates. Therefore, $N(\lambda) = N(\alpha)^m\, N(\mu_1)$, and $N(\mu_1) = \prod_{k=1}^{d} \mu_k$ is a form with rational coefficients of degree d in d integer variables (the coordinates of $\mathbb{Z}^d$). The denominators of the coefficients of the form $N(\mu_1)$ must divide det V so the prime factors of the denominators of the numbers in $N(\Lambda_\alpha)$ must divide $N(\alpha)$ or det V. Therefore, a positive fraction of the numbers in $N(\Lambda_\alpha)$ have numerators that are values of an integral form of degree d in d integer variables. For $d \geq 3$, we obtain a binary form of degree 3 by setting all except 2 of these integer variables to 0 and obtain the lower asymptotic bound for the values of the numerators by a Theorem of Erdös and Mahler [9]. For $d = 2$ we obtain a binary quadratic form and we obtain a lower asymptotic bound given by a Theorem of Odoni ([26], Theorem S).

We refer the reader to Sect. 4 for a discussion of Pontryagin duality. If $\chi \in \widehat{G}$ and $c \in \mathbb{T}_c$, then the zero set $\mathscr{S}(\chi - c) = \{g \in G : \chi(g) - c = 0\}$ has dimension $d - 1$. We call a real-analytic subset of G *simple* if it is contained in a finite union of such sets. Lagarias and Yang conjectured [18] that certain real-analytic subsets of $\mathcal{T}^n$, that arise in the construction of refinable functions of several variables related to tilings and that are analogous to our set $\mathcal{S}(A)$, are simple. We used Lojasiewicz's theorem [21] to prove their conjecture. Thus, we find the following interesting result:

Theorem 7 *If A is nonzero, then $\mathcal{S}(A)$ is not simple.*

Proof The argument used in the proof of Theorem 5 shows that for every $\lambda \in \Lambda_\alpha$ there exists $m \in \mathbb{Z}$ such that $\lambda\alpha^m \in \mathcal{S}(\widehat{a})$. Therefore, Theorem 6 implies that the set of norms $N(\mathcal{S}(\widehat{a}))$ contains a set of rational numbers whose numerators whose modulus is less than L has asymptotic density $> O(L^{1/d})$. If $\mathcal{S}(A)$ were a proper simple subset of G then all but a finite number of points in $\mathcal{S}(\widehat{a})$ would be contained in a finite union of sets having the form $\beta + \delta\mathbb{Z}$ where β and δ are elements in $\mathbb{Q}[\alpha]$. However, $N(\beta + \delta k)$ is a form of degree d with rational coefficients in single integer variable k, and therefore, the set of numerators of the values of this form has asymptotic density $O(L^{1/d})$, thus giving a contradiction.

Remark 1 Integral binary quadratic forms were studied by Gauss [12], who focussed on forms having negative discriminant. Asymptotic estimates for the number of integers represented by integral binary quadratic forms with negative discriminant were obtained in the doctoral dissertation of Bernays [1] and by James [16]. Numerical studies were compiled by Sloane [30].

Inspired by developments in Diophantine geometry and o-minimal theory, we make the following assertion whose validity proves Conjecture 1.

Conjecture 2 Every real-analytic subset of $G\{0\}$ that intersects every homoclinic orbit is simple.

4 Appendix: Pontryagin Duality and Kronecker–Weyl Theorem

A *character* of a locally compact abelian topological group G is a continuous homomorphism $\chi : G \to \mathbb{T}_c$. The *dual group* $\widehat{G}$ consists of all characters under pointwise multiplication and the topology of uniform convergence on compact subsets. The *Pontryagin duality* theorem says that the homomorphism $\gamma : G \to \widehat{\widehat{G}}$

$$\gamma(g)(\chi) = \chi(g), \quad g \in G, \ \chi \in \widehat{G} \tag{26}$$

is a bijective isomorphism. This was proved for second countable groups that are either compact or discrete in 1934 by Lev Semyonovich Pontryagin [28] and extended to general locally compact groups in 1934 by Egbert van Kampen [32]. This theory shows that G is compact; discrete; connected; dimension d iff $\widehat{G}$ is discrete;

compact; torsion free; rank d. For $a \in \mathbb{Z}^n$ define $\chi_a \in \widehat{\mathbb{T}^n}$ by $\chi_a(g) = \exp 2\pi i(a(1)g(1) + \cdots + a(n)g(n))$. The mapping $a \to \chi_a$ is an isomorphism. If H is a closed subgroup of G, then the quotient G/H is locally compact and we have an obvious injective homomorphism $\widehat{G/H} \to \widehat{G}$. Therefore, Pontrygin duality implies that H is proper iff every character on G that vanishes on H vanishes on G. This gives the following classical result [33].

Lemma 2 *(Kronecker–Weyl Theorem) If $v \in \mathbb{R}^n$, then $\mathbb{R}v + \mathbb{Z}^n$ is dense in $\mathbb{T}^n$ iff the entries of v are rationally independent.*

Proof The closure $H = \overline{\mathbb{R}v + \mathbb{Z}^n}$ is a closed subgroup of $\mathbb{T}^n$ and is proper iff there exists $a \in \mathbb{Z}^n \backslash \{0\}$ with $\chi_a(tv + \mathbb{Z}^n) = \exp(2\pi i t a^T v),\ t \in \mathbb{R}$, or equivalently, $a^T v = 0$. This occurs iff the entries of v are rationally dependent.

Acknowledgements This work was done during my visit in the Department of Mathematics and Statistics at Auburn University in Spring 2016. I thank Professor Richard Zalik for his great hospitality during my stay in the department and Keith Mathews and John Robertson for helpful discussions about the representation of integers by integral homogeneous forms.

References

1. P. Bernays, *Über die Darstellung von positiven, ganzen Zahlen durch die primitiven binären quadratischen Formen einer nichtquadratischen Diskriminante*, Dissertation, Universität Göttingen, 1912
2. H. Bohr, Zur Theorie der fastperiodischen Funktionen I. Acta Mathematica **45**, 29–127 (1924)
3. *An Introduction to Diophantine Approximation*, 45th edn. Cambridge Tracts in Mathematics and Mathematical Physics (Cambridge University Press, New York, 1957)
4. A.S. Cavaretta, W. Dahmen, C.A. Micchelli, Stationary subdivision. Memoirs Am. Math. Soc. **93** (1991)
5. X.R. Dai, D.J. Feng, Y. Wang, Refinable functions with non-integer dilations. J. Funct. Anal. **250**, 1–20 (2007)
6. I. Daubechies, Orthonormal bases of compactly supported wavelets. Commun. Pure Appl. Math. **41**(7), 909–966 (1988)
7. R. Dedekind, Über den Zusammenhang zwischen der Theorie der Ideale und der Theorie der höheren Congruenzen. Abhandlungen der Königlichen Gesellschaft der Wissenschaften zu Göttingen **23**(1), 3–37 (1878)
8. P. Erdös, On a family of symmetric Bernoulli convolutions. Am. J. Math. **61**, 974–976 (1939)
9. P. Erdös, K. Mahler, On the number of integers which can be represented by a binary form. J. Lond. Math. Soc. **13**, 134–139 (1938)
10. J.M.G. Fell, A Hausdorff topology for the closed sets of a locally compact non-Hausdorff space. Proc. Am. Math. Soc. **13**, 472–476 (1962)
11. H. Furstenberg, *Recurrence in Ergodic Theory and Combinatorial Number Theory* (Princeton University Press, New Jersey, 1981)
12. C.F. Gauss, *Disquisitiones Arithmeticae* (Yale University Press, London, 1965)
13. J.P. Gazeau, J. Patera, Tau-wavelets of Haar. J. Phys. A: Math. Gen. **29**, 4549–4559 (1996)
14. J.P. Gazeau, V. Spiridonov, Toward discrete wavelets with irrational scaling factor. J. Math. Phys. **37**(6), 3001–3013 (1996)
15. J.S. Geronimo, D.P. Hardin, P.R. Massopust, Fractal functions and wavelet expansions based on several scaling functions. J. Approx. Theory **78**, 373–401 (1994)

16. R.D. James, The distribution of integers represented by quadratic forms. Am. J. Math. **60**(3), 737–744 (1938)
17. S.G. Krantz, H.R. Parks, *A Primer of Real Analytic Functions* (Birkhäuser, Boston, 1992)
18. J.C. Lagarias, Y. Wang, Integral self-affine tiles in R^n, part II: lattice tilings. J. Fourier Anal. Appl. **3**(1), 83–102 (1997)
19. P.F. Lam, On expansive transformation groups. Trans. Am. Math. Soc. **150**, 131–138 (1970)
20. W. Lawton, *The Structure of Compact Connected Groups Which Admit an Expansive Automorphism, Recent advances in Topological Dynamics*. Lecture Notes in Mathematics, vol. 318 (Springer, Berlin, 1973)
21. W. Lawton, Proof of the hyperplane zeros conjecture of Lagarias and Wang. J. Fourier Anal. Appl. **14**(4), 588–605 (2008)
22. W. Lawton, Multiresolution analysis on quasilattices. Poincare J. Anal. Appl. **2**, 37–52 (2015)
23. D.A. Lind, T. Ward, Automorphisms of solenoids and p-adic entropy. Ergodic Theory Dyn. Syst. **8**, 411–419 (1988)
24. S. Lojasiewicz, *Introduction to Complex Analytic Geometry* (Birkhäuser, Boston, 1991)
25. W. Narkiewicz, in *Elementary and Analytic Theory of Algebraic Numbers*, Springer Monographs in Mathematics (Springer, Berlin, 2004)
26. R.W.K. Odoni, Representations of algebraic integers by binary quadratic forms and norm forms from full modules of extension fields. J. Number Theory **10**, 324–333 (1978)
27. C. Pisot, *La répartition modulo 1 et nombres réels algébriques*, Ann. Sc. Norm. Super. Pisa, II, Ser. **7**, 205–248 (1938)
28. L. Pontrjagin, The theory of opological commutative groups. Ann. Math. **35**(2), 361–388 (1934)
29. K. Schmidt, *Dynamical Systems of Algebraic Origin* (Birkhäser, Basel, 1995)
30. N.J.A. Sloane, *Binary Quadratic Forms and OEIS* (2014), https://oeis.org/wiki/Binary_Quadratic_Forms_and_OEIS. Accessed 05–12 June 2014
31. Q.N. Tychonoff, Über die topologische Erweiterung von Räumen. Mathematische Annalen **102**(1), 544–561 (1930)
32. E.R. van Kampen, in Proceedings of the National Academy of Science, *Locally compact abelian groups*, (1934) 434–436
33. H. Weyl, Über die Gleichverteilung von Zahlen mod. Eins. Math. Ann. **77**, 313–352 (1916)

Polyhyperbolic Cardinal Splines

Jeff Ledford

Abstract In this note, we discuss solutions of differential equation $(D^2 - \alpha^2)^k u = 0$ on $\mathbb{R} \setminus \mathbb{Z}$, which we call polyhyperbolic splines. We develop the fundamental function of interpolation and prove various properties related to these splines.

Keywords Interpolation · Splines · Fundamental functions · Band-limited functions

1 Introduction

Beginning with the work of Schoenberg, see for instance [10], splines have been a popular area of research and a useful tool for approximation. This paper will focus on what are here called hyperbolic splines, following the convention of [11], but are often found in the literature as exponential splines [4, 12] and fit into the theory of L-splines [7]. The results of this paper are similar to those found in [1, 2, 5, 6, 8, 9]; in each of these works, various fundamental functions of interpolation are studied. The reader interested in a more general reference for fundamental functions is invited to consult [3] where radial basis functions are used to generate fundamental functions.

Our main result concerns band-limited functions and shows that as the degree tends to infinity the polyhyperbolic spline interpolant tends to the target function. Our methods require the use of the Fourier transform, for which we set the following convention. If $g \in L_1(\mathbb{R})$, then its Fourier transform, written $\hat{g}$, is given by

$$\hat{g}(\xi) = (2\pi)^{-1/2} \int_{\mathbb{R}} g(x) e^{-ix\xi} d\xi.$$

This definition is extended to distributions when necessary.

J. Ledford (✉)
Virginia Commonwealth University, 1015 Floyd Avenue, Richmond, VA 23284, USA
e-mail: jpledford@vcu.edu

G.E. Fasshauer and L.L. Schumaker (eds.), *Approximation Theory XV: San Antonio 2016*, Springer Proceedings in Mathematics & Statistics 201,
DOI 10.1007/978-3-319-59912-0_9

2 Polyhyperbolic Cardinal Splines

Definition 1 Given $\alpha > 0$ and $k \in \mathbb{N}$, the set H_α^k of all *k-hyperbolic cardinal splines* is given by

$$H_\alpha^k = \left\{ f \in C^{2k-2}(\mathbb{R}) : (D^2 - \alpha^2)^k f = 0, \text{ on } \mathbb{R} \setminus \mathbb{Z} \right\} \tag{1}$$

If $k \geq 1$, we say that $f \in H_\alpha^k$ is a *polyhyperbolic cardinal spline*. The justification for the name comes from the fact that $\{\cosh(\alpha\cdot), \sinh(\alpha\cdot)\}$ forms a fundamental set of solutions for the differential equation $(D^2 - \alpha^2)f = 0$. In fact, it is not too hard to see that $\{(\cdot)^j e^{\pm\alpha(\cdot)} : 0 \leq j < k\}$ forms a fundamental set of solutions for the equation $(D^2 - \alpha^2)^k f = 0$.

Most of this paper is analogous to the treatment given to polyharmonic cardinal splines in [6]. We begin with what is called the fundamental solution in [6]. The solution to the equation $(D^2 - \alpha^2)^k E_k(t) = \delta(t)$ is given by

$$E_k(t) = C_{\alpha,k}[e^{-\alpha|\cdot|}]^{*k}(t) \tag{2}$$

where f^{*k} is the convolution of f with itself k times and $C_{\alpha,k}$ is chosen so that the resulting multiple of the Dirac δ–function is 1. The Fourier transform of this function plays an important role in what follows, it is given by

$$\widehat{E_k}(\xi) = (-1)^k(\xi^2 + \alpha^2)^{-k}. \tag{3}$$

From this function, we develop a fundamental function of cardinal interpolation L_k, which is defined by its Fourier transform

$$\widehat{L_k}(\xi) = (2\pi)^{-1/2} \frac{\widehat{E_k}(\xi)}{\sum_{j\in\mathbb{Z}} \widehat{E_k}(\xi - 2\pi j)}. \tag{4}$$

We will also be concerned with the related periodic distribution $\widehat{\Phi_k}$, defined by

$$\widehat{\Phi_k}(\xi) = \left[\widehat{E_k}(\xi)\right]^{-1} \widehat{L_k}(\xi). \tag{5}$$

In order to study L_k and Φ_k further, we make use of some basic analysis in the complex domain. For $A \subset \mathbb{R}$, we define

$$A_\varepsilon = \{\zeta \in \mathbb{C} : \text{Re}(\zeta) \in A \text{ and } |\text{Im}(\zeta)| \leq \varepsilon\}.$$

We may now prove our first lemma.

Lemma 1 *There exists $\varepsilon > 0$ such that both $\widehat{\Phi_k}$ and $\widehat{L_k}$ have extensions to $\mathbb{R}_\varepsilon$ which are analytic.*

Proof We follow along the same lines as the $\alpha = 0$ case which may be found in [6]. We will henceforth assume that $\alpha > 0$. Set $q(\zeta) = -(\zeta^2 + \alpha^2)$ where $\zeta = \xi + i\eta$. We have

$$(2\pi)^{-1/2}[\widehat{\Phi_k}(\xi)]^{-1} = \left\{1 + [q(\xi)]^k F(\xi)\right\}[q(\xi)]^{-k}$$

where

$$F(\zeta) = \sum_{j \neq 0} [q(\zeta - 2\pi j)]^{-k}.$$

Since $\alpha > 0$, q^{-k} is analytic in $\mathbb{R}_\varepsilon$ for any $\varepsilon \in (0, \alpha)$. It is also clear that F is analytic on $(-\pi, \pi]_\varepsilon$. Now since $[q(\xi)]^k F(\xi) \geq 0$ for $\xi \in (-\pi, \pi]$, we can find $\varepsilon > 0$ (which may be smaller than our first choice) such that $1 + [q(\zeta)]^k F(\zeta)$ has no zeros in $(-\pi, \pi]_\varepsilon$. This shows that $\widehat{\Phi_k}$ possesses an extension which is analytic in $(-\pi, \pi]_\varepsilon$. Using periodicity, we may extend this result to $\mathbb{R}_\varepsilon$. The analytic extension of $\widehat{L_k}$ to $\mathbb{R}_\varepsilon$ is given by $\widehat{L_k}(\zeta) = \widehat{\Phi_k}(\zeta)[q(\zeta)]^{-k}$, which as the product of analytic functions is analytic.

Lemma 2 *We have that*

$$\Phi_k(x) = \sum_{j \in \mathbb{Z}} a_j \delta(x - j) \tag{6}$$

where the a_j's depend on k and α. Furthermore, there exists two constants $c, C > 0$ (which also depend on k and α) such that for all $j \in \mathbb{Z}$,

$$|a_j| \leq Ce^{-c|j|}. \tag{7}$$

Proof Periodicity implies that (6) holds with

$$a_j = (2\pi)^{-1/2} \int_{(-\pi,\pi]} \widehat{\Phi_k}(\xi) e^{ij\xi} d\xi.$$

To see that (7) holds, we use Lemma 1 to replace $(-\pi, \pi]$ with $\{\zeta \in \mathbb{C} : \text{Re}(\zeta) \in (-\pi, \pi]$ and $\text{Im}(\zeta) = \text{sgn}(j)c\}$ where $0 < c < \varepsilon$. We have

$$|a_j| \leq (2\pi)^{-1/2} e^{-c|j|} \int_{(-\pi,\pi]} |\widehat{\Phi_k}(\xi)| d\xi \leq Ce^{-c|j|}.$$

Theorem 1 *Let L_k be defined by its Fourier transform* (4) *where $k \in \mathbb{N}$. Then L_k has the following properties:*

1. *L_k is a k-hyperbolic cardinal spline.*
2. *For all $j \in \mathbb{Z}$, $L_k(j) = \delta_{0,j}$ where δ is the Kronecker δ function.*
3. *There are constants $c, C > 0$ (which depend on k and α) such that*

$$|L_k(x)| \leq Ce^{-c|x|}$$

for all $x \in \mathbb{R}$.

4. L_k *has the following representation in terms of* E_k*:*

$$L_k(x) = \sum_{j\in\mathbb{Z}} a_j E_k(x-j) = \Phi_k * E_k(x)$$

where Φ_k *is the function whose Fourier transform is defined by* (5) *and the* a_j*'s are those from Lemma 2.*

Proof Item 1. We see that $(D^2-\alpha^2)^k L_k = \Phi_k$, from examining the equation in the Fourier domain. The smoothness follows from Item 4. Now, Lemma 2 shows that $L_k \in H_\alpha^k$.

Item 2. We have

$$\begin{aligned} L_k(j) =& (2\pi)^{-1/2}\int_{\mathbb{R}} \widehat{L_k}(\xi)e^{ij\xi}d\xi = (2\pi)^{-1/2}\int_{-\pi}^{\pi}\sum_{l\in\mathbb{Z}}\widehat{L_k}(\xi-2\pi l)e^{ij\xi}d\xi \\ =& (2\pi)^{-1}\int_{-\pi}^{\pi} e^{ij\xi}d\xi = \delta_{0,j}. \end{aligned}$$

The interchange of the sum and in the integral in the "periodization trick" is justified by the decay of $\widehat{E_k}$ and the periodicity of the denominator of $\widehat{L_k}$. That the sum is $(2\pi)^{-1/2}$ is clear from (4) and Lemma 1.

Item 3. An argument completely analogous to the one given in Lemma 2 shows that this is true.

Item 4. This is clear from (4) and (5).

We turn briefly to the data that we wish to interpolate.

Definition 2 For $\beta > 0$, let Y^β be the set of all sequences $b = \{b_j : j \in \mathbb{Z}\}$ which satisfy

$$|b_j| \le C(1+|j|)^\beta, \tag{8}$$

for some constant $C > 0$.

Our next result shows what to expect when we interpolate data of this type.

Theorem 2 *Suppose that* $b = \{b_j : j \in \mathbb{Z}\}$ *is a sequence of polynomial growth and define the function* f_b *by*

$$f_b(x) = \sum_{j\in\mathbb{Z}} b_j L_k(x-j). \tag{9}$$

The following are true:

1. *The expansion* (9) *converges absolutely and uniformly in every compact subset of* $\mathbb{R}$.
2. *The function* f_b *is a* k*-hyperbolic spline and* $f_b(j) = b_j$ *for all* $j \in \mathbb{Z}$.

3. If $b \in Y^\beta$, then f_b satisfies

$$|f_b(x)| \leq C(1+|x|)^\beta,$$

for some $C > 0$.

Proof Items 1 and 2 are straightforward consequences of Theorem 1.
For item 3, we let $m_x \in \mathbb{Z}$ satisfy $m_x \leq x < m_x + 1$, then we have

$$\begin{aligned}
|f_b(x)| &\leq \sum_{j\in\mathbb{Z}} |b_j L_k(x-j)| \leq C \sum_{j\in\mathbb{Z}} (1+|j|)^\beta e^{-c|x-k|} \\
&= C(1+|m_x|)^\beta e^{-c|x-m_x|} \left\{ 1 + \sum_{j\neq m_x} \left(\frac{1+|j|}{1+|m_x|}\right)^\beta e^{c|x-m_x|-c|x-j|} \right\} \\
&=: C(1+|m_x|)^\beta e^{-c|x-m_x|} \{1 + A + B\}
\end{aligned}$$

where

$$A = \sum_{j>m_x} \left(\frac{1+|j|}{1+|m_x|}\right)^\beta e^{c|x-m_x|-c|x-j|},$$

and

$$B = \sum_{j<m_x} \left(\frac{1+|j|}{1+|m_x|}\right)^\beta e^{c|x-m_x|-c|x-j|}.$$

The estimates being similar, we show only the estimate for A:

$$\begin{aligned}
A &= e^{c|x-m_x|} \sum_{j=m_x+1}^{\infty} \left(\frac{1+|j|}{1+|m_x|}\right)^\beta e^{-c|x-j|} \\
&\leq e^c \sum_{k=1}^{\infty} \left(\frac{1+|m_x+k|}{1+|m_x|}\right)^\beta e^{-c(k+m_x-x)} \\
&\leq e^{c-c(m_x-x)} \sum_{k=1}^{\infty} (1+|k|)^\beta e^{-ck} \\
&\leq e^{2c} \sum_{k=1}^{\infty} (1+|k|)^\beta e^{-ck} < \infty.
\end{aligned}$$

Here, we have used the triangle inequality and re-indexed the sum. In the second inequality, we have used that $\beta > 0$. Finally, the last inequality comes from the fact that $m_x \leq x < m_x + 1$. Combining this estimate with the analogous estimate for B yields the desired result.

Note that our study has not shown that these splines are unique. We will leave this question to future study and end this note with a result concerning the behavior of polyhyperbolic splines as their degree tends to infinity.

For this, we recall that the classical Paley–Wiener space PW is defined by

$$PW = \{g \in L_2(\mathbb{R}) : \hat{g}(\xi) = 0 \text{ a.e. } \xi \notin [-\pi, \pi]\}.$$

A function $g \in PW$ is often called *band-limited*. The Paley–Wiener theorem shows that band-limited functions are entire functions of exponential type whose restrictions to the real line are in $L_2(\mathbb{R})$.

In light of Theorem 2, we may interpolate a function $g \in PW$ with a k-hyperbolic spline as follows

$$I_k[g](x) = \sum_{j \in \mathbb{Z}} g(j) L_k(x - j). \tag{10}$$

Theorem 3 *Suppose that* $g \in PW$. *The function* $I_k[g]$ *defined in* (10) *satisfies*

(i)

$$\lim_{k \to \infty} \|g - I_k[g]\|_{L_2(\mathbb{R})} = 0, \textit{ and} \tag{11}$$

(ii)

$$\lim_{k \to \infty} |g(x) - I_k[g](x)| = 0, \textit{ uniformly on } \mathbb{R}. \tag{12}$$

Proof We begin by noting that if $y = \{y_j : j \in \mathbb{Z}\} \in \ell_p$ for $1 \le p \le \infty$ then $f_y \in L_p(\mathbb{R})$ where f_y is defined by (9). Since L_k has exponential decay, the estimates of the form

$$\|f_y\|_{L_1(\mathbb{R})} \le C_k \|y\|_{\ell_1} \text{ and}$$
$$\|f_y\|_{L_\infty(\mathbb{R})} \le C_k' \|y\|_{\ell_\infty}$$

are easily established. Now, the Riesz–Thorin interpolation theorem provides a constant $C_{k,p} > 0$ such that

$$\|f_y\|_{L_p(\mathbb{R})} \le C_{k,p} \|y\|_{\ell_p}.$$

It is well known that if $g \in PW$, then $\{g(j) : j \in \mathbb{Z}\} \in \ell_2$. Hence, we can see that

$$\|I_k[g]\|_{L_2(\mathbb{R})} \le C_k \|\{g(j) : j \in \mathbb{Z}\}\|_{\ell_2}$$

for some constant $C_k > 0$. This allows us to make use of the Fourier transform. We have

$$\widehat{I_k[g]}(\xi) = \widehat{L_k}(\xi) \sum_{j \in \mathbb{Z}} g(j) e^{ij\xi}.$$

In particular, for $\xi \in (-\pi, \pi]$, $\widehat{I_k[g]}(\xi) = \sqrt{2\pi}\widehat{L_k}(\xi)\hat{g}(\xi)$. Now, we have the following estimates using Plancherel's identity and the fact that g is band-limited:

$$\begin{aligned}
\|g - I_k[g]\|^2_{L_2(\mathbb{R})} &= \|\hat{g} - \widehat{I_k[g]}\|_{L_2(\mathbb{R})} \\
&= \int_{-\pi}^{\pi} |\hat{g}(\xi)|^2 \left(1 - \sqrt{2\pi}\widehat{L_k}(\xi)\right)^2 d\xi + \sum_{\ell \neq 0} \sqrt{2\pi} \int_{-\pi}^{\pi} |\hat{g}(\xi)\widehat{L_k}(\xi - 2\pi\ell)|^2 d\xi \\
&= \sqrt{2\pi} \int_{-\pi}^{\pi} |\hat{g}(\xi)|^2 \left(\sum_{\ell \neq 0} \widehat{L_k}(\xi - 2\pi\ell)\right)^2 d\xi \\
&\quad + \sqrt{2\pi} \int_{-\pi}^{\pi} |\hat{g}(\xi)|^2 \sum_{\ell \neq 0} |\widehat{L_k}(\xi - 2\pi\ell)|^2 d\xi. \\
&\leq 2\sqrt{2\pi} \int_{-\pi}^{\pi} |\hat{g}(\xi)|^2 \left(\sum_{\ell \neq 0} \widehat{L_k}(\xi - 2\pi\ell)\right)^2 d\xi
\end{aligned}$$

The second equality follows from periodization, while the third follows from Tonelli's theorem and manipulating (4). The final inequality follows from the positivity of the terms in the sum. In light of (3), we have the following elementary estimate for $\xi \in [-\pi, \pi]$ and $\ell \neq 0$:

$$|L_k(\xi - 2\pi\ell)| \leq (2\pi)^{-1/2} \left(\frac{\pi^2 + \alpha^2}{(2|\ell| - 1)^2\pi^2 + \alpha^2}\right)^k.$$

This allows us to use the dominated convergence theorem since the corresponding series is bounded. For $|\xi| < \pi$, the estimate above works with ξ^2 replacing π^2 in the numerator, thus each term in parentheses is strictly less that 1, hence tends to 0 as $k \to \infty$. This completes the proof of (i). The proof of (ii) follows from (i) together with the Cauchy–Schwarz inequality. Using the inversion formula and the triangle inequality, we have

$$\begin{aligned}
|g(x) - I_k[g](x)| &= (2\pi)^{-1/2} \left| \int_{\mathbb{R}} \left(\hat{g}(\xi) - \widehat{I_k[g]}(\xi)\right) e^{ix\xi} d\xi \right| \\
&\leq (2\pi)^{-1/2} \int_{\mathbb{R}} \left|\hat{g}(\xi) - \widehat{I_k[g]}(\xi)\right| d\xi.
\end{aligned}$$

Now, we may use periodization and the Cauchy–Schwarz inequality and reasoning similar to that used in (i) to see that

$$|g(x) - I_k[g](x)| \leq 2\int_{-\pi}^{\pi} |\hat{g}(\xi)|^2 \left(\sum_{\ell\neq 0} \widehat{L_k}(\xi - 2\pi\ell)\right)^2 d\xi,$$

which tends to 0 independently of $x \in \mathbb{R}$.

References

1. B.J.C. Baxter, The asymptotic cardinal function of the multiquadratic $\varphi(r) = (r^2 + c^2)^{1/2}$ as $c \to \infty$. Comput. Math. Appl. **24**(12), 1–6 (1992)
2. B.J.C. Baxter, N. Sivakumar, On shifted cardinal interpolation by Gaussians and multiquadrics. J. Approx. Theory **87**, 36–59 (1996)
3. M.D. Buhmann, *Radial Basis Functions* (Cambridge University Press, Cambridge, 2003)
4. O. Christensen, P. Massopust, Exponential B-splines and the partition of unity property. Adv. Comput. Math. **37**(3), 301–318 (2012)
5. J. Ledford, On the convergence of regular families of cardinal interpolators. Adv. Comput. Math. **41**(2), 357–371 (2015)
6. W. Madych, S. Nelson, Polyharmonic cardinal splines. J. Approx. Theory **60**, 141–156 (1990)
7. C.A. Micchelli, Cardinal L-splines, *Studies in Spline Functions and Approximation Theory* (Academic Press, New York, 1976), pp. 203–250
8. M.J.D. Powell, Univariate multiquadric approximation: reproduction of linear polynomials, in *Multivariate Approximation and Interpolation*, ed. by W. Haussman, K. Jetter (Birkhäuser, Basel, 1990), pp. 227–240
9. S. Riemenschneider, N. Sivakumar, On the cardinal-interpolation operator associated with the one-dimensional multiquadric east. J. Approx. **7**(4), 485–514 (2001)
10. I.J. Schoenberg, Cardinal spline interpolation, in *Conference Board of the Mathematical Sciences Regional Conference*. Series in Applied Mathematics, vol. 12 (SIAM, Philadelphia, 1973)
11. L. Schumaker, On hyperbolic splines. J. Approx. Theory **38**, 144–166 (1983)
12. M. Unser, T. Blu, Cardinal exponential splines. I. theory and filtering algorithms. IEEE Trans. Signal Process **53**(4), 1425–1438 (2005)

Adaptive Computation with Splines on Triangulations with Hanging Vertices

Shiying Li and Larry L. Schumaker

Abstract It is shown how computational methods based on Bernstein–Bézier methods for polynomial splines on triangulations can be carried over to compute with splines on triangulations with hanging vertices. Allowing triangulations with hanging vertices provides much more flexibility than using ordinary triangulations and allows for simple adaptive algorithms based on local refinements. The use of these techniques is illustrated for two application areas of splines—namely, function fitting and the solution of boundary value problems.

Keywords Spline functions · Hanging vertices · Adaptive methods for fitting · Adaptive finite elements for PDE

1 Introduction

Polynomial splines defined on ordinary triangulations are a well-known and highly effective tool in numerical mathematics and are used in a variety of settings for approximating data and solving partial differential equations numerically, among other things. The recent book [17] explains in detail how to compute with such splines efficiently using Bernstein–Bézier techniques, and even includes an extensive MATLAB library.

Polynomial splines defined on triangulations with hanging vertices (H-triangulations) have been used for some time by engineers to solve boundary value problems (BVP), see [2, 3, 20] to cite just a couple of examples. The main advantage of allowing hanging vertices is that it allows much simpler local adaptive refinement techniques than is possible in the framework of ordinary triangulations.

S. Li (✉) · L.L. Schumaker
Department of Mathematics, Vanderbilt University, Nashville, TN 37240, USA
e-mail: shiying.li@vanderbilt.edu

L.L. Schumaker
e-mail: larry.schumaker@vanderbilt.edu

G.E. Fasshauer and L.L. Schumaker (eds.), *Approximation Theory XV: San Antonio 2016*, Springer Proceedings in Mathematics & Statistics 201,
DOI 10.1007/978-3-319-59912-0_10

On the other hand, splines on triangulations with hanging vertices have been treated in the mathematical literature only recently—see [18, 19], where questions such as dimension, construction of stable local bases, and approximation power are dealt with using Bernstein–Bézier techniques. The purpose of this paper is to show how these same techniques can effectively be used for computing with such splines, thus avoiding the use of parametric maps as is common in the engineering literature.

All of the numerical algorithms described in [17] for working with splines on ordinary triangulations can be carried over to the case of H-triangulations. These include a variety of scattered data fitting methods such as minimal energy methods, local macro-element-based methods, and local methods based on derivative estimation. They also include least-squares fitting and penalized least-squares for fitting noisy data. However, in this paper, we focus on only two specific applications of splines on H-triangulations and show how adaptivity can be effectively exploited: (1) function approximation and (2) solution of boundary value problems for partial differential equations. Our aim is to describe simple algorithms and to illustrate them with specific numerical examples.

The paper is organized as follows. In Sect. 2, we introduce various tools for working with splines on H-triangulations. Section 3 is devoted to function approximation assuming we can sample the function at arbitrary points—we have chosen this application as a way to illustrate the capabilities of adaptive methods coupled with H-triangulations. More practical scattered data interpolation and fitting methods will be treated elsewhere. The use of splines on H-triangulations in the Ritz–Galerkin approach to solving boundary value problems is the subject of Sect. 5. Sections 4 and 6 are devoted to numerical examples. We conclude in Sect. 7 with an extensive list of remarks.

2 Splines on Triangulations with Hanging Vertices

In this section, we introduce the spline spaces of interest. We will follow the terminology and notation introduced in [19], see also [13]. We begin by discussing triangulations with hanging vertices.

2.1 H-Triangulations

Definition 1 Let $\triangle := \{T_i\}_{i=1}^{n_t}$ be a collection of triangles such that the interior of the domain $\Omega := \cup T_i$ is connected. In addition, suppose that any pair of distinct triangles can intersect each other only at points on their edges. Then, we call $\triangle$ an *H-triangulation* of Ω.

This definition allows triangulations to have *hanging vertices*, i.e., a vertex v of a triangle may lie in the interior of an edge of another triangle. Figure 1 shows two examples of H-triangulations. We refer to H-triangulations with no hanging vertices

Fig. 1 Two H-triangulations

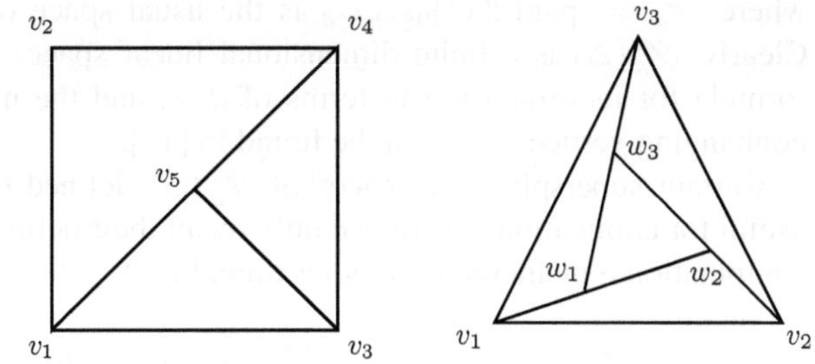

(a) Without cycles (b) With a cycle

as *ordinary triangulations*. They are the ones used in the classical spline theory, see [13, 17]. H-triangulations allow for domains with one or more holes.

Definition 2 We say that an H-triangulation $\triangle$ is *regular* provided that for every vertex v of $\triangle$, the set of all triangles containing v has a connected interior.

Throughout this paper whenever we refer to an H-triangulation, we assume it is regular. For examples of nonregular triangulations, see Remark 2. We now clarify what we mean by an edge of an H-triangulation. If $e := \langle v, w \rangle$ is a line segment of $\triangle$ connecting two vertices v, w of $\triangle$ such that there are no vertices lying in the interior of e, then e is said to be an *edge segment*. An edge segment can be an actual edge of a triangle, but it need not be. If e is a line segment connecting two vertices such that all vertices lying in the interior of e are hanging vertices, and if e cannot be extended to a longer line segment with the same property, then e is said to be a *composite edge* of $\triangle$. Composite edges can consist of one or more edge segments. The triangulation on the left in Fig. 1 has six composite edges and two noncomposite edges.

Definition 3 Suppose $w_1, \ldots, w_n$ is a collection of hanging vertices in an H-triangulation $\triangle$ such that for each $i = 1, \ldots, n$, the vertex w_i lies on a composite edge with one endpoint at w_{i+1}, where we set $w_{n+1} = w_1$. Then $w_1, \ldots, w_n$ is said to form a *cycle*.

The H-triangulation on the right in Fig. 1 contains the cycle w_1, w_2, w_3. In this paper, we will restrict ourselves to H-triangulations without cycles.

2.2 Splines on H-Triangulations

Suppose $\triangle = \{T_i\}_{i=1}^{n_t}$ is an H-triangulation of a polygonal domain Ω, and let $0 \leq r < d$ be given integers. Then, we define the associated *space of splines of degree d and smoothness r* to be

$$\mathcal{S}_d^r(\triangle) := \{s \in C^r(\Omega) : s|_{T_i} \in \mathcal{P}_d \text{ for all } i = 1, \ldots, n_t\},$$

where $\mathscr{P}_d := \mathrm{span}\{x^i y^j\}_{0 \le i+j \le d}$ is the usual space of polynomials of degree d. Clearly, $\mathscr{S}_d^r(\Delta)$ is a finite dimensional linear space. For $d \ge 4r+1$, an explicit formula for its dimension in terms of d, r, and the number of both hanging and nonhanging vertices of Δ can be found in [19].

Certain superspline subspaces of $\mathscr{S}_d^r(\Delta)$ defined on H-triangulations are also useful for applications. Here, we only recall their definition, leaving a discussion of computational methods for another time. Fix $0 \le 2r \le \rho < d$. Then

$$\mathscr{S}_d^{r,\rho}(\Delta) := \{s \in \mathscr{S}_d^r(\Delta) : s \in C^\rho(v) \text{ for all vertices of } \Delta\}, \tag{1}$$

is called a *superspline* space, where $s \in C^\rho(v)$ means that the polynomial pieces $s|_T$ associated with triangles T with a vertex at v have common derivatives up to order ρ at the point v. They were introduced in [19] for H-triangulations.

For a detailed theoretical treatment of supersplines on ordinary triangulations, see [13], and for computational methods based on them, see [15, 17]. Superspline spaces like $\mathscr{S}_d^{r,\rho}(\Delta)$ have several advantages. In particular, they have the same approximation properties as $\mathscr{S}_d^0(\Delta)$, but with fewer degrees of freedom, and often they are easier to work with than the spaces $\mathscr{S}_d^r(\Delta)$.

2.3 The Bernstein–Bézier Representation

It turns out that for splines defined on H-triangulations, we can use the same Bernstein–Bézier techniques that are the key tools for dealing with splines on ordinary triangulations. Given a triangle T, suppose

$$\mathscr{D}_{d,T} := \left\{ \xi_{ijk}^T := \frac{iv_1 + jv_2 + kv_3}{d}, \quad i+j+k = d \right\}$$

is the associated set of *domain points of degree d*, see e.g., [13, 15, 19]. Let $\{B_\xi^T\}_{\xi \in \mathscr{D}_{d,T}}$ be the $n_d := (d+1)(d+2)/2$ Bernstein basis polynomials of degree d associated with T.

Given an H-triangulation $\Delta = \{T_i\}_{i=1}^{n_t}$, we write $\mathbb{P}$ for the linear space of all piecewise polynomials defined on Δ. Then, for any $s \in \mathbb{P}$ and any $T \in \Delta$, we can write

$$s|_T = \sum_{\xi \in \mathscr{D}_{d,T}} c_\xi B_\xi^T.$$

The set of coefficients c_ξ are called the *B-coefficients* of s. It is easy to see that $\mathbb{P}$ is in 1-1 correspondence with the set of domain points

$$\tilde{\mathscr{D}}_{d,\Delta} := \bigcup_{T \in \Delta} \mathscr{D}_{d,T}$$

associated with Δ, where here the union is to be understood in the sense that multiple appearances of the same point are allowed. This implies immediately that $\dim \mathbb{P} = n_t n_d$.

2.4 The Space $\mathscr{S}_d^0(\Delta)$

For most applications, we prefer to work with spaces of splines that are at least continuous, i.e., that are subspaces of $\mathscr{S}_d^0(\Delta)$. Indeed, for many applications, it suffices to work with $\mathscr{S}_d^0(\Delta)$ itself. This space is well understood in the case of ordinary triangulations, but has to be treated with care when working on H-triangulations.

First, we describe an efficient scheme for storing splines $s \in \mathscr{S}_d^0(\Delta)$. Rather than storing s as a member of $\mathbb{P}$, which would involve $n_t n_d$ coefficients, we can take advantage of the continuity of s and store a smaller set of coefficients. Let $\hat{D}_{d,\Delta}$ be the subset of $\tilde{\mathscr{D}}_{d,\Delta}$ obtained by choosing just one point at each vertex, $d-1$ points on each edge, and the $\binom{d-1}{2}$ points of $\mathscr{D}_{d,T}$ that lie inside of T for each triangle $T \in \Delta$. It is clear the cardinality of $\hat{D}_{d,\Delta}$ is

$$n_c = n_v + (d-1)n_e + \binom{d-1}{2} n_t,$$

where n_v, n_e, and n_t are the numbers of vertices, edges, and triangles of Δ.

The number n_c is not generally equal to the dimension of $\mathscr{S}_d^0(\Delta)$. Indeed, if the coefficients of s are known for domain points on a composite edge e, then the C^0 continuity of s determines all of its coefficients associated with domain points on subedges of e. For example, consider a spline $s \in \mathscr{S}_1^0(\Delta)$ on the triangulation in Fig. 1a. Then, since s reduces to a linear polynomial on the edge $\langle v_1, v_4 \rangle$, it follows that the value of s at the vertex v_5 is determined by its values at the vertices v_1 and v_4.

To get the dimension of $\mathscr{S}_d^0(\Delta)$, we now look for a smallest set $\mathscr{M}$ of points in $\hat{D}_{d,\Delta}$ such that setting the corresponding coefficients of a spline $s \in \mathscr{S}_d^0(\Delta)$ uniquely determines all other coefficients. Such a set is called a *minimal determining set* (MDS), see e.g., [13, 17, 19]. Let $\mathscr{M}$ be the subset of $\hat{D}_{d,\Delta}$ which is obtained from $\hat{D}_{d,\Delta}$ by dropping the domain points at hanging vertices along with those lying in the interior of edges which are proper subedges of composite edges.

Theorem 1 *For any H-triangulation Δ, the set $\mathscr{M}$ is a MDS for $\mathscr{S}_d^0(\Delta)$, and*

$$\dim \mathscr{S}_d^0(\Delta) = \hat{n}_v + (d-1)\hat{n}_e + \binom{d-1}{2} n_t, \tag{2}$$

where $\hat{n}_v$ is the number of nonhanging vertices of Δ, and $\hat{n}_e$ is the number of composite edges of Δ.

Proof It is easy to verify that the $\mathcal{M}$ described above is a minimal determining set for $\mathcal{S}_d^0(\triangle)$. It follows that the dimension of $\mathcal{S}_d^0(\triangle)$ is just the cardinality of $\mathcal{M}$, which is given by the formula in (2). □

To give an example of this theorem, consider the triangulation shown in Fig. 1a consisting of three triangles with one hanging vertex. It has five vertices, one of which is a hanging vertex. Moreover, it has eight edges, of which six are composite edges. It follows that the number of coefficients needed to store $s \in \mathcal{S}_1^0(\triangle)$ is five, while the dimension of the space is four. Similarly, the number of coefficients needed to store $s \in \mathcal{S}_2^0(\triangle)$ is 13, while the dimension of the space is 10.

The minimal determining set $\mathcal{M}$ plays another important role in working with $\mathcal{S}_d^0(\triangle)$ since it provides a simple way to create a *basis* for $\mathcal{S}_d^0(\triangle)$. Indeed, given $\xi \in \mathcal{M}$, suppose ϕ_ξ is the spline obtained by setting the coefficient $c_\xi = 1$, and $c_\eta = 0$ for all other $\eta \in \mathcal{M}$. Then, as is the case for splines on ordinary triangulations, it is easy to see that $\{\phi_\xi\}_{\xi \in \mathcal{M}}$ are linearly independent and thus form a basis for $\mathcal{S}_d^0(\triangle)$, see [19]. On ordinary triangulations, these splines have small supports: either the star of a vertex, the set of all triangles sharing an edge, or a single triangle, see [13]. The situation is slightly more complicated when working on H-triangulations, since some of the ϕ_ξ can have somewhat larger supports due to the hanging vertices. For a detailed discussion of supports of these dual basis functions in the case of H-triangulations, see [19].

While we often need a basis for obtaining theoretical results for splines, in practice we generally do not need to construct them explicitly. It is enough to compute the list of domain points forming a minimal set. We call the corresponding B-coefficients the *degrees of freedom (DOF)* of the spline space. These coefficients are the ones we actually compute in solving a practical problem. To work with splines computationally, we will also make use of the so-called *transformation matrix A* which provides a way to compute the vector c of all Bernstein–Bézier coefficients of s from those in the set $\{c_\xi\}_{\xi \in \mathcal{M}}$, i.e., from the degrees of freedom. The matrix A will be of size n_c by m, where $m = \dim \mathcal{S}_d^0(\triangle)$. It can be constructed for any given spline space by using the smoothness conditions defining the space to write down equations connecting coefficients associated with domain points not in $\mathcal{M}$ with the degrees of freedom. Assembling the degrees of freedom in a vector $\tilde{c}$, we can write $c = A\tilde{c}$.

2.5 Smooth Spaces of Splines

In some applications, it is desirable to work with spaces of splines with one or more continuous derivatives, for example, in solving fourth order partial differential equations by the Galerkin method. As discussed in [19], given a spline $s \in \mathcal{S}_d^0(\triangle)$, it is easy to write down linear conditions on its set of B-coefficients to guarantee C^r smoothness across an edge of $\triangle$. Some care is required in working with coefficients associated with domain points in disks around hanging vertices, or with domain points on edges that are contained in larger composite edges. This complicates the construction of minimal determining sets, although in principle the approach is the same as in the case of ordinary triangulations.

Minimal determining sets have been described in detail in [19] for the spaces $\mathscr{S}_d^r(\Delta)$ for all $d \geq 4r+1$, see Lemma 4.2 in that paper. An explicit formula for the dimension of $\mathscr{S}_d^r(\Delta)$ in this case is also given in Theorem 5.3 of [19], along with the construction of dual basis functions whose support properties can be precisely determined. Analogous results were also obtained there for the superspline spaces $\mathscr{S}_d^{r,\rho}(\Delta)$ introduced above.

2.6 Approximation Power of Splines on H-Triangulations

The approximation power of spline spaces defined on H-triangulations was investigated in [19]. Suppose $\mathscr{S}_d^{r,\rho}(\Delta)$ is the superspline space defined in (1) with $0 \leq 2r \leq \rho < d$. Let $|\Delta| = \max_{T \in \Delta} |T|$ be the mesh size of Δ, and for any $m \geq 0$ and $1 \leq q \leq \infty$, let $W_q^{m+1}(\Omega)$ be the usual Sobolev space defined on the domain Ω. Let $\|\cdot\|_q$ be the usual q-norm on Ω, and let $|\cdot|_{m+1,q,\Omega}$ be the seminorm measuring $m+1$-st order derivatives in the q-norm.

Theorem 2 *Suppose $f \in W_q^{m+1}(\Omega)$ with $0 \leq m \leq d$ and $1 \leq q \leq \infty$. Then, there exists a spline $s \in \mathscr{S}_d^{r,\rho}(\Delta)$ such that*

$$\|D_x^\nu D_y^\mu (f - s)\|_q \leq K |\Delta|^{m+1-\nu-\mu} \, |f|_{m+1,q,\Omega},$$

for all $0 \leq \nu + \mu \leq m$.

The constant K depends on d, the size of the smallest angle in Δ, the length of the longest chain of hanging vertices (see Remark 6), and the constant

$$\alpha_\Delta := \max_{e \in \mathscr{E}_c} \max_{\tilde{e} \subset e} \frac{|e|}{|\tilde{e}|}, \tag{3}$$

where $\mathscr{E}_c$ is the set of all composite edges containing two or more edge segments. If Ω is not convex, then K also depends on the Lipschitz constant of the boundary of Ω.

2.7 Locally Refining an H-Triangulation

Given an H-triangulation Δ, we can locally refine it by choosing a triangle T and splitting it into two or more subtriangles. This can be done in several different ways. Here are two commonly used approaches:

1. *S1: Edge split.* Insert an edge segment connecting one vertex v of T to the midpoint w of the edge e opposite to v. This splits T into two subtriangles.
2. *S2: Midpoint refinement.* Connect the midpoints of the edges of T to each other to split T into four subtriangles.

For other refinement possibilities, see Remark 9. Each of these methods has advantages and disadvantages. Method S1 may reduce the size of the smallest angle in the triangulation, but this can be somewhat mitigated by always splitting the longest edge of the triangle. Method S2 leaves the smallest angle unchanged.

In both of these methods, if a new vertex is introduced, it will be either a boundary vertex, or a new hanging interior vertex of the refined triangulation. If a split point falls on an existing vertex, then that vertex becomes a nonhanging vertex.

3 Function Approximation

Suppose f is a given continuous function defined on a domain Ω. We propose to use an adaptive algorithm to compute a spline approximant of f. In this section, we choose an initial triangulation of Ω and work with splines of degree d on $\triangle$ and refinements of it. Here is a general algorithm for carrying this out. Suppose `Appx` is some approximation process producing an approximation in a spline space $\mathcal{S}(\triangle)$. Suppose `Err` is a process to compute a vector $err = (err_1, \ldots, err_{n_t})$, where n_t is the number of triangles in $\triangle$. Choose a maximal number of iterations n_r.

Algorithm 1
For $i = 1$ *until* n_r

1. *Pick a type of spline space* $\mathcal{S}(\triangle)$
2. *Pick an initial triangulation* $\triangle$
3. *Use* `Appx` *to compute a spline* s *in* $\mathcal{S}(\triangle)$
4. *Use* `Err` *to compute the error vector* err
5. *Sort* err
6. *Split the triangle with the largest error*
7. *Replace* $\triangle$ *by the new triangulation and repeat this process*

Refining one triangle at a time can be slow since at each step, we have to reconstruct the approximating spline and compute the vector err of triangle errors. Thus, in practice, we suggest using the following variant of this algorithm where we refine groups of triangles. Pick a number $0 < p < 100$.

Algorithm 2 *Replace step 4 in Algorithm 1 by*

4'. Choose the triangles corresponding to the largest p percent of the errors, and split them.

For the experiments below, we will primarily use Algorithm 2 with $p = 5$. The choice of the approximation process and error computation will depend on the application. Here, we define the approximation scheme `Appx` as follows: Find a spline

$s \in \mathcal{S}(\triangle)$ that interpolates f at all of the domain points $\mathcal{D}_{d,\triangle}$. Next, we have to choose a scheme `Err` for computing the error on each triangle. Here are some possible choices. Let m be an integer greater than d, and define

$$e1(T) := A(T) \sum_{\xi \in \mathcal{D}_{m,T}} |s(\xi) - f(\xi)|, \tag{4}$$

$$e2(T) := A(T) \sum_{\xi \in \mathcal{D}_{m,T}} |s(\xi) - f(\xi)|^2, \tag{5}$$

$$em(T) := A(T) \max_{\xi \in \mathcal{D}_{m,T}} |s(\xi) - f(\xi)|, \tag{6}$$

$$E1(T) := \int_T |s - f|, \tag{7}$$

$$E2(T) := \int_T |s - f|^2, \tag{8}$$

where $A(T)$ is the area of the triangle T. In practice, we typically take $m = d + 2$.

4 Examples of Adaptive Approximation of a Function

In this section, we give two examples showing the performance of the method of the previous section.

Example 1 Fit the function $f_1 = e^{-500[(x-.375)^2+(y-.375)^2]}$ on the unit square by a spline in $\mathcal{S}_d^0(\triangle)$.

Discussion: A plot of the function f_1 is shown in Fig. 2b. We begin with the type-2 triangulation of $[0, 1]^2$ shown in Fig. 2a. It has 64 triangles. Then, for fixed d, we apply Algorithm 2 using the local error indicator defined in (6) with $m = d + 2$. In each pass through the algorithm we identify the triangles corresponding to the top $p = 5$ percent of the errors and split each such triangle uniformly into four subtriangles, see method S2 in Sect. 2.7. Table 1 illustrates the behavior of the algorithm for $d = 4$, where

nr := number of passes through the algorithm,
nt := number of triangles of the final triangulation,
$ndof$:= number of degrees of freedom of the associated spline space,
$emax$:= maximum error on a 1001×1001 grid,
RMS := the root mean square error over that grid,
$time$:= number of seconds to perform the computation.

With $nr = 5$, we get the triangulation shown in Fig. 2c. It has 136 triangles and gives maximum and RMS errors of 1.54(−3) and 4.96(−5). We do not show a plot

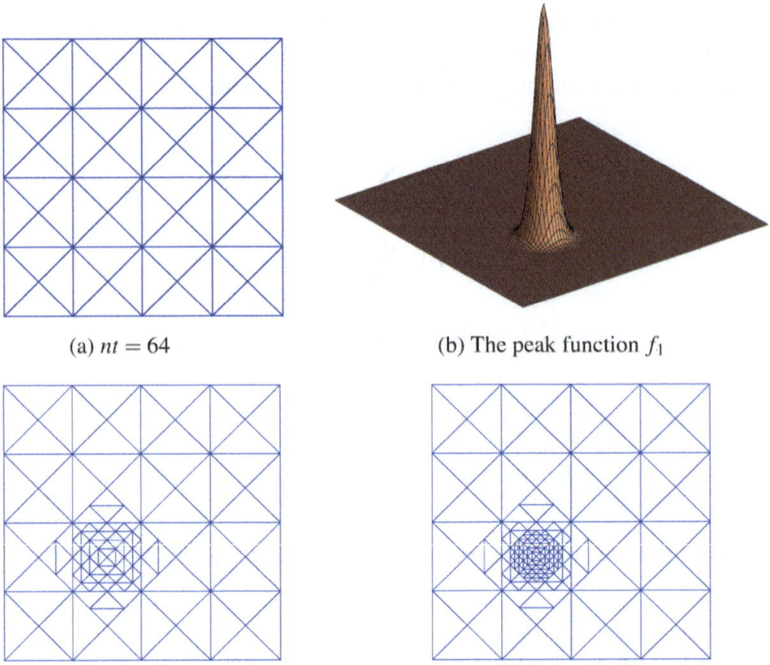

(a) $nt = 64$ (b) The peak function f_1

(c) $nt = 136$ (d) $nt = 220$

Fig. 2 Results for Example 1 with $d = 4$

Table 1 Table of errors for Example 1 with $d = 4$

nr	nt	ndof	emax	RMS	Time
5	136	1089	1.54e-03	4.96e-05	0.14
10	220	1737	8.64e-05	3.49e-06	0.40
15	388	3017	3.65e-05	1.16e-06	0.85
20	508	3993	2.04e-05	4.84e-07	1.56
30	808	6369	2.97e-06	1.04e-07	3.34

of the associated spline surface since it is visually indistinguishable from a plot of f_1 itself. With $nr = 10$, we get the triangulation shown in Fig. 2d which has 220 triangles. Now, the errors are 8.64(−5) and 3.49(−6). As we can see from the table, as nr increases, we get more triangles, but the accuracy improves rapidly in both the maximum and RMS norms.

Table 2 shows the case $d = 5$. We get essentially the same accuracy with 340 triangles and 4161 degrees of freedom as we do with $d = 4$ and 808 triangles and 6369 degrees of freedom. For these examples, we have used the local error indicator given in (6). Using the other local error indicators defined above give similar results. □

Table 2 Table of errors for Example 1 with $d = 5$

nr	nt	ndof	emax	RMS	Time
5	136	1701	6.12e-04	1.22e-05	0.15
10	244	3001	1.53e-05	4.82e-07	0.47
15	340	4161	6.04e-06	1.52e-07	0.95
20	532	6561	1.71e-06	4.01e-08	1.71
30	952	11701	2.53e-07	5.47e-09	4.93

To illustrate the amount of compression achieved by the adaptive algorithm, one can compare with the interpolants from $\mathscr{S}_4^0(\triangle)$ based on large type-2 triangulations. For example, using a type-2 triangulation defined on a 17×17 grid gives an RMS error of $7.52(-5)$, but with 1024 triangles and 8321 degrees of freedom. We achieve a better fit with only 136 triangles and 1089 degrees of freedom. Similarly, using a type-2 triangulation on a 33×33 grid gives an RMS error of $2.42(-6)$, but with 4096 triangles and 33025 degrees of freedom. We achieve a better fit with only 388 triangles and 3017 degrees of freedom.

As a second example, we take a test function which we shall use later in our study of boundary value problems.

Example 2 Fit the function $f_2 := \tanh(40y - 80x^2) - \tanh(40x - 80y^2)$ on the unit square by a spline in $\mathscr{S}_d^0(\triangle)$.

Discussion: A plot of the function f_2 is shown in Fig. 3d. This time, we begin with the type-1 triangulation of $[0, 1]^2$ on a 5×5 grid shown in Fig. 3a. It has 32 triangles. We then apply Algorithm 2 using the local error indicator defined in (6) with $m = d + 2$. In each run of the algorithm, we identify the triangles corresponding to the top $p = 5$ percent of the errors and split each such triangle uniformly into four subtriangles, see method S2 in Sect. 2.7. Tables 3, 4 and 5 give numerical results for $d = 3, 5, 7$. They contain the same information as given in Tables 1 and 2. As to be expected, larger values of d give smaller errors. We have also experimented with the other local error indicators defined earlier and get similar results.

For comparison purposes, we note that interpolation with $\mathscr{S}_7^0(\triangle)$ on a type-1 triangulation of a 17×17 grid has 512 triangles and an RMS error of $3.85(-3)$. This can be compared with our adapted triangulation with 230 triangles which has an RMS error $8.71(-4)$. Interpolation with $\mathscr{S}_7^0(\triangle)$ on a type-1 triangulation of a 33×33 grid has 2048 triangles and an RMS error of $1.2(-4)$. This can be compared with our adapted triangulation with 1079 triangles which has an RMS error of $1.32(-6)$. □

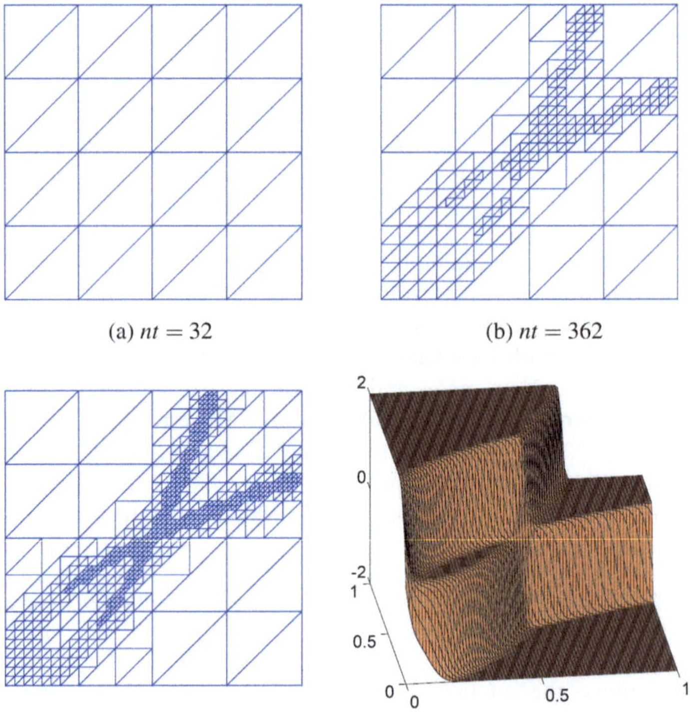

(a) $nt = 32$ (b) $nt = 362$

(c) $nt = 1079$ (d) The function f_2

Fig. 3 Results for Example 2 with $d = 7$

Table 3 Table of errors for Example 2 with $d = 3$

nr	nt	ndof	emax	RMS	Time
10	134	598	3.94e-01	2.89e-02	0.43
20	248	1072	1.60e-01	1.03e-02	0.70
30	422	1819	1.03e-01	3.93e-03	1.59
40	632	2704	5.71e-02	1.58e-03	3.08
50	860	3688	1.80e-02	8.20e-04	5.23
60	1100	4756	1.37e-02	5.08e-04	8.02
70	1448	6262	1.16e-02	3.24e-04	11.68
80	1928	8332	5.52e-03	1.61e-04	17.07

Table 4 Table of errors for Example 2 with $d = 5$

nr	nt	ndof	emax	RMS	Time
5	68	876	7.14e-01	4.67e-02	0.10
10	128	1586	2.90e-01	1.02e-02	0.28
20	254	3101	8.40e-02	2.54e-03	1.01
30	398	4851	2.56e-02	5.81e-04	2.32
40	578	6951	1.01e-02	2.18e-04	4.49
50	806	9771	3.66e-03	6.75e-05	7.97
60	980	11931	1.44e-03	2.73e-05	12.70

Table 5 Table of errors for Example 2 with $d = 7$

nr	nt	ndof	emax	RMS	Time
10	98	2395	3.53e-01	1.34e-02	0.31
20	230	5496	4.34e-02	8.71e-04	1.31
30	362	8723	1.17e-02	1.42e-04	3.27
40	536	12818	2.86e-03	4.10e-05	6.39
50	686	16430	9.43e-04	1.35e-05	11.82
60	884	21218	2.67e-04	3.80e-06	19.64
70	1079	26013	1.71e-04	1.32e-06	29.69

5 Solution of Elliptic Boundary Value Problems

In [15, 17], it was shown how various spline spaces can be used in conjunction with the Ritz–Galerkin method to compute approximations to the solutions of elliptic boundary value problems. The discussion there was for spline spaces on ordinary triangulations, but there is no essential difficulty in carrying out the same program for splines defined on H-triangulations. We just need to work with spline spaces for which we know a minimal determining set, and for which, we can compute the transformation matrix A. In this section, we illustrate the method using the spaces $\mathscr{S}_d^0(\triangle)$ along with local refinement to find good triangulations.

Here is the problem to be attacked. We follow the notation of [17]. Suppose we are given functions f and κ defined on a domain Ω, and suppose g is a function defined on the boundary $\partial\Omega$ of Ω.

Problem 1 Find a function u defined on Ω such that

$$Lu := -\nabla \cdot (\kappa \nabla u) = f, \qquad \text{on } \Omega, \tag{9}$$

$$u = g, \qquad \text{on } \partial\Omega. \tag{10}$$

Here, ∇ denotes the vector-valued differential operator $[D_x, D_y]^T$, and the dot denotes the vector inner product. Now, given a spline space $\mathscr{S}(\triangle)$, suppose $\{\phi_1, \ldots, \phi_{n_0}\}$ is a basis for $U_0 := \{s \in \mathscr{S}(\triangle) : s \equiv 0 \text{ on } \partial\Omega\}$. Then, we look for an approximate solution of Problem 1 in the form

$$s = \sum_{i=1}^{n_0} c_i \phi_i + s_b, \tag{11}$$

where s_b is a spline in $\mathscr{S}(\Delta)$ such that s_b is approximately equal to g on the boundary $\partial\Omega$.

5.1 The Ritz–Galerkin Method

Definition 4 Suppose the spline s in (11) is such that

$$\int_\Omega [Ls(x,y) - f(x,y)]\phi_i(x,y)dxdy = 0, \qquad i = 1, \ldots, n_0. \tag{12}$$

Then s is called the *Ritz–Galerkin approximation* to the solution u of Problem 1.

We recall some notation from [17]. For each triangle $T \in \Delta$, let

$$\langle \phi, \psi \rangle_{2,T} := \int_T \phi\psi dxdy,$$

and

$$\begin{aligned}\langle \phi, \psi \rangle_{G,T} &:= \int_T \kappa(x,y)\nabla\phi(x,y) \cdot \nabla\psi(x,y)dxdy \\ &= \int_T \kappa(x,y)[\phi_x(x,y)\psi_x(x,y) + \phi_y(x,y)\psi_y(x,y)]dxdy.\end{aligned}$$

Let

$$\langle \phi, \psi \rangle_2 := \sum_{T\in\Delta} \langle \phi, \psi \rangle_{2,T}, \qquad \langle \phi, \psi \rangle_G := \sum_{T\in\Delta} \langle \phi, \psi \rangle_{G,T}.$$

Then, as shown in Theorem 9.5 of [17], the coefficients of the spline s in (11) approximating u via the Ritz–Galerkin method can be computed by solving the linear system of equations

$$Mc = r,$$

with $M = [\langle \phi_i, \phi_j \rangle_G]_{i,j=1}^{n_0}$ and

$$r_i = \langle f, \phi_i \rangle_2 - \langle s_b, \phi_i \rangle_G, \qquad i = 1, \ldots, n_0.$$

To use this approach with the space $\mathscr{S}_d^0(\triangle)$ on an H-triangulation, we can follow the algorithm outlined in Sect. 9.3.1 of [17] to set up the stiffness matrix M and right-hand side vector r. This algorithm makes use of the transformation matrix A for the spline space.

5.2 A Posteriori Error Estimators

In order to couple adaptive methods with the Ritz–Galerkin method for solving our boundary value Problem 1, we make use of either Algorithm 1 or 2. To use them, we need to have a way of computing an error value associated with each triangle in $\triangle$. Since we do not know the true solution of the boundary value problem, this error will have to be constructed numerically without using u. Computed error estimates of this type are usually called *a posteriori error estimates* and have been studied extensively in the PDE literature, see e.g., [1, 4–6, 10, 14]. Here are some possibilities in our setting, based on estimating the size of the residual $Ls - f$. Given a triangle T and a spline s along with an integer $m > d$, let

$$r_1(T) := A(T) \sum_{\xi \in \mathscr{D}_{m,T}} |Ls(\xi) - f(\xi)|, \tag{13}$$

$$r_2(T) := A(T) \sum_{\xi \in \mathscr{D}_{m,T}} |Ls(\xi) - f(\xi)|^2, \tag{14}$$

$$R_1(T) := \int_T |Ls - f|, \tag{15}$$

$$R_2(T) := \int_T |Ls - f|^2, \tag{16}$$

where $A(T)$ is the area of T. The last two of these are integrals, and need to be computed by numerical quadrature, see e.g., Sect. 4.6 of [17]. In practice, we typically take $m = d + 2$.

6 Examples of the Adaptive Ritz–Galerkin Method

In this section, we give two examples of the use of the Ritz–Galerkin approach coupled with our algorithm for locally refining triangulations. Here is an algorithm.

Algorithm 3

1. *Pick a type of spline space* $\mathscr{S}(\triangle)$
2. *Pick an initial triangulation* $\triangle$

3. *Use the Ritz–Galerkin method to find a spline s in $\mathcal{S}(\triangle)$ giving an approximate solution of the boundary value problem*
4. *Apply Algorithm 2 to successively refine $\triangle$ either a fixed number of times, or until some error measure is sufficiently small*

For the examples in this section, we will work with the a posteriori error estimator R_2 defined in (16). Our first example deals with the Poisson equation where the differential operator is the Laplace operator.

Example 3 Use the spline spaces $\mathcal{S}_d^0(\triangle)$ to solve the boundary value Problem 1 on $\Omega := [0,1]^2$ with $\kappa = 1$, $f := -\Delta u$, and $g = u|_{\partial\Omega}$, where $u := \tanh(40y - 80x^2) - \tanh(40x - 80y^2)$.

Discussion: The true solution of this boundary value problem u is the same as the function f_2 used in Example 2, see Fig. 3d. In running Algorithm 2, we sort the triangles according to the size of the error estimates, then refine the top 5% using the refinement method S2 described above. We repeat this nr times to get our solution.

Table 6 gives numerical results for the case $d = 3$, starting with a type-1 triangulation with 25 vertices. It has the same columns as in our previous tables, where again we use a 1001×1001 grid on Ω to compute the max and RMS errors. The cases $d = 5, 7$ are shown in Tables 7 and 8.

For comparison purposes, we note that working with $\mathcal{S}_7^0(\triangle)$ on a type-1 triangulation of a 33×33 grid with 2048 triangles and 50625 degrees of freedom gives an RMS error of 8.26(−5). This can be compared with using our adapted triangulation with 488 triangles and 11614 degrees of freedom which leads to an RMS error 4.54(−5).

We do not give any plots of the splines produced in this example or their associated triangulations since they are very similar to those obtained in Example 2 where we

Table 6 Table of errors for Example 3 with $d = 3$

nr	nt	ndof	emax	RMS	Time
10	104	445	8.85e-01	1.16e-01	2.96
20	218	907	2.00e-01	1.99e-02	9.72
30	392	1612	8.93e-02	1.03e-02	22.32
40	518	2137	8.15e-02	7.74e-03	41.68
50	626	2587	4.55e-02	3.18e-03	66.32
60	872	3598	4.28e-02	2.43e-03	102.46

Table 7 Table of errors for Example 3 with $d = 5$

nr	nt	ndof	emax	RMS	Time
10	92	1136	7.89e-01	7.55e-02	3.70
20	182	2181	6.40e-02	5.57e-03	12.31
30	284	3381	4.13e-02	2.19e-03	27.53
40	374	4446	2.34e-02	1.11e-03	51.45
50	458	5466	1.78e-02	8.01e-04	82.55
60	596	7136	6.59e-03	3.20e-04	123.81

Table 8 Table of errors for Example 3 with $d = 7$

nr	nt	ndof	emax	RMS	Time
10	104	2500	1.98e-01	1.65e-02	5.87
20	188	4467	6.61e-02	2.88e-03	20.09
30	248	5909	7.96e-03	6.06e-04	44.11
40	374	8856	4.84e-03	2.28e-04	81.46
50	488	11614	1.02e-03	4.54e-05	139.34

computed an interpolating spline for u. For example, for $d = 7$ using $nr = 40$ here we get a triangulation with 374 triangles and an associated RMS error of $2.28(-04)$, while in Table 5 with $nr = 30$ we had 362 triangles, and an RMS error of $1.42(-4)$. Thus, surprisingly the Ritz–Galerkin solution has virtually the same accuracy as the interpolating spline fit to the true solution. □

In this example, we have worked with the space $\mathscr{S}_7^0(\triangle)$ along with the a posteriori estimator R_1 given in (16). The performance is not as good with other error

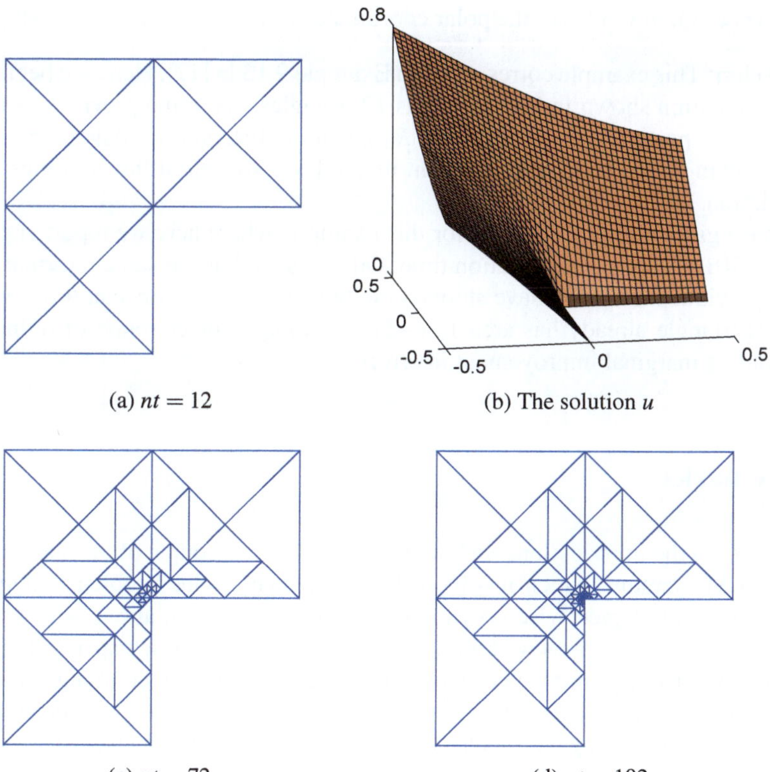

(a) $nt = 12$ (b) The solution u

(c) $nt = 72$ (d) $nt = 102$

Fig. 4 Results for Example 4

Table 9 Table of errors for Example 4 with $d = 3$

nr	nt	ndof	emax	RMS	Time
0	12	67	2.79e-02	2.85e-03	0.00
5	42	178	1.16e-02	1.07e-03	0.69
10	72	307	2.45e-03	1.40e-04	1.73
15	102	433	2.68e-04	6.37e-05	3.44

estimators listed above, see Remark 14. Our second example is a boundary value problem defined on an L-shaped domain with a reentrant corner.

Example 4 Use the spline space $\mathscr{S}_3^0(\triangle)$ to solve the boundary value Problem 1 on the L-shaped domain shown in Fig. 4a with $\kappa \equiv 1$, $f \equiv 0$, and Dirichlet boundary values taken from the function

$$u(x, y) = r(x, y)^{2/3} \sin\left(\frac{2}{3}\theta(x, y)\right),$$

where $(r(x, y), \theta(x, y))$ are the polar coordinates of (x, y) for all $\theta \in [0, \frac{3}{2}\pi]$.

Discussion: This example corresponds to Example 9.13 in [17]. Here, we begin with the triangulation shown in Fig. 4a. It has 12 triangles. As in the previous example, we use the a posteriori error estimator R_2 given in (16), and run Algorithm 2 with $p = 5$, again using the local refinement method S2. We repeat this nr times to get our solution.

Table 9 gives numerical results for this example. where here we report errors on a 101×101 grid to save evaluation time. Otherwise, it has the same information as in our previous tables. We have stopped the table at $nr = 15$ since at this level the smallest triangle already has area 1.5(−5) and doing another round of refinement gives only a marginal improvement in errors. □

7 Remarks

Remark 1 In the introduction, we have referenced only a couple of the many papers in the FEM literature which involve solving boundary value problems adaptively with some kind of piecewise polynomials defined on triangulations with hanging vertices. However, as far as we know, the approach in all of these papers follows the standard engineering approach of working with a single reference triangle which is then mapped to the triangles of an H-triangulation of the physical domain. Here, we are able to avoid working with such parametric maps by employing Bernstein–Bézier techniques.

Remark 2 The reason for restricting our attention to H-triangulations that are regular is to prevent the situation where two triangles touch at a common vertex, while no other triangles share that vertex, see for example Fig. 4.1 in [13]. We need connectivity to connect the polynomial pieces on adjoining triangles.

Remark 3 It can be shown that if Δ is an H-triangulation with no cycles, then the H-triangulation $\widetilde{\Delta}$ obtained applying either of the refinement methods discussed in Sect. 2.7 does not have any cycles, see [19].

Remark 4 The construction of minimal determining sets for $\mathcal{S}_d^r(\Delta)$ on H-triangulations was carried out in [19] only for $d \geq 4r + 1$. It is possible to extend this to all $d \geq 3r + 2$ by arguments similar to those used for splines on ordinary triangulations, see [13].

Remark 5 For splines on ordinary triangulations, the columns of the transformation matrix A give the coefficients of the individual dual basis functions $\{\phi_\xi\}_{\xi \in \mathcal{M}}$, see [15]. The same holds here.

Remark 6 Suppose e is a composite edge of Δ, and suppose $e_1, \ldots, e_m$ is a maximal sequence of composite edges such that for each $i = 1, \ldots, m$, one end of e_i is in the interior of e_{i+1}, where $e_{m+1} = e$. Following [19], we call $e_1, \ldots, e_m$ a *chain* ending at e. We refer to m as the *length* of the chain.

Remark 7 Theorem 2 gives an error bound on the entire domain Ω. There is an analogous local result which gives a bound on a single triangle. For a constructive proof, see [19]. We have stated the global form here for simplicity.

Remark 8 If we start with an ordinary triangulation and repeatedly locally refine it using the refinement method S2, the constant α_Δ in (3) can be bounded by $1/m$, where m is the maximum number of times a given edge is refined. We also note that if we use this refinement process, the smallest angle in the resulting triangulation cannot be smaller than the smallest angle in the original triangulation.

Remark 9 Here, we have focused only on two possible refinement strategies. As is well known in the literature, see e.g., [13], and references therein, in some applications it may be desirable to split triangles in different ways—e.g., with the so-called Powell–Sabin split into either 6 or 12 subtriangles, or the Clough–Tocher split into 3 subtriangles.

Remark 10 In using local refinement with H-triangulations, it will happen that eventually very small triangles will be introduced. Fortunately, if we use the refinement strategy S2, we do not have to worry about the angles, but if the triangles are too small it is hard to keep track of the triangle lists or even to decide whether a point is in a triangle or not. Thus, in our test codes, we have avoided splitting any triangle whose area is below a threshold (which we took to be 10^{-7}).

Remark 11 All experiments here were run in MATLAB on a typical desktop. The times reported are just for comparison purposes—they of course would be different for other machines.

Remark 12 In the interest of saving time, our experiments have all been run with Algorithm 2 where we process the top 5% of triangles in turn, before recomputing the approximation and associated error. If we work with a larger percentage, the algorithms will run faster, but with slightly less accurate results. Conversely, if we work with a smaller percentage, computational times will go up, but we will get somewhat better errors. The best would be to do only one triangle at a time.

Remark 13 For both function approximation and solution of boundary value problems, the approximation process involves computing a minimal determining set. When an H-triangulation is refined locally, it is not necessary to recompute the entire MDS—one can simply update it in the affected triangulations. This would save time, but was not done in the reported results. The same is true of the second stage of the process, the computation of the approximation itself. For example, when solving boundary value problems, instead of computing the entire stiffness matrix anew, we can locally update it each time we have completed a refinement cycle. This would also save time, but again was not implemented for the results reported here.

Remark 14 The performance of the adaptive method of Sect. 5 for solving boundary value problems is somewhat sensitive to which a posteriori error estimator is used. Using the one given in (14), which while quite similar to the one in (16) used here, gives significantly different results. The others do not seem to function as well either.

Remark 15 The experiments in this paper were all conducted using the midpoint refinement method S2. We get similar results for all of the examples if we use S1 instead, although since it only splits a triangle into two subtriangles instead of four, we have to work with larger numbers nr of iterations to get comparable numbers of triangles in the final triangulation.

Remark 16 The experiments in Sect. 3 were carried out with C^0 spline spaces. We get very similar results when working with piecewise polynomials, i.e., splines with no global continuity. In fact, the algorithms run faster, and as soon as we get a reasonably accurate result, the lack of continuity cannot be detected visually.

Remark 17 In refining a single triangle from a triangulation $\triangle$ with either of subdivision methods discussed here, we have chosen the new vertices to be midpoints of existing edges of $\triangle$. This helps maintain the shape of the triangles, but also has the effect that as we continue to refine, these new hanging vertices have a tendency to disappear as we refine a neighboring triangle. The practical effect is that our final H-triangulations do not usually include a large percentage of hanging vertices. This feature also helps avoid long chains of hanging vertices (see [19]).

Remark 18 The Ritz–Galerkin method can also be used to solve higher order boundary value problems. A typical example is the biharmonic equation which is of order four, see Sect. 9.6 of [17]. For conformality, this problem requires working with splines with global smoothness C^1. This is a good example where the space $\mathscr{S}_5^{1,2}(\triangle)$ can be of use, and we can apply the same adaptive approach in this setting.

Remark 19 The key to computing with splines on H-triangulations using Bernstein–Bézier techniques is to work only with spaces where we know a minimal determining set and can construct the transformation matrix A. This is the case for the spaces $\mathcal{S}_d^0(\triangle)$ for all d, as well as for the superspline spaces $\mathcal{S}_d^{r,\rho}(\triangle)$ with $0 \le 2r \le \rho < d$ with $d > 2\rho$. It also holds for the analogs of all of the macro-element spaces discussed in [13] defined over Powell–Sabin refinements, Clough–Tocher refinements, etc.

Remark 20 In this paper, we have focused on splines on H-triangulations in the plane. It is also possible to define splines on H-triangulations on the sphere, see [16]. Indeed, everything done here can just as easily be done on the sphere.

Remark 21 Splines on triangulations can also be used to approximate images, for example using piecewise linear polynomials with C^0 or C^{-1} global continuity, see e.g., [7–9, 11]. Typically such algorithms involve starting with a coarse triangulation and adding new vertices adaptively, after which a Delaunay triangulation is constructed using the resulting vertices. Then, the image is stored (and transmitted) by giving only the locations of the vertices and the associated coefficients. It is also possible to approximate images with splines on H-triangulations using the ideas of Sect. 3. With our approach, we need only store an ordered list of the triangles which have been split, along with the final set of coefficients. We have done some preliminary experimentation with this approach, but to compare with previous methods, one has to properly encode this information. We plan to look into this further in later work.

Remark 22 A completely different approach to solving boundary value problems adaptively with splines has been proposed recently in [12]. Their method works with ordinary triangulations, and the refinement algorithm is global rather than local. To make it easier to compare the results in [12] (which focus on the Poisson problem) with ours, in Example 3 we solve the same BVP as they do in Example 1 of their paper. For this BVP, their Table 2 shows an RMS error on a 1001×1001 grid of 3.31(−3) for a spline $s \in \mathcal{S}_7^0(\triangle)$, based on an adapted triangulation with 1568 triangles. For this problem, we got a spline in $\mathcal{S}_7^0(\triangle)$ on an H-triangulation with only 200 triangles with an RMS accuracy of 2.64(−3), see our Table 6.

References

1. M. Ainsworth, J.T. Oden, *A Posteriori Error Estimation in Finite Element Analysis* (Wiley-Interscience, New York, 2000)
2. M. Ainsworth, R. Rankin, Constant free error bounds for nonuniform order discontinuous Galerkin finite-element approximation on locally refined meshes with hanging nodes. IMA J. Numer. Anal. **31**, 254–280 (2011)
3. M. Ainsworth, L. Demkowicz, C.-W. Kim, Analysis of the equilibrated residual method for a posteriori error estimation on meshes with hanging nodes. Comput. Methods Appl. Mech. Eng. **196**, 3493–3507 (2007)
4. I. Babuška, W.C. Rheinboldt, Error estimates for adaptive finite element computations. SIAM J. Numer. Anal. **15**, 736–754 (1978)

5. R.E. Bank, R.K. Smith, A posteriori error estimates based on hierarchical bases. SIAM J. Numer. Anal. **30**(4), 921–935 (1993)
6. C. Carstensen, J. Hu, A. Orlando, Framework for the a posteriori error analysis of nonconforming finite elements. SIAM J. Numer. Anal. **45**(1), 68–82 (2007)
7. A. Cohen, N. Dyn, F. Hecht, J.-M. Mirebeau, Adaptive multiresolution analysis based on anisotropic triangulations. Math. Comp. **81**, 789–810 (2012)
8. L. Demaret, N. Dyn, M.S. Floater, A. Iske, Adaptive Thinning for Terrain Modelling and Image Compression, *Advances in Multiresolution for Geometric Modelling*, Math. Vis. (Springer, Berlin, 2005)
9. L. Demaret, N. Dyn, A. Iske, Image compression by linear splines over adaptive triangulations. Signal Process **86**, 1604–1616 (2006)
10. C. Erath, D. Praetorius, A posteriori error estimate and adaptive mesh refinement for the cell-centered finite volume method for elliptic boundary value problems. SIAM J. Numer. Anal. **47**(1), 109–135 (2008)
11. A. Iske, *Multiresolution Methods in Scattered Data Modelling* (Springer, Berlin, 2004)
12. M.-J. Lai, C. Mersmann, An Adaptive Triangulation Method for Bivariate Spline Solutions of PDEs, in *Approximation Theory XV: San Antonio 2016*, Springer Proc. in Math. and Stat. vol. 201, ed. by G.E. Fasshauer, L.L. Schumaker (Springer-Verlag, 2017) pp. 155–175
13. M.-J. Lai, L.L. Schumaker, *Spline Functions on Triangulations* (Cambridge University Press, Cambridge, 2007)
14. W.F. Mitchell, A comparison of adaptive refinement techniques for elliptic problems. ACM Trans. Math. Softw. **15**, 326–347 (1989)
15. L.L. Schumaker, Computing bivariate splines in scattered data fitting and the finite element method. Numer. Algorithms **48**, 237–260 (2008)
16. L.L. Schumaker, Splines on spherical triangulations with hanging vertices. Comput. Aided Geom. Design **30**, 263–275 (2013)
17. L.L. Schumaker, *Spline Functions: Computational Methods* (SIAM, Philadelphia, 2015)
18. L.L. Schumaker, L. Wang, Spline spaces on TR-meshes with hanging vertices. Numer. Math. **118**, 531–548 (2011)
19. L.L. Schumaker, L. Wang, Splines on triangulations with hanging vertices. Constr. Approx. **36**, 487–511 (2012)
20. P. Šolń, J. Červený, I. Doležel, Arbitrary-level hanging nodes and automatic adaptivity in the hp-FEM. Math. Comput. Simul. **77**(1), 117–132 (2008)

Scaling Limits of Polynomials and Entire Functions of Exponential Type

D.S. Lubinsky

Abstract The connection between polynomials and entire functions of exponential type is an old one, in some ways harking back to the simple limit

$$\lim_{n\to\infty}\left(1+\frac{z}{n}\right)^n = e^z.$$

On the left-hand side, we have $P_n\left(\frac{z}{n}\right)$, where P_n is a polynomial of degree n, and on the right, an entire function of exponential type. We discuss the role of this type of scaling limit in a number of topics: Bernstein's constant for approximation of $|x|$; universality limits for random matrices; asymptotics of L_p Christoffel functions and Nikolskii inequalities; and Marcinkiewicz–Zygmund inequalities. Along the way, we mention a number of unsolved problems.

Keywords Scaling limits · Universality limits · Christoffel functions · Bernstein's constant

1 Introduction

The classical limit

$$\lim_{n\to\infty}\left(1+\frac{z}{n}\right)^n = e^z, \tag{1}$$

plays a role in many areas of mathematics, expressing very simply the scaling limit of a sequence of a polynomials as an entire function of exponential type 1. Recall that an entire function f has exponential type A if for every $\varepsilon > 0$,

$$|f(z)| = O\left(e^{(A+\varepsilon)|z|}\right), \quad \text{as } |z| \to \infty,$$

Research supported by NSF grant DMS1362208.

D.S. Lubinsky (✉)
School of Mathematics, Georgia Institute of Technology, Atlanta, GA 30332-0160, USA
e-mail: lubinsky@math.gatech.edu

G.E. Fasshauer and L.L. Schumaker (eds.), *Approximation Theory XV: San Antonio 2016*, Springer Proceedings in Mathematics & Statistics 201,
DOI 10.1007/978-3-319-59912-0_11

and A is the smallest number with this property. Within approximation theory, this connection has long been recognized, perhaps most notably in relating asymptotics of errors of polynomial approximation to errors of approximation by entire functions of exponential type. Indeed, the classical monograph of Timan (see [58] for a translation of the Russian original) presents some of this theory, and this topic continues to be explored to this day, notably in the works of Michael Ganzburg [21, 23]. In this survey, we explore this connection in a number of topics in approximation theory.

A less trivial example than (1) involves Lagrange interpolation at roots of unity. Given $n \geq 1$, the fundamental polynomials of Lagrange interpolation at the nth roots of unity are

$$\ell_{jn}(z) = \frac{1}{n} \frac{z^n - 1}{ze^{-2\pi ij/n} - 1}, \quad 0 \leq j \leq n-1,$$

satisfying

$$\ell_{jn}\left(e^{2\pi ik/n}\right) = \delta_{jk}.$$

Let us fix t and take the scaling limit: As $n \to \infty$,

$$\begin{aligned} \ell_{jn}\left(e^{2\pi it/n}\right) &= \frac{1}{n} \frac{e^{2\pi it} - 1}{e^{2\pi i(t-j)/n} - 1} \\ &= \frac{1}{n} \frac{e^{\pi it}}{e^{\pi i(t-j)/n}} \frac{\sin \pi t}{\sin \frac{\pi(t-j)}{n}} \\ &\to e^{i\pi t} \frac{\sin \pi t}{\pi (t-j)}. \end{aligned}$$

Thus,

$$\lim_{n\to\infty} \ell_{jn}\left(e^{2\pi it/n}\right) = e^{i\pi t} (-1)^j \,\mathbb{S}(t-j), \tag{2}$$

where $\mathbb{S}$ is the classical sinc kernel

$$\mathbb{S}(t) = \begin{cases} \frac{\sin \pi t}{\pi t}, & t \neq 0, \\ 1, & t = 0. \end{cases}$$

If

$$\sum_{j=0}^{\infty} |c_j|^2 < \infty,$$

and for $n \geq 1$, we let

$$P_n(z) = \sum_{j=0}^{n-1} (-1)^j c_j \ell_{jn}(z),$$

then one can use (2) to show

$$\lim_{n\to\infty} P_n\left(e^{2\pi it/n}\right) = e^{i\pi t}\sum_{j=0}^{\infty} c_j \mathbb{S}(t-j) = e^{i\pi t} f(t), \tag{3}$$

where

$$f(t) = \sum_{j=0}^{\infty} c_j \mathbb{S}(t-j)$$

is entire of exponential type $\leq \pi$, and its restriction to the real line lies in $L_2(\mathbb{R})$. The space of all such entire functions with the usual L_2 norm on the real line is L_2 Paley-Wiener space PW_2. Its remarkable reproducing kernel $\mathbb{S}$ plays a role in everything from sampling theory to random matrices, satisfying the orthonormality relation

$$\int_{-\infty}^{\infty} \mathbb{S}(t-j)\,\mathbb{S}(t-k)\,dt = \delta_{jk}.$$

The reproducing kernel relation is

$$f(z) = \int_{-\infty}^{\infty} f(t)\,\mathbb{S}(t-z)\,dt, \quad z \in \mathbb{C}, \quad f \in PW_2.$$

The paper is organized as follows: in the next section, we discuss approximation of $|x|$ by polynomials. Section 3 deals with universality limits for random matrices, Sect. 4 deals with asymptotics of Christoffel functions, Sect. 5 with Nikolskii inequalities, and Sect. 6 deals with Marcinkiewicz–Zygmund inequalities.

2 Bernstein's Constant in Approximation of $|x|$

Let $f : [-1, 1] \to \mathbb{R}$. For $n \geq 1$, let $E_n[f]$ denote the error in best uniform approximation of $|x|$ by polynomials of degree $\leq n$, so that

$$E_n[f] = \inf\left\{\| f - P \|_{L_\infty[-1,1]}: \deg(P) \leq n\right\}.$$

In a 1913 paper [6], S. N. Bernstein established that the error in approximating $f(x) = |x|$ decays exactly like $\frac{1}{n}$, that is,

$$\Lambda_1 = \lim_{n\to\infty} nE_n[|x|]$$

exists and is finite and positive. Λ_1 is often called the Bernstein constant. The proof that the limit exists is long and difficult and is unclear in parts. Twenty-five years later [7], he presented a much simpler proof that works for the more general function $|x|^\alpha$ for all $\alpha > 0$ that is not an even integer. It involves dilations of the interval, making essential use of the homogeneity of $|x|^\alpha$, namely that for $\lambda > 0$,

$$|\lambda x|^{\alpha} = \lambda^{\alpha} |x|^{\alpha} .$$

This enabled Bernstein to relate the error in approximation on $[-\lambda, \lambda]$ to that on $[-1, 1]$. It also yielded a formulation of the limit as the error in approximation on the whole real axis by entire functions of exponential type, namely

$$\begin{aligned} \Lambda_{\alpha} &:= \lim_{n\to\infty} n^{\alpha} E_n \left[|x|^{\alpha} \right] \\ &= \inf \left\{ \| \, |x|^{\alpha} - f(x) \, \|_{L_{\infty}(\mathbb{R})} : f \text{ is entire of exponential type } \le 1 \right\} . \end{aligned} \tag{4}$$

You might ask: what is the connection of all this to scaling limits? Well, Bernstein related the errors of polynomial approximation by polynomials of degree $\le n$ on $[-1, 1]$ and $[-n, n]$. Thus, if we let P_n^* denote the unique polynomial of degree $\le n$ that best approximates $|x|^{\alpha}$ in the uniform norm on $[-n, n]$, Bernstein's proof essentially involved scaling $P_n^*(x)$ to $P_n^*(x/n)$. Since 0 is the place where $|x|^{\alpha}$ is least smooth, it is not surprising that we scale about 0. One of the classical unsolved problems of approximation theory is

Problem 1 Give an explicit representation for Λ_1.

Of course, this is a little imprecise, but something such as Λ_1 is a root of an explicit equation, or given by some explicit series, would be a real achievement. Bernstein obtained upper and lower bounds for Λ_1, and using these, speculated that possibly

$$\Lambda_1 = \frac{1}{2\sqrt{\pi}} = 0.28209\,47917\ldots .$$

Some 70 years later, this was disproved by Varga and Carpenter [64, 65] using high precision scientific computation. They showed that

$$\Lambda_1 = 0.28016\,94990\ldots$$

They also showed numerically that the normalized error $2nE_{2n}[|x|]$ should admit an asymptotic expansion in negative powers of n. Further, numerical explorations for approximation of $|x|^{\alpha}$ have been provided by Carpenter and Varga [9].

Bernstein also showed that for $\alpha > 1$, [7], [9, p. 194]

$$\frac{\left|\sin \frac{\alpha\pi}{2}\right|}{\pi} \Gamma(\alpha) \left(1 - \frac{1}{\alpha - 1}\right) < \Lambda_{\alpha} < \frac{\left|\sin \frac{\alpha\pi}{2}\right|}{\pi} \Gamma(\alpha) .$$

Surprisingly, the much deeper analogous problem of rational approximation has already been solved, by the great Herbert Stahl [54]. He proved, using sophisticated methods of potential theory and other complex analytic tools, that

$$\lim_{n\to\infty} e^{\pi\sqrt{n}} R_n [|x|] = 8,$$

where $R_n[|x|]$ denotes the error in best L_∞ approximation of $|x|$ on $[-1, 1]$ by rational functions with numerator and denominator degree $\leq n$. Later [55], he extended this to $|x|^\alpha$, establishing

$$\lim_{n\to\infty} e^{\pi\sqrt{\alpha n}} R_n\left[|x|^\alpha\right] = 4^{1+\alpha/2}\left|\sin\frac{\pi\alpha}{2}\right|.$$

Although Λ_α is not known explicitly, the ideas of Bernstein have been refined, and greatly extended. They are covered in the monograph of Timan [58, p. 48 ff.]. M. Ganzburg has shown limit relations of this type for large classes of functions, in one and several variables, even when weighted norms are involved [21–24]. Nikolskii [46] and Raitsin [48, 49] considered not only uniform, but also L_p norms. They and later Ganzburg [22] showed that for $1 \leq p \leq \infty$, there exists

$$\Lambda_{p,\alpha} = \lim_{n\to\infty} n^{\alpha+\frac{1}{p}} \inf\left\{\| \, |x|^\alpha - P(x) \, \|_{L_p[-1,1]}: \deg(P) \leq n\right\}.$$

More explicitly, Nikolskii [46] proved that at least for odd integers α,

$$\Lambda_{1,\alpha} = \frac{\left|\sin\frac{\alpha\pi}{2}\right|}{\pi} 8\Gamma(\alpha+1)\sum_{k=0}^{\infty}(-1)^k(2k+1)^{-\alpha-2}.$$

He also established an integral representation valid for all $\alpha > -1$, and Bernstein later noted that this implies the above series representation for all $\alpha > -1$. Raitsin [49] proved that for $\alpha > -\frac{1}{2}$,

$$\Lambda_{2,\alpha} = \frac{\left|\sin\frac{\alpha\pi}{2}\right|}{\pi} 2\Gamma(\alpha+1)\sqrt{\pi/(2\alpha+1)}.$$

These are the only known explicit values of $\Lambda_{p,\alpha}$. The extremal entire functions associated with these constants were given in explicit form in [25].

Vasiliev [63] extended Bernstein's results in another direction, replacing the interval $[-1, 1]$ by fairly general compact sets E. Totik [62] has put Vasiliev's results in final form, using sophisticated estimates for harmonic measures. In this more general setting, Λ_α still appears, multiplied by a quantity involving the equilibrium density of potential theory for E. The Bernstein constant was discussed in the recent book of Finch on mathematical constants [17, p. 257 ff.] in different areas of mathematics.

Another recent mode of attack on the Bernstein problem involves sophisticated properties of conformal maps of comb domains: instead of approximating directly on $[-1, 1]$, one solves the asymptotic problem on the symmetric interval $[-1, -a] \cup [a, 1]$. Reknowned complex analysts such as Eremenko, Nazarov, Peherstorfer, Volberg, and Yuditskii have been involved in this effort [16, 43]

The author's own attempts at this problem directly involve scaling limits: as above, let P_n^* denote the unique polynomial of degree $\leq n$ that best approximates $|x|^\alpha$ in the uniform norm. In [35], we proved a pointwise limit, namely that uniformly in compact subsets of $\mathbb{C}$,

$$\lim_{n\to\infty} n^{\alpha} P_n^* (z/n) = H_\alpha^* (z) , \tag{5}$$

where H_α^* is the unique entire function of exponential type 1 attaining the inf in (4). That paper also contained L_p analogues and implicit integral representations of H_α^*. Closely related to Problem 1 is:

Problem 2 Give an explicit description of the function H_α^*.

3 Universality Limits in Random Matrices

Although they have much older roots, the theory of random matrices rose to prominence in the 1950s, when the physicist Eugene Wigner found them an indispensable tool in analyzing scattering theory for neutrons off heavy nuclei. The mathematical context of the unitary case may be briefly described as follows. Let $\mathcal{M}(n)$ denote the space of n by n, Hermitian matrices $M = \left(m_{ij}\right)_{1\le i,j\le n}$. Consider a probability distribution on $\mathcal{M}(n)$,

$$\begin{aligned} P^{(n)}(M) &= cw(M)\, dM \\ &= cw(M) \left(\prod\nolimits_{j=1}^{n} dm_{jj}\right) \left(\prod\nolimits_{j<k} d\left(\mathrm{Re}\, m_{jk}\right) d\left(\mathrm{Im}\, m_{jk}\right)\right). \end{aligned}$$

Here, $w(M)$ is a function defined on $\mathcal{M}(n)$, and c is a normalizing constant. The most important case is

$$w(M) = \exp\left(-2n \text{ tr } Q(M)\right),$$

involving the trace tr, for appropriate functions Q defined on $\mathcal{M}(n)$. In particular, the choice

$$Q(M) = M^2,$$

leads to the Gaussian unitary ensemble (apart from scaling) that was considered by Wigner. One may identify $P^{(n)}$ above with a probability density on the eigenvalues $x_1 \le x_2 \le \cdots \le x_n$ of M,

$$P^{(n)}(x_1, x_2, \ldots, x_n) = c \left(\prod_{j=1}^{m} w(x_j)\right) \left(\prod\nolimits_{i<j} \left(x_i - x_j\right)^2\right).$$

See [10, p. 102 ff.]. Again, c is a normalizing constant.

Orthogonal polynomials enable one to explicitly represent $P^{(n)}$ and a number of other statistical quantities. Let μ be a finite positive Borel measure with support in the real line, with infinitely many points in the support, and all finite power moments.

Define orthonormal polynomials

$$p_n(x) = \gamma_n x^n + \cdots, \quad \gamma_n > 0,$$

$n = 0, 1, 2, \ldots$, satisfying the orthonormality conditions

$$\int p_j p_k d\mu = \delta_{jk}. \tag{6}$$

The nth reproducing kernel for μ is

$$K_n(x, y) = \sum_{k=0}^{n-1} p_k(x) p_k(y), \tag{7}$$

and the normalized kernel is

$$\tilde{K}_n(x, y) = \mu'(x)^{1/2} \mu'(y)^{1/2} K_n(x, y), \tag{8}$$

where μ' denotes the Radon-Nikodym derivative of μ.

There is the basic formula for the probability distribution $P^{(n)}$ [10, p. 112]:

$$P^{(n)}(x_1, x_2, \ldots, x_n) = \frac{1}{n!} \det\left(\tilde{K}_n\left(x_i, x_j\right)\right)_{1 \leq i,j \leq n}.$$

One may use this to compute a host of statistical quantities—for example, the probability that a fixed number of eigenvalues of a random matrix lie in a given interval. One particularly important quantity is the m-point correlation function for $M(n)$ [10, p. 112]:

$$\begin{aligned} R_m(x_1, x_2, \ldots, x_m) &= \frac{n!}{(n-m)!} \int \cdots \\ &\quad \int P^{(n)}(x_1, x_2, \ldots, x_n)\, dx_{m+1}\, dx_{m+2} \ldots dx_n \\ &= \det\left(\tilde{K}_n\left(x_i, x_j\right)\right)_{1 \leq i,j \leq m}. \end{aligned}$$

This last remarkable identity is due to Freeman Dyson.

The *universality limit in the bulk* asserts that for fixed $m \geq 2$, ξ in the interior of the support of μ, and real $a_1, a_2, \ldots, a_m$, we have

$$\begin{aligned} &\lim_{n \to \infty} \frac{1}{\tilde{K}_n(\xi, \xi)^m} R_m\left(\xi + \frac{a_1}{\tilde{K}_n(\xi, \xi)}, \xi + \frac{a_2}{\tilde{K}_n(\xi, \xi)}, \ldots, \xi + \frac{a_m}{\tilde{K}_n(\xi, \xi)}\right) \\ &= \det\left(\mathbb{S}\left(a_i - a_j\right)\right)_{1 \leq i,j \leq m}. \end{aligned}$$

Because m is fixed in this limit, this reduces to the case $m = 2$, namely

$$\lim_{n\to\infty} \frac{\tilde{K}_n\left(\xi + \frac{a}{\tilde{K}_n(\xi,\xi)}, \xi + \frac{b}{\tilde{K}_n(\xi,\xi)}\right)}{\tilde{K}_n(\xi,\xi)} = \mathbb{S}(a-b), \tag{9}$$

for real a, b. Thus, an assertion about the distribution of eigenvalues of random matrices has been reduced to a scaling limit involving orthogonal polynomials. The term universal is quite justified: the limit on the right-hand side of (9) is independent of ξ, but more importantly is independent of the underlying measure. Since in many cases,

$$\lim_{n\to\infty} \frac{1}{n}\tilde{K}_n(\xi,\xi) = \omega(\xi),$$

where ω is an appropriate "equilibrium density", we can also often recast (9) as

$$\lim_{n\to\infty} \frac{\tilde{K}_n\left(\xi + \frac{a}{n\omega(\xi)}, \xi + \frac{b}{n\omega(\xi)}\right)}{n\omega(\xi)} = \mathbb{S}(a-b). \tag{10}$$

For example, if $\mu' > 0$ is positive a.e. in $(-1, 1)$,

$$\omega(x) = \frac{1}{\pi\sqrt{1-x^2}},$$

the ubiquitous *arcsine distribution.*

Typically, the limit (9) is established uniformly for a, b in compact subsets of the real line, but if we remove the normalization from the outer K_n, we can also establish its validity for complex a, b, that is,

$$\lim_{n\to\infty} \frac{K_n\left(\xi + \frac{a}{\tilde{K}_n(\xi,\xi)}, \xi + \frac{b}{\tilde{K}_n(\xi,\xi)}\right)}{K_n(\xi,\xi)} = \mathbb{S}(a-b). \tag{11}$$

There are a variety of methods to establish (9–11). Perhaps the deepest methods are the Riemann-Hilbert methods, which yield far more than universality. See [5, 10–12, 27–29, 42] for Riemann-Hilbert references.

Inspired by the 60th birthday conference for Percy Deift, the author came up with a new comparison method to establish universality. Let μ be a measure supported on $(-1, 1)$, and assume for example that $\mu' > 0$ a.e. in $(-1, 1)$. Let μ be absolutely continuous in a neighborhood of some given $\xi \in (-1, 1)$ and assume that μ' is positive and continuous at ξ. Then, in [38] we established (10). This result was soon extended to a far more general setting by Findley, Simon, and Totik [18, 51, 52, 60], [61]. In particular, when μ is a measure with compact support that is regular, and $\log \mu'$ is integrable in a subinterval of the support (c, d), then Totik established that the universality (10) holds a.e. in (c, d). Totik used the method of polynomial pullbacks to go first from one to finitely many intervals, and then used the latter to approximate general compact sets. In contrast, Simon used the theory of Jost functions.

The drawback of this comparison method is that it requires regularity of the measure μ. Although the latter is a weak global condition, it is nevertheless most probably an unnecessary restriction. To circumvent this, the author developed a different method, based on tools of classical complex analysis, such as normal families, and the theory of entire functions of exponential type. In [39], this was used to show that universality holds in linear Lebesgue measure, *meas*, without any local or global conditions, in the set

$$\{\mu' > 0\} := \{\xi : \mu'(\xi) > 0\}.$$

Theorem 1 *Let μ be a measure with compact support and with infinitely many points in the support. Let $\varepsilon > 0$ and $r > 0$. Then*

$$meas\left\{\xi \in \{\mu' > 0\} : \sup_{|u|,|v|\le r}\left|\frac{K_n\left(\xi + \frac{u}{\tilde{K}_n(\xi,\xi)}, \xi + \frac{v}{\tilde{K}_n(\xi,\xi)}\right)}{K_n(\xi,\xi)} - \mathbb{S}(u-v)\right| \ge \varepsilon\right\} \to 0 \text{ as } n \to \infty. \tag{12}$$

The method of proof of this result is instructive, because it contains ideas often used in establishing scaling limits:

Step 1. Let

$$f_n(u,v) := \frac{K_n\left(\xi + \frac{u}{\tilde{K}_n(\xi,\xi)}, \xi + \frac{v}{\tilde{K}_n(\xi,\xi)}\right)}{K_n(\xi,\xi)},$$

a polynomial in u, v. For $R > 0$ and "most" ξ, one can use tools such as the Bernstein-Walsh growth lemma to show that for $n \ge n_0(R)$ and $|u|, |v| \le R$,

$$|f_n(u,v)| \le C_1 e^{C_2(|u|+|v|)}. \tag{13}$$

Here, C_1 and C_2 are independent of n, u, v.

Step 2. The uniform boundedness in (13) allows us to choose a subsequence $\{f_n\}_{n\in S}$ that converges uniformly for u, v in compact subsets of the plane to an entire function f satisfying the bound

$$|f(u,v)| \le C_1 e^{C_2(|u|+|v|)}, \quad u, v \in \mathbb{C}.$$

Thus, f is of exponential type in each variable.

Step 3. Inasmuch as each K_n is a reproducing kernel for polynomials of degree $\le n-1$, one expects its subsequential limit f to be a reproducing kernel for some space of functions. Indeed, it is, and this is the hard part: to show that f is a reproducing kernel for Paley-Wiener space PW. As reproducing kernels are unique, it follows that

$$f(u, v) = \mathbb{S}(u - v).$$

Since the limit is independent of the subsequence, we have the result for the full sequence.

It is unlikely that convergence in measure in (12) can be replaced by convergence a.e., but nevertheless we pose:

Problem 3 Does universality hold a.e. in $\{\mu' > 0\}$?

We emphasize that this is a tiny slice of a major topic. At the endpoints of compactly supported μ (the "edge of the spectrum"), one scales not with x/n but with x/n^2 and the limiting kernel is a Bessel kernel. For moving boundaries, one scales with $x/n^{2/3}$ and the limiting kernel is an Airy kernel. Other kernels arise when there are jump or other discontinuities, and there are several other more complex scaling limits associated with other universality limits. See, for example, [1–4, 15, 19, 20, 27, 31, 37, 57].

4 L_p Christoffel Functions

Let μ denote a finite positive Borel measure on $[-1, 1]$. Its L_2 Christoffel function is

$$\lambda_n(\mu, x) = \inf_{\deg(P) \leq n-1} \frac{\int_{-1}^{1} |P(t)|^2 \, d\mu(t)}{|P(x)|^2}. \tag{14}$$

If $\{p_j\}$ are the orthonormal polynomials for μ, then it follows from Cauchy-Schwarz' inequality and orthogonality that

$$\lambda_n(\mu, x) = 1 / \sum_{j=0}^{n-1} p_j^2(x), \tag{15}$$

while a minimizing polynomial for a given x is

$$P(t) = K_n(x, t).$$

From these formulas, it is fairly clear why $\lambda_n(\mu, x)$ is useful: bounds on λ_n are essentially bounds on averages of the orthonormal polynomials. Moreover, the extremal or variational property (14) allows comparison of λ_n for different measures.

As is often the case with orthogonal polynomials, it is easier first to start on the unit circle and then later map to $[-1, 1]$. Accordingly, let ω denote a finite positive

Borel measure on the unit circle, or equivalently $[-\pi, \pi]$. Its L_2 Christoffel function is

$$\lambda_n(\omega, z) = \inf_{\deg(P) \le n-1} \frac{\int_{-\pi}^{\pi} \left| P\left(e^{i\theta}\right) \right|^2 d\omega(\theta)}{|P(z)|^2},$$

and yes again, there is a connection to orthogonal polynomials on the unit circle.

Asymptotics of Christoffel functions have been studied for a very long time. Paul Nevai's 1986 ode to them [45] is still very relevant, while Barry Simon's books [50, 53] contain the most recent research. In a breakthrough 1991 paper, Maté, Nevai, and Totik [41] proved that when ω is *regular* and satisfies in some subinterval I of $[-\pi, \pi]$

$$\int_I \log \omega'\left(e^{i\theta}\right) d\theta > -\infty,$$

then for a.e. $\theta \in I$,

$$\lim_{n\to\infty} n\lambda_n\left(\omega, e^{i\theta}\right) = \omega'(\theta).$$

Here, ω is *regular* if

$$\lim_{n\to\infty} \left(\inf_{\deg(P)\le n} \frac{\int_{-\pi}^{\pi} |P|^2 d\omega}{\|P\|^2_{L_\infty(|z|=1)}} \right)^{1/n} = 1.$$

A sufficient condition for regularity, the so-called Erdős-Turán condition, is that $\omega' > 0$ a.e. in $[-\pi, \pi]$. However, there are pure jump measures, and pure singularly continuous measures that are regular [56].

That 1991 paper of Maté, Nevai, and Totik also addresses measures on $[-1, 1]$. If μ is regular on $[-1, 1]$ and satisfies on some subinterval I,

$$\int_I \log \mu' > -\infty,$$

then

$$\lim_{n\to\infty} n\lambda_n(\mu, x) = \pi\sqrt{1-x^2}\mu'(x), \tag{16}$$

for a.e. $x \in I$. Totik subsequently extended this to measures μ with arbitrary compact support [59].

The extension of Christoffel functions from L_2 to L_p also goes back a long way—in some contexts, back to Szegő. For ω as above, define its L_p Christoffel function

$$\lambda_{n,p}(\omega, z) := \inf_{\deg(P)\le n-1} \frac{\int_{-\pi}^{\pi} \left| P\left(e^{i\theta}\right) \right|^p d\omega(\theta)}{|P(z)|^p}. \tag{17}$$

By a compactness argument, there is a polynomial $P^*_{n,p,z}$ of degree $\le n-1$ with $P^*_{n,p,z}(z) = 1$ and

$$\lambda_{n,p}(\omega, z) = \int_{-\pi}^{\pi} \left|P_{n,p,z}^{*}\left(e^{i\theta}\right)\right|^{p} d\omega(\theta). \tag{18}$$

When $p \geq 1$, this polynomial is unique.

The classical Szegő theory provides asymptotics for $\lambda_{n,p}(\omega, z)$ when $|z| < 1$. For example, if ω is absolutely continuous, then [50, p. 153] for $|z| < 1$,

$$\lim_{n\to\infty} \lambda_{n,p}(\omega, z) = \inf\left\{\int |f|^{p} d\omega : f \in H^{\infty} \text{ and } f(z) = 1\right\}.$$

Here, H^{∞} is the usual Hardy space for the unit disk. Moreover, for general measures, there is an alternative expression involving the Poisson kernel for the unit disk [50, p. 154].

On the unit circle, and for measures on $[-1, 1]$, bounds for L_p Christoffel functions have been known for a long time, notably those in Paul Nevai's landmark memoir [44]. However, limits for L_p Christoffel functions on the circle or interval, were first established by Eli Levin and the author [33]. The asymptotic involves an extremal problem for the L_p Paley-Wiener space PW_p. This is the set of all entire functions f of exponential type at most π, whose restriction to the real lies in $L_p(\mathbb{R})$. We define

$$\mathscr{E}_p = \inf\left\{\int_{-\infty}^{\infty} |f(t)|^{p} dt : f \in PW_p \text{ and } f(0) = 1\right\}. \tag{19}$$

Moreover, we let $f_p^{*} \in PW_p$ be a function attaining the infimum in (19), so that $f_p^{*}(0) = 1$ and

$$\mathscr{E}_p = \int_{-\infty}^{\infty} \left|f_p^{*}(t)\right|^{p} dt.$$

When $p \geq 1$, f_p^{*} is unique. For $p < 1$, uniqueness is apparently unresolved. For $p > 1$, we may give an alternate formulation:

$$\mathscr{E}_p = \inf \int_{-\infty}^{\infty} \left|\mathbb{S}(t) - \sum_{j=-\infty, j\neq 0}^{\infty} c_j \mathbb{S}(t-j)\right|^{p} dt, \tag{20}$$

where the inf is taken over all $\{c_j\} \in \ell_p$, that is, over all $\{c_j\}$ satisfying

$$\sum_{j=-\infty, j\neq 0}^{\infty} \left|c_j\right|^{p} < \infty. \tag{21}$$

When $p = 2$, the orthonormality of the integer translates $\{\mathbb{S}(t-j)\}$ shows that $f_2^{*} = \mathbb{S}$, and

$$\mathscr{E}_2 = \int_{-\infty}^{\infty} \mathbb{S}(t)^2 \, dt = 1.$$

The precise value of $\mathscr{E}_p$ is apparently not known for $p \neq 2$. The estimate

$$\mathscr{E}_p > p^{-1}$$

goes back to 1949, to Korevaar's thesis [8, p. 102], [26].

We proved [33]:

Theorem 2 *Let $p > 0$, let ω be a finite positive measure supported on the unit circle, and assume that ω is regular. Let $|z_0| = 1$, and assume that z_0 is a Lebesgue point of ω, while the derivative ω' of the absolutely continuous part of ω is lower semi-continuous at z_0.*

(a) Then

$$\lim_{n\to\infty} n\lambda_{n,p}(\omega, z_0) = 2\pi \mathscr{E}_p \omega'(z_0). \tag{22}$$

(b) If also $\omega'(z_0) > 0$ and $p > 1$, we have

$$\lim_{n\to\infty} P^*_{n,p,z_0}\left(z_0 e^{2\pi i z/n}\right) = e^{i\pi z} f_p^*(z), \tag{23}$$

uniformly for z in compact subsets of the plane.

The proof of this theorem very heavily depends on scaling limits. Here are some ideas when $p > 1$, when ω is Lebesgue measure on the unit circle and when $z_0 = 1$: fix any $f \in PW_p$ with $f(0) = 1$. It admits the expansion

$$f(z) = \sum_{j=-\infty}^{\infty} f(j)\, \mathbb{S}(z-j) \tag{24}$$

that converges locally uniformly in the plane. This allows us to construct polynomials along the lines in the introduction: fix $m \geq 1$ and let

$$S_n(z) = \sum_{|j|\leq m} f(j)(-1)^j \ell_{jn}(z).$$

Here $\{\ell_{jn}\}$ are the fundamental polynomials of Lagrange interpolation at the roots of unity, as in Sect. 1. Since $S_n(1) = f(0) = 1$, we have

$$\lambda_{n,p}(\omega, 1) \leq \int_{-\pi}^{\pi} |S_n(z)|^p \, d\theta.$$

Here, for each $r > 0$, the limit (3) shows

$$\lim_{n\to\infty} n \int_{-2\pi r/n}^{2\pi r/n} |S_n(z)|^p \, d\theta = 2\pi \lim_{n\to\infty} \int_{-r}^{r} \left| S_n\left(e^{2\pi i t/n}\right) \right|^p dt$$
$$= 2\pi \int_{-r}^{r} \left| \sum_{|j|\le m} f(j) \mathbb{S}(t-j) \right|^p dt.$$

We estimate the tails of both sides, let $r \to \infty$, and deduce that

$$\limsup_{n\to\infty} n\lambda_{n,p}(\omega, 1) \le \limsup_{n\to\infty} n \int_{-\pi}^{\pi} |S_n(z)|^p \, d\theta$$
$$\le 2\pi \int_{-\infty}^{\infty} \left| \sum_{|j|\le m} f(j) \mathbb{S}(t-j) \right|^p dt.$$

Next, we let $m \to \infty$, and obtain

$$\limsup_{n\to\infty} n\lambda_{n,p}(\omega, 1) \le 2\pi \int_{-\infty}^{\infty} |f(t)|^p \, dt.$$

As we may choose any $f \in L_\pi^p$, with $f(0) = 1$, we obtain

$$\limsup_{n\to\infty} n\lambda_{n,p}(\omega, 1) \le 2\pi \mathcal{E}_p.$$

The converse inequality is more difficult, but also involves scaling limits.

There are also analogous results on $[-1, 1]$. Let μ be a finite positive measure with support $[-1, 1]$. It was probably Paul Nevai, who first systematically studied for measures on $[-1, 1]$, the general L_p Christoffel function

$$\lambda_{n,p}(\mu, x) = \inf_{\deg(P)\le n-1} \frac{\int_{-1}^{1} |P(t)|^p \, d\mu(t)}{|P(x)|^p}, \tag{25}$$

in his 1979 memoir [44]. It was useful in establishing Bernstein and Nikolskii inequalities, in estimating quadrature sums, and in studying convergence of Lagrange interpolation and orthogonal expansions. Let $P^*_{n,p,\xi}$ denote a polynomial of degree $\le n-1$ with $P^*_{n,p,\xi}(\xi) = 1$, that attains the inf in (25).

Let us say that μ is *regular on* $[-1, 1]$, or just *regular*, if

$$\lim_{n\to\infty} \left(\inf_{\deg(P)\le n} \frac{\int_{-1}^{1} P^2 d\mu}{\|P\|^2_{L_\infty[-1,1]}} \right)^{1/n} = 1.$$

As for the unit circle, a simple sufficient condition for regularity is that $\mu' > 0$ a.e. in $[-1, 1]$, although it is far from necessary. We proved [33]:

Theorem 3 *Let $p > 0$, and let μ be a finite positive measure supported on $[-1, 1]$, and assume that μ is regular. Let $\xi \in (-1, 1)$ be a Lebesgue point of μ, and let the derivative of its absolutely continuous part μ' be lower semi-continuous at ξ.*

(a) Then

$$\lim_{n\to\infty} n\lambda_{n,p}(\mu, \xi) = \pi\sqrt{1-\xi^2}\mathscr{E}_p\mu'(\xi).$$

(b) If also $\mu'(\xi) > 0$ and $p > 1$, we have

$$\lim_{n\to\infty} P^*_{n,p,\xi}\left(\xi + \frac{\pi\sqrt{1-\xi^2}z}{n}\right) = f^*_p(z), \tag{26}$$

uniformly for z in compact subsets of the plane, where f^*_p is the function attaining the inf in (19).

Observe that for both the unit circle and $[-1, 1]$, the only difference between the L_2 and L_p asymptotics is the constant $\mathscr{E}_p$. This suggests:

Problem 4 Evaluate $\mathscr{E}_p$, or at least estimate it, for $p \neq 2$.

Problem 5 Characterize the entire function f^*_p attaining the inf in (19).

5 Nikolskii Inequalities

Nikolskii inequalities compare the norms of polynomials in different L_p spaces. Accordingly, define

$$\|P\|_p = \left(\frac{1}{2\pi}\int_{-\pi}^{\pi} \left|P\left(e^{i\theta}\right)\right|^p d\theta\right)^{1/p}, \quad \text{if } p < \infty$$

and

$$\|P\|_\infty = \sup_{|z|=1} |P(z)|.$$

The simplest Nikolskii inequalities assert that given $q > p > 0$, there exists C depending on p, q, such that for $n \geq 1$ and polynomials P of degree $\leq n$,

$$\frac{\|P\|_p}{\|P\|_q} \geq Cn^{\frac{1}{q}-\frac{1}{p}}. \tag{27}$$

They are useful in studying convergence of orthonormal expansions and Lagrange interpolation, and in analyzing quadrature and discretization of integrals. A proof for trigonometric polynomials, which includes this case, appears in [13, Theorem 2.6, p. 102]. The converse sharp inequality, namely

$$\frac{\|P\|_p}{\|P\|_q} \leq 1$$

follows from Hölder's inequality. It is a longstanding problem to determine the sharp constant in (27). Accordingly define

$$\Lambda_{n,p,q} = \inf_{\deg(P)\leq n-1} \frac{\|P\|_p}{\|P\|_q}. \tag{28}$$

Our results from the previous section resolve the case $q = \infty$: as $n \to \infty$,

$$\lim_{n\to\infty} \Lambda_{n,p,\infty} n^{1/p} = \inf \left\{ \frac{\|f\|_{L_p(\mathbb{R})}}{\|f\|_{L_\infty(\mathbb{R})}} : f \in PW_p \right\}.$$

This suggests a generalization that might work for Nikolskii inequalities. Define

$$\mathscr{E}_{p,q} = \inf \left\{ \frac{\|f\|_{L_p(\mathbb{R})}}{\|f\|_{L_q(\mathbb{R})}} : f \in PW_p \right\}. \tag{29}$$

Using precisely the sort of scaling limits discussed in the introduction, and in the previous section, Eli Levin and I proved [32] that if $q > p > 0$, then

$$\limsup_{n\to\infty} \Lambda_{n,p,q} n^{\frac{1}{p}-\frac{1}{q}} \leq \mathscr{E}_{p,q}. \tag{30}$$

Despite repeated attempts, we were unable to prove the limit. Accordingly, we pose:

Problem 6 Prove

$$\lim_{n\to\infty} \Lambda_{n,p,q} n^{\frac{1}{p}-\frac{1}{q}} = \mathscr{E}_{p,q}. \tag{31}$$

Problem 7 Characterize, or describe, the entire functions attaining the inf in $\mathscr{E}_{p,q}$.

6 Marcinkiewicz–Zygmund Inequalities

The Plancherel-Polya inequalities [30, p. 152] assert that for $1 < p < \infty$, and entire functions f of exponential type at most π,

$$A_p \sum_{j=-\infty}^{\infty} |f(j)|^p \leq \int_{-\infty}^{\infty} |f|^p \leq B_p \sum_{j=-\infty}^{\infty} |f(j)|^p, \tag{32}$$

provided the integral in the middle is finite. For $0 < p \leq 1$, the left-hand inequality is still true, but the right-hand inequality requires additional restrictions [8, 47]. Here,

A_p and B_p are independent of f. The Marcinkiewicz–Zygmund inequalities assert [66, Vol. II, p. 30] that for $p > 1, n \geq 1$, and polynomials P of degree $\leq n - 1$,

$$\frac{A'_p}{n} \sum_{j=1}^{n} \left| P\left(e^{2\pi i j/n}\right) \right|^p \leq \int_0^1 \left| P\left(e^{2\pi i t}\right) \right|^p dt \leq \frac{B'_p}{n} \sum_{j=1}^{n} \left| P\left(e^{2\pi i j/n}\right) \right|^p. \tag{33}$$

Here too, A'_p and B'_p are independent of n and P, and the left-hand inequality is also true for $0 < p \leq 1$ [34]. We assume that A_p, B_p, A'_p, B'_p are the sharp constants, so that A_p and A'_p are as large as possible, while B_p and B'_p are as small as possible. These inequalities are useful in studying convergence of Fourier series, Lagrange interpolation, in number theory, and weighted approximation. See [14, 34, 35]. Of course if $p = 2$, then $A_2 = B_2 = A'_2 = B'_2 = 1$.

In [40], I proved that the sharp constants in (32) and (33) are the same:

Theorem 4 *For* $0 < p < \infty$,

$$A_p = A'_p$$

and for $1 < p < \infty$,

$$B_p = B'_p.$$

Moreover if $p \neq 2$, *then* $A_p < 1 < B_p$.

In [8, p. 101, Thm. 6.7.15], it is proven that $A_p \geq \frac{\pi}{4e^{\pi\pi/2}}$ (this was recorded incorrectly in [40]).

Problem 8 Evaluate or estimate A_p and B_p.

Acknowledgements The author would like to thank the organizers of Approximation Theory XV for the opportunity to take part in a very successful and stimulating conference. In addition, the author thanks the referees for their thorough reports.

References

1. G. Akemann, J. Baik, P. Di Francesco (eds.), *The Oxford Handbook of Random Matrix Theory* (Oxford University Press, Oxford, 2011)
2. G. Anderson, A. Guionnet, O. Zeitouni, *An Introduction to Random Matrices, Cambridge Studies in Advanced Mathematics, 118* (Cambridge University Press, Cambridge, 2010)
3. J. Baik, T. Kriecherbauer, K. McLaughlin, P. Miller, *Uniform Asymptotics for Polynomials Orthogonal with Respect to a General Class of Discrete Weights and Universality Results for Associated Ensembles*. Annals of Math. Studies, vol. 164 (Princeton University Press, Princeton, 2007)

4. J. Baik, L. Li, T. Kriecherbauer, K. McLaughlin, C. Tomei, *Proceedings of the Conference on Integrable Systems, Random Matrices and Applications*. Contemporary Mathematics, vol. 458 (American Mathematical Society, Providence, 2008)
5. P. Bleher, A. Its, *Random Matrix Models and their Applications* (Cambridge University Press, Cambridge, 2001)
6. S.N. Bernstein, Sur la meilleure approximation de $|x|$ par des polynômes de degré donnés. Acta Math. **37**, 1–57 (1913)
7. S.N. Bernstein, Sur la meilleure approximation de $|x|^p$ par des polynô mes de degrés très élevés. Bull. Acad. Sc. USSR, Ser. Math., **2** 181–190 (1938)
8. R.P. Boas, *Entire Functions* (Academic Press, New York, 1954)
9. A.J. Carpenter, R.S. Varga, Some Numerical Results on Best Uniform Polynomial Approximation of x^α on [0, 1]. Springer Lecture Notes in Mathematics, vol. 1550 (1993), pp. 192–222
10. P. Deift, *Orthogonal Polynomials and Random Matrices: A Riemann-Hilbert Approach, Courant Institute Lecture Notes*, vol. 3 (New York University Press, New York, 1999)
11. P. Deift, D. Gioev, *Random Matrix Theory: Invariant Ensembles and Universality, Courant Institute Lecture Notes*, vol. 18 (New York University Press, New York, 2009)
12. P. Deift, T. Kriecherbauer, K. McLaughlin, S. Venakides, X. Zhou, *Uniform Asymptotics for Polynomials Orthogonal with respect to Varying Exponential Weights and Applications to Universality Questions in Random Matrix Theory*. Communications on Pure and Applied Mathematics, vol. L11 (1999), pp. 1335–1425
13. R. DeVore G. Lorentz, *Constructive Approximation*, vol. 1 (Springer, Berlin, 1993)
14. F. Filbir, H.N. Mhaskar, Marcinkiewicz-Zygmund measures on manifolds. J. Complex. **27**, 568–596 (2011)
15. L. Erdős, Universality of Wigner random matrices: a survey of recent results. Russian Math. Surv. **66**, 507–626 (2011)
16. A. Eremenko, P. Yuditskii, Polynomials of the best uniform approximation to sgn x on two intervals. J. d'Analyse Mathématique **114**, 285–315 (2011)
17. S.R. Finch, *Mathematical Constants* (Cambridge University Press, Cambridge, 2003)
18. E. Findley, Universality for regular measures satisfying Szegő's condition. J. Approx. Theory **155**, 136–154 (2008)
19. A.F. Moreno, A. Martinez-Finkelshtein, V. Sousa, Asymptotics of orthogonal polynomials for a weight with a jump on $[-1, 1]$. Constr. Approx. **33**, 219–263 (2011)
20. P. Forrester, *Log-Gases and Random Matrices* (Princeton University Press, Princeton, 2010)
21. M. Ganzburg, *Limit Theorems and Best Constants of Approximation Theory* (in), Handbook on Analytic Computational Methods in Applied Mathematics, ed. by G. Anastassiou (CRC Press, Boca Raton, FL 2000)
22. M. Ganzburg, The Bernstein constant and polynomial interpolation at the Chebyshev nodes. J. Approx. Theory **119**, 193–213 (2002)
23. M. Ganzburg, Limit Theorems of Polynomial Approximation. Memoirs Am. Math. Soc. **192**(897) (2008)
24. M. Ganzburg, *Polynomial interpolation and asymptotic representations for zeta functions*. Dissertationes Math. (Rozprawy Mat.) **496**, 117 (2013)
25. M. Ganzburg, D.S. Lubinsky, Best approximating entire functions to $|x|^\alpha$ in L_2. Contemp. Math. **455**, 93–107 (2008)
26. J. Korevaar, An inequality for entire functions of exponential type. Nieuw. Arch. Wiskunde **23**, 55–62 (1949)
27. A. Kuijlaars, *Universality, Chapter 6 in "The Oxford Handbook on Random Matrix Theory*, ed. by G. Akemann, J. Baik, P. Di Francesco (Oxford University Press, Oxford, 2011), pp. 103–134
28. A. Kuijlaars, M. Vanlessen, Universality for eigenvalue correlations at the origin of the spectrum. Commun. Math. Phys. **243**, 163–191 (2003)
29. A. Kuijlaars, K.T.-R. McLaughlin, W. Van Assche, M. Vanlessen, The Riemann-Hilbert approach to strong asymptotics for orthogonal polynomials on [–1,1]. Adv. Math. **188**, 337–398 (2004)

30. B. Ja Levin, *Lectures on Entire Functions*, Translations of Mathematical Monographs (American Mathematical Society, Providence, 1996)
31. E. Levin, D.S. Lubinsky, *Universality Limits at the Soft Edge of the Spectrum via Classical Complex Analysis*. International Maths. Research Notices (2010), https://doi.org/10.1093/imrn/rnq185
32. E. Levin, D.S. Lubinsky, Asymptotic behavior of Nikolskii constants for polynomials on the unit circle. Comput. Methods Funct. Theory **15**, 459–468 (2015)
33. E. Levin, D.S. Lubinsky, L_p Christoffel Functions, L_p Universality, and Paley-Wiener Spaces. J. d'Analyse Mathématique, **125** 243–283 (2015)
34. D.S. Lubinsky, *Marcinkiewicz-Zygmund Inequalities: Methods and Results*, (in) Recent Progress in Inequalities, ed. by G.V. Milovanovic et al. (Kluwer Academic Publishers, Dordrecht, 1998), pp. 213–240
35. D.S. Lubinsky, On the Bernstein constants of polynomial approximation. Constr. Approx. **25**, 303–366 (2007)
36. D.S. Lubinsky, Universality limits in the bulk for arbitrary measures on compact sets. J. d'Analyse Mathématique **106**, 373–394 (2008)
37. D. S. Lubinsky, *Universality Limits at the Hard Edge of the Spectrum for Measures with Compact Support*, International Maths. Research Notices, International Maths. Research Notices (2008), Art. ID rnn 099, 39 pp
38. D.S. Lubinsky, A new approach to universality limits involving orthogonal polynomials. Ann. Math. **170**, 915–939 (2009)
39. D.S. Lubinsky, Bulk universality holds in measure for compactly supported measures. J d'Analyse Mathématique **116**, 219–253 (2012)
40. D.S. Lubinsky, On sharp constants in Marcinkiewicz-Zygmund and Plancherel-Polya inequalities. Proc. Am. Math. Soc. **142**, 3575–3584 (2014)
41. A. Maté, P. Nevai, V. Totik, Szegő's extremum problem on the unit circle. Ann. Math. **134**, 433–453 (1991)
42. K.T.-R. McLaughlin, P.D. Miller, *The $\bar{\partial}$-Steepest Descent Method and the Asymptotic Behavior of Polynomials Orthogonal on the Unit Circle with Fixed and Exponentially Varying Nonanalytic Weights*, International Maths. Research Notices (2006), Article ID 48673, pp. 1–78
43. F. Nazarov, F. Peherstorfer, A. Volberg, P. Yuditskii, Asymptotics of the best polynomial approximation of $|x|^p$ and of the best Laurent polynomial approximation of sgn(x) on two symmetric intervals. Constr. Approx. **29**, 23–39 (2009)
44. P. Nevai, *Orthogonal Polynomials*. Memoirs of the AMS, vol. 213 (1979)
45. P. Nevai, Geza Freud, orthogonal polynomials and Christoffel functions: a case study. J. Approx. Theory **48**, 3–167 (1986)
46. S.M. Nikolskii, On the best mean approximation by polynomials of the functions $|x-c|^s$. Izvestia Akad. Nauk SSSR **11**, 139–180 (1947). (in Russian)
47. M. Plancherel, G. Polya, Fonctions entierers et integrales de Fourier multiples. Comment. Math. Helvet. **10**, 110–163 (1937)
48. R.A. Raitsin, S. N. Bernstein limit theorem for the best approximation in the mean and some of its applications. Izv. Vysch. Uchebn. Zaved. Mat. **12** 81–86(1968)
49. R.A. Raitsin, On the best approximation in the mean by polynomials and entire functions of finite degree of functions having an algebraic singularity. Izv. Vysch. Uchebn. Zaved. Mat. **13**, 59–61 (1969)
50. B. Simon, *Orthogonal Polynomials on the Unit Circle, Parts 1 and 2* (American Mathematical Society, Providence, 2005)
51. B. Simon, Two extensions of Lubinsky's universality theorem. Journal d'Analyse de Mathématique **105**, 345–362 (2008)
52. B. Simon, Weak convergence of CD kernels and applications. Duke Math. J. **146**, 305–330 (2009)
53. B. Simon, *Szegö's theorem and its Descendants: Spectral Theory for L_2 Perturbations of Orthogonal Polynomials* (Princeton University Press, Princeton, 2011)

54. H. B. Stahl, *Best Uniform Rational Approximations of* $|x|$ *on* $[-1, 1]$, Mat. Sb. **183**(1992), 85–118. (Translation in Russian Acad. Sci. Sb. Math., **76**(1993), 461–487)
55. H.B. Stahl, Best uniform rational approximation of x^{α} on $[0, 1]$. Acta Math. **190**, 241–306 (2003)
56. H. Stahl, V. Totik, *General Orthogonal Polynomials* (Cambridge University Press, Cambridge, 1992)
57. T. Tao, *Topics in Random Matrix Theory, Graduate Studies in Mathematics*, vol. 132 (American Mathematical Society, Providence, 2012)
58. A.F. Timan, *Theory of Approximation of Functions of a Real Variable (translated by J Berry)* (Dover, New York, 1994)
59. V. Totik, Asymptotics for Christoffel functions for general measures on the real line. J. d'Analyse Mathématique **81**, 283–303 (2000)
60. V. Totik, Universality and fine zero spacing on general sets. Arkiv för Matematik **47**, 361–391 (2009)
61. V. Totik, Universality under Szegő's condition. Canad. Math. Bull. **59**, 211–224 (2016)
62. V. Totik, *Metric properties of harmonic measure*. Memoirs Am. Math. Soc. **184** 867 (2006)
63. R.K. Vasiliev, *Chebyshev Polynomials and Approximation*, Theory on Compact Subsets of the Real Axis (Saratov University Publishing House, 1998)
64. R.S. Varga, *Scientific Computation on Mathematical Problems and Conjectures*. CBMS-NSF Regional Conference Series in Applied Mathematics, SIAM, Vermont (1990)
65. R.S. Varga, A.J. Carpenter, On the Bernstein conjecture in approximation theory. Constr. Approx. **1**, 333–348 (1985)
66. A. Zygmund, *Trigonometric Series*, vols. I, II (Cambridge University Press, Cambridge, II, 1990)

Generalized B-Splines in Isogeometric Analysis

Carla Manni, Fabio Roman and Hendrik Speleers

Abstract In this paper, we survey the use of generalized B-splines in isogeometric Galerkin and collocation methods. Generalized B-splines are a special class of Tchebycheffian B-splines and form an attractive alternative to standard polynomial B-splines and NURBS in both modeling and simulation. We summarize their definition and main properties, and we illustrate their use in a selection of numerical examples in the context of isogeometric analysis. For practical applications, we mainly focus on trigonometric and hyperbolic generalized B-splines.

Keywords Generalized B-splines · Isogeometric analysis · Galerkin and collocation methods

1 Introduction

Isogeometric analysis (IgA) is a technology introduced nearly a decade ago in a seminal paper by Hughes et al. [24] that unifies computer-aided design (CAD) and finite element analysis (FEA). CAD software, used in industry for geometric modeling, typically describes physical domains by means of tensor-product B-splines or non-uniform rational B-splines (NURBS). Such geometries are then further processed in the analysis phase. One of the key concepts in IgA is to use the same discretization and representation tools for the design as well as for the analysis (in an isoparametric environment), providing a true design-through-analysis methodology [13, 24]. The isogeometric approach based on B-splines/NURBS shows important advan-

C. Manni (✉) · H. Speleers
Department of Mathematics, University of Rome "Tor Vergata",
Via della Ricerca Scientifica, 00133 Rome, Italy
e-mail: manni@mat.uniroma2.it

H. Speleers
e-mail: speleers@mat.uniroma2.it

F. Roman
Department of Mathematics, University of Turin, Via Carlo Alberto 10,
10123 Turin, Italy
e-mail: fabio.roman@unito.it

G.E. Fasshauer and L.L. Schumaker (eds.), *Approximation Theory XV: San Antonio 2016*, Springer Proceedings in Mathematics & Statistics 201,
DOI 10.1007/978-3-319-59912-0_12

tages over standard finite element approaches. The geometry of the physical domain is exactly described, so the interaction with the CAD system during any further refinement process in the analysis phase is eliminated. In addition, B-spline and NURBS spaces possess an inherent higher smoothness than those in standard FEA, leading to a higher accuracy per degree of freedom. The concept of IgA leads to improved convergence and smoothness properties of the PDE solutions and faster overall simulations. Thanks to the successful application in various engineering fields (from structural mechanics to fluid–structure interaction), IgA is rapidly becoming a mainstream analysis methodology and a new paradigm for geometric design.

Even though NURBS are the de facto standard in CAD systems [38], they suffer from a few major drawbacks. For example, their parameterization of conics does not correspond to the arc length, and they behave poorly with respect to differentiation and integration, which are crucial operators in analysis. On the other hand, in the literature there exist alternatives to overcome such problems as we will briefly review in the following.

Standard polynomial splines of degree p are smooth piecewise algebraic polynomials of degree less than or equal to p. This concept can be generalized in two directions, see [43, Chap. 11]: (a) the spaces from which the pieces are drawn can be more general than algebraic polynomials, and (b) the derivative conditions defining smoothness at each knot can be replaced by more general linear functionals. Some special generalizations have been independently studied by several authors primarily as tools in computer-aided geometric design.

Extended Tchebycheff (ET-) spaces of dimension $p + 1$ on a given interval J are spaces such that any non-trivial element has at most p zeros (counting multiplicity) in J; these are a natural generalization of algebraic polynomial spaces. Similarly, Tchebycheffian splines, where the pieces are drawn from ET-spaces, are a natural and elegant generalization of polynomial splines, see [43, Chap. 9]. They offer several advantages over standard polynomial splines, mainly due to the wide variety of ET-spaces. Despite this flexibility, many results of the polynomial framework extend in an elegant way to the broader Tchebycheff framework. As it is difficult to trace all the works on Tchebycheffian splines, we limit ourselves to mention [25] for locally supported bases, [36] for a blossoming approach, and [29] for a knot insertion procedure. We refer the reader to [43] for an extended bibliography on the topic.

In this paper, we focus on a special class of Tchebycheffian splines, the so-called *generalized splines*. Following [27], these are splines where the pieces are drawn from spaces of the form $\langle 1, x, \ldots, x^{p-2}, \overline{U}(x), \overline{V}(x)\rangle$, called section spaces, with suitable assumptions on the functions $\overline{U}$, $\overline{V}$ and with the usual derivative conditions defining smoothness at each knot. Therefore, the generalized splines considered here are less general than those described in [43, Chap. 11]. They can be seen as the minimal extension of standard polynomial splines still offering a wide variety of additional flexibility in geometric modeling (see, e.g., [10, 11, 26, 27, 30, 37]) and numerical simulation (see, e.g., [32, 33, 35]). This compromise between “minimal distance” from the polynomial case and flexibility is the motivation of our choice. Generalized splines possess many fundamental features of standard polynomial splines. In particular, they admit a representation in terms of basis functions with similar properties to polynomial B-splines. Such functions are called generalized B-splines (GB-splines).

GB-splines with a suitable selection of their section spaces — typically including polynomial and hyperbolic/trigonometric functions — allow for an exact representation of polynomial curves, conics, helices, and other profiles of salient interest in applications. In particular, conics are well parameterized by hyperbolic/trigonometric GB-splines. Moreover, in contrast to NURBS, they behave completely similar to B-splines with respect to differentiation and integration. Thanks to their structural similarity to polynomial B-splines, GB-splines are "plug-to-plug" compatible with B-splines in modeling and simulation software.

In this paper, we survey some recent results on the use of GB-splines as bases for the discretization spaces in numerical simulation according to the isogeometric paradigm. The goal of our survey is twofold. First, we want to highlight the structural similarity between B-splines and GB-splines and illustrate this by means of their behavior with respect to local refinement (see [6, 8, 34]) and the spectral properties of the matrices obtained in Galerkin/collocation discretizations (see [41]). Second, we show how a fine-tuning of the section spaces—according to a problem-oriented strategy taking into account the geometrical and/or analytical peculiar issues of the specific problem—generally results in a gain from the accuracy point of view, see [32–35].

The remainder of the paper is divided into 4 sections. Section 2 presents the definition and main properties of GB-splines. A short summary of isogeometric Galerkin and collocation methods is provided in Sect. 3. Section 4 compares numerical discretizations based on B-splines and GB-splines according to the isogeometric paradigm for three model differential problems taken from the literature. Finally, Sect. 5 collects some concluding remarks.

2 Generalized B-Splines

In this section, we present the definition and main properties of GB-splines. We start by defining the almost-polynomial spaces we are interested in. Throughout the paper, we denote by $\langle f_0, \ldots, f_p \rangle$ the space spanned by the set of functions $\{f_0, \ldots, f_p\}$.

2.1 Almost-Polynomial Spaces

Let $\overline{U}, \overline{V}$ be two univariate functions in $C^p([a,b])$ such that the space $\langle D^{(p-1)}\overline{U}, D^{(p-1)}\overline{V} \rangle$ is an ET-space of dimension 2 on $[a,b]$; i.e., any non-trivial element in the space spanned by the $(p-1)$th derivatives of $\overline{U}, \overline{V}$ has at most one zero (including multiplicity) in $[a,b]$. Then, we consider the following almost-polynomial space:

$$\mathbb{P}_p^{\overline{U},\overline{V}}([a,b]) := \langle 1, x, \ldots, x^{p-2}, \overline{U}(x), \overline{V}(x)\rangle, \quad x \in [a,b]. \tag{1}$$

The functions $\overline{U}$, $\overline{V}$ can be carefully selected in order to incorporate specific features in the space (1). Popular choices are as follows:

$$\mathbb{P}_p([a,b]) := \langle 1, x, \ldots, x^{p-2}, x^{p-1}, x^p\rangle, \tag{2}$$

$$\mathbb{H}_{p,\alpha}([a,b]) := \langle 1, x, \ldots, x^{p-2}, \cosh(\alpha x), \sinh(\alpha x)\rangle, \quad 0 < \alpha \in \mathbb{R}, \tag{3}$$

$$\mathbb{T}_{p,\alpha}([a,b]) := \langle 1, x, \ldots, x^{p-2}, \cos(\alpha x), \sin(\alpha x)\rangle, \quad 0 < \alpha(b-a) < \pi. \tag{4}$$

We refer to [12] for a detailed analysis of the properties of spaces of the form (1). In particular, they are ET-spaces of dimension $p+1$ and admit a basis which is a natural generalization of standard Bernstein polynomials. Moreover, for fixed values of the involved parameters, the spaces (3) and (4) have the same approximation power as the polynomial space in (2), see [12, Sect. 3]. The spaces (3) and (4) have been widely used in geometric modeling and constrained interpolation/approximation because the phase parameters α can be exploited as shape-parameters [26, 30, 47].

In the following, we are interested in spaces of smooth piecewise functions with sections in spaces of the form (1). Therefore, the space (1) will be referred to as *section space*. In Sect. 2.2, we describe how to build basis functions for such piecewise spaces that possess the same nice properties as standard polynomial B-splines.

2.2 GB-Splines: Definition and Properties

B-splines are the most famous basis for the space of polynomial splines. They can be defined in different ways, see [14, 43], but probably their most popular definition is given in terms of a recurrence relation. For $n \geq 1$ and $p \geq 0$, let $\Xi := \{\xi_i\}_{i=1}^{n+2p+1}$ be a non-decreasing sequence of knots. Without loss of generality, we assume Ξ to be open; i.e., its end knots have a multiplicity $p+1$,

$$\xi_1 = \cdots = \xi_{p+1} < \cdots \leq \xi_i \leq \xi_{i+1} \leq \cdots < \xi_{n+p+1} = \cdots = \xi_{n+2p+1}. \tag{5}$$

The ith B-spline $B_{i,p,\Xi}$ of degree p over the knot sequence Ξ is defined recursively as follows. For $p = 0$,

$$B_{i,0,\Xi}(x) := \begin{cases} 1, & \text{if } x \in [\xi_i, \xi_{i+1}), \\ 0, & \text{elsewhere,} \end{cases}$$

and for $p \geq 1$,

$$B_{i,p,\Xi}(x) := \frac{x - \xi_i}{\xi_{i+p} - \xi_i} B_{i,p-1,\Xi}(x) + \frac{\xi_{i+1+p} - x}{\xi_{i+1+p} - \xi_{i+1}} B_{i+1,p-1,\Xi}(x),$$

where fractions with zero denominators are considered to be zero. The B-spline $B_{i,p,\Xi}$ is a nonnegative piecewise polynomial supported on the interval $[\xi_i, \xi_{i+p+1}]$. Moreover, it has a very elegant differentiation formula for $p \geq 1$,

$$D_+ B_{i,p,\Xi}(x) = p\left(\frac{B_{i,p-1,\Xi}(x)}{\xi_{i+p} - \xi_i} - \frac{B_{i+1,p-1,\Xi}(x)}{\xi_{i+p+1} - \xi_{i+1}}\right),$$

where D_+ denotes the usual right derivative, and integration formula,

$$\delta_{i,p,\Xi} := \int_{\xi_i}^{\xi_{i+p+1}} B_{i,p,\Xi}(s)\,\mathrm{d}s = \frac{\xi_{i+p+1} - \xi_i}{p+1}.$$

By combining the above formulas, we arrive at an alternative recurrence relation for B-splines of degree $p \geq 1$:

$$B_{i,p,\Xi}(x) = \int_{\xi_i}^{x} \frac{B_{i,p-1,\Xi}(s)}{\delta_{i,p-1,\Xi}}\,\mathrm{d}s - \int_{\xi_{i+1}}^{x} \frac{B_{i+1,p-1,\Xi}(s)}{\delta_{i+1,p-1,\Xi}}\,\mathrm{d}s, \tag{6}$$

where we use the convention that if $\delta_{i,k,\Xi} = 0$, then

$$\int_{\xi_i}^{x} \frac{B_{i,k,\Xi}(s)}{\delta_{i,k,\Xi}}\,\mathrm{d}s := \begin{cases} 1, & \text{if } x \geq \xi_{i+k+1}, \\ 0, & \text{otherwise.} \end{cases}$$

A generalization of (6) leads to B-spline-like functions with sections in spaces of the form (1), see [27, 31] and references therein. They will be called *GB-splines of degree p*. Note that the term *degree* is used here to stress the similarity with the polynomial case. Given a knot sequence Ξ as in (5), we select for each $\xi_i < \xi_{i+1}$ an almost-polynomial space $\mathbb{P}_p^{\overline{U}_i,\overline{V}_i}([\xi_i, \xi_{i+1}])$ generated by the functions $\overline{U}_i, \overline{V}_i$. We denote by $\overline{u}_i, \overline{v}_i$ the unique elements in $\langle D^{(p-1)}\overline{U}_i, D^{(p-1)}\overline{V}_i\rangle$ satisfying

$$\overline{u}_i(\xi_i) = 1, \quad \overline{u}_i(\xi_{i+1}) = 0, \quad \overline{v}_i(\xi_i) = 0, \quad \overline{v}_i(\xi_{i+1}) = 1.$$

For notational convenience, we also define the piecewise functions U, V by

$$U(x) := \overline{U}_i(x), \quad V(x) := \overline{V}_i(x), \quad x \in [\xi_i, \xi_{i+1}).$$

Due to the structure of spaces of the form (1), it is natural to consider a recurrence relation starting from GB-splines of degree 1.

Definition 1 Given a knot sequence Ξ and a sequence of almost-polynomial spaces $\mathbb{P}_p^{\overline{U}_i,\overline{V}_i}([\xi_i, \xi_{i+1}])$, the ith GB-spline $B_{i,p,\Xi}^{U,V}$ of degree p over Ξ is defined recursively as follows. For $p = 1$,

$$B^{U,V}_{i,1,\Xi}(x) := \begin{cases} \overline{v_i}(x), & \text{if } x \in [\xi_i, \xi_{i+1}), \\ \overline{u}_{i+1}(x), & \text{if } x \in [\xi_{i+1}, \xi_{i+2}), \\ 0, & \text{elsewhere,} \end{cases}$$

and for $p \geq 2$,

$$B^{U,V}_{i,p,\Xi}(x) := \int_{\xi_i}^{x} \frac{B^{U,V}_{i,p-1,\Xi}(s)}{\delta^{U,V}_{i,p-1,\Xi}}\, \mathrm{d}s - \int_{\xi_{i+1}}^{x} \frac{B^{U,V}_{i+1,p-1,\Xi}(s)}{\delta^{U,V}_{i+1,p-1,\Xi}}\, \mathrm{d}s,$$

where

$$\delta^{U,V}_{i,p,\Xi} := \int_{\xi_i}^{\xi_{i+p+1}} B^{U,V}_{i,p,\Xi}(s)\, \mathrm{d}s,$$

and we use the convention that if $\delta^{U,V}_{i,k,\Xi} = 0$, then

$$\int_{\xi_i}^{x} \frac{B^{U,V}_{i,k,\Xi}(s)}{\delta^{U,V}_{i,k,\Xi}}\, \mathrm{d}s := \begin{cases} 1, & \text{if } x \geq \xi_{i+k+1}, \\ 0, & \text{otherwise.} \end{cases}$$

GB-splines with only section spaces in (2)–(4) will be referred to as *standard B-splines*, *hyperbolic (or exponential) GB-splines*, and *trigonometric GB-splines*, respectively. Two GB-splines of degree 1 are depicted in Fig. 1, and some sets of cubic GB-splines are illustrated in Fig. 2.

GB-splines share many fundamental features with standard B-splines (see, e.g., [27, 47]). We collect some properties in the next proposition.

Proposition 1 *Let* $\{B^{U,V}_{i,p,\Xi} : i = 1, \ldots, n + p\}$ *be a set of GB-splines of degree* $p \geq 2$ *over the knot sequence* Ξ*. The following properties hold:*

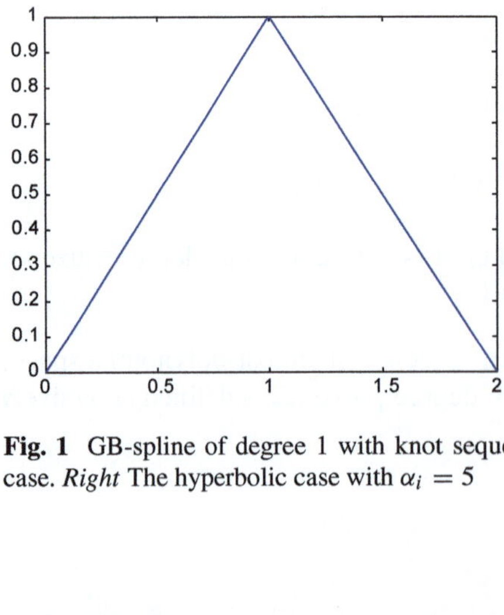

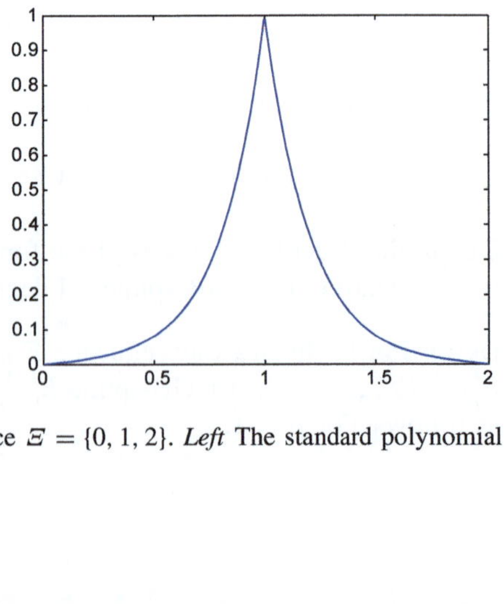

Fig. 1 GB-spline of degree 1 with knot sequence $\Xi = \{0, 1, 2\}$. *Left* The standard polynomial case. *Right* The hyperbolic case with $\alpha_i = 5$

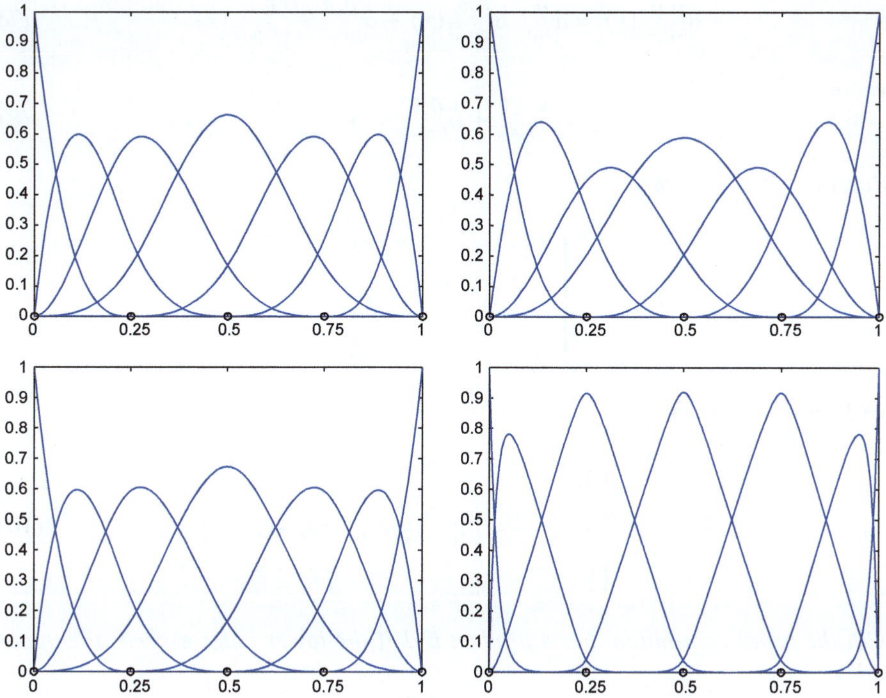

Fig. 2 Examples of GB-splines of degree 3 defined on the knot sequence $\Xi = \{0, 0, 0, 0, 1/4, 1/2, 3/4, 1, 1, 1, 1\}$. *Top* Trigonometric GB-splines with $\alpha_i = \frac{2}{3}\pi$ (*left*) and $\alpha_i = 3\pi$ (*right*). *Bottom* Hyperbolic GB-splines with $\alpha_i = 3$ (*left*) and $\alpha_i = 50$ (*right*)

- *Piecewise structure:* $B^{U,V}_{i,p,\Xi} \in \mathbb{P}^{\overline{U}_m,\overline{V}_m}_p([\xi_m, \xi_{m+1}))$;
- *Positivity:* $B^{U,V}_{i,p,\Xi}(x) > 0,\ x \in (\xi_i, \xi_{i+p+1})$;
- *Local support:* $B^{U,V}_{i,p,\Xi}(x) = 0,\ x \notin [\xi_i, \xi_{i+p+1}]$;
- *Smoothness:* $B^{U,V}_{i,p,\Xi}$ *is* $p - \mu_j$ *times continuously differentiable at* ξ_j*, being* μ_j *the multiplicity of* ξ_j *in* $\{\xi_i, \ldots, \xi_{i+p+1}\}$;
- *Local partition of unity:* $\sum_{i=m-p}^{m} B^{U,V}_{i,p,\Xi}(x) = 1,\ x \in [\xi_m, \xi_{m+1})$;
- *Local linear independence:* $\left\{B^{U,V}_{i,p,\Xi} : i = m-p, \ldots, m\right\}$ *are linearly independent on* $[\xi_m, \xi_{m+1})$;
- *Differentiation:* $D_+ B^{U,V}_{i,p,\Xi}(x) = \frac{B^{U,V}_{i,p-1,\Xi}(x)}{\delta^{U,V}_{i,p-1,\Xi}} - \frac{B^{U,V}_{i+1,p-1,\Xi}(x)}{\delta^{U,V}_{i+1,p-1,\Xi}}$.

In complete analogy with the polynomial case, there is also a knot insertion procedure available for GB-splines, see [29, 47].

Proposition 2 *Let* $\widehat{\Xi}$ *be the knot sequence obtained by inserting a new knot* $\hat{\xi}$ *in the knot sequence* Ξ *in the interval* (ξ_{p+1}, ξ_{n+p+1})*, and set* $\widehat{U} := U$, $\widehat{V} := V$*. For fixed* i*, let* μ_j *be the multiplicity of* ξ_j *in* $\{\xi_i, \ldots, \xi_{i+p+1}\}$*. Then, for* $p \geq 2$ *we have*

$$B_{i,p,\Xi}^{U,V}(x) = \widehat{\omega}_{i,p}^{\widehat{U},\widehat{V}} B_{i,p,\widehat{\Xi}}^{\widehat{U},\widehat{V}}(x) + \widehat{\sigma}_{i,p}^{\widehat{U},\widehat{V}} B_{i+1,p,\widehat{\Xi}}^{\widehat{U},\widehat{V}}(x), \tag{7}$$

where

$$\widehat{\omega}_{i,p}^{\widehat{U},\widehat{V}} + \widehat{\sigma}_{i-1,p}^{\widehat{U},\widehat{V}} = 1. \tag{8}$$

Moreover, if $\mu_i > 1$ *then*

$$\widehat{\omega}_{i,p}^{\widehat{U},\widehat{V}} := \begin{cases} 1, & \text{if } \xi_{i+p} \leq \hat{\xi}, \\ \prod_{\ell=\mu_i-1}^{p-1} \frac{\delta_{i,\ell,\widehat{\Xi}}^{\widehat{U},\widehat{V}}}{\delta_{i,\ell,\Xi}^{U,V}}, & \text{if } \xi_i < \hat{\xi} < \xi_{i+p}, \\ 0, & \text{if } \hat{\xi} \leq \xi_i, \end{cases} \tag{9}$$

while, if $\mu_i = 1$ *then*

$$\widehat{\omega}_{i,p}^{\widehat{U},\widehat{V}} := \begin{cases} 1, & \text{if } \xi_{i+p} \leq \hat{\xi}, \\ \frac{D_+ B_{i,1,\Xi}(\xi_i)}{D_+ B_{i,1,\widehat{\Xi}}(\xi_i)} \prod_{\ell=1}^{p-1} \frac{\delta_{i,\ell,\widehat{\Xi}}^{\widehat{U},\widehat{V}}}{\delta_{i,\ell,\Xi}^{U,V}}, & \text{if } \xi_i < \hat{\xi} < \xi_{i+p}, \\ 0, & \text{if } \hat{\xi} \leq \xi_i, \end{cases} \tag{10}$$

with the usual convention that a product is 1 if the lower index exceeds the upper index.

The expressions in (9) and (10) are obtained by comparing the first non-vanishing (right) derivative at ξ_i of the left and the right term in (7). A similar expression for $\widehat{\sigma}_{i,p}^{\widehat{U},\widehat{V}}$ can be obtained by comparing the first non-vanishing (left) derivative at ξ_{i+p+1} of the left and the right term in (7). The equality (8) can be deduced by the local partition of unity property.

The space spanned by a set of GB-splines defined over a fixed knot sequence will be referred to as a *generalized spline space*. For notational convenience, we denote by $\mathbb{S}_{p,\Xi}$, $\mathbb{S}_{p,\Xi}^{\mathbb{H}_\alpha}$, and $\mathbb{S}_{p,\Xi}^{\mathbb{T}_\alpha}$ the spaces spanned by standard B-splines and hyperbolic and trigonometric GB-splines, respectively.

It is clear that hyperbolic and trigonometric GB-splines allow for an exact representation of conics as well as some transcendental curves (helix, cycloid, ...). They are very attractive from the geometrical point of view. Indeed, in contrast to NURBS, they are able to provide parameterizations of conics close to the arc length so that equally spaced points in the parameter domain correspond to almost equally spaced points on the described curve. Since the section spaces (1) may be chosen differently on each knot interval, GB-splines allow for an exact representation of profiles composed by a sequence of curve segments of different kind: arcs of ellipses, hyperbolas, polynomial curves, etc.

Furthermore, the differential operator acts on trigonometric and hyperbolic generalized splines in the same way as on standard polynomial splines. More precisely,

$$D_+\mathbb{S}_{p,\Xi} = \mathbb{S}_{p-1,\Xi}, \quad D_+\mathbb{S}_{p,\Xi}^{\mathbb{H}_\alpha} = \mathbb{S}_{p-1,\Xi}^{\mathbb{H}_\alpha}, \quad D_+\mathbb{S}_{p,\Xi}^{\mathbb{T}_\alpha} = \mathbb{S}_{p-1,\Xi}^{\mathbb{T}_\alpha}, \quad p \geq 2. \tag{11}$$

NURBS do not possess a property like (11). This property makes the structural similarity between hyperbolic/trigonometric GB-splines and standard B-splines even stronger.

Remark 1 Hyperbolic/trigonometric GB-splines tend to standard B-splines of the same degree and over the same knot sequence if the local phase parameters α tend to 0. The same is true if, for fixed values of α, the length of each knot interval $\xi_{i+1} - \xi_i$ tends to zero.

2.3 GB-Splines on Uniformly Spaced Knots

In many practical applications, GB-splines on uniformly spaced knots are of particular interest. As for the standard polynomial case, they enjoy special properties. In this section, we summarize some of them.

Given the almost-polynomial space $\mathbb{P}_p^{U,V}([0,1])$, we denote by $\overline{u}, \overline{v}$ the unique elements in $\langle D^{(p-1)}U, D^{(p-1)}V \rangle$ satisfying

$$\overline{u}(0) = 1, \quad \overline{u}(1) = 0, \quad \overline{v}(0) = 0, \quad \overline{v}(1) = 1. \tag{12}$$

Definition 2 The (normalized) cardinal GB-spline of degree $p \geq 1$ over the uniform knot sequence $\{0, 1, \ldots, p+1\}$ with sections belonging to integer translations of $\mathbb{P}_p^{U,V}([0,1])$ is denoted by $\phi_p^{U,V}$ and is defined recursively as follows. For $p = 1$,

$$\phi_1^{U,V}(t) := \frac{1}{\delta_1^{U,V}} \begin{cases} \overline{v}(t), & \text{if } t \in [0,1), \\ \overline{u}(t-1), & \text{if } t \in [1,2), \\ 0, & \text{elsewhere}, \end{cases}$$

where

$$\delta_1^{U,V} := \int_0^1 \overline{v}(s)\,\mathrm{d}s + \int_1^2 \overline{u}(s-1)\,\mathrm{d}s.$$

For $p \geq 2$,

$$\phi_p^{U,V}(t) := \int_0^t \left(\phi_{p-1}^{U,V}(s) - \phi_{p-1}^{U,V}(s-1)\right) \mathrm{d}s.$$

If $\mathbb{P}_p^{U,V}([0,1]) = \mathbb{P}_p([0,1])$, then the function defined in Definition 2 is the standard (polynomial) cardinal B-spline of degree p, denoted by ϕ_p. In the next proposition, we list some properties of cardinal GB-splines (see, e.g., [41, Sect. 3.1]).

Proposition 3 *Let $\phi_p^{U,V}$ be the cardinal GB-spline of degree $p \geq 2$. The following properties hold:*

- *Smoothness:* $\phi_p^{U,V} \in C^{p-1}$;
- *Positivity:* $\phi_p^{U,V}(t) > 0,\ t \in (0, p+1)$;
- *Local support:* $\phi_p^{U,V}(t) = 0,\ t \notin (0, p+1)$;
- *Partition of unity:* $\sum_{k=1}^{p} \phi_p^{U,V}(k) = 1$;
- *Differentiation:* $D\,\phi_p^{U,V}(t) = \phi_{p-1}^{U,V}(t) - \phi_{p-1}^{U,V}(t-1)$;
- *Integration:* $\int_0^{p+1} \phi_p^{U,V}(s)\,\mathrm{d}s = 1$;
- *Convolution relation:* $\phi_p^{U,V}(t) = \int_0^1 \phi_{p-1}^{U,V}(t-s)\,\mathrm{d}s$.

The hyperbolic cardinal GB-spline is denoted by $\phi_p^{\mathbb{H}_\alpha}$ and is defined by taking $U(t) := \cosh(\alpha t)$ and $V(t) := \sinh(\alpha t)$. In this case, we have

$$\overline{u}(t) = \frac{\sinh(\alpha(1-t))}{\sinh(\alpha)}, \quad \overline{v}(t) = \frac{\sinh(\alpha t)}{\sinh(\alpha)},$$

satisfying (12). The trigonometric cardinal GB-spline is denoted by $\phi_p^{\mathbb{T}_\alpha}$ and is defined by taking $U(t) := \cos(\alpha t)$ and $V(t) := \sin(\alpha t)$. In this case, we have

$$\overline{u}(t) = \frac{\sin(\alpha(1-t))}{\sin(\alpha)}, \quad \overline{v}(t) = \frac{\sin(\alpha t)}{\sin(\alpha)},$$

satisfying (12).

We now focus on the following sequence of uniformly spaced knots,

$$\{\xi_1, \dots, \xi_{n+2p+1}\} = \Big\{\underbrace{0, \dots, 0}_{p+1}, \frac{1}{n}, \frac{2}{n}, \dots, \frac{n-1}{n}, \underbrace{1, \dots, 1}_{p+1}\Big\}, \tag{13}$$

and consider the same almost-polynomial space (1) on each knot interval. The corresponding GB-splines of degree p according to Definition 1 are compactly denoted by

$$\Big\{B_{i,p}^{U,V} : i = 1, \dots, n+p\Big\}, \tag{14}$$

and their span is denoted by $\mathbb{S}_{n,p}^{U,V}$. It turns out that

$$\mathbb{S}_{n,p}^{U,V} = \left\{f \in C^{p-1}([0,1]) : f \in \mathbb{P}_p^{\overline{U},\overline{V}}\left(\left[\frac{i}{n}, \frac{i+1}{n}\right)\right),\ i = 0, \dots, n-1\right\}. \tag{15}$$

In particular, let

$$\Big\{B_{i,p}^{\mathbb{H}_\alpha} : i = 1, \dots, n+p\Big\}, \quad \Big\{B_{i,p}^{\mathbb{T}_\alpha} : i = 1, \dots, n+p\Big\} \tag{16}$$

be the sets of hyperbolic and trigonometric GB-splines of degree p defined over the knot sequence (13) with sections in (3) and (4), respectively. Then, it is easy to check that for $i = p+1, \dots, n$ and $p \geq 2$,

$$B_{i,p}^{\mathbb{H}_\alpha}(x) = \phi_p^{\mathbb{H}_{\alpha/n}}(nx - i + p + 1), \quad B_{i,p}^{\mathbb{T}_\alpha}(x) = \phi_p^{\mathbb{T}_{\alpha/n}}(nx - i + p + 1), \tag{17}$$

in a complete analogy with the polynomial case. The spans of (16) are denoted by $\mathbb{S}_{n,p}^{\mathbb{H}_\alpha}$ and $\mathbb{S}_{n,p}^{\mathbb{T}_\alpha}$, respectively, while $\mathbb{S}_{n,p}$ is the standard polynomial spline space over the knots (13).

Remark 2 A main issue in working with GB-splines is the construction of efficient algorithms for their evaluation. We refer to [42] for a discussion and the description of an approximation procedure in case of uniform knots based on cardinal B-splines.

2.4 Multivariate GB-Splines

Multivariate versions of the univariate GB-splines in Definition 1 can be easily obtained by the usual tensor-product approach (in d dimensions):

$$B_{\mathbf{i},\mathbf{p},\boldsymbol{\Xi}}^{\mathbf{U},\mathbf{V}} = B_{i_1,p_1,\Xi_1}^{U_1,V_1} \otimes B_{i_2,p_2,\Xi_2}^{U_2,V_2} \otimes \cdots \otimes B_{i_d,p_d,\Xi_d}^{U_d,V_d},$$

where different knot sequences, degrees and section spaces can be considered in the different directions. The structural simplicity makes tensor-product splines a powerful tool for both modeling and simulation.

Adaptive local refinement is important in applications. Unfortunately, a simple tensor-product spline structure lacks adequate local refinement. This triggered the interest in alternative spline structures supporting local refinement but still retaining a local tensor-product structure. On this concern, we mention (analysis-suitable) T-splines [28, 45], hierarchical splines [18, 20, 21], and locally refined (LR-) splines [16]. All of them can be seen as special cases of polynomial splines over T-meshes [15, 44]. In the more recent literature, we also find some specific extensions to the generalized spline setting. For example, generalized T-splines [5, 7], hierarchical generalized splines [34], and generalized splines on T-meshes [6, 8, 9] have been addressed.

In this section, we limit ourselves to hierarchical/multilevel bases, and we outline the construction of hierarchical bases in terms of a hierarchy of tensor-product GB-splines. This is a simplified version of the construction proposed in [21, 34, 46].

Let $\widehat{\Omega}$ be a hyper-rectangle in $\mathbb{R}^d$. We consider a sequence of nested tensor-product d-variate generalized spline spaces defined on $\widehat{\Omega}$,

$$\mathbb{W}^0 \subset \mathbb{W}^1 \subset \mathbb{W}^2 \subset \cdots. \tag{18}$$

Any element of $\mathbb{W}^\ell$ is a piecewise function defined over a partition of $\widehat{\Omega}$ consisting of hyper-rectangles, which will be called *cells* of level ℓ. We denote by

$$\mathscr{B}^\ell := \{b_{i,\ell} : i = 1, \ldots, \dim(\mathbb{W}^\ell)\} \tag{19}$$

the tensor-product GB-spline basis of $\mathbb{W}^\ell$, where the knot sequences are assumed to be open in each direction. Finally, we consider a nested sequence of closed subsets of $\widehat{\Omega}$,

$$\widehat{\Omega} =: \Omega^0 \supseteq \Omega^1 \supseteq \Omega^2 \supseteq \cdots, \tag{20}$$

where each Ω^ℓ in (20) is the union of a selection of cells of level ℓ. The set

$$\boldsymbol{\Omega}_N := \{\Omega^0, \Omega^1, \ldots, \Omega^{N-1}\}, \quad N \geq 1, \tag{21}$$

will be referred to as a *hierarchy of subsets of* $\widehat{\Omega}$ *of depth* N.

Given a sequence of spaces and bases as in (18)–(19) and a hierarchy of subsets as in (21), we can now construct the corresponding set of so-called hierarchical basis functions as follows. We first take all the basis elements in $\mathscr{B}^0$. Then, we apply an iterative procedure which selects at each level ℓ all the basis functions obtained in the previous step whose support is not entirely contained in Ω^ℓ and all the basis functions in $\mathscr{B}^\ell$ whose support is entirely contained in Ω^ℓ. More precisely, we state the following definition.

Definition 3 The hierarchical set of basis functions $\mathscr{H}_{\boldsymbol{\Omega}_N}$ associated with a hierarchy of subsets of $\widehat{\Omega}$ as in (21) is recursively constructed as follows:

(i) $\mathscr{H}^0 := \mathscr{B}^0$;
(ii) for $\ell = 0, \ldots, N-2$:

$$\mathscr{H}^{\ell+1} := \mathscr{H}_C^{\ell+1} \cup \mathscr{H}_F^{\ell+1},$$

where

$$\mathscr{H}_C^{\ell+1} := \{b_{i,j} \in \mathscr{H}^\ell : \operatorname{supp}(b_{i,j}) \nsubseteq \Omega^{\ell+1}\},$$
$$\mathscr{H}_F^{\ell+1} := \{b_{i,\ell+1} \in \mathscr{B}^{\ell+1} : \operatorname{supp}(b_{i,\ell+1}) \subseteq \Omega^{\ell+1}\};$$

(iii) $\mathscr{H}_{\boldsymbol{\Omega}_N} := \mathscr{H}^{N-1}$.

The space $\mathbb{S}_{\boldsymbol{\Omega}_N} := \langle b_{i,j} : b_{i,j} \in \mathscr{H}_{\boldsymbol{\Omega}_N} \rangle$ is called the *hierarchical space associated with* $\boldsymbol{\Omega}_N$. Since the elements in $\mathscr{B}^\ell$ are locally linearly independent, Definition 3 provides a set of linearly independent functions [21]. Therefore, the elements in $\mathscr{H}_{\boldsymbol{\Omega}_N}$ form a basis of $\mathbb{S}_{\boldsymbol{\Omega}_N}$, which will be referred to as the *hierarchical basis* of $\mathbb{S}_{\boldsymbol{\Omega}_N}$. Figure 3 shows the construction of the hierarchical basis for hierarchical cubic hyperbolic GB-splines on the interval [0, 1] associated with a hierarchy of depth 2.

The elements of the hierarchical basis are obviously nonnegative, but they do not form a partition of unity. Nevertheless, it is possible to construct an alternative basis of $\mathbb{S}_{\boldsymbol{\Omega}_N}$, the so-called *truncated hierarchical basis* whose elements form a convex partition of unity [21].

Remark 3 In Definition 3, we assumed a sequence of nested linear spaces $\mathbb{W}^\ell$ following the classical definition of hierarchical bases [21]. However, the construction

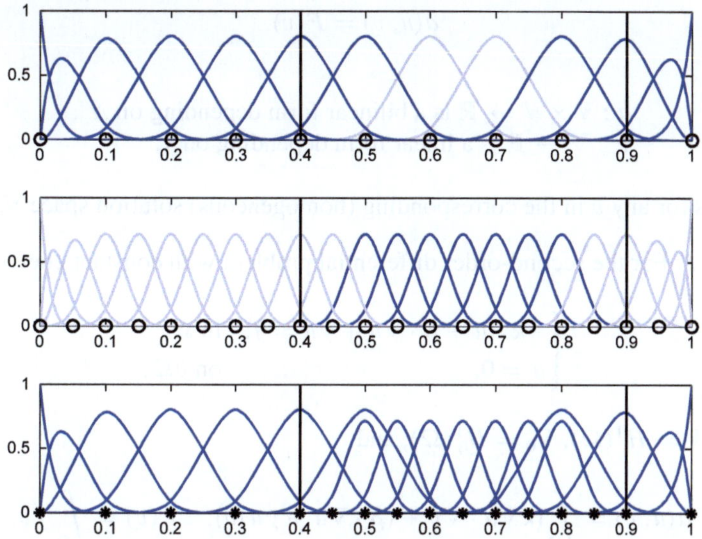

Fig. 3 Illustration of the construction of hierarchical cubic hyperbolic GB-splines with $\alpha = 50$, $N = 2$, $\Omega_0 = [0, 1]$, $\Omega_1 = [0.4, 0.9]$. The light colored functions are removed in the construction of the hierarchical basis

also works when the spaces $\mathbb{W}^\ell$ do not form a nested sequence. Giving up nestedness allows for more freedom in the construction and leads to the so-called *multilevel bases* [34]. Under the assumption of local linear independence for the basis functions considered in each space $\mathbb{W}^\ell$, the resulting set of multilevel functions are still linearly independent.

3 Isogeometric Galerkin and Collocation Methods

In this section, we outline the Galerkin and collocation approximations of PDE solutions in the IgA framework. For the sake of simplicity, we illustrate the isogeometric approach for second-order elliptic differential problems with homogeneous Dirichlet boundary conditions. A more comprehensive presentation can be found in [13].

Let $\mathscr{L}$ be a linear second-order elliptic differential operator on the domain $\Omega \subset \mathbb{R}^d$ with Lipschitz boundary $\partial\Omega$. We consider the differential problem

$$\begin{cases} \mathscr{L}u = f, & \text{in } \Omega, \\ u = 0, & \text{on } \partial\Omega, \end{cases} \tag{22}$$

for the unknown $u : \Omega \to \mathbb{R}$. This is the so-called *strong form* of the problem. The *weak form* of (22) is given by

$$a(u,v) = F(v), \tag{23}$$

where

$$a : \mathbb{V} \times \mathbb{V} \to \mathbb{R} \text{ is a bilinear form depending on } \mathscr{L},$$
$$F : \mathbb{V} \to \mathbb{R} \text{ is a linear form depending on } f,$$

and holds for any v in the corresponding (homogeneous) solution space $\mathbb{V}_0 \subset \mathbb{V}$.

Example 1 For the second-order differential problem with constant coefficients,

$$\begin{cases} -\kappa \Delta u + \boldsymbol{\beta} \cdot \nabla u + \gamma u = f, & \text{in } \Omega, \\ u = 0, & \text{on } \partial\Omega, \end{cases}$$

we have $\mathbb{V} = H^1(\Omega)$, $\mathbb{V}_0 = H_0^1(\Omega)$, and

$$a(u,v) = \int_\Omega \big(\kappa \nabla u \cdot \nabla v + (\boldsymbol{\beta} \cdot \nabla u + \gamma u)v\big), \quad F(v) = \int_\Omega f v.$$

3.1 Isogeometric Galerkin Methods

The Galerkin approach to approximate the solution of (22) is based on the weak form (23). First, we select a finite-dimensional approximation space on Ω,

$$\mathbb{V}_h := \langle \varphi_1, \varphi_2, \ldots, \varphi_{m_h} \rangle \subset \mathbb{V}, \quad \dim(\mathbb{V}_h) = m_h, \tag{24}$$

related to a fineness parameter $h > 0$, and

$$\mathbb{V}_{h,0} := \{ v_h \in \mathbb{V}_h : v_{h|\partial\Omega} = 0 \} \subset \mathbb{V}_0, \quad \dim(\mathbb{V}_{h,0}) = n_h < m_h. \tag{25}$$

Then, we look for $u_h^G \in \mathbb{V}_{h,0}$ as the solution of

$$a(u_h^G, v_h) = F(v_h), \quad \forall v_h \in \mathbb{V}_{h,0}. \tag{26}$$

Assuming $\mathbb{V}_{h,0} = \langle \varphi_1, \varphi_2, \ldots, \varphi_{n_h} \rangle$ and writing $u_h^G = \sum_{i=1}^{n_h} c_i \varphi_i$, condition (26) gives rise to a linear system $A\mathbf{c} = \mathbf{F}$, where

$$A := [a(\varphi_j, \varphi_i)]_{i,j=1}^{n_h} \in \mathbb{R}^{n_h \times n_h} \text{ is the stiffness matrix,}$$
$$\mathbf{F} := [F(\varphi_i)]_{i=1}^{n_h} \in \mathbb{R}^{n_h} \text{ is the load vector.}$$

Different Galerkin methods correspond to different choices of the subspace $\mathbb{V}_h$. Since the weak form (23) is expressed in terms of integrals (see Example 1), the construction of the stiffness matrix and load vector requires the use of numerical integration. Therefore, the efficiency of a Galerkin method deeply depends on the efficiency of the quadrature rules used in the construction of the corresponding linear systems.

In the standard formulation of IgA, the physical domain Ω is represented by means of a *global geometry map*. We define the geometry map $\mathbf{G}$ from the parametric domain $\widehat{\Omega} := [0, 1]^d$ to the closure of Ω as (see Fig. 4)

$$\mathbf{G}(\hat{\mathbf{x}}) := \sum_{i=1}^{m_h} \mathbf{P}_i \hat{\varphi}_i(\hat{\mathbf{x}}), \quad \mathbf{P}_i \in \mathbb{R}^d, \quad \hat{\mathbf{x}} \in \widehat{\Omega}, \tag{27}$$

where the basis functions

$$\{\hat{\varphi}_1, \ldots, \hat{\varphi}_{m_h}\} \tag{28}$$

have to be selected so as to produce an exact representation of the geometry. Following the isoparametric approach, the space $\mathbb{V}_h$ in (24) is then defined as

$$\varphi_i(\mathbf{x}) := \hat{\varphi}_i \circ \mathbf{G}^{-1}(\mathbf{x}) = \hat{\varphi}_i(\hat{\mathbf{x}}), \quad i = 1, \ldots, m_h, \quad \mathbf{x} = \mathbf{G}(\hat{\mathbf{x}}). \tag{29}$$

Usually, the functions in (28) are chosen to be tensor-product B-splines or NURBS. In this paper, we consider GB-splines whose section spaces will be selected according to a problem-oriented strategy.

The Galerkin formulation has been intensively and successfully employed in the isogeometric context. In contrast to the finite element context, where elementwise Gauss quadrature is known to be optimal, it is not yet completely clear how to construct efficient IgA quadrature rules, see [23] and references therein.

The quadrature issue motivated the idea of taking advantage of the high smoothness of IgA basis functions to construct efficient and geometrically flexible collocation methods, see [2] or the recent survey [40]. They will be discussed in the next section.

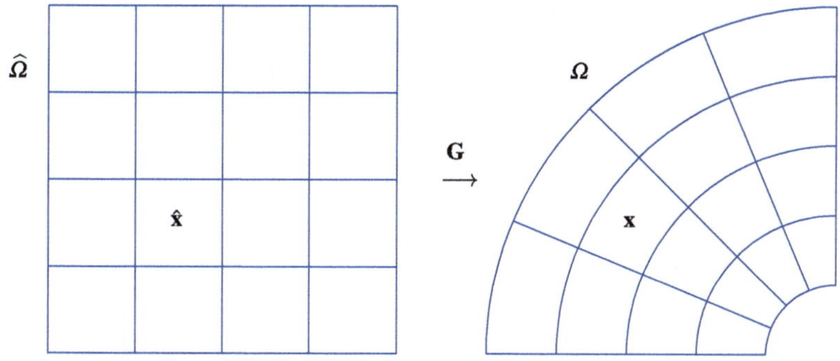

Fig. 4 Isogeometric Galerkin. Description of the physical domain Ω by means of a global geometry map $\mathbf{G}$

3.2 Isogeometric Collocation Methods

The collocation approach to approximate the solution of (22) is based on the strong form of the problem: We look for $u_h^C \in \mathbb{V}_{h,0}$ (see (25)) as the solution of

$$\mathscr{L}u_h^C(\tau_i) = f(\tau_i), \quad i = 1, \ldots, n_h, \tag{30}$$

where $\tau_i \in \Omega$ are the so-called *collocation points*. Writing $u_h^C = \sum_{i=1}^{n_h} c_i \varphi_i$, condition (30) gives rise to a linear system $A\mathbf{c} = \mathbf{F}$, where

$$A := [\mathscr{L}\varphi_j(\tau_i)]_{i,j=1}^{n_h} \in \mathbb{R}^{n_h \times n_h} \text{ and } \mathbf{F} := [f(\tau_i)]_{i=1}^{n_h} \in \mathbb{R}^{n_h}.$$

Isogeometric collocation emanates from the combination of the isogeometric technology and the standard collocation method. Following the isoparametric approach, we consider the approximation spaces $\mathbb{V}_{h,0}$ and $\mathbb{V}_h$ spanned by the functions in (29), and we select the collocation points as

$$\tau_i := \mathbf{G}(\hat{\tau}_i), \quad i = 1, \ldots, n_h, \tag{31}$$

where $\hat{\tau}_i$ are suitable points belonging to the parametric domain $\widehat{\Omega}$, see Fig. 5.

Usually, in collocation IgA, the functions $\hat{\varphi}_i$ in (28) are chosen to be tensor-product B-splines or NURBS. The major advantage of isogeometric collocation over Galerkin-type methods is the minimal computational effort with respect to quadrature, since for each degree of freedom only one point evaluation at the collocation point is required. This property leads to extremely easy and fast constructions of the corresponding linear systems.

On the other hand, in contrast to isogeometric Galerkin methods, the theoretical understanding of isogeometric collocation methods is not so mature yet. The optimal choice of the collocation points, the unisolvency of the resulting system, and the

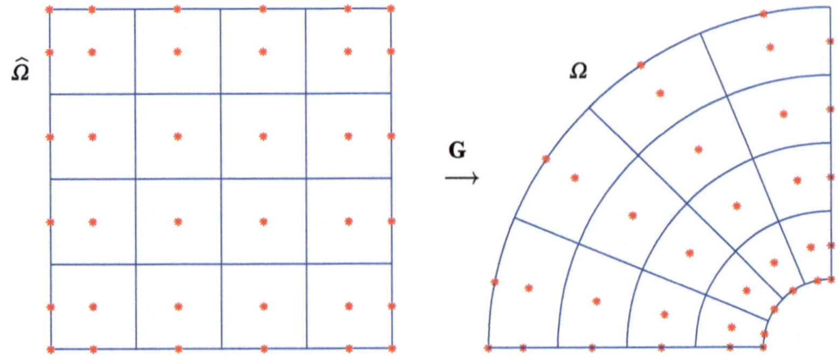

Fig. 5 Isogeometric collocation. Description of the physical domain Ω by means of a global geometry map $\mathbf{G}$ and related collocation points

accuracy performance of the obtained approximate solution are still main open issues. Some results in this direction are known in the one-dimensional setting for a special choice of the points $\hat{\tau}_i$ in (31), the so-called *Tchebycheff–Demko points*, see [2]. However, in practical applications, the most popular choice for the points $\hat{\tau}_i$ is the so-called *Greville points* corresponding to the selected spline space. In case of the polynomial spline space $\mathbb{S}_{p,\Xi}$, they can be computed as knot averages; e.g., the ith Greville point corresponding to $\mathbb{S}_{p,\Xi}$ is

$$\frac{\xi_{i+1} + \cdots + \xi_{i+p}}{p};$$

there are tensor-product variations in case of tensor-product spline spaces.

In the following, we focus on isogeometric methods based on tensor-product GB-splines. The above-mentioned Greville and Tchebycheff–Demko points can be defined for generalized spline spaces (if $p \geq 3$) in a similar way as for polynomial spline spaces. The latter points can be computed by a similar iterative algorithm as the one proposed for B-splines in [14, Chap. XIII], see [35]. Nevertheless, in our numerical experiments, we take the collocation points as (the images of) the Greville points of the standard B-splines of the same degree. This choice greatly simplifies the computation without sacrificing any accuracy [35].

4 Examples

In this section, we illustrate the use of GB-splines as basis functions (28) in isogeometric Galerkin and collocation methods. We will show that a proper selection of the section spaces (1) according to a problem-dependent strategy may result in an accuracy gain compared with standard B-splines. To this end, we present three examples. The first one is an eigenvalue problem for the one-dimensional Laplace operator. In the second example, we address an advection-dominated advection–diffusion problem on the unit square. Finally, in our third example, we consider a vector problem (linear elasticity) on a bivariate domain Ω whose description requires a non-trivial geometry map. In the first and last example, a Galerkin discretization is used, while for the advection–diffusion problem the isogeometric collocation approach is applied.

4.1 An Eigenvalue Problem

In our first example, we consider the following eigenvalue problem for the univariate Laplace operator:

$$\begin{cases} -\,D^2u(x) = \omega^2 u(x), & x \in (0,1), \\ u(0) = 0, \quad u(1) = 0, \end{cases} \tag{32}$$

whose non-trivial exact solutions are

$$u_k(x) := \sin(\omega_k x), \quad \omega_k := k\pi, \quad k = 1, 2, \ldots. \tag{33}$$

The weak form of problem (32) reads as follows: Find non-trivial $u \in H_0^1(0, 1)$ and ω such that

$$\int_0^1 Du(s)Dv(s)\,\mathrm{d}s - \omega^2 \int_0^1 u(s)v(s)\,\mathrm{d}s = 0, \quad \forall v \in H_0^1(0, 1).$$

Following the Galerkin approach, we choose a subspace $\mathbb{V}_{h,0}$ of $H_0^1(0, 1)$ spanned by the basis $\{\varphi_1, \ldots, \varphi_{n_h}\}$ and we find approximate values ω_h^G to ω by solving

$$K\mathbf{u} = \left(\omega_h^G\right)^2 M\mathbf{u},$$

where the matrices K and M consist of the elements

$$K_{i,j} := \int_0^1 D\varphi_j(s)D\varphi_i(s)\,\mathrm{d}s, \quad M_{i,j} := \int_0^1 \varphi_j(s)\varphi_i(s)\,\mathrm{d}s, \quad i, j = 1, \ldots, n_h.$$

This means that each $(\omega_h^G)^2$ is an eigenvalue of the matrix $L := M^{-1}K$. Thus, the n_h eigenvalues of the matrix L are an approximation of the first n_h eigenvalues of the problem (32), namely

$$\omega_k^2 = (k\pi)^2, \quad k = 1, \ldots, n_h.$$

Similarly, an approximation of the eigenfunction u_k is obtained by considering

$$\sum_{i=1}^{n_h} u_{k,i}\varphi_i, \tag{34}$$

where $\mathbf{u}_k := (u_{k,1}, \ldots, u_{k,n_h})$ is the kth eigenvector of L. Of course, a proper normalization is needed. Taking into account the expression of u_k in (33), we normalize $\mathbf{u}_k$ so that the infinity-norm of the function in (34) is equal to 1. More information on this eigenvalue problem can be found in [4].

We now choose the approximation space $\mathbb{V}_{h,0}$ to be a space of smooth generalized spline functions vanishing at the two ends of the unit interval. More precisely, for $p \geq 2$, $n \geq 1$, we set

$$\mathbb{V}_{h,0} := \left\{ f \in \mathbb{S}_{n,p}^{U,V} : f(0) = f(1) = 0 \right\},$$

where we recall $\mathbb{S}_{n,p}^{U,V}$ from (15). This space has dimension $n + p - 2$ and is spanned by the set of GB-splines $\left\{ B_{2,p}^{U,V}, \ldots, B_{n+p-1,p}^{U,V} \right\}$. Furthermore, we set

$$nK_{n,p}^{U,V} := \left[\int_0^1 DB_{j+1,p}^{U,V}(s)\, DB_{i+1,p}^{U,V}(s)\, \mathrm{d}s \right]_{i,j=1}^{n+p-2}, \tag{35}$$

$$\frac{1}{n}M_{n,p}^{U,V} := \left[\int_0^1 B_{j+1,p}^{U,V}(s)\, B_{i+1,p}^{U,V}(s)\, \mathrm{d}s \right]_{i,j=1}^{n+p-2}. \tag{36}$$

It is clear that the matrices $K_{n,p}^{U,V}$ and $M_{n,p}^{U,V}$ are symmetric, and due to the compact support of the GB-spline basis, they have a $(2p+1)$-band structure. As discussed before, in view of the eigenvalue problem (32), we need to determine the eigenvalues of the matrix

$$L_{n,p}^{U,V} := n^2 \left(M_{n,p}^{U,V}\right)^{-1} K_{n,p}^{U,V}. \tag{37}$$

Considering the form (33) of the solution, the natural choice of the section spaces for the GB-splines is (4) with parameter $n\alpha$. Note that the corresponding spaces $\mathbb{S}_{n,p}^{\mathbb{T}_{n\alpha}}$ are not nested for increasing values of n. We denote by $K_{n,p}^{\mathbb{T}_{n\alpha}}$, $M_{n,p}^{\mathbb{T}_{n\alpha}}$ the matrices in (35) and (36) obtained with such a choice. Taking into account (17), we see that the central parts of the matrices $K_{n,p}^{\mathbb{T}_{n\alpha}}$, $M_{n,p}^{\mathbb{T}_{n\alpha}}$ have a Toeplitz structure (i.e., the entries are constant along each diagonal). More precisely, the nonzero entries in the rows $i = 2p, \ldots, n-p-1$ are given by

$$\left(K_{n,p}^{\mathbb{T}_{n\alpha}}\right)_{i,i\pm k} = \int_0^{p+1} D\phi_p^{\mathbb{T}_\alpha}(t \mp k) D\phi_p^{\mathbb{T}_\alpha}(t)\, \mathrm{d}t,$$

$$\left(M_{n,p}^{\mathbb{T}_{n\alpha}}\right)_{i,i\pm k} = \int_0^{p+1} \phi_p^{\mathbb{T}_\alpha}(t \mp k) \phi_p^{\mathbb{T}_\alpha}(t)\, \mathrm{d}t,$$

for $k = 0, 1, \ldots, p$. These central parts agree with Toeplitz matrices generated by the functions

$$f_p^{\mathbb{T}_\alpha}(\theta) := \sum_{k=-p}^{p} \left(\int_0^{p+1} D\phi_p^{\mathbb{T}_\alpha}(t-k)\, D\phi_p^{\mathbb{T}_\alpha}(t)\, \mathrm{d}t \right) \cos(k\theta),$$

$$h_p^{\mathbb{T}_\alpha}(\theta) := \sum_{k=-p}^{p} \left(\int_0^{p+1} \phi_p^{\mathbb{T}_\alpha}(t-k)\, \phi_p^{\mathbb{T}_\alpha}(t)\, \mathrm{d}t \right) \cos(k\theta),$$

respectively, see [41]. The above functions are defined on $[-\pi, \pi]$, but since they are symmetric, we can restrict them to the range $[0, \pi]$ for visualization purposes. The selection of the parameter α is crucial in this approximation strategy, and we refer to [41] for a discussion of different selection criteria.

For the sake of comparison, we also consider the Galerkin approximation of the solution of (33) by using standard B-splines of degree p over the same knot sequence (13). The resulting matrices (35) and (36) have the same structure as $K_{n,p}^{\mathbb{T}_{n\alpha}}$ and $M_{n,p}^{\mathbb{T}_{n\alpha}}$. In particular, their central parts are Toeplitz matrices generated by the functions

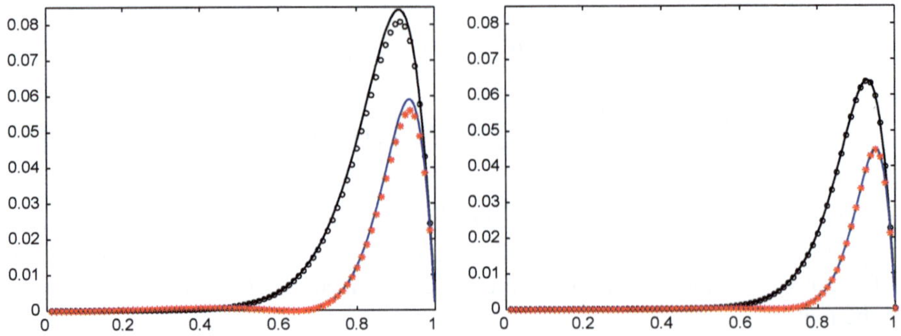

Fig. 6 Eigenvalue problem. Plot of the relative spectral error values in the trigonometric case (∗) and the polynomial case (∘) for $n = 80$, together with the scaled graphs of $e_p^{\mathbb{T}_\alpha}$ (*blue line*) and e_p (*black line*). *Left* $p = 3, \alpha = \frac{7}{11}\pi$. *Right* $p = 4, \alpha = \frac{9}{13}\pi$

$$f_p(\theta) := \sum_{k=-p}^{p} \left(\int_0^{p+1} D\phi_p(t-k)\, D\phi_p(t)\, \mathrm{d}t \right) \cos(k\theta),$$

$$h_p(\theta) := \sum_{k=-p}^{p} \left(\int_0^{p+1} \phi_p(t-k)\, \phi_p(t)\, \mathrm{d}t \right) \cos(k\theta),$$

respectively, see [19].

In Fig. 6, we plot the relative spectral error values

$$\left(\frac{k}{n}, \left(\frac{\omega_k^{\mathbb{T}_{n\alpha}}}{k\pi} \right)^2 - 1 \right), \quad k = 1, \ldots, n, \tag{38}$$

where $\omega_k^{\mathbb{T}_{n\alpha}}$ denotes the trigonometric GB-spline approximation for ω_k. In the same figure, we also show the graph of the function

$$e_p^{\mathbb{T}_\alpha}(\theta) := \frac{f_p^{\mathbb{T}_\alpha}(\theta)}{h_p^{\mathbb{T}_\alpha}(\theta)} \frac{1}{\theta^2} - 1,$$

which gives an asymptotic description of the quantity in (38), see [41]. For comparison, we also plot the relative spectral error values and the graph of the corresponding function e_p in case of standard B-splines of the same degree over the same knot sequence, i.e.,

$$e_p(\theta) := \frac{f_p(\theta)}{h_p(\theta)} \frac{1}{\theta^2} - 1.$$

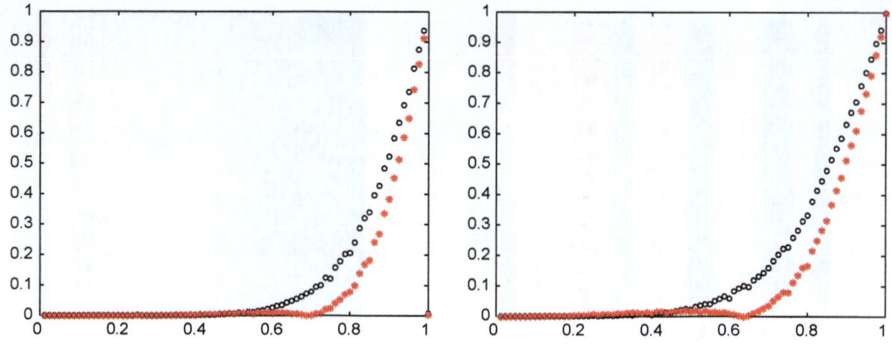

Fig. 7 Eigenvalue problem. Plot of the infinity-norm of the error for all the eigenfunctions in the trigonometric case ($*$) and the polynomial case ($\circ$) for $n = 80$. *Left* $p = 3, \alpha = \frac{7}{11}\pi$. *Right* $p = 4$, $\alpha = \frac{9}{13}\pi$

Both the functions are rescaled over the interval [0, 1] in the figure. The values of α are selected according to a heuristic rule, see [41, Sect. 5.3]. Figure 7 depicts the infinity-norm of the error for all the obtained eigenfunctions tabulated at 500 equally spaced points in [0, 1].

Two remarks are in order. First, there is a very good match between the functions $e_p^{\mathbb{T}_\alpha}$ obtained by the theoretical spectral analysis of the involved matrices and the computed relative error in (38) of the Galerkin approximation of the spectrum based on trigonometric spline spaces $\mathbb{S}_{n,p}^{\mathbb{T}_{n\alpha}}$. The same holds for the approximation based on standard B-splines. Second, trigonometric GB-splines present improvements over standard B-splines in the global approximation of both the eigenvalues and the eigenfunctions.

4.2 *An Advection–Diffusion Problem with Boundary Layers*

In our second example, we consider the following advection–diffusion problem defined on the bivariate domain $\Omega = (0, 1)^2$:

$$-\kappa \Delta u + \boldsymbol{\beta} \cdot \nabla u = f, \quad \text{in } \Omega,$$

complemented by homogeneous Dirichlet boundary conditions. The parameters are taken to be $\kappa = 10^{-3}$ and $\boldsymbol{\beta} = [1 \quad 0]^T$, while the right-hand side has a constant value $f = 1$. The exact solution of such a problem is a ramp of unit slope along the x-axis, showing two layers at $y = 0$ and $y = 1$, and a third, sharper layer at $x = 1$.

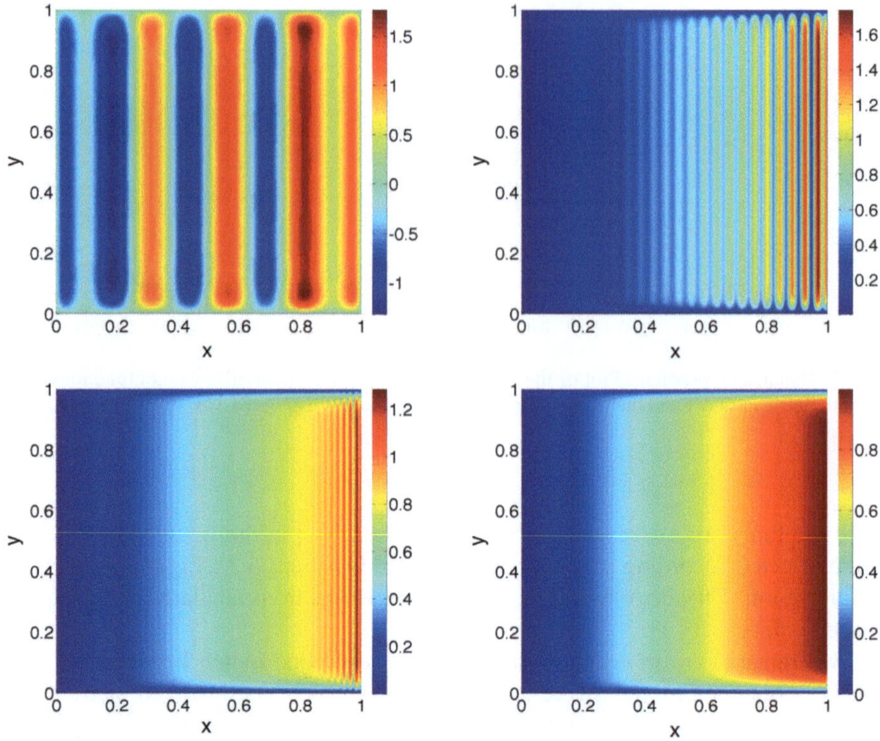

Fig. 8 Advection–diffusion problem. Contour plots of the numerical solution and corresponding color bars depicting the attained values of the function. Collocation is performed at standard tensor-product B-spline Greville points, using degree $p = 4$ in both directions and standard tensor-product B-splines (*top, left* 10×10 degrees of freedom; *top, right* 50×50; *bottom, left* 100×100; *bottom, right* 300×300)

The solution of the problem is approximated by collocation IgA considering both tensor-product polynomial B-splines and tensor-product GB-splines. The physical domain and the parametric domain agree so there is no need of a geometry map. In both cases, we use standard tensor-product B-spline Greville points as collocation points and we consider uniformly spaced knot sequences as in (13). The solution presents strong gradients and thin layers, i.e., very small regions where the function exhibits a strong variation. Therefore, it is natural to use hyperbolic GB-splines due to their ability of describing strong variations without oscillations.

We choose the parameter α to be equal to the global Péclet number, i.e., $\alpha = \|\boldsymbol{\beta}\|/\kappa = 10^3$. In Figs. 8 and 9, we present some results for the same degree $p = 4$ in both directions and different choices of uniformly spaced knot sequences. We can see that, as expected, standard B-splines lead to spurious oscillations until the discretization is fine enough to resolve the boundary layers. A discretization consisting of 300×300 degrees of freedom ($n = 298$) is needed to get a solution where the

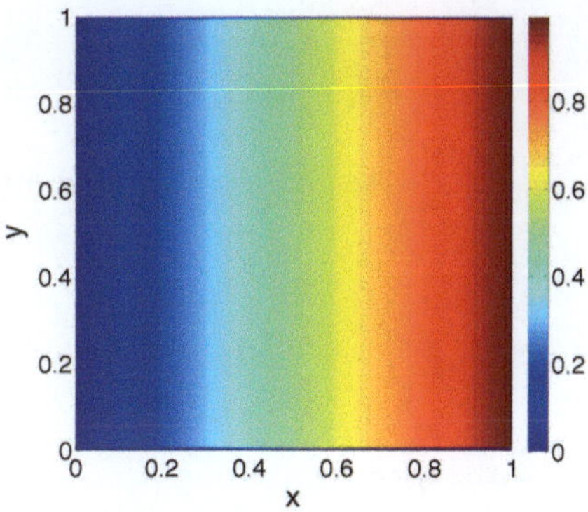

Fig. 9 Advection–diffusion problem. Contour plots of the numerical solution and corresponding color bars depicting the attained values of the function. Collocation is performed at standard tensor-product B-spline Greville points, using degree $p = 4$ in both directions, and tensor-product hyperbolic B-splines with $\alpha = \|\boldsymbol{\beta}\|/\kappa = 10^3$ (*left* 10×10 degrees of freedom; *right* 50×50)

oscillations (indeed still present) are negligible. Instead, with hyperbolic GB-splines and the adopted selection of the parameter α, already a very coarse discretization (10×10 degrees of freedom) leads to acceptable results. However, we notice that in this case the two boundary layers at $y = 0$ and $y = 1$ appear to be too sharp, as compared with the behavior of the 300×300 B-spline case. A less coarse discretization (50×50 degrees of freedom) gives a solution that, in practice, cannot be distinguished from the 300×300 B-spline case. Finally, we remark that, in contrast to the 300×300 B-spline discretization, no oscillations (not even very small) are observed in this case. Other problems discretized by isogeometric collocation using GB-splines can be found in [35].

4.3 An Elastic Plate Problem with a Circular Hole

In our last example, we address a model problem in solid mechanics. We consider an infinite plate with a circular hole of radius r, subject to an in-plane uniform tension T_x in x-direction, see Fig. 10 (left). For a homogeneous and isotropic material, this problem features an exact solution which can be found in [22, Sect. 7.6]. The infinite plate is modeled by a finite circular domain with radius R. Due to the symmetry, the computational domain Ω is restricted to a quarter, see Fig. 10 (right). We study the linear elastic behavior of the displacement field $\mathbf{u} : \Omega \to \mathbb{R}^2$ described by

$$\operatorname{div} \sigma(\mathbf{u}) = 0, \quad \text{in } \Omega.$$

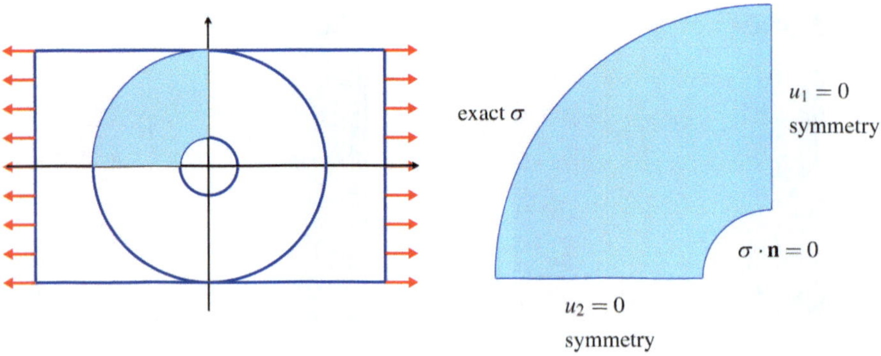

Fig. 10 Elastic plate problem with a circular hole

The boundary conditions are shown in Fig. 10 (right) where we apply the exact solution as Neumann boundary condition. For the sake of completeness, we recall that $\sigma(\mathbf{u}) := \{\sigma_{ij}(\mathbf{u})\}_{i,j=1,2}$ with

$$\sigma_{ij}(\mathbf{u}) := \lambda \operatorname{div} \mathbf{u}\, \delta_{ij} + 2\mu\varepsilon_{ij}(\mathbf{u}), \quad \varepsilon_{ij}(\mathbf{u}) := \frac{1}{2}\left(\frac{\partial u_i}{\partial x_j} + \frac{\partial u_j}{\partial x_i}\right), \quad i, j = 1, 2,$$

and

$$\mathbf{u} := (u_1, u_2), \quad (x, y) := (x_1, x_2), \quad \lambda := \frac{E\nu}{(1+\nu)(1-2\nu)}, \quad \mu := \frac{E}{2(1+\nu)},$$

where δ_{ij} denotes the Kronecker delta, E the Young's modulus, and ν the Poisson's ratio. In our computed example, we have taken

$$r = 1, \quad R = 4, \quad E = 10^5, \quad \nu = 0.3, \quad T_x = 10.$$

Without the hole, the stress would be uniform:

$$\sigma_{1,1} = T_x, \quad \sigma_{1,2} = \sigma_{2,2} = 0.$$

This distribution will change only in the vicinity of the hole. More precisely, we get a peak stress concentration at the upper side of the hole, that is at the point (0, 1), see [22] and Fig. 11 (right). Due to the behavior of the solution, a numerical approximation strategy using local refinement around this point is preferable.

The solution of the problem is approximated by Galerkin IgA. We refer to [39, Chap. 3] for the weak form of the problem. Since the physical domain Ω is a quarter of an annulus, see Fig. 10 (right), it is natural to use (quadratic) trigonometric

Table 1 Elastic plate problem. Control points of the geometry function

k	$\mathbf{P}_{k,1}$	$\mathbf{P}_{k,2}$	$\mathbf{P}_{k,3}$
1	(−1, 0)	(−2.5, 0)	(−4, 0)
2	(−1, 1)	(−2.5, 2.5)	(−4, 4)
3	(0, 1)	(0, 2.5)	(0, 4)

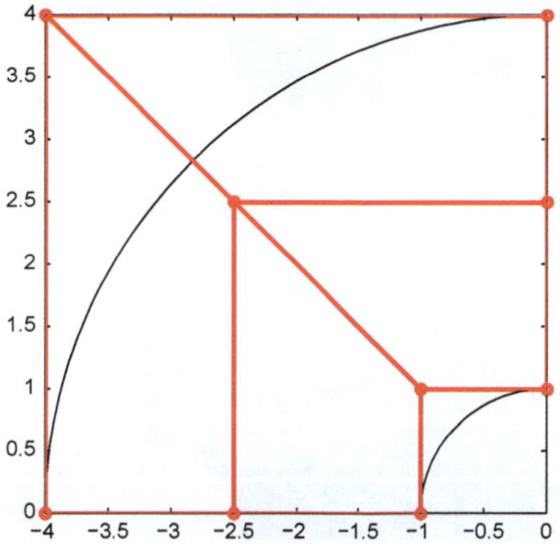

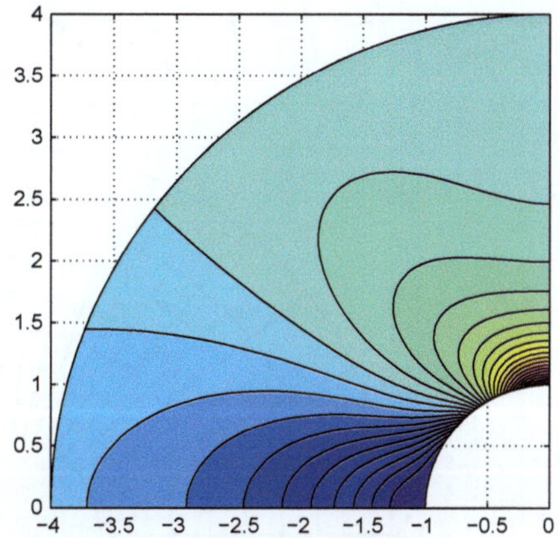

Fig. 11 Elastic plate problem. *Left* control net of the geometry function. *Right* exact $\sigma_{1,1}$

GB-splines for the representation of the circular arcs. Hence, we construct a global geometry function as in (27) considering the tensor-product space $\mathbb{S}_{n,2}^{\mathbb{T}_\alpha} \otimes \mathbb{S}_{n,2}$ with $\alpha = \frac{\pi}{2}$. The control points $\mathbf{P}_i = \mathbf{P}_{k,j}$, $k, j = 1, 2, 3$, are depicted in Table 1 for a coarse grid consisting of one interval per edge, see Fig. 11 (left). Then, we approximate both components of the displacement in the bivariate tensor-product spaces $\mathbb{S}_{n,2}^{\mathbb{T}_\alpha} \otimes \mathbb{S}_{n,2}$ with $\alpha = \frac{\pi}{2}$ and different values of n. Standard quadratic B-splines $\mathbb{S}_{n,2} \otimes \mathbb{S}_{n,2}$ have also been considered for the sake of comparison. Next, we consider bivariate spaces spanned by hierarchical bases built from the same tensor-product spaces mentioned before, defined on the hierarchical meshes shown in Fig. 12 (left column) according to the construction outlined in Sect. 2.4. Figure 13 shows the reduction of the computed error versus the number of degrees of freedom, in the cases of uniform and local hierarchical refinements with the different considered spaces. The presented hierarchical meshes are constructed manually. For automatic refinement strategies, we refer to the literature, see e.g., [3, 17]. Further details on this example can be found in [34].

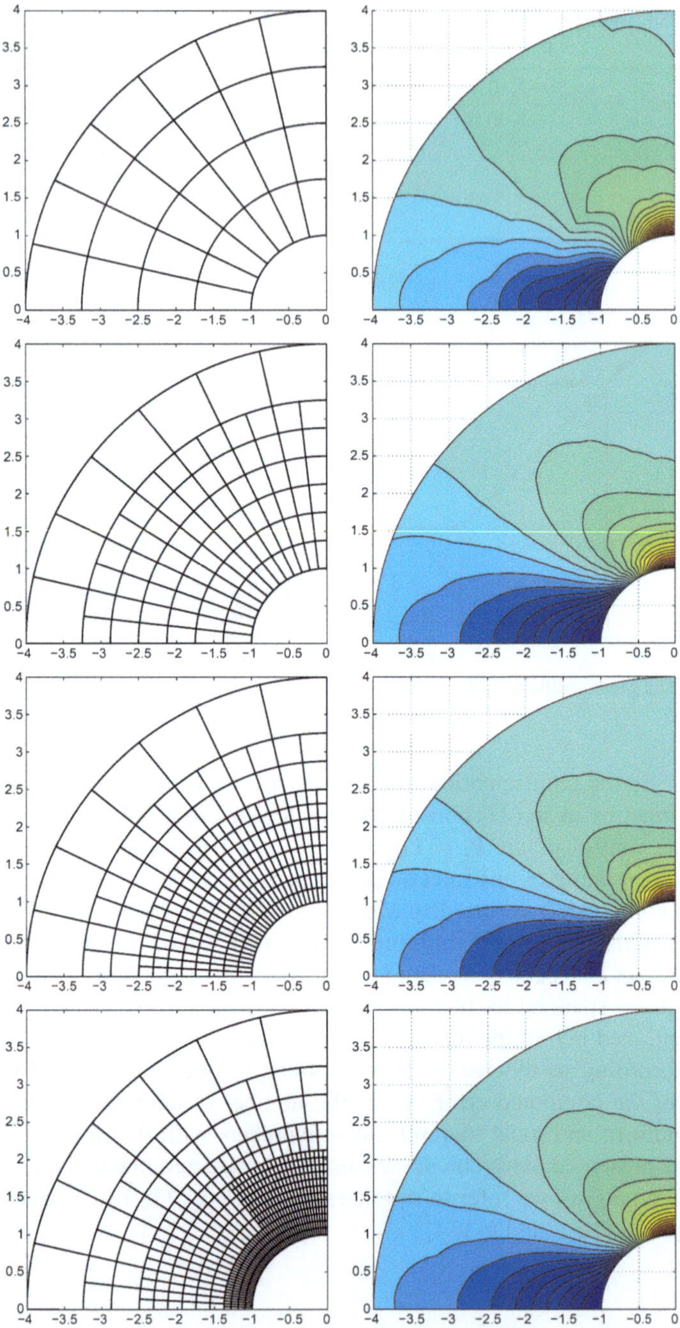

Fig. 12 Elastic plate problem. Hierarchical meshes and corresponding hierarchical quadratic trigonometric GB-spline approximations of $\sigma_{1,1}$ with $\alpha = \pi/2$

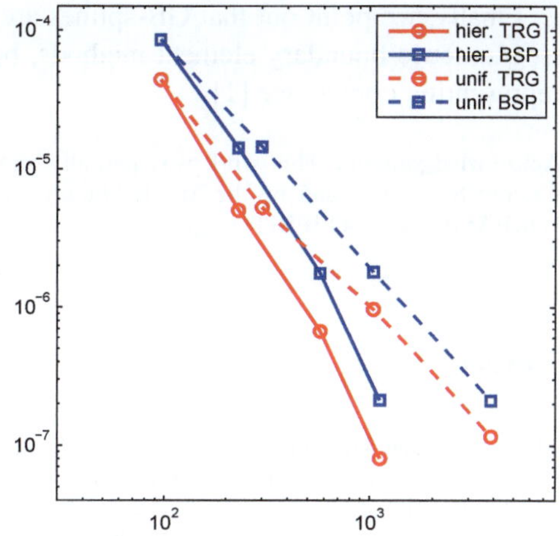

Fig. 13 Elastic plate problem. L_2-norm of the displacement error using quadratic trigonometric GB-splines (*red line*) and standard quadratic B-splines (*blue line*) versus the number of degrees of freedom, computed on locally refined hierarchical meshes (*solid*) and uniformly refined meshes (*dashed*)

5 Conclusions

In this paper, we have reviewed the definition and main properties of GB-splines, and we have illustrated their use in isogeometric Galerkin and collocation methods as a possible alternative to standard B-splines/NURBS.

GB-splines are a special case of the larger and richer class of Tchebycheffian B-splines, and they can be seen as the minimal extension of the standard polynomial B-splines sharing similar properties. In particular, GB-splines support local refinement based on the hierarchical/multilevel approach. The complete structural similarity between B-splines and GB-splines makes them "plug-to-plug" compatible in isogeometric Galerkin and collocation methods.

Trigonometric and hyperbolic GB-splines are of particular interest in IgA because they offer some relevant advantages over NURBS whenever the section spaces are selected according to a problem-dependent strategy. In particular, trigonometric/hyperbolic GB-splines

- allow for the exact representation of conics by means of a parameterization which is close to the arc length, without requiring rational forms;
- behave completely similar to standard B-splines with respect to differentiation/integration;
- give rise to matrices with spectral properties analogous to standard B-splines when used in Galerkin/collocation methods;
- have the same approximation properties as standard B-splines of the same degree, for fixed values of the local phase parameters α; and
- tend to standard B-splines of the same degree as the local phase parameters (or equivalently the knot spacing) tend to zero.

Finally, we point out that GB-splines are also a promising tool in the context of isogeometric boundary element methods, because of the advantages they offer for representing conics, see [1].

Acknowledgements This work was partially supported by INdAM Gruppo Nazionale per il Calcolo Scientifico and by the MIUR "Futuro in Ricerca 2013" Program through the project "DREAMS" (RBFR13FBI3).

References

1. A. Aimi, M. Diligenti, M.L. Sampoli, A. Sestini, Non-polynomial spline alternatives in isogeometric symmetric Galerkin BEM. Appl. Numer. Math. **116**, 10–23 (2017)
2. F. Auricchio, L. Beirão da Veiga, T.J.R. Hughes, A. Reali, G. Sangalli, Isogeometric collocation. Math. Models Methods Appl. Sci. **20**, 2075–2107 (2010)
3. R.E. Bank, R.K. Smith, A posteriori error estimates based on hierarchical bases. SIAM J. Numer. Anal. **30**, 921–935 (1993)
4. D. Boffi, Finite element approximation of eigenvalue problems. Acta Numer. **19**, 1–120 (2010)
5. C. Bracco, D. Cho, Generalized T-splines and VMCR T-meshes. Comput. Methods Appl. Mech. Eng. **280**, 176–196 (2014)
6. C. Bracco, F. Roman, Spaces of generalized splines over T-meshes. J. Comput. Appl. Math. **294**, 102–123 (2016)
7. C. Bracco, D. Berdinsky, D. Cho, M.-J. Oh, T.-W. Kim, Trigonometric generalized T-splines. Comput. Methods Appl. Mech. Eng. **268**, 540–556 (2014)
8. C. Bracco, T. Lyche, C. Manni, F. Roman, H. Speleers, Generalized spline spaces over T-meshes: dimension formula and locally refined generalized B-splines. Appl. Math. Comput. **272**, 187–198 (2016)
9. C. Bracco, T. Lyche, C. Manni, F. Roman, H. Speleers, On the dimension of Tchebycheffian spline spaces over planar T-meshes. Comput. Aided Geom. Design **45**, 151–173 (2016)
10. J.M. Carnicer, E. Mainar, J.M. Peña, Critical length for design purposes and extended Chebyshev spaces. Constr. Approx. **20**, 55–71 (2004)
11. P. Costantini, Curve and surface construction using variable degree polynomial splines. Comput. Aided Geom. Design **17**, 419–446 (2000)
12. P. Costantini, T. Lyche, C. Manni, On a class of weak Tchebycheff systems. Numer. Math. **101**, 333–354 (2005)
13. J.A. Cottrell, T.J.R. Hughes, Y. Bazilevs, *Isogeometric Analysis: Toward Integration of CAD and FEA* (Wiley, Chichester, 2009)
14. C. de Boor, *A Practical Guide to Splines*, Revised edn. (Springer, New York, 2001)
15. J. Deng, F. Chen, Y. Feng, Dimensions of spline spaces over T-meshes. J. Comput. Appl. Math. **194**, 267–283 (2006)
16. T. Dokken, T. Lyche, K.F. Pettersen, Polynomial splines over locally refined box-partitions. Comput. Aided Geom. Design **30**, 331–356 (2013)
17. M. Dörfel, B. Jüttler, B. Simeon, Adaptive isogeometric analysis by local h-refinement with T-splines. Comput. Methods Appl. Mech. Eng. **199**, 264–275 (2010)
18. D.R. Forsey, R.H. Bartels, Hierarchical B-spline refinement. Comput. Graph. **22**, 205–212 (1988)
19. C. Garoni, C. Manni, F. Pelosi, S. Serra-Capizzano, H. Speleers, On the spectrum of stiffness matrices arising from isogeometric analysis. Numer. Math. **127**, 751–799 (2014)
20. C. Giannelli, B. Jüttler, H. Speleers, THB-splines: the truncated basis for hierarchical splines. Comput. Aided Geom. Design **29**, 485–498 (2012)
21. C. Giannelli, B. Jüttler, H. Speleers, Strongly stable bases for adaptively refined multilevel spline spaces. Adv. Comp. Math. **40**, 459–490 (2014)

22. P.L. Gould, *Introduction to Linear Elasticity* (Springer, Berlin, 1999)
23. R.R. Hiemstra, F. Calabrò, D. Schillinger, T.J.R. Hughes, Optimal and reduced quadrature rules for tensor product and hierarchically refined splines in isogeometric analysis. Comput. Methods Appl. Mech. Eng. **316**, 966–1004 (2017)
24. T.J.R. Hughes, J.A. Cottrell, Y. Bazilevs, Isogeometric analysis: CAD, finite elements, NURBS, exact geometry and mesh refinement. Comput. Methods Appl. Mech. Eng. **194**, 4135–4195 (2005)
25. J.W. Jerome, L.L. Schumaker, Local support bases for a class of spline functions. J. Approx. Theory **16**, 16–27 (1976)
26. P.E. Koch, T. Lyche, Interpolation with exponential B-splines in tension, in *Geometric Modelling*, ed. by G. Farin, H. Hagen, H. Noltemeier, W. Knödel (Springer, Berlin, 1993), pp. 173–190
27. B.I. Kvasov, P. Sattayatham, GB-splines of arbitrary order. J. Comput. Appl. Math. **104**, 63–88 (1999)
28. X. Li, M.A. Scott, Analysis-suitable T-splines: characterization, refineability, and approximation. Math. Models Methods Appl. Sci. **24**, 1141–1164 (2014)
29. L. Lyche, A recurrence relation for Chebyshevian B-splines. Constr. Approx. **1**, 155–173 (1985)
30. E. Mainar, J.M. Peña, J. Sánchez-Reyes, Shape preserving alternatives to the rational Bézier model. Comput. Aided Geom. Design **18**, 37–60 (2001)
31. C. Manni, H. Speleers, Standard and non-standard CAGD tools for isogeometric analysis: a tutorial, in *IsoGeometric Analysis: A New Paradigm in the Numerical Approximation of PDEs*, ed. by A. Buffa, G. Sangalli. Lecture Notes in Mathematics, vol. 2161 (2016), pp. 1–69
32. C. Manni, F. Pelosi, M.L. Sampoli, Generalized B-splines as a tool in isogeometric analysis. Comput. Methods Appl. Mech. Eng. **200**, 867–881 (2011)
33. C. Manni, F. Pelosi, M.L. Sampoli, Isogeometric analysis in advection–diffusion problems: tension splines approximation. J. Comput. Appl. Math. **236**, 511–528 (2011)
34. C. Manni, F. Pelosi, H. Speleers, Local hierarchical h-refinements in IgA based on generalized B-splines, in *Mathematical Methods for Curves and Surfaces 2012*, ed. by M. Floater, et al. Lecture Notes in Computer Science, vol. 8177 (2014), pp. 341–363
35. C. Manni, A. Reali, H. Speleers, Isogeometric collocation methods with generalized B-splines. Comput. Math. Appl. **70**, 1659–1675 (2015)
36. M.L. Mazure, Four properties to characterize Chebyshev blossoms. Constr. Approx. **17**, 319–334 (2001)
37. M.L. Mazure, How to build all Chebyshevian spline spaces good for geometric design? Numer. Math. **119**, 517–556 (2011)
38. L. Piegl, W. Tiller, *The NURBS Book (Monographs in Visual Communication)*, 2nd edn. (Springer, New York, 1997)
39. A. Quarteroni, *Numerical Models for Differential Problems* (Springer, Mailand, 2009)
40. A. Reali, T.J.R. Hughes, An introduction to isogeometric collocation methods, in *Isogeometric Methods for Numerical Simulation*, ed. by G. Beer, S.P. Bordas (Springer, Wien, 2015)
41. F. Roman, C. Manni, H. Speleers, Spectral analysis of matrices in Galerkin methods based on generalized B-splines with high smoothness. Numer. Math. **135**, 169–216 (2017)
42. F. Roman, C. Manni, H. Speleers, Numerical approximation of GB-splines with a convolutional approach. Appl. Numer. Math. **116**, 273–285 (2017)
43. L.L. Schumaker, *Spline Functions: Basic Theory*, 3rd edn. (Cambridge University Press, Cambridge, 2007)
44. L.L. Schumaker, L. Wang, Approximation power of polynomial splines on T-meshes. Comput. Aided Geom. Design **29**, 599–612 (2012)
45. T. Sederberg, J. Zheng, A. Bakenov, A. Nasri, T-splines and T-NURCCs. ACM Trans. Graph. **22**, 477–484 (2003)
46. H. Speleers, C. Manni, Effortless quasi-interpolation in hierarchical spaces. Numer. Math. **132**, 155–184 (2016)
47. G. Wang, M. Fang, Unified and extended form of three types of splines. J. Comput. Appl. Math. **216**, 498–508 (2008)

On Polynomials with Vanishing Hessians and Some Density Problems

Tom McKinley and Boris Shekhtman

Abstract We propose a conjecture regarding homogeneous polynomials with vanishing Hessian and indicate some evidence for its validity. The conjecture is related to a question of Allan Pinkus and Bronislaw Wajnryb regarding density of certain classes of polynomials.

Keywords Homogeneous polynomial · Algebraic dependency · Density

1 Introduction

In this article, we propose a conjecture regarding homogeneous polynomials with vanishing Hessian and indicate some evidence for its validity. The conjecture is related to a question of Allan Pinkus and Bronislaw Wajnryb regarding density of certain classes of polynomials.

In what follows, the symbol $\mathbb{k}$ will stand either for the field $\mathbb{R}$ of real numbers or the field $\mathbb{C}$ of complex numbers. We use $\mathbb{k}[x_1, \ldots, x_n]$ to denote the algebra of polynomials of n variables with coefficients in $\mathbb{k}$. For every $p \in \mathbb{k}[x_1, \ldots, x_n]$ we let $p(D_1, \ldots, D_n)$ be the differential operator obtained from p by formally replacing the variables $x_1, \ldots, x_n$ with the differential operators $D_1, \ldots, D_n$, where D_k is differentiation with respect to x_k.

Conjecture 1 Let p, f be homogeneous polynomials in $\mathbb{k}[x_1, \ldots, x_n]$. Then,

$$p(D_1 f, \ldots, D_n f) = 0 \Rightarrow p(D_1, \ldots, D_n) f = 0. \tag{1}$$

The conjecture is obviously true if p is a linear form, since, in this case, $p(D_1 f, \ldots, D_n f) = p(D_1, \ldots, D_n) f$. It is also obvious if $n = 1$, since $\left(f'\right)^m = 0$ implies $f' = 0$, hence $f^{(m)} = 0$.

T. McKinley · B. Shekhtman (✉)
University of South Florida, Tampa, FL 33620-5700, USA
e-mail: shekhtma@usf.edu

G.E. Fasshauer and L.L. Schumaker (eds.), *Approximation Theory XV: San Antonio 2016*, Springer Proceedings in Mathematics & Statistics 201,
DOI 10.1007/978-3-319-59912-0_13

The condition $p(D_1 f, \ldots, D_n f) = 0$ means that the partial derivatives of f are algebraically dependent which is equivalent [2, 9] to the Jacobian determinant of the map $\nabla f : \mathbb{k}^n \to \mathbb{k}^n$ vanishing identically, i.e., to the vanishing of the Hessian determinant of f.

The characterization of homogeneous polynomials with vanishing Hessian started by Hesse [5] and continued in Gordan and Noether [4] as well as by Perazzo [8]. Recently, the interest in the topic had been renewed [1, 3, 6, 10]. In particular, Gordan and Noether verified the claim of Hesse for polynomials in four variables (Hesse believed that the result below holds independently of the number of variables) and constructed a counterexample to the claim in 5 variables:

Theorem 1 (Gordan and Noether [3]) *If $n \leq 4$ and the Hessian determinant of f vanishes then $D_1 f, \ldots, D_n f$ are linearly dependent.*

A simple counter example in five variables is

$$f(x_1, x_2, x_3, x_4, x_5) := x_1 x_4^2 + x_2 x_4 x_5 + x_3 x_5^2 \tag{2}$$

It is easy to check that the first partial derivatives in this case are linearly independent yet satisfy the algebraic equation

$$(D_1 f)(D_3 f) - (D_2 f)^2 = 0 \tag{3}$$

as well as the corresponding linear differential equation

$$D_1 D_3 f - D_2^2 f = 0 \tag{4}$$

Notice that while f is a polynomial in five variables, the polynomial $p = x_1 x_3 - x_2^2$ that satisfies our assumption (??) depends only on three variables. This matches the special case of the conjecture proved in the next section.

2 The Main Result

Theorem 2 *Suppose that*

$$p(y_1, \ldots, y_n) = q(l_1(y_1, \ldots, y_n), l_2(y_1, \ldots, y_n), l_3(y_1, \ldots, y_n))$$

where l_1, l_2, l_3 are linear forms. Then the conjecture holds.

We start with a lemma; notice that in it f does not need to be homogeneous but p does:

Lemma 1 *Let $f, p \in \mathbb{k}[x_1, \ldots, x_n]$ and p homogeneous. If $n \leq 3$ then $p(D_1 f, \ldots, D_n f) = 0$ implies that $D_1 f, \ldots, D_n f$ are linearly dependent.*

Proof Let $\widetilde{f}(x_0, x_1, \ldots, x_n)$ be the homogenization of the polynomial f, i.e.,

$$\widetilde{f}(x_0, x_1, \ldots, x_n) = x_0^m f\left(\frac{x_1}{x_0}, \ldots, \frac{x_n}{x_0}\right)$$

where $m = \deg f$. Then, from homogeneity of p, we conclude that $p\left(D_1\widetilde{f}, \ldots, D_n\widetilde{f}\right) = 0$ for all $x_0, x_1, \ldots, x_n$ with $x_0 \neq 0$ and hence, by continuity for all $x_0, x_1, \ldots, x_n$. Since $\widetilde{f}$ is a function of at most four variables, the Gordan–Noether Theorem 1 implies that $D_0\widetilde{f}, D_1\widetilde{f}, \ldots, D_n\widetilde{f}$ are linearly dependent, i.e., there exists a nontrivial linear combination

$$\sum_{k=0}^{n} c_k D_k \widetilde{f} = 0.$$

Assume that $D_1 f, \ldots, D_n f$ are linearly independent. Then, without loss of generality we can assume that

$$\sum_{k=1}^{n} c_k D_k \widetilde{f} = D_0 \widetilde{f}. \tag{5}$$

Now $x_0\widetilde{f}$ is also a homogeneous polynomials satisfying $p(D_1(x_0 f), \ldots, D_n(x_0 f)) = 0$. Using Theorem 1 one more time and linear independence of $D_1 f, \ldots, D_n f$, we obtain coefficients (b_k) such that

$$x_0\left(\sum_{k=0}^{n} b_k D_k \widetilde{f}\right) = D_0\left(x_0\widetilde{f}\right) = x_0 D_0\widetilde{f} + \widetilde{f}. \tag{6}$$

Setting $x_0 = 1$ and subtracting Eq. (5) from Eq. (6) we obtain

$$\sum_{k=1}^{n} (b_k - c_k) D_k f = f$$

which is not possible since the degree of the polynomial on the left is one less than the one on the right.

Proof (*Proof of Theorem* 2) First observe that the conjecture is invariant with respect to linear change of variables. Hence, it sufficient to assume that $f \in \Bbbk[x_1, \ldots, x_n]$ is homogeneous, and p is a homogeneous polynomial in two or three variables. We begin with two: that is we assume that

$$p(D_1 f, D_2 f) = 0 \tag{7}$$

and conclude that

$$p(D_1, D_2) f = 0. \tag{8}$$

To simplify the notation, we use $\mathbf{u}$ to denote the variables $x_3, x_4, \ldots, x_n$; hence, the homogeneous polynomial $f(x_1, \ldots, x_n)$ will be written as $f(x_1, x_2, \mathbf{u})$. Now, let us fix $\mathbf{u}$. Then, $f(x_1, x_2, \mathbf{u})$ is a polynomial (not necessarily homogeneous) for which (7) holds. By Lemma 1 the polynomials $D_1 f$ and $D_2 f$ are linearly dependent. Without loss of generality we can find a scalar $a(\mathbf{u})$ such that

$$D_1 f = a(\mathbf{u}) D_2 f. \tag{9}$$

From (9), we easily conclude that

$$D_1^k f = a^k(\mathbf{u}) D_2^k f. \tag{10}$$

If $p(y_1, y_2) = \sum\limits_{k=0}^{m} c_k y_1^k y_2^{m-k}$ then (8) says that

$$\sum_{k=0}^{m} c_k (D_1 f)^k (D_2 f)^{m-k} = 0. \tag{11}$$

By (9), we have

$$\sum_{k=0}^{m} c_k a^k(\mathbf{u}) (D_2 f)^m = 0 \tag{12}$$

hence either

$$\sum_{k=0}^{m} c_k a^k(\mathbf{u}) = 0 \tag{13}$$

or

$$(D_2 f)^m = 0. \tag{14}$$

In either case, using (10), we obtain

$$\sum_{k=0}^{m} c_k \left(D_1^k D_2^{m-k} f\right) = \sum_{k=0}^{m} c_k \left(D_2^{m-k} D_1^k f\right) = \sum_{k=0}^{m} c_k a^k(\mathbf{u}) \left(D_2^m f\right) = 0, \tag{15}$$

where the last equality follows from (13) or (14) and holds for every $\mathbf{u}$.

The proof for $n = 3$ is similar. Assuming

$$p(D_1 f, D_2 f, D_3 f) = 0, \tag{16}$$

we want to conclude

$$p(D_1, D_2, D_3) f = 0. \tag{17}$$

Once again we write $f(x_1, \ldots, x_n)$ as $f(x_1, x_2, x_3, \mathbf{u})$ where $\mathbf{u}$ this time denotes the remaining $n-3$ variables. Hence for every fixed $\mathbf{u}$, the polynomial $f(x_1, x_2, x_3, \mathbf{u})$ is a polynomial of three variables satisfying (16). Lemma 1 gives coefficients $b(\mathbf{u})$ and $c(\mathbf{u})$ such that

$$D_3 f = a(\mathbf{u}) D_1 f + b(\mathbf{u}) D_2 f. \tag{18}$$

Let $p(y_1, y_2) = \sum_{k=0}^{m} c_k(y_1, y_2) y_3^k$ where $c_k(x_1, x_2)$ are homogeneous polynomials of degree $m-k$. Then (16) and (18) give

$$q(D_1 f, D_2 f) := \left(\sum_{k=0}^{m} c_k(D_1 f, D_2 f)(a(\mathbf{u}) D_1 f + b(\mathbf{u}) D_2 f) \right)^k = 0.$$

In other words, it satisfies equation of type (7). By what was proved previously for the case of two derivatives it follows that

$$\left(\sum_{k=0}^{m} c_k(D_1, D_2)(a(\mathbf{u}) D_1 + b(\mathbf{u}) D_2) \right)^k f = 0. \tag{19}$$

By (18) $D_3^k f = (a(\mathbf{u}) D_1 + b(\mathbf{u}) D_2)^k f$ hence

$$\begin{aligned} p(D_1, D_2, D_3) f &= \sum_{k=0}^{m} c_k(D_1, D_2) D_3^k f \\ &= \sum_{k=0}^{m} c_k(D_1, D_2)(a(\mathbf{u}) D_1 + b(\mathbf{u}) D_2)^k f = 0 \end{aligned}$$

by (19).

Theorem 3 *Conjecture 1 also holds for $n = 4$.*

We will skip the proof since the Gordan–Noether Theorem 1 implies the linear dependency of four partials. The rest of the calculations are the same as in the proof of the previous theorem reducing the case to three variables and hence Theorem 2.

3 Applications to Density Problem

For a fixed polynomial $f \in \mathbb{k}[x_1, \ldots, x_n]$, we let

$$\mathscr{P}(f) := \operatorname{span}\left\{ (f(\cdot + \mathbf{b}))^k : \mathbf{b} \in \mathbb{R}^d, k \in \mathbb{Z}_+ \right\}. \tag{20}$$

Allan Pinkus and Bronislaw Wajnryb in their survey paper [9] introduced and studied density properties of this class of polynomials. In particular, they proved that if

f is homogeneous then $\mathscr{P}(f) \neq \Bbbk[x_1, \ldots, x_n]$ implies that the Hessian determinant of f is equal to zero. They also proved the converse in case $n \leq 4$ and asked if it is true for arbitrary n:

Problem 1 Let f be a homogeneous polynomial in $\mathbb{R}[x_1, \ldots, x_n]$ with vanishing Hessian. Does this imply that $\overline{\mathscr{P}(f)} \neq C(\mathbb{R}^d)$?

Proposition 1 *If Conjecture 1 holds then the answer to Problem 1 is affirmative.*

To prove this, we will site the following theorem proved in [7]:

Theorem 4 *For any n and any $f \in \mathbb{R}[x_1, \ldots, x_n]$, the following are equivalent:*

(i) $\mathscr{P}(f) \neq \mathbb{R}[x_1, \ldots, x_n]$
(ii) There exists a nonzero polynomial $p \in \mathbb{R}[x_1, \ldots, x_n]$ such that

$$p(D_1, \ldots, D_n)\left(f^k\right) = 0 \tag{21}$$

for all $k \in \mathbb{Z}_+$.
(iii) $\overline{\mathscr{P}(f)} \neq C\left(\mathbb{R}^d\right)$.

Proof (*Proof of Proposition* 1) Observe that the vanishing of the Hessian is equivalent to the existence of a polynomial $p \in \mathbb{R}[x_1, \ldots, x_n]$ such that

$$p(D_1 f, \ldots, D_n f) = 0. \tag{22}$$

Since f is homogeneous, every homogeneous component of p satisfies (22) so we may, as well, assume that p in (22) is homogeneous of some degree m. By the chain rule,

$$\begin{aligned}
&p\left(D_1(f)^k, \ldots, D_n\left(f^k\right)\right) \\
&= p\left(kf^{k-1}(D_1 f, \ldots, D_n f)\right) \\
&= \left(kf^{k-1}\right)^m p(D_1 f, \ldots, D_n f) = 0.
\end{aligned}$$

Now, the validity of the conjecture implies

$$p(D_1, \ldots, D_n)\left(f^k\right) = 0 \tag{23}$$

for all $k \in \mathbb{Z}_+$. By Theorem 4 we conclude that $\mathscr{P}(f) \neq \mathbb{R}[x_1, \ldots, x_n]$.

As a corollary, we immediately obtain the following partial answer to the question of Pinkus and Wajnryb:

Theorem 5 *Let f be a homogeneous polynomial in $\mathbb{R}[x_1, \ldots, x_d]$ such that there exists a homogeneous polynomial*

$$p(y_1, \ldots, y_n) = q(l_1(y_1, \ldots, y_n), l_2(y_1, \ldots, y_n), l_3(y_1, \ldots, y_n))$$

with l_1, l_2, l_3 linear forms satisfying $p(D_1 f, \ldots, D_n f) = 0$. *Then* $\mathscr{P}(f) \neq \mathbb{R}[x_1, \ldots, x_n]$.

The assumption of the theorem may seem too restrictive (and it is) but most of the known cubic polynomials f in six variables with vanishing Hessians [1, 3, 10] either have partial derivatives that are linearly dependent or have a polynomial p that depends on three variables. In the case of linear dependency, the polynomial p in (22) is linear and (23) simply follows from the chain rule. Hence, the conclusion of the theorem is still valid.

4 A Converse to the Main Conjecture

It is clear that the reverse implication in Conjecture 1 is not valid. For instance $D_1^2 x_1 = 0$ yet $(D_1 x_1)^2 = 1$. In the previous section, it was observed that if the conjecture holds, then

$$p(D_1 f, \ldots, D_n f) = 0 \Rightarrow p(D_1, \ldots, D_n)\left(f^k\right) = 0$$

for all $k \in \mathbb{Z}_+$. It turns out that in this stronger form, the inverse implication holds even without the assumption of homogeneity of f:

Theorem 6 *Let* $p, f \in \Bbbk[x_1, \ldots, x_n]$ *and* p *be homogeneous. If* $p(D_1, \ldots, D_n)\left(f^k\right) = 0$ *for all* $k \in \mathbb{Z}_+$ *then* $p(D_1 f, \ldots, D_n f) = 0$.

Proof Let p be a homogeneous polynomial of degree m and $k \geq m$. It is easy to see (by induction on m) that if $m_1 + m_2 + \cdots + m_n = m$, then

$$D_1^{m_1} D_2^{m_2} \ldots D_n^{m_n}\left(f^k\right)$$

$$= \frac{k!}{(k-m)!} f^{k-m} (D_1 f)^{m_1} (D_2 f)^{m_2} \ldots (D_n f)^{m_n} + \sum_{j=0}^{m-1} \frac{k!}{(k-j)!} f^{k-j} f_j$$

where the polynomials f_j do not depend on k. Hence, for any homogeneous polynomial p of degree m, we have

$$p(D_1, \ldots, D_n)\left(f^k\right) = \frac{k!}{(k-m)!} f^{k-m} p(D_1 f, \ldots, D_n f) + \sum_{j=0}^{m-1} \frac{k!}{(k-j)!} f^{k-j} g_j \tag{24}$$

for some g_j that do not depend on k. If $p(D_1, \ldots, D_n)\left(f^k\right) = 0$ then (24) implies

$$\frac{k!}{(k-m)!} f^{k-m} p(D_1 f, \ldots, D_n f) + \sum_{j=0}^{m-1} \frac{k!}{(k-j)!} f^{k-j} g_j = 0.$$

Factoring out f^{k-m} and dividing out by $\frac{k!}{(k-m)!}$, we have

$$p(D_1 f, \ldots, D_n f) = -\sum_{j=0}^{m-1} \frac{(k-m)!}{(k-j)!} f^{m-j} g_j = 0. \tag{25}$$

Normalizing f so that the uniform norm of f is one on the unit ball of $\Bbbk^n$ and, letting $k \to \infty$ the right-hand side in (25) tends to zero, hence $p(D_1 f, \ldots, D_n f) = 0$ on the unit ball of $\Bbbk^n$ and thus is identically zero.

The theorem gives an alternative proof of the result of Pinkus and Wajnryb mentioned at the beginning of the previous section:

Corollary 1 *Let f be a homogeneous polynomial in $\mathbb{R}[x_1, \ldots, x_n]$ and $\mathscr{P}(f) \neq \mathbb{R}[x_1, \ldots, x_n]$. Then, the Hessian determinant of f vanishes identically.*

Proof If $\mathscr{P}(f) \neq \mathbb{R}[x_1, \ldots, x_n]$, then by Theorem 4 there exists a nonzero polynomial $p \in \mathbb{R}[x_1, \ldots, x_n]$ such that

$$p(D_1, \ldots, D_n)\left(f^k\right) = 0 \tag{26}$$

for all $k \in \mathbb{Z}_+$. Since f is homogeneous, (26) holds for every homogeneous component of p. Hence, we can assume that p is homogeneous which, by Theorem 6, implies that the first partial derivatives of f are algebraically dependent, i.e., the Hessian of f vanishes.

Acknowledgements The authors wish to thank the anonymous referees for their kind words and useful suggestions.

References

1. C. Ciliberto, F. Russo, A. Simis, Homaloidal hypersurfaces and hypersurfaces with vanishing Hessian. Adv. Math. **218**(6), 1759–1805 (2008), http://dx.doi.org/10.1016/j.aim.2008.03.025
2. R. Ehrenborg, G.C. Rota, Apolarity and canonical forms for homogeneous polynomials. European J. Combin. **14**(3), 157–181 (1993), http://dx.doi.org/10.1006/eujc.1993.1022
3. R. Gondim, F. Russo, On cubic hypersurfaces with vanishing hessian. J. Pure Appl. Algebra **219**(4), 779–806 (2015), http://dx.doi.org/10.1016/j.jpaa.2014.04.030
4. P. Gordan, M. Nöther, Ueber die algebraischen Formen, deren Hesse'sche Determinante identisch verschwindet. Math. Ann. **10**(4), 547–568 (1876), http://dx.doi.org/10.1007/BF01442264
5. O. Hesse, Zur Theorie der ganzen homogenen Functionen. J. Reine Angew. Math. **56**, 263–269 (1859), http://dx.doi.org/10.1515/crll.1859.56.263
6. C. Lossen, When does the Hessian determinant vanish identically? (On Gordan and Noether's proof of Hesse's claim). Bull. Braz. Math. Soc. (N.S.) **35**(1), 71–82 (2004), http://dx.doi.org/10.1007/s00574-004-0004-0
7. T. McKinley, B. Shekhtman, On a problem of Pinkus and Wajnryb regarding density of multivariate polynomials. Proc. Am. Math. Soc. 6 (2016), http://dx.doi.org/10.1090/proc/13196. (Published electronically)
8. U. Perazzo, Sulle variet'a cubiche la cui hessiana svanisce identicamente. Giornale di Matematiche (Battaglini) **38**, 337–354 (1900)

9. A. Pinkus, Wajnryb, B.: On a problem of approximation by means of multidimensional polynomials. Uspekhi Mat. Nauk 50, 89–110 (1995). (Russian Math. Surveys **50** 319–340 (1995). (in English))
10. F. Russo, *Hypersurfaces with Vanishing Hessian*. Lecture Notes of the Unione Matematica Italiana. (Springer International Publishing AG, Switzerland 2016), pp. 177–220

Batched Stochastic Gradient Descent with Weighted Sampling

Deanna Needell and Rachel Ward

Abstract We analyze a batched variant of Stochastic gradient descent (SGD) with weighted sampling distribution for smooth and non-smooth objective functions. We show that by distributing the batches computationally, a significant speedup in the convergence rate is provably possible compared to either batched sampling or weighted sampling alone. We propose several computationally efficient schemes to approximate the optimal weights and compute proposed sampling distributions explicitly for the least squares and hinge loss problems. We show both analytically and experimentally that substantial gains can be obtained.

Keywords Stochastic gradient descent · Batched SGD · Kaczmarz method

1 Mathematical Formulation

We consider minimizing an objective function of the form

$$F(\boldsymbol{x}) = \frac{1}{n}\sum_{i=1}^{n} f_i(\boldsymbol{x}) = \mathbb{E} f_i(\boldsymbol{x}). \tag{1}$$

One important such objective function is the least squares objective for linear systems. Given an $n \times m$ matrix $\boldsymbol{A}$ with rows $\boldsymbol{a}_1, \ldots, \boldsymbol{a}_n$ and a vector $\boldsymbol{b} \in \mathbb{R}^n$, one searches for the least squares solution $\boldsymbol{x}_{LS}$ given by

D. Needell (✉)
Claremont McKenna College, Claremont, CA 91711, USA
e-mail: dneedell@cmc.edu

R. Ward
University of Texas, Austin, TX 78712, USA
e-mail: rachel@math.utexas.edu

G.E. Fasshauer and L.L. Schumaker (eds.), *Approximation Theory XV: San Antonio 2016*, Springer Proceedings in Mathematics & Statistics 201,
DOI 10.1007/978-3-319-59912-0_14

$$x_{LS} \stackrel{\text{def}}{=} \operatorname{argmin}_{x\in\mathbb{R}^m} \frac{1}{2}\|Ax - b\|_2^2 = \operatorname{argmin}_{x\in\mathbb{R}^m} \frac{1}{n}\sum_{i=1}^{n}\frac{n}{2}(b_i - \langle a_i, x\rangle)^2 = \operatorname{argmin}_{x\in\mathbb{R}^m} \mathbb{E} f_i(x), \tag{2}$$

where the functionals are defined by $f_i(x) = \frac{n}{2}(b_i - \langle a_i, x\rangle)^2$.

Another important example is the setting of support vector machines where one wishes to minimize the hinge loss objective given by

$$x_{HL} \stackrel{\text{def}}{=} \operatorname{argmin}_{w\in\mathbb{R}^m} \frac{1}{n}\sum_{i=1}^{n}[y_i\langle w, x_i\rangle]_+ + \frac{\lambda}{2}\|w\|_2^2. \tag{3}$$

Here, the data is given by the matrix X with rows $x_1, \ldots, x_n$ and the labels $y_i \in \{-1, 1\}$. The function $[z]_+ \stackrel{\text{def}}{=} \max(0, z)$ denotes the positive part. We view the problem (3) in the form (1) with $f_i(w) = [1 - y_i\langle w, x_i\rangle]_+$ and regularizer $\frac{\lambda}{2}\|w\|_2^2$.

The stochastic gradient descent (SGD) method solves problems of the form (1) by iteratively moving in the gradient direction of a randomly selected functional. SGD can be described succinctly by the update rule:

$$x_{k+1} \leftarrow x_k - \gamma\nabla f_{i_k}(x_k),$$

where index i_k is selected randomly in the kth iteration, and an initial estimation x_0 is chosen arbitrarily. Typical implementations of SGD select the functionals uniformly at random, although if the problem at hand allows a one-pass preprocessing of the functionals, certain *weighted* sampling distributions preferring functionals with larger variation can provide better convergence (see e.g., [1, 2] and references therein). In particular, Needell et al. show that selecting a functional with probability proportional to the Lipschitz constant of its gradient yields a convergence rate depending on the *average* of all such Lipschitz constants, rather than the supremum [1]. An analogous result in the same work shows that for non-smooth functionals, the probabilities should be chosen proportional to the Lipschitz constant of the functional itself.

Another variant of SGD utilizes so-called *mini-batches*; in this variant, a batch of functionals is selected in each iteration rather than a single one [3–6]. The computations over the batches can then be run in parallel, and speedups in the convergence are often quite significant.

Contribution. We propose a weighted sampling scheme to be used with mini-batch SGD. We show that when the batches can be implemented in parallel, significant speedup in convergence is possible. In particular, we analyze the convergence using *efficiently computed distributions* for the least squares and hinge loss objectives, the latter being especially challenging since it is non-smooth. We demonstrate theoretically and empirically that weighting the distribution and utilizing batches of functionals per iteration together form a complementary approach to accelerating convergence, highlighting the precise improvements, and weighting schemes for these settings of practical interest.

Organization. We next briefly discuss some related work on SGD, weighted distributions, and batching methods. We then combine these ideas into one cohesive framework and discuss the benefits in various settings. Section 2 focuses on the impact of weighting the distribution. In Sect. 3, we analyze SGD with weighting and batches for smooth objective functions, considering the least squares objective as a motivating example. We analyze the non-smooth case along with the hinge loss objective function in Sect. 4. We display experimental results for the least squares problem in Sect. 5 that serve to highlight the relative trade-offs of using both batches and weighting, along with different computational approaches. We conclude in Sect. 6.

Related work. Stochastic gradient descent, stemming from the work [7], has recently received renewed attention for its effectiveness in treating large-scale problems arising in machine learning [8–11]. Importance sampling in stochastic gradient descent, as in the case of mini-batching (which we also refer to simply as *batching* here), also leads to variance reduction in stochastic gradient methods and, in terms of theory, leads to improvement of the leading constant in the complexity estimate, typically via replacing the maximum of certain data-dependent quantities by their average. Such theoretical guarantees were shown for the case of solving least squares problems where stochastic gradient descent coincides with the randomized Kaczmarz method in [12]. This method was extended to handle noisy linear systems in [13]. Later, this strategy was extended to the more general setting of smooth and strongly convex objectives in [1], building on an analysis of stochastic gradient descent in [14]. Later, [2] considered a similar importance sampling strategy for convex but not necessarily smooth objective functions. Importance sampling has also been considered in the related setting of stochastic coordinate descent/ascent methods [15–18]. Other papers exploring advantages of importance sampling in various adaptations of stochastic gradient descent include but are not limited to [19–22].

Mini-batching in stochastic gradient methods refers to pooling together several random examples in the estimate of the gradient, as opposed to just a single random example at a time, effectively reducing the variance of each iteration [23]. On the other hand, each iteration also increases in complexity as the size of the batch grows. However, if parallel processing is available, the computation can be done concurrently at each step, so that the "per-iteration cost" with batching is not higher than without batching. Ideally, one would like the consequence of using batch size b to result in a convergence rate speedup by factor of b, but this is not always the case [24]. Still, [6] showed that by incorporating parallelization or multiple cores, this strategy can only improve on the convergence rate over standard stochastic gradient and can improve the convergence rate by a factor of the batch size in certain situations, such as when the matrix has nearly orthonormal rows. Other recent papers exploring the advantages of mini-batching in different settings of stochastic optimization include [3, 5, 25–27].

The recent paper [28] also considered the combination of importance sampling and mini-batching for a stochastic dual coordinate ascent algorithm in the general setting of empirical risk minimization, wherein the function to minimize is smooth and convex. There the authors provide a theoretical optimal sampling strategy that is not practical to implement but can be approximated via alternating minimization.

They also provide a computationally efficient formula that yields better sample complexity than uniform mini-batching, but without quantitative bounds on the gain. In particular, they do not provide general assumptions under which one achieves provable speedup in convergence depending on an average Lipschitz constant rather than a maximum.

For an overview of applications of stochastic gradient descent and its weighted/ batched variants in large-scale matrix inversion problems, we refer the reader to [29].

2 SGD with Weighting

Recall the objective function (1). We assume in this section that the function F and the functionals f_i satisfy the following convexity and smoothness conditions:

Convexity and smoothness conditions

1. Each f_i is continuously differentiable and the gradient function ∇f_i has Lipschitz constant bounded by L_i: $\|\nabla f_i(\boldsymbol{x}) - \nabla f_i(\boldsymbol{y})\|_2 \leq L_i \|\boldsymbol{x} - \boldsymbol{y}\|_2$ for all vectors $\boldsymbol{x}$ and $\boldsymbol{y}$.
2. F has strong convexity parameter μ; that is, $\langle \boldsymbol{x} - \boldsymbol{y}, \nabla F(\boldsymbol{x}) - \nabla F(\boldsymbol{y})\rangle \geq \mu\|\boldsymbol{x} - \boldsymbol{y}\|_2^2$ for all vectors $\boldsymbol{x}$ and $\boldsymbol{y}$.
3. At the unique minimizer $\boldsymbol{x}_* = \operatorname{argmin} F(\boldsymbol{x})$, the average gradient norm squared $\|\nabla f_i(\boldsymbol{x}_*)\|_2^2$ is not too large, in the sense that

$$\frac{1}{n}\sum_{i=1}^{n}\|\nabla f_i(\boldsymbol{x}_*)\|_2^2 \leq \sigma^2.$$

An unbiased gradient estimate for $F(\boldsymbol{x})$ can be obtained by drawing i uniformly from $[n] \stackrel{\text{def}}{=} \{1, 2, \ldots, n\}$ and using $\nabla f_i(\boldsymbol{x})$ as the estimate for $\nabla F(\boldsymbol{x})$. The standard SGD update with fixed step size γ is given by

$$\boldsymbol{x}_{k+1} \leftarrow \boldsymbol{x}_k - \gamma \nabla f_{i_k}(\boldsymbol{x}_k) \tag{4}$$

where each i_k is drawn uniformly from $[n]$. The idea behind weighted sampling is that, by drawing i from a weighted distribution $\mathcal{D}^{(p)} = \{p(1), p(2), \ldots, p(n)\}$ over $[n]$, the weighted sample $\frac{1}{p(i_k)}\nabla f_{i_k}(\boldsymbol{x}_k)$ is still an unbiased estimate of the gradient $\nabla F(\boldsymbol{x})$. This motivates the weighted SGD update

$$\boldsymbol{x}_{k+1} \leftarrow \boldsymbol{x}_k - \frac{\gamma}{np(i_k)}\nabla f_{i_k}(\boldsymbol{x}_k), \tag{5}$$

In [1], a family of distributions $\mathcal{D}^{(p)}$ whereby functions f_i with larger Lipschitz constants are more likely to be sampled was shown to lead to an improved convergence rate in SGD over uniform sampling. In terms of the distance $\|\boldsymbol{x}_k - \boldsymbol{x}_*\|_2^2$ of the

kth iterate to the unique minimum, starting from initial distance $\varepsilon_0 = \|\boldsymbol{x}_0 - \boldsymbol{x}_*\|_2^2$, Corollary 3.1 in [1] is as follows.

Proposition 1 *Assume the convexity and smoothness conditions are in force. For any desired* $\varepsilon > 0$*, and using a step size of*

$$\gamma = \frac{\mu\varepsilon}{4(\varepsilon\mu\frac{1}{n}\sum_{i=1}^{n} L_i + \sigma^2)},$$

we have that after

$$k = \left\lceil 4\log(2\varepsilon_0/\varepsilon)\left(\frac{\frac{1}{n}\sum_{i=1}^{n} L_i}{\mu} + \frac{\sigma^2}{\mu^2\varepsilon}\right)\right\rceil \tag{6}$$

iterations of weighted SGD (5) *with weights*

$$p(i) = \frac{1}{2n} + \frac{1}{2n}\cdot\frac{L_i}{\frac{1}{n}\sum_i L_i}, \tag{7}$$

the following holds in expectation with respect to the weighted distribution (7)*:* $\mathbb{E}^{(p)}\|\mathbf{x}_k - \mathbf{x}_*\|_2^2 \le \varepsilon$.

Remark 1 Note that one can obtain a guarantee with the same number of iterations as (6), same weights (7), and step sizes which depend on the index selected by cleverly re-writing the objective function as the sum of scaled functionals f_i each repeated an appropriate number of times. We state the version in Proposition 1 derived from [1] here for simplicity and convenience, and note that it improves upon classical results even in the uniformly bounded Lipschitz case.

Remark 2 This should be compared to the result for uniform sampling SGD [1]: using step size $\gamma = \frac{\mu\varepsilon}{4(\varepsilon\mu(\sup_i L_i)+\sigma^2)}$, one obtains the comparable error guarantee $\mathbb{E}\|\mathbf{x}_k - \mathbf{x}_*\|_2^2 \le \varepsilon$ after a number of iterations

$$k = \left\lceil 2\log(2\varepsilon_0/\varepsilon)\left(\frac{\sup_i L_i}{\mu} + \frac{\sigma^2}{\mu^2\varepsilon}\right)\right\rceil. \tag{8}$$

Since the average Lipschitz constant $\frac{1}{n}\sum_i L_i$ is always at most $\sup_i L_i$, and can be up to n times smaller than $\sup_i L_i$, SGD with weighted sampling requires twice the number of iterations of uniform SGD in the worst case, but can potentially converge much faster, specifically, in the regime where

$$\frac{\sigma^2}{\mu^2\varepsilon} \le \frac{\frac{1}{n}\sum_{i=1}^{n} L_i}{\mu} \ll \frac{\sup_i L_i}{\mu}.$$

3 Mini-batch SGD with Weighting: The Smooth Case

Here, we present a weighting and mini-batch scheme for SGD based on Proposition 1. For practical purposes, we assume that the functions $f_i(\boldsymbol{x})$ such that $F(\boldsymbol{x}) = \frac{1}{n}\sum_{i=1}^{n} f_i(\boldsymbol{x})$ are initially partitioned into fixed batches of size b and denote the partition by $\{\tau_1, \tau_2, \ldots \tau_d\}$ where $|\tau_i| = b$ for all $i < d$ and $d = \lceil n/b \rceil$ (for simplicity we will henceforth assume that $d = n/b$ is an integer). We will randomly select from this pre-determined partition of batches; however, our analysis extends easily to the case where a batch of size b is randomly selected each time from the entire set of functionals. With this notation, we may re-formulate the objective given in (1) as follows:

$$F(\boldsymbol{x}) = \frac{1}{d}\sum_{i=1}^{d} g_{\tau_i}(\boldsymbol{x}) = \mathbb{E} g_{\tau_i}(\boldsymbol{x}),$$

where now, we write $g_{\tau_i}(\boldsymbol{x}) = \frac{1}{b}\sum_{j\in\tau_i} f_j(\boldsymbol{x})$. We can apply Proposition 1 to the functionals g_{τ_i} and select batch τ_i with probability proportional the Lipschitz constant of ∇g_{τ_i} (or of g_{τ_i} in the non-smooth case, see Sect. 4). Note that

- The strong convexity parameter μ for the function F remains invariant to the batching rule.
- The residual error σ_τ^2 such that $\frac{1}{d}\sum_{i=1}^{d} \|\nabla g_{\tau_i}(\boldsymbol{x}_*)\|_2^2 \le \sigma_\tau^2$ can only **decrease** with increasing batch size, since for $b \ge 2$,

$$\sigma_\tau^2 = \frac{1}{d}\sum_{i=1}^{d} \left\| \frac{1}{b}\nabla\left(\sum_{k\in\tau_i} f_k(\boldsymbol{x}_*)\right)\right\|_2^2 \le \frac{1}{n}\sum_{i=1}^{n} \|\nabla f_i(\boldsymbol{x}_*)\|_2^2 \le \sigma^2.$$

  Note that for certain objective functions, this bound can be refined with a dependence on the block size b, in which case even further improvements can be gained by batching, see e.g., (19) and surrounding discussions.
- The average Lipschitz constant $\overline{L}_\tau = \frac{1}{d}\sum_{i=1}^{d} L_{\tau_i}$ of the gradients of the batched functions g_{τ_i} can only **decrease** with increasing batch size, since by the triangle inequality, $L_{\tau_i} \le \frac{1}{b}\sum_{k\in\tau_i} L_k$, and thus

$$\frac{1}{d}\sum_{i=1}^{d} L_{\tau_i} \le \frac{1}{n}\sum_{k=1}^{n} L_k = \overline{L}.$$

Incorporating these observations, applying Proposition 1 in the batched weighted setting implies that incorporating weighted sampling and mini-batching in SGD results in a convergence rate that equals or improves on the rate obtained using weights alone:

Theorem 1 *Assume that the convexity and smoothness conditions on* $F(\boldsymbol{x}) = \frac{1}{n}\sum_{i=1}^{n} f_i(\boldsymbol{x})$ *are in force. Consider the* $d = n/b$ *batches* $g_{\tau_i}(\boldsymbol{x}) = \frac{1}{b}\sum_{k\in\tau_i} f_k(\boldsymbol{x})$, *and the batched weighted SGD iteration*

$$\boldsymbol{x}_{k+1} \leftarrow \boldsymbol{x}_k - \frac{\gamma}{d \cdot p(\tau_{i_k})} \nabla g_{\tau_{i_k}}(\boldsymbol{x}_k)$$

where batch τ_i is selected at iteration k with probability

$$p(\tau_i) = \frac{1}{2d} + \frac{1}{2d} \cdot \frac{L_{\tau_i}}{\overline{L}_\tau}. \tag{9}$$

For any desired ε, and using a step size of

$$\gamma = \frac{\mu\varepsilon}{4(\varepsilon\mu\overline{L}_\tau + \sigma_\tau^2)},$$

we have that after a number of iterations

$$k = \left\lceil 4\log(2\varepsilon_0/\varepsilon)\left(\frac{\overline{L}_\tau}{\mu} + \frac{\sigma_\tau^2}{\mu^2\varepsilon}\right)\right\rceil \le \left\lceil 4\log(2\varepsilon_0/\varepsilon)\left(\frac{\overline{L}}{\mu} + \frac{\sigma^2}{\mu^2\varepsilon}\right)\right\rceil, \tag{10}$$

the following holds in expectation with respect to the weighted distribution (9)*: $\mathbb{E}^{(p)}\|\mathbf{x}_k - \mathbf{x}_*\|_2^2 \le \varepsilon$.*

Remark 3 The inequality in (10) implies that batching and weighting can only improve the convergence rate of SGD compared to weighting alone. As a reminder, this is under the assumption that the batches can be computed in parallel, so depending on the number of cores available, one needs to weigh the computational trade-off between iteration complexity and improved convergence rate. We investigate this trade-off as well as other computational issues in the following sections.

To completely justify the strategy of batching + weighting, we must also take into account the precomputation cost in computing the weighted distribution (9), which increases with the batch size b. In the next section, we refine Theorem 1 precisely this way in the case of the least squares objective, where we can quantify more precisely the gain achieved by weighting and batching. We give several explicit bounds and sampling strategies on the Lipschitz constants in this case that can be used for computationally efficient sampling.

3.1 Least Squares Objective

Although there are of course many methods for solving linear systems, methods like SGD for least squares problems have attracted recent attention due to their ability to utilize small memory footprints even for very large systems. In settings for example where the matrix is too large to be stored in memory, iterative approaches like the

Kaczmarz method (a variant of SGD) are necessary. With this motivation, we spend this section analyzing the least squares problem using weights and batching.

Consider the least squares objective

$$F(\boldsymbol{x}) = \frac{1}{2}\|\boldsymbol{A}\boldsymbol{x} - \boldsymbol{b}\|_2^2 = \frac{1}{n}\sum_{i=1}^{n} f_i(\boldsymbol{x}),$$

where $f_i(\boldsymbol{x}) = \frac{n}{2}(b_i - \langle \boldsymbol{a}_i, \boldsymbol{x}\rangle)^2$. We assume the matrix $\boldsymbol{A}$ has full column-rank, so that there is a unique minimizer $\boldsymbol{x}_*$ to the least squares problem:

$$\boldsymbol{x}_{LS} = \boldsymbol{x}_* = \arg\min_{\boldsymbol{x}} \|\boldsymbol{A}\boldsymbol{x} - \boldsymbol{b}\|_2^2.$$

Note that the convexity and smoothness conditions are satisfied for such functions. Indeed, observe that $\nabla f_i(\boldsymbol{x}) = n(\langle \boldsymbol{a}_i, \boldsymbol{x}\rangle - b_i)\boldsymbol{a}_i$, and

1. The individual Lipschitz constants are bounded by $L_i = n\|\boldsymbol{a}_i\|_2^2$, and the average Lipschitz constant by $\frac{1}{n}\sum_i L_i = \|\boldsymbol{A}\|_F^2$ (where $\|\cdot\|_F$ denotes the Frobenius norm),
2. The strong convexity parameter is $\mu = \sigma_{\min}^{-1}(\boldsymbol{A})$, the reciprocal of the smallest singular value of $\boldsymbol{A}$,
3. The residual is $\sigma^2 = n\sum_i \|\boldsymbol{a}_i\|_2^2 |\langle \boldsymbol{a}_i, \boldsymbol{x}_*\rangle - b_i|^2$.

In the batched setting, we compute

$$g_{\tau_i}(\boldsymbol{x}) = \frac{1}{b}\sum_{k\in\tau_i} f_k(\boldsymbol{x}) = \frac{n}{2b}\sum_{k\in\tau_i}(b_k - \langle \boldsymbol{a}_k, \boldsymbol{x}\rangle)^2 = \frac{d}{2}\|\boldsymbol{A}_{\tau_i}\boldsymbol{x} - \boldsymbol{b}_{\tau_i}\|_2^2, \tag{11}$$

where we have written $\boldsymbol{A}_{\tau_i}$ to denote the submatrix of $\boldsymbol{A}$ consisting of the rows indexed by τ_i.

Denote by σ_τ^2 the residual in the batched setting. Since $\nabla g_{\tau_i}(\boldsymbol{x}) = d\sum_{k\in\tau_i} (\langle \boldsymbol{a}_k, \boldsymbol{x}\rangle - b_k)\boldsymbol{a}_k$,

$$\begin{aligned}\sigma_\tau^2 &= \frac{1}{d}\sum_{i=1}^{d}\|\nabla g_{\tau_i}(\boldsymbol{x}_*)\|_2^2 = d\sum_{i=1}^{d}\|\sum_{k\in\tau_i}(\langle \boldsymbol{a}_k, \boldsymbol{x}_*\rangle - b_k)\boldsymbol{a}_k\|_2^2\\ &= d\sum_{i=1}^{d}\|\boldsymbol{A}_{\tau_i}^*(\boldsymbol{A}_{\tau_i}\boldsymbol{x}_* - \boldsymbol{b}_{\tau_i})\|_2^2 \le d\sum_{i=1}^{d}\|\boldsymbol{A}_{\tau_i}\|^2\|\boldsymbol{A}_{\tau_i}\boldsymbol{x}_* - \boldsymbol{b}_{\tau_i}\|_2^2,\end{aligned}$$

where we have written $\|\boldsymbol{B}\|$ to denote the spectral norm of the matrix $\boldsymbol{B}$, and $\boldsymbol{B}^*$ the adjoint of the matrix. Denote by L_{τ_i} the Lipschitz constant of ∇g_{τ_i}. Then, we also have

$$
\begin{aligned}
L_{\tau_i} &= \sup_{x,y} \frac{\|\nabla g_{\tau_i}(x) - \nabla g_{\tau_i}(y)\|_2}{\|x-y\|_2} \\
&= \frac{n}{b} \sup_{x,y} \frac{\|\sum_{k\in\tau_i} \left[(\langle a_k, x\rangle - b_k)a_k - (\langle a_k, y\rangle - b_k)a_k\right]\|_2}{\|x-y\|_2} \\
&= \frac{n}{b} \sup_{z} \frac{\|\sum_{k\in\tau_i} \langle a_k, z\rangle a_k\|_2}{\|z\|_2} \\
&= \frac{n}{b} \sup_{z} \frac{\|A^*_{\tau_i} A_{\tau_i} z\|_2}{\|z\|_2} \\
&= \frac{n}{b} \|A^*_{\tau_i} A_{\tau_i}\| \\
&= d\|A_{\tau_i}\|^2.
\end{aligned}
$$

We see thus that *if there exists a partition such that $\|A_{\tau_i}\|$ are as small as possible (e.g., within a constant factor of the row norms) for all τ_i in the partition, then both* σ_τ^2 and $L_\tau = \frac{1}{d}\sum_i L_{\tau_i}$ *are decreased by a factor of the batch size b compared to the unbatched setting.* These observations are summed up in the following corollary of Theorem 1 for the least squares case.

Corollary 1 *Consider* $F(x) = \frac{1}{2}\|Ax - b\|_2^2 = \frac{1}{2}\sum_{i=1}^d \|A_{\tau_i}x - b_{\tau_i}\|_2^2$. *Consider the batched weighted SGD iteration*

$$x_{k+1} \leftarrow x_k - \frac{\gamma}{p(\tau_i)} \sum_{j\in\tau_i} (\langle a_j, x_k\rangle - b_j)a_j, \tag{12}$$

with weights

$$p(\tau_i) = \frac{b}{2n} + \frac{1}{2} \cdot \frac{\|A_{\tau_i}\|^2}{\sum_{i=1}^d \|A_{\tau_i}\|^2}. \tag{13}$$

For any desired ε, and using a step size of

$$\gamma = \frac{\frac{1}{4}\varepsilon}{\varepsilon \sum_{i=1}^d \|A_{\tau_i}\|^2 + d\sigma_{\min}^{-2}(A)\sum_{i=1}^d \|A_{\tau_i}\|^2 \|A_{\tau_i}x_* - b_{\tau_i}\|_2^2}, \tag{14}$$

we have that after

$$k = \left\lceil 4\log(2\varepsilon_0/\varepsilon)\left(\sigma_{\min}^{-2}(A)\sum_{i=1}^d \|A_{\tau_i}\|^2 + \frac{d\sigma_{\min}^{-4}(A)\sum_{i=1}^d \|A_{\tau_i}\|^2\|A_{\tau_i}x_* - b_{\tau_i}\|_2^2}{\varepsilon}\right)\right\rceil \tag{15}$$

iterations of (12), $\mathbb{E}^{(p)}\|\mathbf{x}_k - \mathbf{x}_*\|_2^2 \le \varepsilon$ *where* $\mathbb{E}^{(p)}[\cdot]$ *means the expectation with respect to the index at each iteration drawn according to the weighted distribution* (13).

This corollary suggests a heuristic for batching and weighting in SGD for least squares problems, in order to optimize the convergence rate. Note of course that, like other similar results for SGD, it is only a heuristic since in particular to compute the step size and required number of iterations in (14) and (15), one needs an estimate on the size of the system residual $\|\boldsymbol{A}\boldsymbol{x}_\star - \boldsymbol{b}\|_2$ (which is of course zero in the consistent case). We summarize the desired procedure here:

1. Find a partition $\tau_1, \tau_2, \ldots, \tau_d$ that roughly minimizes $\sum_{i=1}^d \|\boldsymbol{A}_{\tau_i}\|^2$ among all such partitions.
2. Apply the weighted SGD algorithm (12) using weights given by (13).

We can compare the results of Corollary 1 to the results for weighted SGD when a single functional is selected in each iteration, where the number of iterations to achieve expected error ε is

$$k = \left\lceil 4\log(2\varepsilon_0/\varepsilon)\left(\sigma_{\min}^{-2}(\boldsymbol{A})\sum_{i=1}^n \|\boldsymbol{a}_i\|^2 + \frac{n\sigma_{\min}^{-4}(\boldsymbol{A})\sum_{i=1}^n \|\boldsymbol{a}_i\|^2\|\langle \boldsymbol{a}_i, \boldsymbol{x}_*\rangle - b_i\|_2^2}{\varepsilon}\right)\right\rceil. \tag{16}$$

That is, the ratio between the standard weighted number of iterations k_{stand} in (16) and the batched weighted number of iterations k_{batch} in (15) is

$$\frac{k_{stand}}{k_{batch}} = \frac{\varepsilon\sum_{i=1}^n \|\boldsymbol{a}_i\|^2 + n\sigma_{\min}^{-2}(\boldsymbol{A})\sum_{i=1}^n \|\boldsymbol{a}_i\|^2\|\langle \boldsymbol{a}_i, \boldsymbol{x}_*\rangle - b_i\|_2^2}{\varepsilon\sum_{i=1}^d \|\boldsymbol{A}_{\tau_i}\|^2 + d\sigma_{\min}^{-2}(\boldsymbol{A})\sum_{i=1}^d \|\boldsymbol{A}_{\tau_i}\|^2\|\boldsymbol{A}_{\tau_i}\boldsymbol{x}_* - \boldsymbol{b}_{\tau_i}\|_2^2}. \tag{17}$$

In case the least squares residual error is uniformly distributed over the n indices, that is, $\|\langle \boldsymbol{a}_i, \boldsymbol{x}_*\rangle - b_i\|_2^2 \approx \frac{1}{n}\|\boldsymbol{A}\boldsymbol{x}_* - \boldsymbol{b}\|^2$ for each $i \in [n]$, this factor reduces to

$$\frac{k_{stand}}{k_{batch}} = \frac{\|\boldsymbol{A}\|_F^2}{\sum_{i=1}^d \|\boldsymbol{A}_{\tau_i}\|^2}. \tag{18}$$

It follows thus that the combination of batching and weighting in this setting always reduces the iteration complexity compared to weighting alone and can result in up to a factor of b speedup:

$$1 \le \frac{k_{stand}}{k_{batch}} \le b.$$

In the remainder of this section, we consider several families of matrices where the maximal speedup is achieved, $\frac{k_{stand}}{k_{batch}} \approx b$. We also take into account the computational cost of computing the norms $\|\boldsymbol{A}_{\tau_i}\|^2$ which determine the weighted sampling strategy.

Orthonormal systems It is clear that the advantage of mini-batching is the strongest when the rows of $\boldsymbol{A}$ in each batch are orthonormal. In the extreme case where $\boldsymbol{A}$ has orthonormal rows, we have

$$\overline{L}_\tau = \sum_{i=1}^{d} \|A_{\tau_i}^* A_{\tau_i}\| = \frac{n}{b} = \frac{1}{b}\overline{L}. \tag{19}$$

Thus for orthonormal systems, we gain a factor of b by using mini-batches of size b. However, there is little advantage to weighting in this case as all Lipschitz constants are the same.

Incoherent systems More generally, the advantage of mini-batching is strong when the rows a_i within any particular batch are *nearly* orthogonal. Suppose that each of the batches is well-conditioned in the sense that

$$\sum_{i=1}^{n} \|a_i\|_2^2 \geq C'n, \qquad \|A_{\tau_i}^* A_{\tau_i}\| = \|A_{\tau_i} A_{\tau_i}^*\| \leq C, \qquad i = 1, \ldots, d, \tag{20}$$

For example, if A^* has the *restricted isometry property* [30] of level δ at sparsity level b, (20) holds with $C \leq 1 + \delta$. Alternatively, if A has unit-norm rows and is incoherent, i.e., $\max_{i \neq j} |\langle a_i, a_j \rangle| \leq \frac{\alpha}{b-1}$, then (20) holds with constant $C \leq 1 + \alpha$ by Gershgorin circle theorem.
If the incoherence condition (20) holds, we gain a factor of b by using weighted mini-batches of size b:

$$\overline{L}_\tau = \sum_{i=1}^{d} \|A_{\tau_i}^* A_{\tau_i}\| \leq C\frac{n}{b} \leq \frac{C}{C'}\frac{\overline{L}}{b}.$$

Incoherent systems, variable row norms More generally, consider the case where the rows of A are nearly orthogonal to each other, but not normalized as in (20). We can then write $A = D\Psi$, where D is an $n \times n$ diagonal matrix with entry $d_{ii} = \|a_i\|_2$, and Ψ with normalized rows satisfies

$$\|\Psi_{\tau_i}^* \Psi_{\tau_i}\| = \|\Psi_{\tau_i} \Psi_{\tau_i}^*\| \leq C, \qquad i = 1, \ldots, d,$$

as is the case if, e.g., Ψ has the restricted isometry property or Ψ is incoherent.

In this case, we have

$$\begin{aligned} \|A_{\tau_i}^* A_{\tau_i}\| = \|A_{\tau_i} A_{\tau_i}^*\| &= \|D_{\tau_i} \Psi_{\tau_i} \Psi_{\tau_i}^* D_{\tau_i}\| \\ &\leq \max_{k \in \tau_i} \|a_k\|_2^2 \|\Psi_{\tau_i} \Psi_{\tau_i}^*\| \\ &\leq C \max_{k \in \tau_i} \|a_k\|_2^2, \qquad i = 1, \ldots, d. \end{aligned} \tag{21}$$

Thus,

$$\overline{L}_\tau = \sum_{i=1}^{d} \|A_{\tau_i}^* A_{\tau_i}\| \leq C \sum_{i=1}^{d} \max_{k \in \tau_i} \|a_k\|_2^2. \tag{22}$$

In order to minimize the expression on the right-hand side over all partitions into blocks of size b, we partition the rows of A according to the order of the decreasing rearrangement of their row norms. This batching strategy results in a factor of b gain in iteration complexity compared to weighting without batching:

$$\begin{aligned}\overline{L}_\tau &\leq C \sum_{i=1}^{d} \|a_{((i-1)b+1)}\|_2^2 \\ &\leq \frac{C}{b-1} \sum_{i=1}^{n} \|a_i\|_2^2 \\ &\leq \frac{C'}{b} \overline{L}.\end{aligned} \tag{23}$$

We now turn to the practicality of computing the distribution given by the constants L_{τ_i}. We propose several options to efficiently compute these values given the ability to parallelize over b cores.

Max-norm The discussion above suggests the use of the maximum row norm of a batch as a proxy for the Lipschitz constant. Indeed, (21) shows that the row norms give an upper bound on these constants. Then, (23) shows that up to a constant factor, such a proxy still has the potential to lead to an increase in the convergence rate by a factor of b. Of course, computing the maximum row norm of each batch costs on the order of mn flops (the same as the non-batched weighted SGD case).

Power method In some cases, we may utilize the power method to approximate $\|A_{\tau_i}^* A_{\tau_i}\|$ efficiently. Suppose that for each batch, we can approximate this quantity by $\hat{Q}_{\tau_i}$. Classical results on the power method allow one to approximate the norm to within an arbitrary additive error, with a number of iterations that depends on the spectral gap of the matrix. An alternative approach, that we consider here, can be used to obtain approximations leading to a *multiplicative* factor difference in the convergence rate, without dependence on the eigenvalue gaps λ_1/λ_2 within batches. For example, [31, Lemma 5] shows that with high probability with respect to a randomized initial direction to the power method, after $T \geq \varepsilon^{-1} \log(\varepsilon^{-1} b)$ iterations of the power method, one can guarantee that

$$\|A_{\tau_i}^* A_{\tau_i}\| \geq \hat{Q}_{\tau_i} \geq \frac{\|A_{\tau_i}^* A_{\tau_i}\|}{1+\varepsilon}.$$

At b^2 computations per iteration of the power method, the total computational cost (to compute all quantities in the partition), shared over all b cores, is $b\varepsilon^{-1} \log(\varepsilon^{-1} \log(b))$. This is actually potentially much *lower* than the cost to compute all row norms $L_i = \|\boldsymbol{a}_i\|_2^2$ as in the standard non-batched weighted method. In this case, the power method yields

$$\overline{L}_\tau \geq \frac{b}{n}\sum_{i=1}^{d}\frac{n}{b}\hat{Q}_{\tau_i} \geq \frac{\overline{L}_\tau}{1+\varepsilon},$$

for a constant ε.

4 Mini-batch SGD with Weighting: The Non-smooth Case

We next present analogous results to the previous section for objectives which are strongly convex but lack the smoothness assumption. Like the least squares objective in the previous section, our motivating example here will be the support vector machine (SVM) with hinge loss objective.

A classical result (see e.g., [32–34]) for SGD establishes a convergence bound of SGD with non-smooth objectives. In this case, rather than taking a step in the gradient direction of a functional, we move in a direction of a subgradient. Instead of utilizing the Lipschitz constants of the gradient terms, we utilize the Lipschitz constants of the actual functionals themselves. Note that in the non-smooth case, one cannot guarantee convergence of the iterates $\boldsymbol{x}_k$ to a unique minimizer $\boldsymbol{x}_\star$ so instead one seeks convergence of the objective value itself. Concretely, a classical bound is of the following form.

Proposition 2 *Let the objective $F(\boldsymbol{x}) = \mathbb{E}g_i(\boldsymbol{x})$ with minimizer $\boldsymbol{x}_\star$ be a μ-strongly convex (possibly non-smooth) objective. Run SGD using a subgradient h_i of a randomly selected functional g_i at each iteration. Assume that $\mathbb{E}h_i \in \partial F(\boldsymbol{x}_k)$ (expectation over the selection of subgradient h_i) and that*

$$\max_{\boldsymbol{x},\boldsymbol{y}} \frac{\|g_i(\boldsymbol{x}) - g_i(\boldsymbol{y})\|}{\|\boldsymbol{x}-\boldsymbol{y}\|} \leq \max_{\boldsymbol{x}} \|h_i(\boldsymbol{x})\| \leq G_i.$$

Set $\overline{G^2} = \mathbb{E}(G_i^2)$. Using step size $\gamma = \gamma_k = 1/(\mu k)$, we have

$$\mathbb{E}\left[F(\boldsymbol{x}_k) - F(\boldsymbol{x}_\star)\right] \leq \frac{C\overline{G^2}(1+\log k)}{\mu k}, \tag{24}$$

where C is an absolute constant.

Such a result can be improved by utilizing averaging of the iterations; for example, if $\boldsymbol{x}_k^\alpha$ denotes the average of the last αk iterates, then the convergence rate bound (24) can be improved to:

$$\mathbb{E}\left[F(\boldsymbol{x}_k) - F(\boldsymbol{x}_\star)\right] \leq \frac{C\overline{G^2}\left(1 + \log \frac{1}{\min(\alpha,(1+1/k)-\alpha)}\right)}{\mu k} \leq \frac{C\overline{G^2}\left(1 + \log \frac{1}{\min(\alpha,1-\alpha)}\right)}{\mu k}.$$

Setting $m_\alpha = 1 + \log \frac{1}{\min(\alpha,1-\alpha)}$, we see that to obtain an accuracy of $\mathbb{E}\left[F(\boldsymbol{x}_k) - F(\boldsymbol{x}_\star)\right] \leq \varepsilon$, it suffices that

$$k \geq \frac{C\overline{G^2} m_\alpha}{\mu \varepsilon}.$$

In either case, it is important to notice the dependence on $\overline{G^2} = \mathbb{E}(G_i^2)$. By using weighted sampling with weights $p(i) = G_i / \sum_i G_i$, we can improve this dependence to one on $(\overline{G})^2$, where $\overline{G} = \mathbb{E}G_i$ [1, 2]. Since $\overline{G^2} - (\overline{G})^2 = \mathrm{Var}(G_i)$, this improvement reduces the dependence by an amount equal to the variance of the Lipschitz constants G_i. Like in the smooth case, we now consider not only weighting the distribution, but also by batching the functionals g_i. This yields the following result, which we analyze for the specific instance of SVM with hinge loss below.

Theorem 2 *Instate the assumptions and notation of Proposition 2. Consider the $d = n/b$ batches $g_{\tau_i}(\boldsymbol{x}) = \frac{1}{b}\sum_{j\in\tau_i} g_j(\boldsymbol{x})$, and assume each batch g_{τ_i} has Lipschitz constant G_{τ_i}. Write $\overline{G}_\tau = \mathbb{E}G_{\tau_i}$. Run the weighted batched SGD method with averaging as described above, with step size $\gamma/p(\tau_i)$. For any desired ε, it holds that after*

$$k = \frac{C(\overline{G}_\tau)^2 m_\alpha}{\mu\varepsilon}$$

iterations with weights

$$p(\tau_i) = \frac{G_{\tau_i}}{\sum_j G_{\tau_j}}, \tag{25}$$

we have $\mathbb{E}^{(p)}\left[F(\boldsymbol{x}_k) - F(\boldsymbol{x}_\star)\right] \leq \varepsilon$ where $\mathbb{E}^{(p)}[\cdot]$ means the expectation with respect to the index at each iteration drawn according to the weighted distribution (25).

Proof Applying weighted SGD with weights $p(\tau_i)$, we re-write the objective $F(\boldsymbol{x}) = \mathbb{E}\left(g_i(\boldsymbol{x})\right)$ as $F(\boldsymbol{x}) = \mathbb{E}^{(p)}\left(\hat{g}_{\tau_i}(\boldsymbol{x})\right)$, where

$$\hat{g}_{\tau_i}(x) = \left(\frac{1}{n}\sum_j G_{\tau_j}\right)\left(\frac{1}{G_{\tau_i}}\sum_{j\in\tau_i} g_j(\boldsymbol{x})\right) = \left(\frac{b}{n}\sum_j G_{\tau_j}\right)\left(\frac{g_{\tau_i}(\boldsymbol{x})}{G_{\tau_i}}\right).$$

Then, the Lipschitz constant $\hat{G}_i$ of $\hat{g}_{\tau_i}$ is bounded above by $\hat{G}_i = \frac{b}{n}\sum_j G_{\tau_j}$, and so

$$\mathbb{E}^{(p)}\hat{G}_i^2 = \sum_i \frac{G_{\tau_i}}{\sum_j G_{\tau_j}} \left(\frac{b}{n}\sum_j G_{\tau_j}\right)^2 = \left(\frac{b}{n}\sum_j G_{\tau_j}\right)^2 = (\mathbb{E}G_{\tau_i})^2 = (\overline{G}_\tau)^2.$$

The result follows from an application of Proposition 2. □

We now formalize these bounds and weights for the SVM with hinge loss objective. Other objectives such as L1 regression could also be adapted in a similar fashion, e.g., utilizing an approach as in [35].

4.1 SVM with Hinge Loss

We now consider the SVM with hinge loss problem as a motivating example for using batched weighted SGD for non-smooth objectives. Recall the SVM with hinge loss objective is

$$F(\boldsymbol{x}) := \frac{1}{n}\sum_{i=1}^{n}[y_i\langle \boldsymbol{x}, \boldsymbol{a}_i\rangle]_+ + \frac{\lambda}{2}\|\boldsymbol{x}\|_2^2 = \mathbb{E}g_i(\boldsymbol{x}), \tag{26}$$

where $y_i \in \{\pm 1\}$, $[u]_+ = \max(0, u)$, and

$$g_i(\boldsymbol{x}) = [y_i\langle \boldsymbol{x}, \boldsymbol{a}_i\rangle]_+ + \frac{\lambda}{2}\|\boldsymbol{x}\|_2^2.$$

This is a key example where the components are (λ-strongly) convex but no longer smooth. Still, each g_i has a well-defined subgradient:

$$\nabla g_i(\boldsymbol{x}) = \chi_i(\boldsymbol{x}) y_i \boldsymbol{a}_i + \lambda \boldsymbol{x},$$

where $\chi_i(\boldsymbol{x}) = 1$ if $y_i\langle \boldsymbol{x}, \boldsymbol{a}_i\rangle < 1$ and 0 otherwise. It follows that g_i is Lipschitz and its Lipschitz constant is bounded by

$$G_i := \max_{\boldsymbol{x},\boldsymbol{y}} \frac{\|g_i(\boldsymbol{x}) - g_i(\boldsymbol{y})\|}{\|\boldsymbol{x} - \boldsymbol{y}\|} \leq \max_{\boldsymbol{x}} \|\nabla g_i(\boldsymbol{x})\| \leq \|\boldsymbol{a}_i\|_2 + \lambda.$$

As shown in [1, 2], in the setting of non-smooth objectives of the form (26), where the components are not necessarily smooth, but each g_i is G_i-Lipschitz, the performance of SGD depends on the quantity $\overline{G^2} = \mathbb{E}[G_i^2]$. In particular, the iteration complexity depends linearly on $\overline{G^2}$.

For the hinge loss example, we have calculated that

$$\overline{G^2} = \frac{1}{n}\sum_{i=1}^{n}(\|\boldsymbol{a}_i\|_2 + \lambda)^2 \leq 2\lambda^2 + \frac{2}{n}\sum_{i=1}^{n}\|\boldsymbol{a}_i\|_2^2.$$

Incorporating (non-batch) weighting to this setting, as discussed in [1], reduces the iteration complexity to depend linearly on $(\overline{G})^2 = (\mathbb{E}[G_i])^2$, which is at most $\overline{G^2}$ and can be as small as $\frac{1}{n}\overline{G^2}$. For the hinge loss example, we have

$$(\overline{G})^2 = \left(\lambda + \frac{1}{n}\sum_{i=1}^{n}\|\boldsymbol{a}_i\|_2\right)^2.$$

We note here that one can incorporate the dependence on the regularizer term $\frac{\lambda}{2}\|\boldsymbol{x}\|_2^2$ in a more optimal way by bounding the functional norm only over the iterates themselves, as in [6, 34]; however, we choose a crude upper bound on the Lipschitz constant here in order to maintain a dependence on the *average* constant rather than the *maximum*, and only sacrifice a constant factor.

4.1.1 Batched Sampling

The paper [6] considered batched SGD for the hinge loss objective. For batches τ_i of size b, let $g_{\tau_i} = \frac{\lambda}{2}\|\boldsymbol{x}\|_2^2 + \frac{1}{b}\sum_{k\in\tau_i}[y_k\langle\boldsymbol{x}, \boldsymbol{a}_k\rangle]_+$ and observe

$$F(\boldsymbol{x}) := \frac{1}{n}\sum_{i=1}^{n}[y_i\langle\boldsymbol{x}, \boldsymbol{a}_i\rangle]_+ + \frac{\lambda}{2}\|\boldsymbol{x}\|_2^2 = \mathbb{E}g_{\tau_i}(\boldsymbol{x}).$$

We now bound the Lipschitz constant G_τ for a batch. Let $\chi = \chi_k(\boldsymbol{x})$ and $\boldsymbol{A}_\tau$ have rows $y_k\boldsymbol{a}_k$ for $k \in \tau$. We have

$$\begin{aligned}
\max_{\boldsymbol{x}}\left\|\frac{1}{b}\sum_{k\in\tau_i}\chi_k(\boldsymbol{x})y_k\boldsymbol{a}_k\right\|_2 &= \max_{\boldsymbol{x}}\sqrt{\left\langle\frac{1}{b}\sum_{k\in\tau_i}\chi_k(\boldsymbol{x})y_k\boldsymbol{a}_k, \frac{1}{b}\sum_{k\in\tau_i}\chi_k(\boldsymbol{x})y_k\boldsymbol{a}_k\right\rangle} \\
&= \frac{1}{b}\max_{\boldsymbol{x}}\sqrt{\chi^*\boldsymbol{A}_\tau\boldsymbol{A}_\tau^*\chi} \\
&\leq \frac{1}{b}\sqrt{b\|\boldsymbol{A}_\tau\boldsymbol{A}_\tau^*\|} \\
&= \frac{1}{\sqrt{b}}\|\boldsymbol{A}_\tau\|,
\end{aligned} \tag{27}$$

and therefore $G_\tau \leq \frac{1}{\sqrt{b}}\|\boldsymbol{A}_\tau\| + \lambda$. Thus, for batched SGD without weights, the iteration complexity depends linearly on

$$\begin{aligned}\overline{G_\tau^2} &= \frac{b}{n}\sum_{i=1}^{d} G_{\tau_i}^2\\ &\le 2\lambda^2 + \frac{2}{n}\sum_{i=1}^{d}\|A_{\tau_i}\|^2\\ &= 2\lambda^2 + \frac{2}{n}\sum_{i=1}^{d}\|A_{\tau_i}^* A_{\tau_i}\|.\end{aligned}$$

Even without weighting, we already see potential for drastic improvements, as noted in [6]. For example, in the orthonormal case, where $\|A_{\tau_i}^* A_{\tau_i}\| = 1$ for each τ_i, we see that with appropriately chosen λ, $\overline{G_\tau^2}$ is on the order of $\frac{1}{b}$, which is a factor of b times smaller than $\overline{G^2} \approx 1$. Similar factors are gained for the incoherent case as well, as in the smooth setting discussed above. Of course, we expect even more gains by utilizing both batching and weighting.

4.1.2 Weighted Batched Sampling

Incorporating weighted batched sampling, where we sample batch τ_i with probability proportional to G_{τ_i}, the iteration complexity is reduced to a linear dependence on $(\overline{G_\tau})^2$, as in Theorem 2. For hinge loss, we calculate

$$(\overline{G_\tau})^2 = \left(\frac{b}{n}\sum_{i=1}^{d} G_{\tau_i}\right)^2 \le \left(\frac{b}{n}\sum_{i=1}^{d}\frac{1}{\sqrt{b}}\|A_{\tau_i}\| + \lambda\right)^2 = \left(\lambda + \frac{\sqrt{b}}{n}\sum_{i=1}^{d}\|A_{\tau_i}\|\right)^2.$$

We thus have the following guarantee for the hinge loss objective.

Corollary 2 *Instate the notation of Theorem 2. Consider* $F(x) = \frac{1}{n}\sum_{i=1}^{n}[y_i\langle x, a_i\rangle]_+ + \frac{\lambda}{2}\|x\|_2^2$. *Consider the batched weighted SGD iteration*

$$x_{k+1} \leftarrow x_k - \frac{1}{\mu k p(\tau_i)}\left(\lambda x_k + \frac{1}{b}\sum_{j\in\tau_i}\chi_j(x_k) y_j a_j\right), \tag{28}$$

where $\chi_j(x) = 1$ *if* $y_j\langle x, a_j\rangle < 1$ *and* 0 *otherwise. Let* A_τ *have rows* $y_j a_j$ *for* $j \in \tau$. *For any desired* ε, *we have that after*

$$k = \frac{C m_\alpha \left(\lambda + \frac{\sqrt{b}}{n}\sum_{i=1}^{d}\|A_{\tau_i}\|\right)^2}{\lambda\varepsilon} \tag{29}$$

iterations of (28) *with weights*

$$p(\tau_i) = \frac{\|\boldsymbol{A}_{\tau_i}\| + \lambda\sqrt{b}}{\frac{n}{\sqrt{b}}\lambda + \sum_j \|\boldsymbol{A}_{\tau_j}\|}, \tag{30}$$

it holds that $\mathbb{E}^{(p)}[F(\mathbf{x}_k) - F(\mathbf{x}_*)] \leq \varepsilon$.

5 Experiments

In this section, we present some simple experimental examples that illustrate the potential of utilizing weighted mini-batching. We consider several test cases as illustration.

Gaussian linear systems The first case solves a linear system $\boldsymbol{Ax} = \boldsymbol{b}$, where $\boldsymbol{A}$ is a matrix with i.i.d. standard normal entries (as is $\boldsymbol{x}$, and $\boldsymbol{b}$ is their product). In this case, we expect the Lipschitz constants of each block to be comparable, so the effect of weighting should be modest. However, the effect of mini-batching in parallel of course still appears. Indeed, Fig. 1 (top) displays the convergence rates in terms of iterations for various batch sizes, where each batch is selected with probability as in (13). When batch updates can be run in parallel, we expect the convergence behavior to mimic this plot (which displays iterations). We see that in this case, larger batches yield faster convergence. In these simulations, the step size γ was set as in (14) (approximations for Lipschitz constants also apply to the step size computation) for the weighted cases and set to the optimal step size as in [1, Corollary 3.2] for the uniform cases. Behavior using uniform selection is very similar (not shown), as expected in this case since the Lipschitz constants are roughly constant. Figure 1 (bottom) highlights the improvements in our proposed weighted batched SGD method versus the classical, single functional and unweighted, SGD method. The power method refers to the method discussed at the end of Sect. 3, and max-norm method refers to the approximation using the maximum row norm in a batch, as in (21). The notation "(opt)" signifies that the optimal step size was used, rather than the approximation; otherwise, in all cases, both the sampling probabilities (13) and step sizes (14) were approximated using the approximation scheme given. Not surprisingly, using large batch sizes yields significant speedup.

Gaussian linear systems with variation We next test systems that have more variation in the distribution of Lipschitz constants. We construct a matrix $\boldsymbol{A}$ of the same size as above, but whose entries in the kth row are i.i.d. normally distributed with mean zero and variance k^2. We now expect a large effect both from batching and from weighting. In our first experiment, we select the fixed batches randomly at the onset and compute the probabilities according to the Lipschitz constants of those randomly selected batches, as in (13). The results are displayed in the top plot of Fig. 2. In the second experiment, we batch sequentially, so that rows with similar Lipschitz constants (row norms) appear in the same batch, and again utilized the weighted sampling. The results are displayed in the center plot

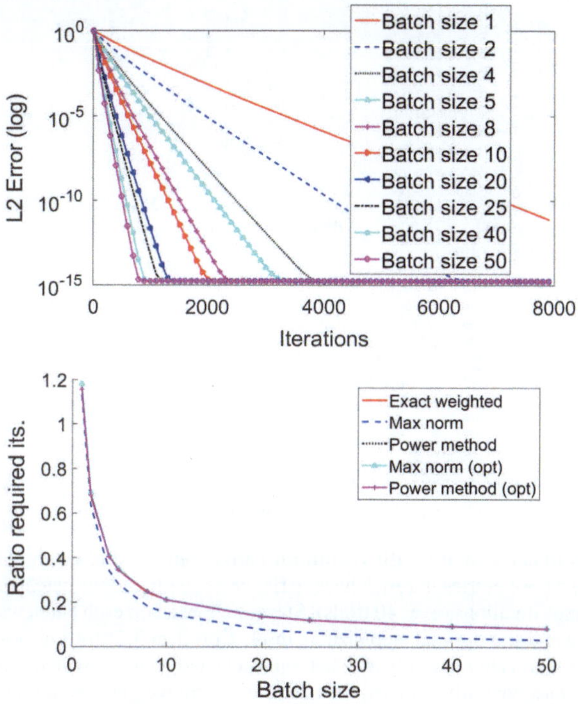

Fig. 1 **(Gaussian linear systems: convergence)** Mini-batch SGD on a Gaussian 1000×50 system with various batch sizes; batches created randomly at onset. Graphs show mean L2-error versus iterations (over 40 trials). Step size γ used on each batch was as given in (14) for the weighted cases and as in [1, Corollary 3.2] for the uniform comparisons, where in all cases corresponding approximations were used to compute the spectral norms. *Top* Batches are selected using proposed weighted selection strategy (13). *Bottom* Ratio of the number of iterations required to reach an error of 10^{-5} for weighted batched SGD versus classical (single functional) uniform (unweighted) SGD. The notation "(opt)" signifies that the optimal step size was used, rather than the approximation

of Fig. 2. Finally, the bottom plot of Fig. 2 shows convergence when batching sequentially and then employing uniform (unweighted) sampling. As our theoretical results predict, batching sequentially yields better convergence, as does utilizing weighted sampling.

Since this type of system nicely highlights the effects of both weighting and batching, we performed additional experiments using this type of system. Figure 3 highlights the improvements gained by using weighting. In the top plot, we see that for all batch sizes improvements are obtained by using weighting, even more so than in the standard normal case, as expected (note that we cut the curves off when the weighted approach reaches machine precision). In the bottom plot, we see that the number of iterations to reach a desired threshold is also less using the various weighting schemes; we compare the sampling method using exact

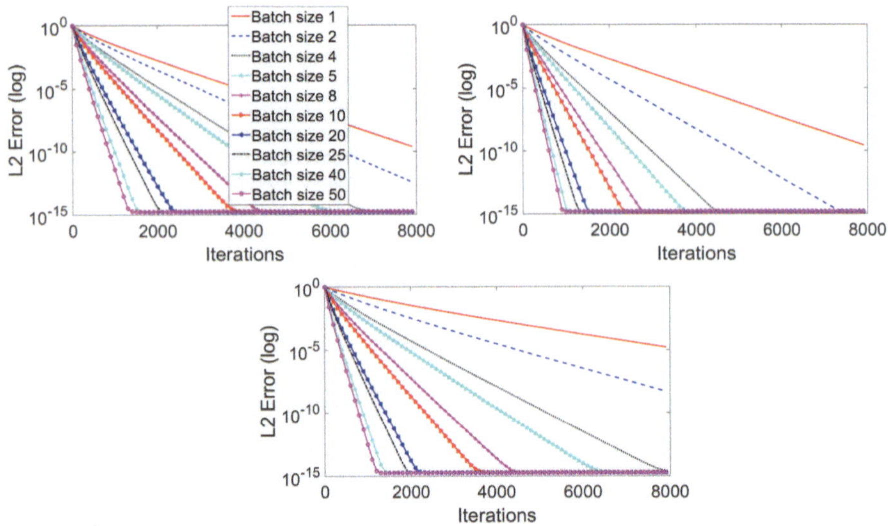

Fig. 2 (Gaussian linear systems with variation: convergence) Mini-batch SGD on a Gaussian 1000×50 system whose entries in row k have variance k^2, with various batch sizes. Graphs show mean L2-error versus iterations (over 40 trials). Step size γ used on each batch was as given in (14) for weighted SGD and the optimal step size as in [1, Corollary 3.2] for uniform sampling SGD. *Top Left* Batches are created randomly at onset, then selected using weighted sampling. *Top Right* Batches are created sequentially at onset, then selected using weighted sampling. *Bottom* Batches are created sequentially at onset, then selected using uniform (unweighted) sampling

computations of the Lipshitz constants (spectral norms), using the maximum row norm as an approximation as in (21), and using the power method (using number of iterations equal to $\varepsilon^{-1}\log(\varepsilon^{-1}b)$ with $\varepsilon = 0.01$). Step size γ used on each batch was again set as in (14) (approximations for Lipschitz constants also apply to the step size computation) for the weighted cases and as in [1, Corollary 3.2] for the uniform cases. For cases when the exact step size computation was used rather than the corresponding approximation, we write "(opt)". For example, the marker "Max-norm (opt)" represents the case when we use the maximum row norm in the batch to approximate the Lipschitz constant, but still use the exact spectral norm when computing the optimal step size. This of course is not practical, but we include these for demonstration. Figure 4 highlights the effect of using batching. The top plot confirms that larger batch sizes yield significant improvement in terms of L2-error and convergence (note that again all curves eventually converge to a straight line due to the error reaching machine precision). The bottom plot highlights the improvements in our proposed weighted batched SGD methods versus the classical, single functional and unweighted, SGD method.

We next further investigate the effect of using the power method to approximate the Lipschitz constants used for the probability of selecting a given batch. We again create the batches sequentially and fix them throughout the remainder of the method. At the onset of the method, after creating the batches, we run the

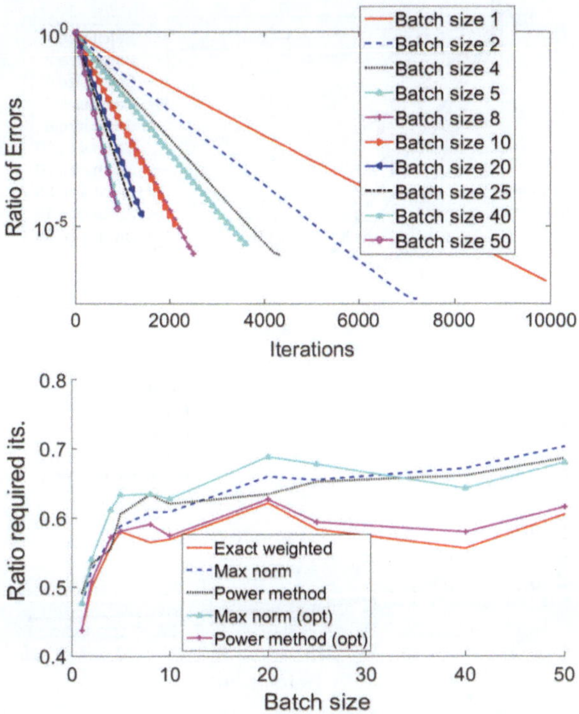

Fig. 3 (Gaussian linear systems with variation: effect of weighting) Mini-batch SGD on a Gaussian 1000×50 system whose entries in row k have variance k^2, with various batch sizes; batches created sequentially at onset. Step size γ used on each batch was set as in (14) (approximations for Lipschitz constants also apply to the step size computation) for the weighted cases and as in [1, Corollary 3.2] for the uniform cases. *Top* Ratio of mean L2-error using weighted versus unweighted random batch selection (improvements appear when plot is less than one). *Bottom* Ratio of the number of iterations required to reach an error of 10^{-5} for various weighted selections versus unweighted random selection. The notation "(opt)" signifies that the optimal step size was used, rather than the approximation

power method using $\varepsilon^{-1}\log(\varepsilon^{-1}b)$ iterations (with $\varepsilon = 0.01$) per batch, where we assume the work can evenly be divided among the b cores. We then determine the number of computational flops required to reach a specified solution accuracy using various batch sizes b. The results are displayed in Fig. 5. The top plot shows the convergence of the method; comparing with the top plot of Fig. 2, we see that the convergence is slightly slower than when using the precise Lipschitz constants, as expected. The bottom plot of Fig. 5 shows the number of computational flops required to achieve a specified accuracy, as a function of the batch size. We see that there appears to be an "optimal" batch size, around $b = 8$ for this case, at which the savings in computational time computing the Lipschitz constants and the additional iterations required due to the inaccuracy are balanced.

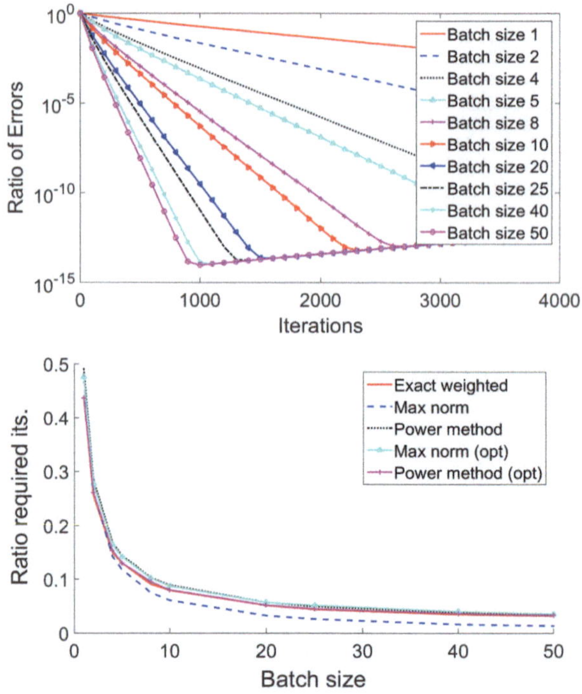

Fig. 4 **(Gaussian linear systems with variation: effect of batching)** Mini-batch SGD on a Gaussian 1000×50 system whose entries in row k have variance k^2, with various batch sizes; batches created sequentially at onset. Step size γ used on each batch was set as in (14) (approximations for Lipschitz constants also apply to the step size computation) for the weighted cases and as in [1, Corollary 3.2] for the uniform cases. *Top* Ratio of mean L2-error using weighted batched SGD versus classical (single functional) weighted SGD (improvements appear when plot is less than one). *Bottom* Ratio of the number of iterations required to reach an error of 10^{-5} for various weighted selections with batched SGD versus classical (single functional) uniform (unweighted) SGD. The notation "(opt)" signifies that the optimal step size was used, rather than the approximation

Correlated linear systems We next tested the method on systems with correlated rows, using a matrix with i.i.d. entries uniformly distributed on [0, 1]. When the rows are correlated in this way, the matrix is poorly conditioned and thus convergence speed suffers. Here, we are particularly interested in the behavior when the rows also have high variance; in this case, row k has uniformly distributed entries on $[0, \sqrt{3}k]$ so that each entry has variance k^2 like the Gaussian case above. Figure 6 displays the convergence results when creating the batches randomly and using weighting (top), creating the batches sequentially and using weighting (center), and creating the batches sequentially and using unweighted sampling (bottom). Like Fig. 2, we again see that batching the rows with larger row norms together and then using weighted sampling produces a speedup in convergence.

Orthonormal systems As mentioned above, we expect the most notable improvement in the case when $\boldsymbol{A}$ is an orthonormal matrix. For this case, we run the

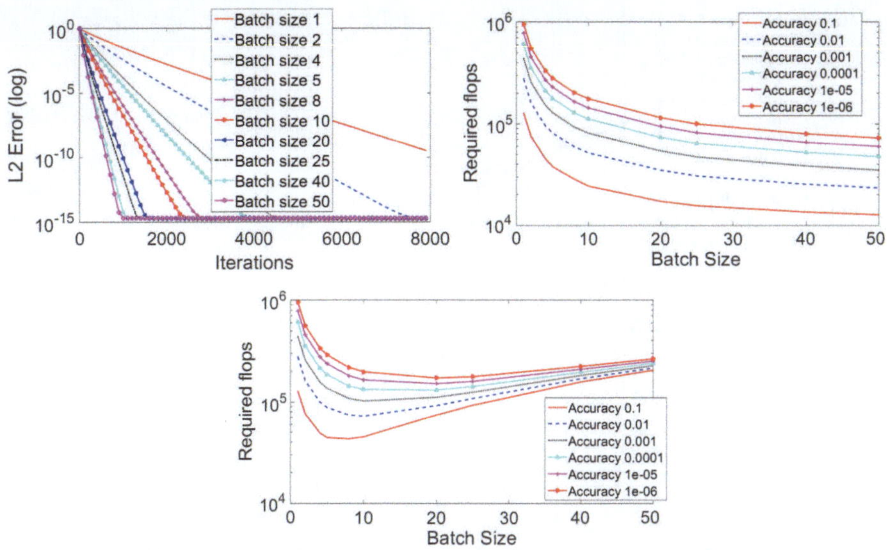

Fig. 5 **(Gaussian linear systems with variation: using power method)** Mini-batch SGD on a Gaussian 1000×50 system whose entries in row k have variance k^2, with various batch sizes; batches created sequentially at onset. Step size γ used on each batch was set as in (14) (approximations for Lipschitz constants also apply to the step size computation) for the weighted cases and as in [1, Corollary 3.2] for the uniform cases. Lipschitz constants for batches are approximated by using $\varepsilon^{-1} \log(\varepsilon^{-1} b)$ (with $\varepsilon = 0.01$) iterations of the power method. *Top Left* Convergence of the batched method. *Next* Required number of computational flops to achieve a specified accuracy as a function of batch size when computation is shared over b cores (*top right*) or done on a single node (*bottom*)

method on a 200×200 orthonormal discrete Fourier transform (DFT) matrix. As seen in the top plot of Fig. 7, we do indeed see significant improvements in convergence with batches in our weighted scheme. Of course, if the matrix is orthonormal, one could also simply apply $\boldsymbol{A}^*$ to solve the system, but we include these experiments for intuition and comparison.

Sparse systems Lastly, we show convergence for the batched weighted scheme on sparse Gaussian systems. The matrix is generated to have 20% nonzero entries, and each nonzero entry is i.i.d. standard normal. Figure 7 (center) shows the convergence results. The convergence behavior is similar to the non-sparse case, as expected, since our method does not utilize any sparse structure.

Tomography data The final system we consider is a real system from tomography. The system was generated using the MATLAB regularization toolbox by P.C. Hansen (http://www.imm.dtu.dk/~pcha/Regutools/) [36]. This creates a 2D tomography problem $\boldsymbol{Ax} = \boldsymbol{b}$ for an $n \times d$ matrix with $n = fN^2$ and $d = N^2$, where $\boldsymbol{A}$ corresponds to the absorption along a random line through an $N \times N$ grid. We set $N = 20$ and the oversampling factor $f = 3$. Figure 7 (bottom) shows the convergence results.

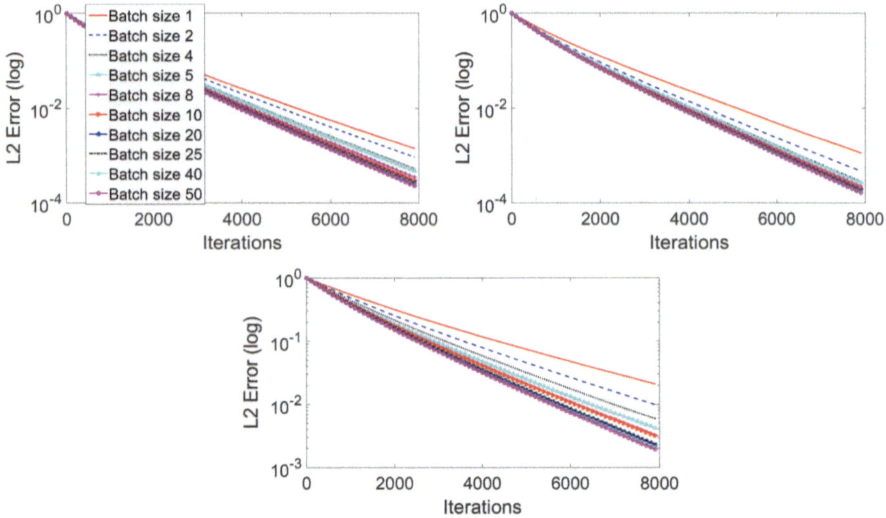

Fig. 6 **(Correlated systems with variation: convergence)** Mini-batch SGD on a uniform 1000×50 system whose entries in row k have variance k^2, with various batch sizes. Graphs show mean L2-error versus iterations (over 40 trials). Step size γ used on each batch was set as in (14) (approximations for Lipschitz constants also apply to the step size computation) for the weighted cases and as in [1, Corollary 3.2] for the uniform cases. *Top Left* Batches are created randomly at onset, then selected using weighted sampling. *Top Right* Batches are created sequentially at onset, then selected using weighted sampling. *Bottom* Batches are created sequentially at onset, then selected using uniform (unweighted) sampling

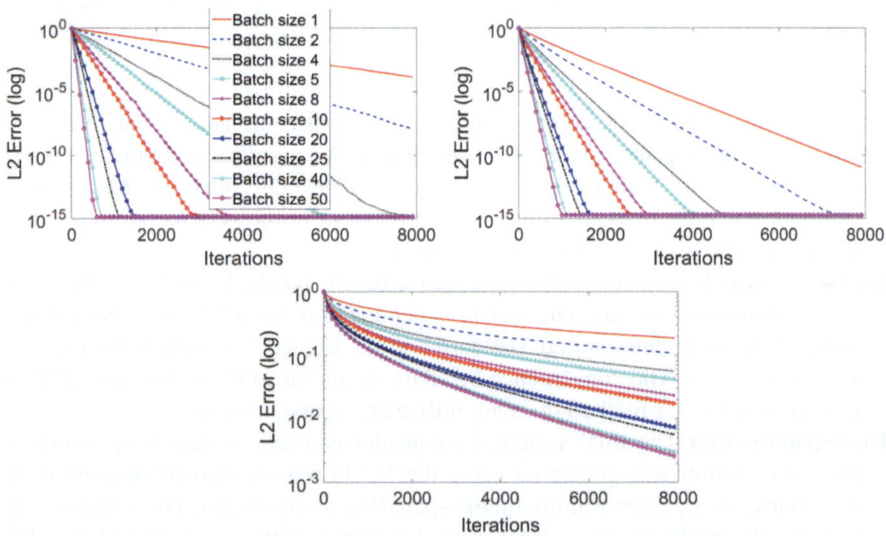

Fig. 7 **(Orthonormal, sparse, and tomography systems: convergence)** Mini-batch SGD on two systems for various batch sizes; batches created randomly at onset. Graphs show mean L2-error versus iterations (over 40 trials). Step size γ used on each batch was set as in (14). *Top Left* Matrix is a 200×200 orthonormal discrete Fourier transform (DFT). *Top Right* 1000×50 matrix is a sparse standard normal matrix with density 20%. *Bottom* Tomography data (1200×400 system)

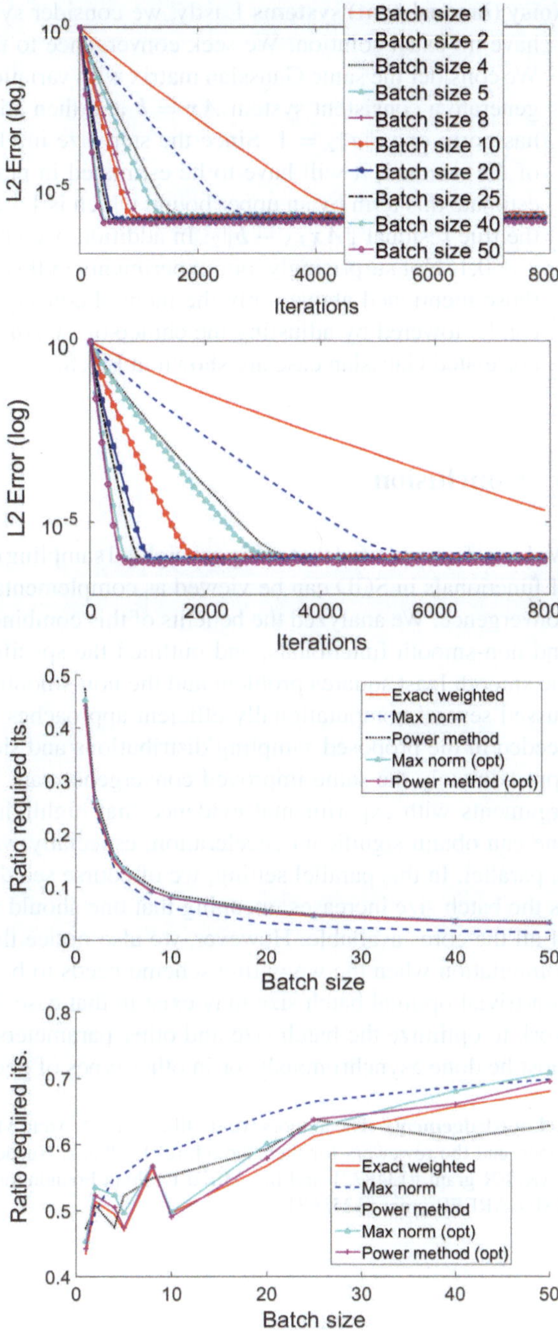

Fig. 8 (Noisy systems: convergence) Mini-batch SGD on a Gaussian 1000×50 system whose entries in row k have variance k^2, with various batch sizes. Noise of norm 1 is added to system to create an inconsistent system. Graphs show mean L2-error versus iterations (over 40 trials). Step size γ used on each batch was set as in (14) for the weighted case and as in [1, Corollary 3.2] for the uniform case; the residual $\boldsymbol{A}\boldsymbol{x}_{LS} - \boldsymbol{b}$ was upper bounded by a factor of 1.1 in all cases. *Top* Batches are created sequentially at onset, then selected using weighted sampling. *Second Plot* Batches are created sequentially at onset, then selected using uniform (unweighted) sampling. *Third Plot* Ratio of the number of iterations required to reach an error of 10^{-5} for various weighted selections with batched SGD versus classical (single functional) uniform (unweighted) SGD. *Bottom* Ratio of the number of iterations required to reach an error of 10^{-5} for various weighted selections with batched SGD versus classical uniform (unweighted) SGD as a function of batch size

Noisy (inconsistent) systems Lastly, we consider systems that are noisy, i.e., they have no exact solution. We seek convergence to the least squares solution $\boldsymbol{x}_{LS}$. We consider the same Gaussian matrix with variation as described above. We first generate a consistent system $\boldsymbol{Ax} = \boldsymbol{b}$ and then add a residual vector $\boldsymbol{e}$ to $\boldsymbol{b}$ that has norm one, $\|\boldsymbol{e}\|_2 = 1$. Since the step size in (14) depends on the magnitude of the residual, it will have to be estimated in practice. In our experiments, we estimate this term by an upper bound which is 1.1 times larger in magnitude than the true residual $\|\boldsymbol{Ax}_{LS} - \boldsymbol{b}\|_2$. In addition, we choose an accuracy tolerance of $\varepsilon = 0.1$. Not surprisingly, our experiments in this case show similar behavior to those mentioned above, only the method convergences to a larger error (which can be lowered by adjusting the choice of ε). An example of such results in the correlated Gaussian case are shown in Fig. 8.

6 Conclusion

We have demonstrated that using a weighted sampling distribution along with batches of functionals in SGD can be viewed as complementary approaches to accelerating convergence. We analyzed the benefits of this combined framework for both smooth and non-smooth functionals, and outlined the specific convergence guarantees for the smooth least squares problem and the non-smooth hinge loss objective. We discussed several computationally efficient approaches to approximating the weights needed in the proposed sampling distributions and showed that one can still obtain approximately the same improved convergence rate. We confirmed our theoretical arguments with experimental evidence that highlight in many important settings, one can obtain significant acceleration, especially when batches can be computed in parallel. In this parallel setting, we of course see that the improvement increases as the batch size increases, meaning that one should unsurprisingly take advantage of all the cores available. However, we also notice that there may be a trade-off in computation when the weighting scheme needs to be calculated a priori, and that a non-trivial optimal batch size may exist in that case. It will be an interesting future work to optimize the batch size and other parameters when the parallel computing must be done asynchronously, or in other types of geometric architectures.

Acknowledgements The authors would like to thank Anna Ma for helpful discussions about this paper, and the reviewers for their thoughtful feedback. Needell was partially supported by NSF CAREER grant #1348721 and the Alfred P. Sloan Foundation. Ward was partially supported by NSF CAREER grant #1255631.

References

1. D. Needell, N. Srebro, R. Ward, Stochastic gradient descent and the randomized Kaczmarz algorithm. Math. Program. Ser. A **155**(1), 549–573 (2016)
2. P. Zhao, T. Zhang, Stochastic optimization with importance sampling for regularized loss minimization, in *Proceedings of the 32nd International Conference on Machine Learning (ICML-15)* (2015)
3. A. Cotter, O. Shamir, N. Srebro, K. Sridharan, Better mini-batch algorithms via accelerated gradient methods, in *Advances in neural information processing systems* (2011), pp. 1647–1655
4. A. Agarwal, J.C. Duchi, Distributed delayed stochastic optimization, in *Advances in Neural Information Processing Systems* (2011), pp. 873–881
5. O. Dekel, R. Gilad-Bachrach, O. Shamir, L. Xiao, Optimal distributed online prediction using mini-batches. J. Mach. Learn. Res. **13**(1), 165–202 (2012)
6. M. Takac, A. Bijral, P. Richtarik, N. Srebro, Mini-batch primal and dual methods for SVMs, in *Proceedings of the 30th International Conference on Machine Learning (ICML-13)*, vol. 3 (2013), pp. 1022–1030
7. H. Robbins, S. Monroe, A stochastic approximation method. Ann. Math. Stat. **22**, 400–407 (1951)
8. L. Bottou, O. Bousquet, The tradeoffs of large-scale learning, in *Optimization for Machine Learning* (2011), p. 351
9. L. Bottou, Large-scale machine learning with stochastic gradient descent, in *Proceedings of COMPSTAT'2010* (Springer, Berlin, 2010), pp. 177–186
10. A. Nemirovski, A. Juditsky, G. Lan, A. Shapiro, Robust stochastic approximation approach to stochastic programming. SIAM J. Optim. **19**(4), 1574–1609 (2009)
11. S. Shalev-Shwartz, N. Srebro, SVM optimization: inverse dependence on training set size, in *Proceedings of the 25th international conference on Machine learning* (2008), pp. 928–935
12. T. Strohmer, R. Vershynin, A randomized Kaczmarz algorithm with exponential convergence. J. Fourier Anal. Appl. **15**(2), 262–278 (2009)
13. D. Needell, Randomized Kaczmarz solver for noisy linear systems. BIT **50**(2), 395–403 (2010)
14. F. Bach, E. Moulines, Non-asymptotic analysis of stochastic approximation algorithms for machine learning, in *Advances in Neural Information Processing Systems (NIPS)* (2011)
15. Y. Nesterov, Efficiency of coordinate descent methods on huge-scale optimization problems. SIAM J. Optim. **22**(2), 341–362 (2012)
16. P. Richtárik, M. Takáč, On optimal probabilities in stochastic coordinate descent methods. Optim. Lett. 1–11 (2015)
17. Z. Qu, P. Richtarik, T. Zhang, Quartz: randomized dual coordinate ascent with arbitrary sampling, in *Advances in neural information processing systems*, vol. 28 (2015), pp. 865–873
18. D. Csiba, Z. Qu, P. Richtarik, Stochastic dual coordinate ascent with adaptive probabilities, in *Proceedings of the 32nd International Conference on Machine Learning (ICML-15)* (2015)
19. Y.T. Lee, A. Sidford, Efficient accelerated coordinate descent methods and faster algorithms for solving linear systems, in *2013 IEEE 54th Annual Symposium on Foundations of Computer Science (FOCS)* (IEEE, 2013), pp. 147–156
20. M. Schmidt, N. Roux, F. Bach, Minimizing finite sums with the stochastic average gradient (2013), arXiv:1309.2388
21. L. Xiao, T. Zhang, A proximal stochastic gradient method with progressive variance reduction. SIAM J. Optim. **24**(4), 2057–2075 (2014)
22. A. Défossez, F.R. Bach, Averaged least-mean-squares: bias-variance trade-offs and optimal sampling distributions, in *AISTATS* (2015)
23. S. Shalev-Shwartz, Y. Singer, N. Srebro, A. Cotter, Pegasos: primal estimated sub-gradient solver for SVM. Math. Program. **127**(1), 3–30 (2011)
24. R.H. Byrd, G.M. Chin, J. Nocedal, Y. Wu, Sample size selection in optimization methods for machine learning. Math. Program. **134**(1), 127–155 (2012)
25. D. Needell, R. Ward, Two-subspace projection method for coherent overdetermined linear systems. J. Fourier Anal. Appl. **19**(2), 256–269 (2013)

26. J. Konecnỳ, J. Liu, P. Richtarik, M. Takac, mS2GD: mini-batch semi-stochastic gradient descent in the proximal setting. IEEE J. Sel. Top. Signal Process. **10**(2), 242–255 (2016)
27. M. Li, T. Zhang, Y. Chen, A.J. Smola, Efficient mini-batch training for stochastic optimization, in *Proceedings of the 20th ACM SIGKDD International Conference on Knowledge Discovery and Data Mining* (ACM, 2014), pp. 661–670
28. D. Csiba, P. Richtarik, Importance sampling for minibatches (2016), arXiv:1602.02283
29. R.M. Gower, P. Richtárik, Randomized quasi-Newton updates are linearly convergent matrix inversion algorithms (2016), arXiv:1602.01768
30. E.J. Candès, T. Tao, Decoding by linear programming. IEEE Trans. Inf. Theory **51**, 4203–4215 (2005)
31. P. Klein, H.-I. Lu, Efficient approximation algorithms for semidefinite programs arising from max cut and coloring, in *Proceedings of the Twenty-Eighth Annual ACM Symposium on Theory of Computing* (ACM, 1996), pp. 338–347
32. Y. Nesterov, *Introductory Lectures on Convex Optimization* (Kluwer, Dordrecht, 2004)
33. O. Shamir, T. Zhang, Stochastic gradient descent for non-smooth optimization: convergence results and optimal averaging schemes (2012), arXiv:1212.1824
34. A. Rakhlin, O. Shamir, K. Sridharan, Making gradient descent optimal for strongly convex stochastic optimization (2012), arXiv:1109.5647
35. J. Yang, Y.-L. Chow, C. Ré, M.W. Mahoney, Weighted SGD for ℓ_p regression with randomized preconditioning, in *Proceedings of the Twenty-Seventh Annual ACM-SIAM Symposium on Discrete Algorithms* (SIAM, 2016), pp. 558–569
36. P.C. Hansen, Regularization tools version 4.0 for matlab 7.3. Numer. Algorithms **46**(2), 189–194 (2007)

A Fractional Spline Collocation Method for the Fractional-order Logistic Equation

Francesca Pitolli and Laura Pezza

Abstract We construct a collocation method based on the fractional B-splines to solve a nonlinear differential problem that involves fractional derivatives, i.e., the fractional-order logistic equation. The use of the fractional B-splines allows us to express the fractional derivatives of the approximating function in an analytical form. Thus, the fractional collocation method is easy to implement, accurate, and efficient. Several numerical tests illustrate the efficiency of the proposed collocation method.

Keywords Fractional-order • Logistic equation • Fractional B-spline • Fractional derivative • Collocation method

1 Introduction

When we speak about *Fractional Derivative*, we are generalizing to positive real order the well-understood notion of derivative of integer order (see, for instance, [14]). The nonlocal behavior of the fractional derivative allows to model physical phenomena where memory processes are relevant. For this reason, in the last decades, fractional-order differential problems were extensively used to describe the anomalous diffusion in porous media, the behavior of viscoelastic materials, the population growth of ecosystems, the intermittent phenomena in plasma, just to cite a few (see, for instance, [6, 9, 15] and references therein).

On the other hand, the nonlocality is a challenge when one wants to construct numerical methods for the solution of fractional-order differential problems. In fact, to approximate the fractional derivative of a function, its values from the beginning of the discretization interval have to be taken into account. Many numerical methods

F. Pitolli (✉) · L. Pezza
Dept. SBAI, Università di Roma "La Sapienza", Via A. Scarpa 16,
00161 Roma, Italy
e-mail: francesca.pitolli@sbai.uniroma1.it

L. Pezza
e-mail: laura.pezza@sbai.uniroma1.it

G.E. Fasshauer and L.L. Schumaker (eds.), *Approximation Theory XV: San Antonio 2016*, Springer Proceedings in Mathematics & Statistics 201,
DOI 10.1007/978-3-319-59912-0_15

that generalize to fractional-order numerical methods well established in the case of integer-order derivatives, have been proposed in the literature (see, for instance, [2, 3, 11, 21] and references therein). Unfortunately, these methods are computationally demanding when a high accuracy in the numerical solution is required.

In this paper, we construct a collocation method that is especially tailored for fractional-order differential problems. The method looks for an approximating function that belongs to the functional space generated by the fractional B-splines and takes advantage of the explicit formula for their fractional derivatives. The resulting method is easy to implement, accurate, and efficient.

We will use the proposed method to solve a nonlinear fractional differential problem, i.e., the fractional-order logistic differential equation, and we will show that the method produces good approximations while keeping the computational cost low.

The organization of the paper is as follows. In Sect. 2, the fractional-order logistic model is presented, and the definition of fractional derivative is given. Section 3 is devoted to the fractional B-splines and their main properties, while the explicit expression of their fractional derivatives is given in Sect. 4. The fractional spline collocation method is introduced in Sect. 5. Finally, in Sect. 6, some numerical tests on the solution of the linear fractional growth model and of the nonlinear fractional logistic model are shown. Some conclusions are drawn in Sect. 7.

2 The Fractional-order Logistic Differential Equation

The *logistic equation* [19]

$$\begin{cases} u'(t) = \rho\, u(t)\,\big(1 - u(t)\big)\,, & t > 0\,, \\ u(0) = u_0\,, \end{cases} \tag{1}$$

where $\rho > 0$ is the growth rate, is extensively used to model the growth of a population in the case of limited resources. It has application in several fields, from ecology to chemical reactions, from medicine to transport of goods. Remarkably, the solution of the logistic equation has a closed form, i.e.,

$$u(t) = \frac{u_0}{u_0 + (1 - u_0)\mathrm{e}^{-\rho t}}\,, \tag{2}$$

also known as the *sigmoid function*. We notice that Eq. (1) is normalized so that $\lim_{t\to\infty} u(t) = 1$.

Recently, it has been observed that many real-life growth phenomena are better described by fractional-order differential problems (see, for instance, [1, 8, 16, 17]), so that it seems natural to generalize to the fractional order the logistic model. The *fractional-order logistic equation*, first analyzed in [4], is

$$\begin{cases} D^\gamma u(t) = \rho^\gamma\, u(t)\,\bigl(1 - u(t)\bigr) & t > 0\,, \quad 0 < \gamma < 1\,, \\ u(0) = u_0\,, \end{cases} \tag{3}$$

where $D_t^\gamma u$ denotes the *Caputo fractional derivative* with respect to the time t.

We recall that the Caputo derivative of fractional-order $\gamma > 0$ of a function $f \in C^{\lceil\gamma\rceil-1}(\mathbb{R}^+)$ having absolute integrable derivative of order $k = \lceil\gamma\rceil$ is defined as

$$D_t^\gamma\, f(t) := \bigl(\mathscr{J}^{(k-\gamma)} f^{(k)}\bigr)(t)\,, \qquad k-1 < \gamma < k\,, \quad k \in \mathbb{N}\backslash\{0\}\,, \qquad t > 0\,, \tag{4}$$

where $\mathscr{J}^{(\gamma)}$ is the *Riemann–Liouville integral operator*

$$\bigl(\mathscr{J}^{(\gamma)} f\bigr)(t) := \frac{1}{\Gamma(\gamma)} \int_0^t f(\tau)\,(t-\tau)^{\gamma-1}\, d\tau\,, \tag{5}$$

and Γ denotes the *Euler's gamma function*

$$\Gamma(\gamma) := \int_0^\infty \tau^{\gamma-1}\, \mathrm{e}^{-\tau}\, d\tau\,. \tag{6}$$

Here, $\mathbb{N}$ denotes the set of nonnegative integers.

For details on fractional calculus and fractional derivatives see, for instance, [14].

The existence and uniqueness of the solution to (3) was proved in [4], while its analytical solution was given in [20],

$$u(t) = \sum_{k=0}^{\infty} \left(\frac{u_0 - 1}{u_0}\right)^k E_\gamma(-k\,\rho^\gamma\, t^\gamma)\,, \tag{7}$$

where $E_\gamma(z)$ is the Mittag–Leffler function defined as

$$E_\gamma(z) = \sum_{k=0}^{\infty} \frac{z^k}{\Gamma(k\,\gamma + 1)}\,. \tag{8}$$

The fractional-order influences the population growth: The lower the derivative order is, the slower the rate of growth becomes.

In the following, we will construct a collocation method based on the fractional B-splines to solve the fractional logistic equation.

3 The Fractional B-Splines

The *fractional B-splines*, i.e., the B-splines of fractional order, were introduced in [18] generalizing to fractional power the classical definition of polynomial B-splines of integer order. We notice that here *'fractional'* actually means *'real.'* In fact, let

$$t_+^\alpha := \begin{cases} t^\alpha , & t \geq 0 , \\ 0 , & \text{otherwise} , \end{cases} \qquad \alpha > -1/2 , \tag{9}$$

be the *fractional truncated power function*, and let Δ_h^γ be the *generalized finite difference operator*

$$\Delta_h^\gamma f(t) := \frac{1}{h^\gamma} \sum_{k=0}^{\infty} (-1)^k \binom{\gamma}{k} f(t - h\,k) , \qquad \gamma \in \mathbb{R}^+ , \tag{10}$$

where

$$\binom{\gamma}{k} := \frac{\Gamma(\gamma+1)}{k!\,\Gamma(\gamma-k+1)} , \qquad k \in \mathbb{N} , \quad \gamma \in \mathbb{R}^+ , \tag{11}$$

are the *generalized binomial coefficients*. When $\gamma \in \mathbb{N}\backslash\{0\}$, $\{\binom{\gamma}{k}\}$ is the usual binomial coefficient sequence so that the series in (10) is a finite sum and Δ_h^γ reduces to the usual finite difference operator. When $\gamma \in \mathbb{R}^+\backslash\mathbb{N}$, the sequence $\{\binom{\gamma}{k}\}$ is no more compactly supported even if it is absolutely summable so that the limit of the series (10) exists under suitable conditions on f.

The *fractional B-spline* B_α of order α is defined as

$$B_\alpha(t) := \frac{\Delta_h^{\alpha+1}\, t_+^\alpha}{\Gamma(\alpha+1)} , \qquad \alpha > -\frac{1}{2} . \tag{12}$$

When $\alpha = n \in \mathbb{N}$, B_n is the well-known polynomial B-spline of integer degree n and compact support $[0, n+1]$. When α is not an integer, the fractional B-spline B_α does not have compact support but it decays as

$$B_\alpha(t) \to \frac{1}{|t|^{-\alpha-2}} , \tag{13}$$

for $t \to \infty$ and belongs to $L^2(\mathbb{R})$.

The fractional B-splines for different values of the parameter α are displayed in Fig. 1 (top left panel). The classical polynomial B-splines are also displayed (dashed lines). The picture shows that even if the fractional B-splines do not have compact support, they decay very fast so that they can be truncated for computational purposes. Moreover, we observe that the fractional B-splines are not always positive, even if the nonnegative part becomes smaller and smaller when α increases.

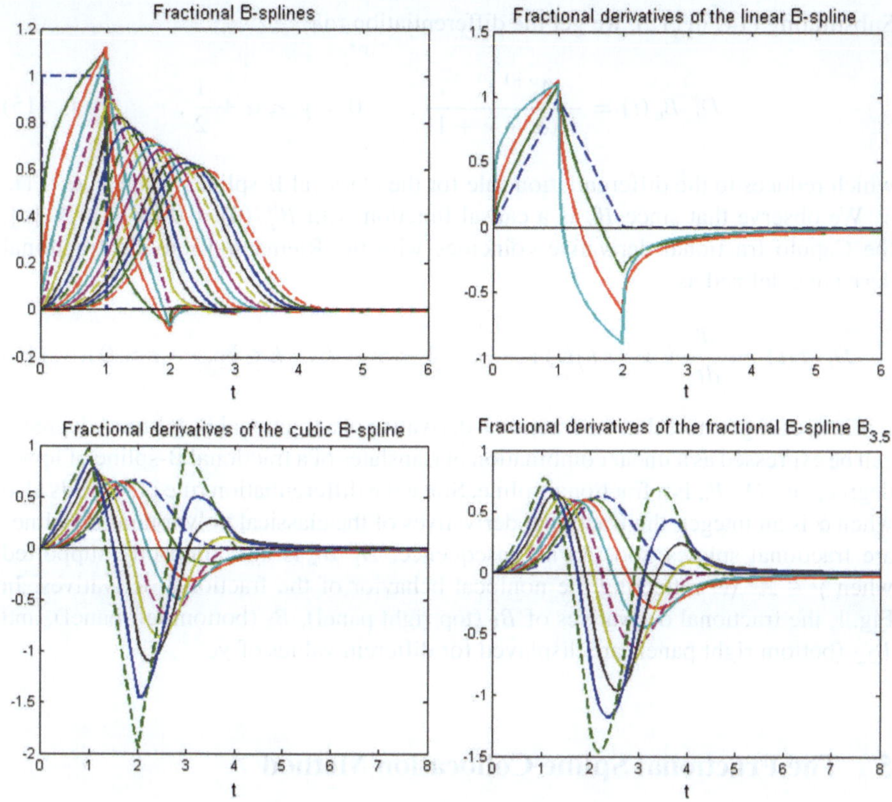

Fig. 1 The fractional B-splines (*solid lines*) and the polynomial B-splines (*dashed lines*) for α ranging from 0 to 4 (*top left panel*). The fractional derivatives of the linear B-spline B_1 for $\gamma = 0.25, 0.5, 0.75$ (*top right panel*). The fractional derivatives of the cubic B-spline B_3 (*bottom left panel*) and of the fractional B-spline $B_{3.5}$ (*bottom right panel*) for γ ranging from 0.25 to 2. Ordinary derivatives are displayed as *dashed lines*

4 The Fractional Derivatives of the Fractional B-Splines

Fractional derivatives of fractional B-splines can be evaluated by differentiating (12). First of all we observe that the fractional derivative of order γ of the fractional truncated power (9) can be evaluated explicitly using definition (4). We get

$$D_t^\gamma t_+^\alpha = \begin{cases} \dfrac{\Gamma(\alpha+1)}{\Gamma(\alpha+1-\gamma)} \, t_+^{\alpha-\gamma} \,, & t > 0 \,, \\ 0 \,, & \text{otherwise} \,, \end{cases} \qquad 0 < \gamma \le \alpha \,. \tag{14}$$

Substituting (14) in (12), we get the differentiation rule

$$D_t^\gamma \, B_\alpha(t) = \frac{\Delta_h^{\alpha+1}\, t_+^{\alpha-\gamma}}{\Gamma(\alpha-\gamma+1)}\,, \qquad 0 < \gamma < \alpha + \frac{1}{2}\,, \tag{15}$$

which reduces to the differentiation rule for the classical B-splines when $\gamma, \alpha \in \mathbb{N}$.

We observe that since B_α is a causal function with $B_\alpha^{(n)}(0) = 0$ for $n \in \mathbb{N}\backslash\{0\}$, the Caputo fractional derivative coincides with the Riemann–Liouville fractional derivative defined as

$$D_t^\gamma \, f(t) := \frac{d^k}{dt^k}\left(\mathscr{I}^{(\gamma)} f\right)(t)\,, \qquad k-1 < \gamma < k\,, \quad k \in \mathbb{N}\,, \qquad t > 0\,. \tag{16}$$

It is interesting to note that the fractional derivative of a fractional B-spline of degree α can be expressed as a linear combination of translates of a fractional B-spline of lower degree, i.e., $D_t^\gamma \, B_\alpha$ is a fractional spline. Since the differentiation rule (15) holds also when α is an integer, the fractional derivatives of the classical polynomial B-splines are fractional splines, too. As a consequence, $D_t^\gamma \, B_n$ is not compactly supported when $\gamma \in \mathbb{R}^+\backslash\mathbb{N}$ reflecting the nonlocal behavior of the fractional derivatives. In Fig. 1, the fractional derivatives of B_1 (top right panel), B_3 (bottom left panel), and $B_{3.5}$ (bottom right panel) are displayed for different values of γ.

5 The Fractional Spline Collocation Method

The *fractional spline collocation method* looks for an approximating function that belongs to the space of the fractional spline B_α. Let $\Delta = \{j\,h, 0 \le j \le N\}$, $N = T/h$, be a partition of equispaced nodes in the interval $[0, T]$ and let $\mathscr{V}_{\alpha,h}$ be the fractional spline space on the partition Δ. As a basis for $\mathscr{V}_{\alpha,h}$, we choose the set of functions

$$\Phi_{\alpha,h,k} = \{\varphi_{\alpha,h,k}(t) = B_\alpha(t - h\,k), k \in \mathscr{N}_h\,, t \in [0, T]\}\,. \tag{17}$$

Here, $\mathscr{N}_h$ is a suitable set of indexes chosen in order to take into account all the translates of B_α that are significantly different from zero in $[0, T]$. When $\alpha = n \in \mathbb{N}$, due to the compact support of B_n we get $\mathscr{N}_h = \{k, -n \le k \le N\}$. When α is not an integer, since B_α decays fast toward infinity, for computational purposes we can assume supp $B_\alpha \approx [0, N_0]$, where $N_0 \ge \lceil\alpha\rceil + 2$ is a truncation parameter (cf. Fig. 1). Thus, $\mathscr{N}_h = \{k, -N_0 + 1 \le k \le N\}$. We observe that the basis $\Phi_{\alpha,h,k}$ has a certain number of boundary functions, depending on the value of α. More precisely, let N_α be an integer equal to n when α has an integer value and equal to N_0 when α is not an integer. Then, the basis $\Phi_{\alpha,h,k}$ has $2N_\alpha$ boundary functions, i.e., the N_α functions $\varphi_{\alpha,h,k}(t)$ with index k ranging from $-N_\alpha$ to -1 (left boundary), and the N_α functions

$\varphi_{\alpha,h,k}(t)$ with index k ranging from $N - N_\alpha + 1$ to N (right boundary). Here, the boundary functions are obtained just restricting $B_\alpha(t - h\,k)$ to the interval $[0, T]$.

For a given partition Δ, we look for an approximating function $u_{\alpha,h}(t) \in \mathscr{V}_{\alpha,h}$, i.e.,

$$u_{\alpha,h}(t) = \sum_{k\in\mathscr{N}_h} \lambda_{\alpha,h,k}\, \varphi_{\alpha,h,k}(t)\,, \tag{18}$$

which solves the differential problem (3) on a set of *collocation points*. Here, we choose as collocation points the set of equispaced nodes $\{t_p = p\,\delta,\, p = 0, \ldots, M\}$, where $M = T/\delta$. Thus, the discretized version of (3) reads

$$\begin{cases} D_t^\gamma\, u_{\alpha,h}(t_p) = \rho^\gamma\, u_{\alpha,h}(t_p)\,\big(1 - u_{\alpha,h}(t_p)\big)\,, & p = 1, \ldots, M\,, \\[2ex] u_{\alpha,h}(0) = u_0\,. \end{cases} \tag{19}$$

Substituting (18) in (19), we get

$$\begin{cases} \displaystyle\sum_{k\in\mathscr{N}_h} \lambda_k\, D_t^\gamma\, \varphi_k(t_p) = \rho^\gamma \sum_{k\in\mathscr{N}_h} \lambda_k \varphi_k(t_p)\,\big(1 - \sum_{k\in\mathscr{N}_h} \lambda_k \varphi_k(t_p)\big)\,, & p = 1, \ldots, M\,, \\[2ex] \displaystyle\sum_{k\in\mathscr{N}_h} \lambda_k \varphi_k(0) = u_0\,, \end{cases} \tag{20}$$

where, for short notation, we dropped the subscripts α and h.

Equations (20) form a nonlinear system of M equations for the $N \le M$ unknowns $\{\lambda_k,\, k \in \mathscr{N}_h\}$ and can be solved, for instance, by the Levenberg–Marquardt algorithm. In the following section, we give some numerical results showing the feasibility of the method.

6 Numerical Tests

In this section, we will show the performance of the proposed collocation method in solving some test problems.

First of all, we test the fractional spline collocation method on the fractional growth equation

$$\begin{cases} D^\gamma u(t) = \rho^\gamma\, u(t)\,, & t > 0\,, \quad 0 < \gamma \le 1\,, \quad \rho > 0\,, \\[2ex] u(0) = u_0\,, \end{cases} \tag{21}$$

which generalizes to the fractional-order derivative the well-known Malthusian model [10]. Its analytical solution, obtained through the Laplace transform, has the following expression (cf. [20])

$$u(t) = u_0\, E_\gamma\big((\rho\, t)^\gamma\big)\,. \tag{22}$$

It reduces to the classical exponential growth law when $\gamma = 1$.

Applying to Eq. (21) the collocation method highlighted in the previous section, we get the linear system

$$\begin{cases} \displaystyle\sum_{k\in\mathcal{N}_h} \lambda_k\, D_t^\gamma \varphi_k(t_p) - \rho^\gamma \sum_{k\in\mathcal{N}_h} \lambda_k\, \varphi_k(t_p) = 0\,, \qquad p = 1, \ldots, M\,, \\ \displaystyle\sum_{k\in\mathcal{N}_h} \lambda_k\, \varphi_k(0) = u_0\,, \end{cases} \tag{23}$$

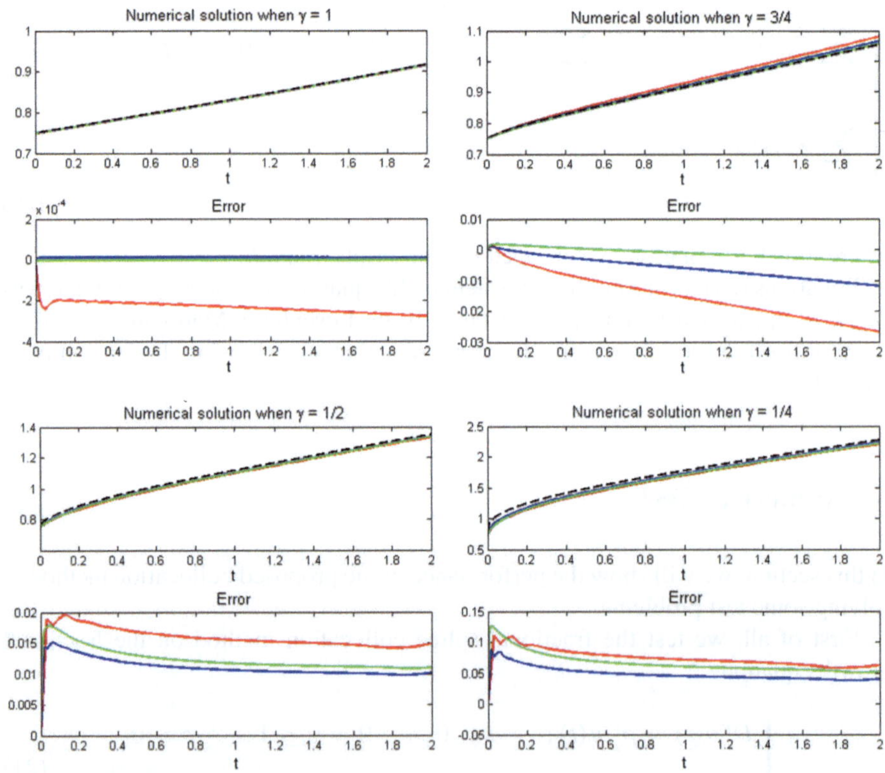

Fig. 2 Fractional growth model: The numerical solution and the error obtained by using the fractional B-spline $B_{3.75}$ for $\gamma = 1$ (*top left panels*), 3/4 (*top right panels*), 1/2 (*bottom left panels*), 1/4 (*bottom right panels*) for increasing values of M and N; i.e., $M = 64$, $N = 37$ (*red*); $M = 128$, $N = 69$ (*blue*); $M = 256$, $N = 133$ (*green*). The analytical solution is shown as a *dashed black line*

where $\{\lambda_k\}$ are the N unknowns. When $N < M$, the linear system is overdetermined and can be solved in the least squares sense.

In the following numerical tests, we set $\rho = 1$.

The numerical solution and the error $u(t) - u_{\alpha,h}(t)$ obtained when $\alpha = 3.75$ are shown in Fig. 2 for different values of γ and decreasing values of the partition step h and of the collocation step δ, i.e., $\gamma = 1, 3/4, 1/2, 1/4$, $h = 2^{-4}, 2^{-5}, 2^{-6}$, and $\delta = 2^{-5}, 2^{-6}, 2^{-7}$. Here, the truncation parameter N_0 is set equal to 6 so that $N = T/h + N_0 - 1 = 37, 69, 133$ while $M = T/\delta = 64, 128, 256$.

Figure 3 displays the numerical solution and the error obtained by using the fractional B-splines B_α with $\alpha = 3, 3.25, 3.5, 3.75, 4$, for the two values $\gamma = 1/4$ and $\gamma = 1$ (Malthus model).

These simple tests show that the numerical solution obtained by the proposed collocation method has a good accuracy even in the case of few nodes and low dimension of the approximating space. Moreover, a suitable choice of the fractional order of the fractional B-spline can improve the approximation accuracy.

Then, we solved the fractional logistic equation using the same parameters as in the previous tests. The numerical solution and the error obtained when $\alpha = 3.75$ are shown in Fig. 4 for different values of γ. Figure 5 displays the numerical solution and the error obtained by using the fractional B-splines B_α with $\alpha = 3, 3.25, 3.5, 3.75, 4$, for the two cases $\gamma = 1/4$ and $\gamma = 1$.

The numerical results show that also for this nonlinear problem, the collocation method provides an accurate solution.

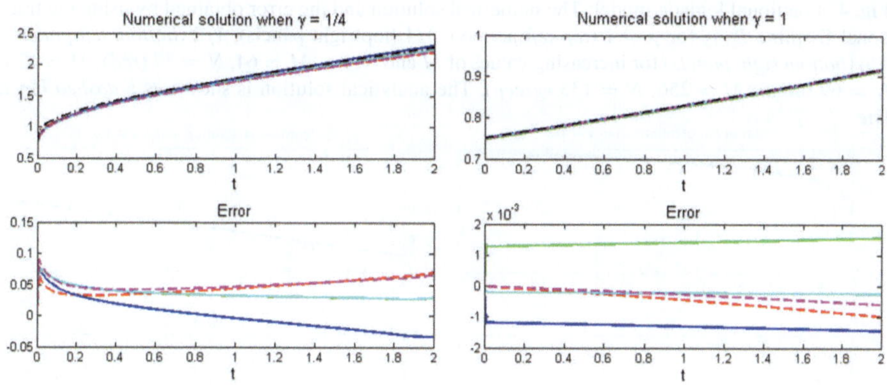

Fig. 3 Fractional growth model: The numerical solution and the error for $\gamma = 1/4$ (left), $\gamma = 1$ (*right*) obtained by using the fractional B-splines $B_{3.25}$ (*blue*), $B_{3.5}$ (*green*), $B_{3.75}$ (*cyan*). *Dashed lines* correspond to the polynomial B-splines B_3 (red) and B_4 (*magenta*). Here, $M = 128$, $N = 69$

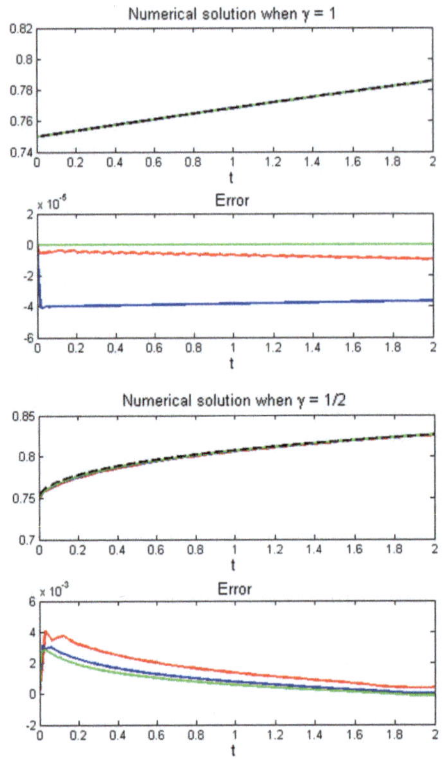

Fig. 4 Fractional logistic model: The numerical solution and the error obtained by using the fractional B-spline $B_{3.75}$ for $\gamma = 1$ (*top left panels*), 3/4 (top right panels), 1/2 (*bottom left panels*), 1/4 (*bottom right panels*) for increasing values of M and N, i.e., $M = 64$, $N = 37$ (*red*); $M = 128$, $N = 69$ (*blue*); $M = 256$, $N = 133$ (*green*). The analytical solution is shown as a *dashed black line*

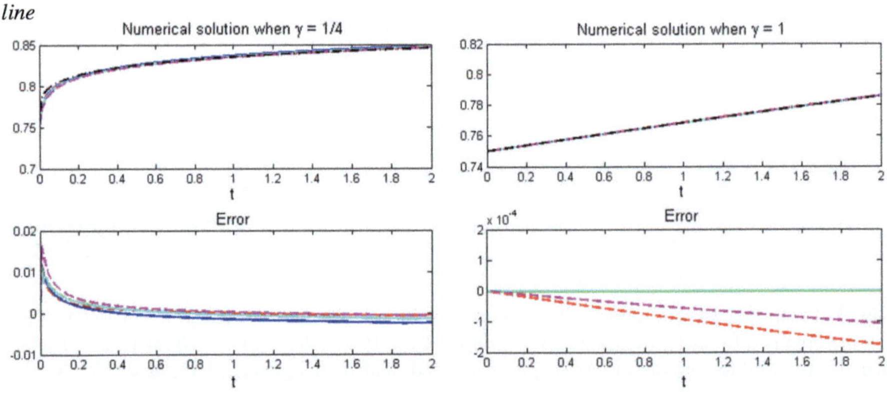

Fig. 5 Fractional logistic model: The numerical solution and the error for $\gamma = 1/4$ (*left panels*), $\gamma = 1$ (*right panels*) obtained by using the fractional B-splines $B_{3.25}$ (*blue*), $B_{3.5}$ (*green*), $B_{3.75}$ (*cyan*). *Dashed lines* corresponds to the polynomial B-splines B_3 (*red*) and B_4 (*magenta*). Here, $M = 128$, $N = 69$

7 Conclusion

We presented a fractional spline collocation method suitable to solve both linear and nonlinear fractional-order differential problems, and we tested their performance on the growth and logistic differential equations of fractional order. The numerical results show that the proposed method produces an accurate approximation with a low computational cost. Moreover, the order of the fractional B-spline to be used in the method can be considered as a free parameter and can be chosen in order to reduce the approximation error.

A theoretical proof of the convergence when h and δ go to zero can be given following classical arguments in approximation theory (cf. [13]) and will be the subject of a forthcoming paper where the use of different approximating operators and different node distributions will be also considered. Moreover, the additional approximation error introduced by the truncation parameter N_0 has to be carefully evaluated.

Another issue to be taken into account is the truncation of the boundary functions since it could generate numerical instabilities due to the possibly high condition number of the corresponding collocation matrix. There exist different approaches to obtain stable bases on finite intervals. Since the fractional B-splines are refinable functions and generate a multiresolution analysis in $L_2(\mathbb{R})$, we can adapt to the fractional B-spline case some methods already used for constructing refinable bases on the interval (see, for instance, [5]).

As a final remark, we observe that refinability is a key ingredient to construct fractional wavelet collocation methods (cf. [12]).

References

1. R. Almeida, N.R.O. Bastos, M.T.T. Monteiro, Modeling some real phenomena by fractional differential equations. Math. Meth. Appl. Sci. **39**, 4846–4855 (2016)
2. D. Baleanu, K. Diethelm, E. Scalas, J.J. Trujillo, *Fractional Calculus, Models and Numerical Methods* (World Scientific, Singapore, 2012)
3. S. Das, P.K. Gupta, Approximate approach to the Das model of fractional logistic population growth. Appl. App. Math. **5**, 605–611 (2010)
4. A.M.A. El-Sayed, A.E.M. El-Mesiry, H.A.A. El-Saka, On the fractional-order logistic equation. Appl. Math. Lett. **20**, 817–823 (2007)
5. L. Gori, L. Pezza, F. Pitolli, Recent results on wavelet bases on the interval generated by GP refinable functions. Appl. Num. Math. **51**, 549–563 (2004)
6. R. Hilfer (ed.), *Applications of Fractional Calculus in Physics* (World Scientific Publishing Co., Inc, River Edge, NJ, 2000)
7. A.A. Kilbas, H.M. Srivastava, J.J. Trujillo, *Theory and Applications of Fractional Differential Equations*. North-Holland Mathematics Studies, vol. 204 (Elsevier Science, Amsterdam, 2006)
8. Y.F. Luchko, M. Rivero, J.J. Trujillo, M.P. Velasco, Fractional models, non-locality, and complex systems. Comput. Math. App. **59**, 1048–1056 (2010)
9. F. Mainardi, *Fractional Calculus and Waves in Linear Viscoelasticity: An Introduction to Mathematical Models* (Imperial College Press, London, 2010)

10. T.R. Malthus, *An Essay on the Principle of Population (1798)* (Macmillan & Co., London, 1926)
11. A. Pedas, E. Tamme, Numerical solution of nonlinear fractional differential equations by spline collocation methods. J. Comput. Appl. Math. **255**, 216–230 (2014)
12. L. Pezza, F. Pitolli, A fractional wavelet Galerkin method for the fractional diffusion problem. IMACS Ser. Comput. Appl. Math. **20**, 1–10 (2017)
13. L. Pezza, F. Pitolli, A multiscale collocation method for fractional differential problems. Math. Comput. Simul
14. I. Podlubny, *Fractional Differential Equations*. Mathematics in Science and Engineering, vol. 198 (Academic Press, Cambridge, 1999)
15. V.E. Tarasov, *Fractional Dynamics. Applications of Fractional Calculus to Dynamics of Particles, Fields and Media*. Nonlinear Physical Science (Springer, Heidelberg, 2010)
16. V.V. Tarasova, V.E. Tarasov, Logistic map with memory from economic model. Chaos, Solitons Fractals **95**, 84–91 (2017)
17. R. Toledo-Hernandez, V. Rico-Ramirez, G.A. Iglesias-Silva, U.M. Diwekar, A fractional calculus approach to the dynamic optimization of biological reactive systems. Part I: Fractional models for biological reactions. Chem. Eng. Sci. **117**, 217–228 (2014)
18. M. Unser, T. Blu, Fractional splines and wavelets. SIAM Rev. **42**, 43–67 (2000)
19. P.F. Verhulst, Notice sur la loi que la population suit dans son accroissement. Correspondance mathématique et physique, Quetelet **10**, 113–121 (1838)
20. B.J. West, Exact solution to fractional logistic equation. Physica A **429**, 103–108 (2015)
21. M. Zayernouri, G.E. Karniadakis, Fractional spectral collocation method. SIAM J. Sci. Comput. **36**, A40–A62 (2014)

The Complete Length Twelve Parametrized Wavelets

David W. Roach

Abstract In this paper, a complete parametrization of the length twelve wavelets is given for the dilation coefficients of the trigonometric polynomials, $m(\omega)$, that satisfy the necessary conditions for orthogonality, that is $m(0) = \sqrt{2}$ and $|m(\omega)|^2 + |m(\omega + \pi)|^2 = 2$. This parametrization has five free parameters and has a simple compatibility with the shorter length parametrizations for some specific choices of the free parameters. These wavelets have varying numbers of vanishing moments and regularity, but continuously transform from one to the other with the perturbation of the free parameters. Finally, we graph some example scaling functions from the parametrization which includes the standard Daubechies wavelets and some new wavelets that perform better than the CDF biorthogonal 9/7 wavelet in an image compression experiment on some fingerprint images.

Keywords Wavelets · Orthogonal · Scaling functions parametrization · Vanishing moments · Image compression

1 Introduction

Wavelets have become a standard approximation theory tool for efficiently representing complex data sets that require approximation at multiple scales. In her seminal work [2], Daubechies developed the theory of compactly supported orthogonal wavelets with dilation factor two. In this theory, she presented a discrete set of compactly supported wavelets with the specific desirable properties of minimal phase, maximum number of vanishing moments, and least asymmetry. She recognized that there were a whole continuum of compactly supported wavelets without those properties and that some of them could be of use in signal processing applications. This paper presents a necessary and sufficient parametrization of the length twelve refinable functions that satisfy the necessary condition for orthogonality,

---

D.W. Roach (✉)
Murray State University, Murray, KY 42071, USA
e-mail: droach@murraystate.edu

G.E. Fasshauer and L.L. Schumaker (eds.), *Approximation Theory XV: San Antonio 2016*, Springer Proceedings in Mathematics & Statistics 201,
DOI 10.1007/978-3-319-59912-0_16

$|m(\omega)|^2 + |m(\omega + \pi)|^2 = 2$. This parametrization includes all the scaling functions of length 12 and consequently all the wavelets of length 12 for any choice of the free parameters. The standard Daubechies wavelets are included along with the continuum of wavelets that have varying numbers of vanishing moments. In fact, the optimal choice of the parametrized wavelets for most test images in a compression scheme was found to be wavelets with only the zeroth vanishing moment. In earlier papers [4, 8, 9], we have presented the complete parametrizations for the shorter length wavelets of 10 or less. In [10], a parametrization of a subclass of the length 12 wavelets was given that only had three free parameters where extra conditions that were not necessary were employed to make the construction easier.

Other researchers have investigated the parametrization of orthogonal wavelets (see [14]). It appears that Schneid and Pittner [12] were the first to give formulas that would lead to the explicit parametrizations for the class of finite length orthogonal scaling functions after finding the Kronecker product of some matrices for wavelet lengths of two through ten, but the explicit parametrizations beyond six were not constructed. Colella and Heil investigated the length four parametrization in [1]. Others have constructed parametrizations for biorthogonal wavelets and multiwavelets (see [3, 7]). Regensburger, in [6], constructed the explicit parametrizations for the orthogonal scaling functions with multiple vanishing moments up to length ten by first solving the linear system of equations that result from the vanishing moment conditions and then solving the necessary condition for orthogonality.

In this current work, we give an explicit parametrization for all compactly supported refinable functions of length 12 that satisfy the necessary conditions for orthogonality, that is $m(0) = \sqrt{2}$ and $|m(\omega)|^2 + |m(\omega + \pi)|^2 = 2$. To illustrate the use of such a parametrization, we give a few example wavelets from the continuum that have a varying number of vanishing moments and yet perform better than the CDF biorthogonal 9/7 wavelet in an image compression scheme with numerous fingerprint images.

2 The Nonlinear Equations

The necessary conditions for orthogonality are well known in the literature (see [2, 5], and others). A scaling function φ that satisfies the dilation equation

$$\varphi(x) = \sum_{k=0}^{N} h_k \varphi(2x - k)$$

has an associated trigonometric polynomial m of degree N which can be be expressed as

$$m(\omega) = \sum_{k=0}^{N} h_k e^{ik\omega}.$$

Moreover, it is well known that m can be written as an infinite product. In order for this product to converge, m must not vanish at the origin, i.e., $m(0) = c \neq 0$. This condition immediately implies that

$$\sum_{k=0}^{N} h_k = \sqrt{2}. \tag{1}$$

Moreover, the necessary condition for the orthonormality of φ with its integer shifts is given by

$$|m(\omega)|^2 + |m(\omega + \pi)|^2 = 2. \tag{2}$$

This condition is equivalent to the dilation coefficients satisfying a system of non-linear equations, specifically

$$\sum_{k=0}^{N-2j} h_k h_{k+2j} = \delta(j), \quad j = 0, \ldots, \frac{N-1}{2}$$

where $\delta(0) = 1$ and $\delta(j) = 0$ for $j \neq 0$.

For the length twelve case (i.e., $N = 11$) that we are currently considering, we have the following underdetermined nonlinear system:

$$\begin{aligned}
h_0^2 + h_1^2 + h_2^2 + h_3^2 + h_4^2 + h_5^2 + h_6^2 + h_7^2 + h_8^2 + h_9^2 + h_{10}^2 + h_{11}^2 &= 1 \\
h_0h_2 + h_1h_3 + h_2h_4 + h_3h_5 + h_4h_6 + h_5h_7 + h_6h_8 + h_7h_9 + h_8h_{10} + h_9h_{11} &= 0 \\
h_0h_4 + h_1h_5 + h_2h_6 + h_3h_7 + h_4h_8 + h_5h_9 + h_6h_{10} + h_7h_{11} &= 0 \\
h_0h_6 + h_1h_7 + h_2h_8 + h_3h_9 + h_4h_{10} + h_5h_{11} &= 0 \\
h_0h_8 + h_1h_9 + h_2h_{10} + h_3h_{11} &= 0 \\
h_0h_{10} + h_1h_{11} &= 0.
\end{aligned}$$

Additionally, these two conditions (1) and (2) imply the zeroth vanishing moment condition $m(\pi) = 0$ or equivalently the linear equations

$$\sum_{k=0}^{(N-1)/2} h_{2k} = \sum_{k=0}^{(N-1)/2} h_{2k+1} = \frac{\sqrt{2}}{2}.$$

Because the products of the coefficients in the system of nonlinear equations have the pattern that the odd indices multiply the other odd indices and similarly for the even indices, it is convenient to separate the even indices from the odd indices in the following fashion:

$$m(\omega) = \sum_{k=0}^{n} a_k e^{2ki\omega} + b_k e^{(2k+1)i\omega}$$

where we let $n = (N - 1)/2$. Note, since there are no odd length scaling functions satisfying the necessary condition for orthogonality, N will always be an odd integer.

As a means of summary with this alternate notation, we conclude with the following statements. Given a scaling function φ and its associated trigonometric polynomial

$$m(\omega) = \sum_{k=0}^{n} a_k e^{2ki\omega} + b_k e^{(2k+1)i\omega}$$

of degree $2n + 1$, the necessary condition for orthogonality,

$$|m(\omega)|^2 + |m(\omega + \pi)|^2 = 2,$$

is equivalent to the following system of nonlinear equations:

$$\sum_{k=0}^{n-j} a_k a_{k+j} + b_k b_{k+j} = \delta(j), \quad j = 0, \ldots, n-1$$

where $\delta(0) = 1$ and $\delta(j) = 0$ for $j \neq 0$.

3 Length Four

Although the length four parametrization is well known (see [1, 8, 14]), it is used in the construction of the length twelve parametrization and is presented here for completeness.

For length four ($N = 3$ and $n = 1$), the nonlinear system of equations is

$$a_0 + a_1 = \frac{\sqrt{2}}{2} \tag{3}$$

$$b_0 + b_1 = \frac{\sqrt{2}}{2} \tag{4}$$

$$a_0^2 + a_1^2 + b_0^2 + b_1^2 = 1 \tag{5}$$

$$a_0 a_1 + b_0 b_1 = 0. \tag{6}$$

Subtracting twice Eq. (6) from Eq. (5) gives

$$(a_0 - a_1)^2 + (b_0 - b_1)^2 = 1.$$

This equation allows the introduction of a free parameter, that is

$$a_0 - a_1 = \cos\left(2\alpha - \frac{\pi}{4}\right) \tag{7}$$

$$b_0 - b_1 = \sin\left(2\alpha - \frac{\pi}{4}\right) \tag{8}$$

where we chose the angle argument so that the parametrization reduces to the length two case whenever $\alpha = \pi/4$. Combining equations (3) and (4) with (7) and (8) followed by some simplification gives the length four parameterization:

$$a_0 = \frac{\sqrt{2}}{2}\cos\alpha\,(\cos\alpha + \sin\alpha)\,, \tag{9}$$

$$b_0 = \frac{\sqrt{2}}{2}\sin\alpha\,(\cos\alpha + \sin\alpha)\,, \tag{10}$$

$$a_1 = \frac{\sqrt{2}}{2}\sin\alpha\,(-\cos\alpha + \sin\alpha)\,, \tag{11}$$

$$b_1 = \frac{\sqrt{2}}{2}\cos\alpha\,(\cos\alpha - \sin\alpha)\,. \tag{12}$$

Other versions of this parametrization are well known (see [1, 8, 14], and others). It should be noted that this parametrization is a necessary representation for the coefficients and upon substituting them back into the system of equations (3)–(6), we see that they are also sufficient.

4 Length Eight

For length eight ($N = 7$ and $n = 3$), the solution is given in [8] and presented here to aid in the understanding of the longer case. For this situation, the nonlinear system of equations is given by

$$a_0 + a_1 + a_2 + a_3 = \frac{\sqrt{2}}{2} \tag{13}$$

$$b_0 + b_1 + b_2 + b_3 = \frac{\sqrt{2}}{2} \tag{14}$$

$$a_0^2 + a_1^2 + a_2^2 + a_3^2 + b_0^2 + b_1^2 + b_2^2 + b_3^2 = 1 \tag{15}$$

$$a_0a_1 + a_1a_2 + a_2a_3 + b_0b_1 + b_1b_2 + b_2b_3 = 0 \tag{16}$$

$$a_0a_2 + a_1a_3 + b_0b_2 + b_1b_3 = 0 \tag{17}$$

$$a_0a_3 + b_0b_3 = 0. \tag{18}$$

Relating it back to the length four case, the sums $a_0 + a_2$, $a_1 + a_3$, $b_0 + b_2$, and $b_1 + b_3$ satisfy the system of equations for length four, i.e.,

$$(a_0 + a_2) + (a_1 + a_3) = \frac{\sqrt{2}}{2}$$
$$(b_0 + a_2) + (b_1 + b_3) = \frac{\sqrt{2}}{2}$$
$$(a_0 + a_2)^2 + (a_1 + a_3)^2 + (b_0 + b_2)^2 + (b_1 + b_3)^2 = 1$$
$$(a_0 + a_2)(a_1 + a_3) + (b_0 + b_2)(b_1 + b_3) = 0.$$

Thus, we can use the length four parameterization for these sums, i.e.,

$$a_0 + a_2 = \frac{\sqrt{2}}{2} \cos\alpha(\cos\alpha + \sin\alpha) \tag{19}$$
$$b_0 + b_2 = \frac{\sqrt{2}}{2} \sin\alpha(\cos\alpha + \sin\alpha) \tag{20}$$
$$a_1 + a_3 = \frac{\sqrt{2}}{2} \sin\alpha(-\cos\alpha + \sin\alpha) \tag{21}$$
$$b_1 + b_3 = \frac{\sqrt{2}}{2} \cos\alpha(\cos\alpha - \sin\alpha). \tag{22}$$

The sum and difference of Equation (15) and twice Equation (18) give the two equations:

$$(a_0 + a_3)^2 + (b_0 + b_3)^2 = 1 - a_1^2 - a_2^2 - b_1^2 - b_2^2 \tag{23}$$
$$(a_0 - a_3)^2 + (b_0 - b_3)^2 = 1 - a_1^2 - a_2^2 - b_1^2 - b_2^2 := q^2. \tag{24}$$

where q^2 is yet to be determined. Using the fact that the right-hand sides of equations (23) and (24) are equivalent, we introduce two new parameters β and γ in the following fashion:

$$a_0 + a_3 = q\cos\beta \tag{25}$$
$$b_0 + b_3 = q\sin\beta \tag{26}$$
$$a_0 - a_3 = q\cos(2\gamma - \beta) \tag{27}$$
$$b_0 - b_3 = q\sin(2\gamma - \beta). \tag{28}$$

Combining these equations with Eqs. (19)–(22), we have

$$a_0 = q\cos(\beta - \gamma)\cos\gamma \tag{29}$$
$$b_0 = q\cos(\beta - \gamma)\sin\gamma \tag{30}$$
$$a_1 = \frac{1}{\sqrt{2}} \sin\alpha(-\cos\alpha + \sin\alpha) + q\sin(\beta - \gamma)\sin\gamma \tag{31}$$
$$b_1 = \frac{1}{\sqrt{2}} \cos\alpha(\cos\alpha - \sin\alpha) - q\sin(\beta - \gamma)\cos\gamma \tag{32}$$

$$a_2 = \frac{1}{\sqrt{2}} \cos\alpha(\cos\alpha + \sin\alpha) - q\cos(\beta - \gamma)\cos\gamma \tag{33}$$

$$b_2 = \frac{1}{\sqrt{2}} \sin\alpha(\cos\alpha + \sin\alpha) - q\cos(\beta - \gamma)\sin\gamma \tag{34}$$

$$a_3 = -q\sin(\beta - \gamma)\sin\gamma \tag{35}$$

$$b_3 = q\sin(\beta - \gamma)\cos\gamma \tag{36}$$

$$q = \frac{1}{\sqrt{2}} \cos(\alpha - \gamma)\left(\cos(\alpha + \beta - \gamma) + \sin(\alpha + \beta - \gamma)\right) \tag{37}$$

where q is computed from the quadratic equation resulting from Eq. (17).

Upon substituting this necessary parametrization back into the system of equations (13)–(18), we see that they are also sufficient.

5 Length Twelve Solution

For the construction of the complete parametrizations for length six, eight, and ten, see [8, 9]. In those constructions, the parametrization describes the complete set of all trigonometric polynomials that satisfy the necessary conditions for the orthogonality of the associated scaling function and their integer shifts. In those works, the coefficients were not normalized and require a multiple of $\sqrt{2}$ when implementing them, but here we have included the normalization in the construction. In [10], a three parameter subclass of the length twelve wavelets is given. For the length twelve case ($N = 11$ and $n = 5$), we will introduce five free parameters to linearize the system of quadratic equations.

Using the necessary condition for orthogonality (2), we have the following nonlinear system of equations:

$$a_0 + a_1 + a_2 + a_3 + a_4 + a_5 = \frac{\sqrt{2}}{2} \tag{38}$$

$$b_0 + b_1 + b_2 + b_3 + b_4 + b_5 = \frac{\sqrt{2}}{2} \tag{39}$$

$$a_0^2 + a_1^2 + a_2^2 + a_3^2 + a_4^2 + a_5^2 + b_0^2 + b_1^2 + b_2^2 + b_3^2 + b_4^2 + b_5^2 = 1 \tag{40}$$

$$a_0a_1 + a_1a_2 + a_2a_3 + a_3a_4 + a_4a_5 + b_0b_1 + b_1b_2 + b_2b_3 + b_3b_4 + b_4b_5 = 0 \tag{41}$$

$$a_0a_2 + a_1a_3 + a_2a_4 + a_3a_5 + b_0b_2 + b_1b_3 + b_2b_4 + b_3b_5 = 0 \tag{42}$$

$$a_0a_3 + a_1a_4 + a_2a_5 + b_0b_3 + b_1b_4 + b_2b_5 = 0 \tag{43}$$

$$a_0a_4 + a_1a_5 + b_0b_4 + b_1b_5 = 0 \tag{44}$$

$$a_0a_5 + b_0b_5 = 0 \tag{45}$$

An important step in the construction is establishing the connection between the sums of the even- and odd-indexed coefficients back to the length four parametrization. Here is the main result.

Theorem 1 *For any real numbers $\alpha, \beta, \gamma, \delta$ and θ, the trigonometric polynomial $m(\omega)$ of the form*

$$m(\omega) = \sum_{k=0}^{5} a_k e^{2ki\omega} + b_k e^{(2k+1)i\omega}$$

with coefficients defined as

$$\begin{aligned}
a_0 &= \frac{p}{2}\Big(\cos\beta + \cos(2\theta - \beta)\Big)\\
b_0 &= \frac{p}{2}\Big(\sin\beta + \sin(2\theta - \beta)\Big)\\
a_1 &= \frac{\sqrt{2}}{4}\sin\alpha(-\cos\alpha + \sin\alpha) + \frac{q}{2}\cos\beta - \frac{r}{2}\sin\beta - \frac{s}{2}\cos(2\theta - \beta) + \frac{t}{2}\sin(2\theta - \beta) - a_5\\
b_1 &= \frac{\sqrt{2}}{4}\cos\alpha(\cos\alpha - \sin\alpha) + \frac{q}{2}\sin\beta + \frac{r}{2}\cos\beta - \frac{s}{2}\sin(2\theta - \beta) - \frac{t}{2}\cos(2\theta - \beta) - b_5\\
a_2 &= \frac{\sqrt{2}}{4}\cos\alpha(\cos\alpha + \sin\alpha) - \frac{q}{2}\cos\beta + \frac{r}{2}\sin\beta - \frac{s}{2}\cos(2\theta - \beta) + \frac{t}{2}\sin(2\theta - \beta)\\
b_2 &= \frac{\sqrt{2}}{4}\sin\alpha(\cos\alpha + \sin\alpha) - \frac{q}{2}\sin\beta - \frac{r}{2}\cos\beta - \frac{s}{2}\sin(2\theta - \beta) - \frac{t}{2}\cos(2\theta - \beta)\\
a_3 &= \frac{\sqrt{2}}{4}\sin\alpha(-\cos\alpha + \sin\alpha) - \frac{q}{2}\cos\beta + \frac{r}{2}\sin\beta + \frac{s}{2}\cos(2\theta - \beta) - \frac{t}{2}\sin(2\theta - \beta)\\
b_3 &= \frac{\sqrt{2}}{4}\cos\alpha(\cos\alpha - \sin\alpha) - \frac{q}{2}\sin\beta - \frac{r}{2}\cos\beta + \frac{s}{2}\sin(2\theta - \beta) + \frac{t}{2}\cos(2\theta - \beta)\\
a_4 &= \frac{\sqrt{2}}{4}\cos\alpha(\cos\alpha + \sin\alpha) + \frac{q}{2}\cos\beta - \frac{r}{2}\sin\beta + \frac{s}{2}\cos(2\theta - \beta) - \frac{t}{2}\sin(2\theta - \beta) - a_0\\
b_4 &= \frac{\sqrt{2}}{4}\sin\alpha(\cos\alpha + \sin\alpha) + \frac{q}{2}\sin\beta + \frac{r}{2}\cos\beta + \frac{s}{2}\sin(2\theta - \beta) + \frac{t}{2}\cos(2\theta - \beta) - b_0\\
a_5 &= \frac{p}{2}\Big(\cos\beta - \cos(2\theta - \beta)\Big)\\
b_5 &= \frac{p}{2}\Big(\sin\beta - \sin(2\theta - \beta)\Big)\\
u &= \cos(\beta - \gamma) + \cos(2\alpha + \beta - \delta - 2\theta)\\
p &= \frac{\sqrt{2}}{8}u\Big(\cos\delta + \sin\delta + \cos\gamma + \sin\gamma\Big)\\
q &= \frac{\sqrt{2}}{4}\Big(-\cos(2\alpha + \beta - 2\theta) - \sin(2\alpha + \beta - 2\theta) + u(\cos\delta + \sin\delta)\Big)\\
r &= \frac{\sqrt{2}}{4}\Big(-\cos\beta + \sin\beta + u(\cos\gamma - \sin\gamma)\Big)\\
s &= \frac{\sqrt{2}}{4}\Big(-\cos\beta - \sin\beta + u(\cos\gamma + \sin\gamma)\Big)\\
t &= \frac{\sqrt{2}}{4}\Big(\cos(2\alpha + \beta - 2\theta) - \sin(2\alpha + \beta - 2\theta) + u(-\cos\delta + \sin\delta)\Big)
\end{aligned}$$

satisfies

$$m(0) = \sqrt{2} \text{ and } |m(\omega)|^2 + |m(\omega + \pi)|^2 = 2.$$

Proof To begin with, the sums $a_0 + a_2 + a_4$, $a_1 + a_3 + a_5$, $b_0 + b_2 + b_4$, and $b_1 + b_3 + b_5$ satisfy the system of equations associated with the length four parametrization, i.e.,

$$\begin{aligned}
(a_0 + a_2 + a_4) + (a_1 + a_3 + a_5) &= \frac{\sqrt{2}}{2} \\
(b_0 + b_2 + b_4) + (b_1 + b_3 + b_5) &= \frac{\sqrt{2}}{2} \\
(a_0 + a_2 + a_4)^2 + (a_1 + a_3 + b_5)^2 + (b_0 + b_2 + b_4)^2 + (b_1 + b_3 + b_5)^2 &= 1 \\
(a_0 + a_2 + a_4)(a_1 + a_3 + b_5) + (b_0 + b_2 + b_4)(b_1 + b_3 + b_5) &= 0.
\end{aligned}$$

The third equation is equivalent to the sum of equations (40), (42), and (44), and the last one is equivalent to the sum of equations (41), (43), and (45). Therefore, we can use the length four parametrization for these sums, i.e.,

$$\begin{aligned}
a_0 + a_2 + a_4 &= \frac{\sqrt{2}}{2} \cos\alpha \, (\cos\alpha + \sin\alpha) \, , \\
b_0 + b_2 + b_4 &= \frac{\sqrt{2}}{2} \sin\alpha \, (\cos\alpha + \sin\alpha) \, , \\
a_1 + a_3 + a_5 &= \frac{\sqrt{2}}{2} \sin\alpha \, (-\cos\alpha + \sin\alpha) \, , \\
b_1 + b_3 + b_5 &= \frac{\sqrt{2}}{2} \cos\alpha \, (\cos\alpha - \sin\alpha) \, .
\end{aligned}$$

In an effort to linearize the system of equations, note that the sum and difference of Equation (40) and twice Equation (45) give the two equations:

$$(a_0 + a_5)^2 + (b_0 + b_5)^2 = \frac{\sqrt{2}}{2} - a_1^2 - a_2^2 - a_3^2 - a_4^2 - b_1^2 - b_2^2 - b_3^2 - b_4^2 \tag{46}$$

$$(a_0 - a_5)^2 + (b_0 - b_5)^2 = \frac{\sqrt{2}}{2} - a_1^2 - a_2^2 - a_3^2 - a_4^2 - b_1^2 - b_2^2 - b_3^2 - b_4^2 := p^2 \tag{47}$$

Although the right-hand side, p^2, has not yet been determined, we use the fact that the right-hand sides of equations (46) and (47) are equivalent and introduce two new free parameters β and γ in the following fashion,

$$a_0 + a_5 = p\cos\beta \tag{48}$$
$$b_0 + b_5 = p\sin\beta \tag{49}$$
$$a_0 - a_5 = p\cos(2\theta - \beta) \tag{50}$$
$$b_0 - b_5 = p\sin(2\theta - \beta), \tag{51}$$

which can be solved directly for a_0, a_5, b_0, and b_5. As before, the angle arguments in (50) and (51) are chosen so that the length 12 solution reduces to the length 10 solution whenever $\theta = \beta$. There are now 8 linear equations that are each a necessary constraint for all trigonometric polynomials of length twelve which satisfy the necessary conditions for orthogonality. It should be noted that the nonlinear equation (45) is satisfied.

Next, we temporarily introduce four parameters $\tilde{q}, \tilde{r}, \tilde{s}$, and $\tilde{t}$ to complete the set of twelve linear equations, i.e.,

$$a_0 - a_2 + a_4 = \tilde{q} \tag{52}$$
$$b_0 - b_2 + b_4 = \tilde{r} \tag{53}$$
$$a_1 - a_3 + a_5 = \tilde{s} \tag{54}$$
$$b_1 - b_3 + b_5 = \tilde{t}. \tag{55}$$

This set of linear equations can be solved for the remaining a_i's and b_i's. After substituting these expressions into the remaining nonlinear equations, we see a tangled involvement of the parameters $\tilde{q}, \tilde{r}, \tilde{s}$, and $\tilde{t}$ with the trigonometric expressions $\sin\beta, \cos\beta, \cos(2\theta - \beta)$, and $\sin(2\theta - \beta)$. With these expressions in mind, we rotate $\tilde{q}, \tilde{r}, \tilde{s}$, and $\tilde{t}$ using an orthogonal matrix to another vector with parameters q, r, s, and t as follows:

$$\begin{bmatrix} \cos\beta & -\sin\beta & \cos(2\theta-\beta) & -\sin(2\theta-\beta) \\ \sin\beta & \cos\beta & \sin(2\theta-\beta) & \cos(2\theta-\beta) \\ \cos\beta & -\sin\beta & -\cos(2\theta-\beta) & \sin(2\theta-\beta) \\ \sin\beta & \cos\beta & -\sin(2\theta-\beta) & -\cos(2\theta-\beta) \end{bmatrix} \begin{bmatrix} q \\ r \\ s \\ t \end{bmatrix} = \begin{bmatrix} \tilde{q} \\ \tilde{r} \\ \tilde{s} \\ \tilde{t} \end{bmatrix} \tag{56}$$

Because of the orthogonality of the matrix, this rotation is invertible and does not overly constrain the values of q, r, s, and t. So, the last four linear equations become

$$a_0 - a_2 + a_4 = q\cos\beta - r\sin\beta + s\cos(2\theta-\beta) - t\sin(2\theta-\beta), \tag{57}$$
$$b_0 - b_2 + b_4 = q\sin\beta + r\cos\beta + s\sin(2\theta-\beta) + t\cos(2\theta-\beta), \tag{58}$$
$$a_1 - a_3 + a_5 = q\cos\beta - r\sin\beta - s\cos(2\theta-\beta) + t\sin(2\theta-\beta), \tag{59}$$
$$b_1 - b_3 + b_5 = q\sin\beta + r\cos\beta - s\sin(2\theta-\beta) - t\cos(2\theta-\beta). \tag{60}$$

After solving for the a_i's and b_i's and substituting them into Eq. (44), we get a necessary expression for p, i.e.,

$$p = \frac{1}{8}\Big(4q + 4s + \sqrt{2}\cos\beta + \sqrt{2}\cos(2\alpha + \beta - 2\theta)$$
$$+\sqrt{2}\sin\beta + \sqrt{2}\sin(2\alpha + \beta - 2\theta)\Big). \quad (61)$$

At this point, the last of the two nonlinear Eqs. (44) and (45) are both satisfied. Additionally, the Eqs. (40) and (42) reduce to the single equation

$$q^2 + r^2 + s^2 + t^2 = \frac{1}{2} \quad (62)$$

and likewise Eqs. (41) and (43) become

$$\begin{aligned} q^2 - r^2 - s^2 + t^2 &= \\ &\frac{\sqrt{2}}{2}(-(r+s)\cos\beta + (r-s)\sin\beta \\ &\quad +(q-t)\cos(2\alpha+\beta-2\theta) + (q+t)\sin(2\alpha+\beta-2\theta)). \end{aligned} \quad (63)$$

Using the quadratic form of Eq. (63), we now have the first two terms in the expressions for q, r, s, and t along with temporary parameters $\bar{q}, \bar{r}, \bar{s}$, and $\bar{t}$, namely

$$q = \frac{\sqrt{2}}{4}\Big(-\cos(2\alpha+\beta-2\theta) - \sin(2\alpha+\beta-2\theta) + \bar{q}\Big) \quad (64)$$
$$r = \frac{\sqrt{2}}{4}\Big(-\cos\beta + \sin\beta + \bar{r}\Big) \quad (65)$$
$$s = \frac{\sqrt{2}}{4}\Big(-\cos\beta - \sin\beta + \bar{s}\Big) \quad (66)$$
$$t = \frac{\sqrt{2}}{4}\Big(\cos(2\alpha+\beta-2\theta) - \sin(2\alpha+\beta-2\theta) + \bar{t}\Big) \quad (67)$$

which transform Equation (62) into

$$\bar{q}^2 + \bar{t}^2 = \bar{r}^2 + \bar{s}^2. \quad (68)$$

Equation (68) can be solved by introducing three free parameters u, γ, and δ where

$$\bar{q} = \sqrt{2}u\cos(\delta - \pi/4) \quad (69)$$
$$\bar{r} = \sqrt{2}u\cos(\gamma - \pi/4) \quad (70)$$
$$\bar{s} = \sqrt{2}u\sin(\gamma - \pi/4) \quad (71)$$
$$\bar{t} = \sqrt{2}u\sin(\delta - \pi/4). \quad (72)$$

With these new parameters, the nonlinear equations (40) and (42) are now satisfied, that is,

$$q = \frac{\sqrt{2}}{4}\Big(-\cos(2\alpha+\beta-2\theta) - \sin(2\alpha+\beta-2\theta) + \sqrt{2}u\cos(\delta-\pi/4)\Big) \quad (73)$$

$$r = \frac{\sqrt{2}}{4}\Big(-\cos\beta + \sin\beta - \sqrt{2}u\sin(\gamma-\pi/4)\Big) \quad (74)$$

$$s = \frac{\sqrt{2}}{4}\Big(-\cos\beta - \sin\beta + \sqrt{2}u\cos(\gamma-\pi/4)\Big) \quad (75)$$

$$t = \frac{\sqrt{2}}{4}\Big(\cos(2\alpha+\beta-2\theta) - \sin(2\alpha+\beta-2\theta) + \sqrt{2}u\sin(\delta-\pi/4)\Big). \quad (76)$$

The shifted angle arguments, $\delta - \pi/4$ and $\gamma - \pi/4$, were chosen to allow for compatibility with the shorter length parametrizations (if $\theta = \beta$, it reduces to the length 10 solution, additionally if $\delta = \gamma$, then length 8, $\gamma = \alpha$ is length 6, and $\beta = \alpha$ together with the other parameter restrictions gives the length four solution). The two remaining nonlinear equations (41) and (43) simplify to a quadratic equation that yields the form of the remaining non-free parameter u, that is

$$u = (\cos(\beta-\gamma) + \cos(2\alpha+\beta-\delta-2\theta)). \quad (77)$$

At each step of the construction, the parametrized constraints were necessary, and upon substituting them back into the nonlinear system, they are also sufficient. This completes the proof. □

6 A Numerical Experiment

A few parametrized wavelets, for example, were selected where the parameters are given in Table 1 with the graphs of their associated scaling functions given in Fig. 1. The first two examples are the parameters needed to get the standard Daubechies scaling functions D12 and LA12 (the least asymmetric Daubechies scaling function of length 12) depicted in the top row of Fig. 1. The scaling function B12 was showcased in [10] as an example from the subclass of the length 12 parametrized wavelets that performed well compared to the 9/7 biorthogonal. The example SF12 was chosen because of its superior performance as compared to the biorthogonal 9/7 in an image compression scheme for a class of fingerprint images. It should be noted that SF12 has only one vanishing moment and is depicted in the lower right of Fig. 1. The linear and quadratic moments of SF12 are fairly small, but the moments above two are quite large.

In order to establish the usefulness of the length twelve parametrization, a parameter search was made in order to find a length 12 wavelet that performed better than the CDF biorthogonal 9/7 wavelet in an image compression scheme for a set of fingerprint images.

Table 1 Parameters associated with some example length twelve parametrized scaling functions

	Free parameters				
Scaling function	α	β	γ	δ	θ
D12	1.95163679104	1.33934218288	1.6336995561	1.7734712041	1.34900022990
LA12	1.04467309738	−0.245858720144	2.37236277960	1.05183797421	0.22291045224
B12	0.67119522483	2.1784697955	−2.3561944906	−2.3561944909	2.2999564636
SF12	1.960084456	4.944302102	2.528015388	4.002884472	2.62299263

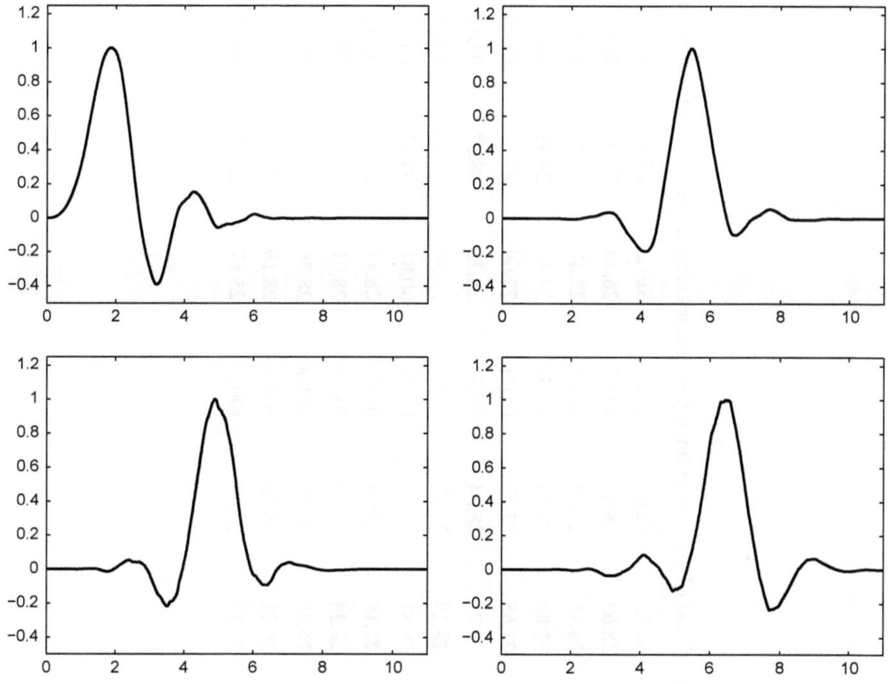

Fig. 1 Graphs for the parametrized length twelve scaling functions given in Table 1. D12 (*upper-left*), LA12 (*upper-right*), B12 (*lower-left*), and SF12 (*lower-right*)

The details of the numerical experiment are as follows:

- Eight-level decomposition with periodic boundaries for SF12 and symmetric boundaries for the CDF 9/7.
- Embedded zero-tree (EZW) compression (see [11, 13]) with a file size compression ratio of 32:1. For this experiment, all of the images are 512×512 with a PGM file size of 256 Kb and a compressed file size of 8 Kb. This particular EZW implementation is not completely optimized and would not necessarily yield the maximum PSNR possible but serves well as a comparative measure of the true compressibility of the wavelet decomposition.

Table 2 Peak signal-to-noise ratio (PSNR) for SF12 and CDF9/7 on 48 fingerprint images compressed at 32:1. Boldface indicates a better PSNR

Image	SF12	CDF9/7	Image	SF12	CDF9/7	Image	SF12	CDF9/7	Image	SF12	CDF9/7
f0001	**26.82**	26.78	f0013	**28.63**	28.62	f0027	**26.48**	26.05	f0041	**27.75**	27.61
f0002	**27.68**	27.58	f0014	**26.45**	26.37	f0028	**27.47**	27.47	f0042	28.33	**28.38**
f0003	30.85	**31.04**	f0015	**27.00**	26.80	f0030	29.37	**29.41**	f0043	**26.05**	25.56
f0004	32.14	**32.36**	f0016	**27.65**	27.47	f0031	**27.69**	27.53	f0044	**27.28**	26.83
f0005	**23.26**	22.96	f0017	28.77	**28.84**	f0032	28.31	**28.32**	f0046	24.53	**24.78**
f0006	**26.93**	26.62	f0018	**28.29**	28.03	f0033	27.48	**27.35**	f0047	28.33	**28.38**
f0007	**28.11**	27.83	f0019	**28.81**	28.71	f0034	30.81	**30.85**	f0048	**29.03**	28.96
f0008	32.74	**32.89**	f0021	**27.15**	26.96	f0035	**28.11**	27.93	f0049	**28.36**	28.25
f0009	**27.62**	27.29	f0023	**28.24**	28.05	f0036	**28.01**	27.78	f0050	31.50	**31.58**
f0010	**27.54**	27.45	f0024	**26.87**	26.73	f0038	**28.76**	28.69	f0051	31.32	**31.45**
f0011	**27.05**	26.69	f0025	26.26	**26.33**	f0039	**28.19**	28.05	f0052	27.60	**27.66**
f0012	30.57	**31.05**	f0026	**27.77**	27.44	f0040	**25.67**	25.47	f0053	28.63	**28.93**

- Eight-level reconstruction followed by a peak signal-to-noise ratio (PSNR), i.e.,

$$RMSE = \sqrt{\frac{1}{512^2}\sum_{i=1}^{512}\sum_{i=1}^{512}|A_{i,j} - \tilde{A}_{i,j}|^2}$$

$$PSNR = 20\log_{10}\left(\frac{255}{RMSE}\right)$$

  where $A_{i,j}$ is the original matrix of grayscale values and $\tilde{A}_{i,j}$ is the compressed version.
- The fingerprint files were obtained from National Institute of Standards and Technology (NIST) 8-Bit Gray Scale Images of Fingerprint Image Groups (FIGS) - NIST Special Database 4. A set of 48 fingerprint images were used in the comparison with the file names as given in Table 2.
- **Experiment Results:** The parametrized length 12 wavelet SF12 outperformed the 9/7 in PSNR value in 67% of the images tested (a higher PSNR is equivalent to a lower RMSE). For the 32 images where SF12 did better, the average difference between their PSNR values was 0.20 whereas the 16 images where 9/7 did better, the average difference was only 0.14. Moreover, on several of the images where 9/7 was better there was significant white space around the fingerprint image. See Table 2 for the comparison details.

7 Conclusions

In this paper, we have constructed an explicit necessary and sufficient parametrization of the class of length twelve orthogonal wavelets with dilation factor two. By perturbing the five free parameters, a continuum of wavelets can be used to analyze data sets as opposed to just using a handful of standard wavelets. Among the parametrization, we were able to find a wavelet that performed better than the CDF biorthogonal 9/7 wavelet in an image compression scheme with a set of 48 finger print images. The parametrized wavelet SF12 did better in terms of PSNR on 67% of the compressed images and had a higher average improvement as well. Although more work needs to be done, one conclusion that must be made is that there is a whole continuum of wavelets available that may have been overlooked due to their lack of vanishing moments. The requirement that a wavelet has a high number of vanishing moments may be too restrictive for many applications, and the optimal choice may be among the parametrized wavelets.

References

1. D. Colella, C. Heil, The characterization of continuous, four-coefficient scaling functions and wavelets, IEEE Trans. Inf. Theory (Special Issue on Wavelet Transforms and Multiresolution Signal Analysis) **38** 876–881 (1992)
2. I. Daubechies, *Ten Lectures on Wavelets* (SIAM, Philadelphia, 1992)
3. Q.T. Jiang, Paramtrization of m-channel orthogonal multifilter banks. Advan. Comput. Math. **12**, 189–211 (2000)
4. M.J. Lai, D.W. Roach, Parameterizations of univariate orthogonal wavelets with short support, in *Approximation Theory X: Splines, Wavelets, and Applications*, ed. by C.K. Chui, L.L. Schumaker, J. Stockler (Vanderbilt University Press, Nashville, 2002), pp. 369–384
5. W. Lawton, Necessary and sufficient conditions for constructing orthonormal wavelet bases. J. Math. Phys. **32**, 57–61 (1991)
6. G. Regensburger, Parametrizing compactly supported orthonormal wavelets by discrete moments. Appl. Algebra Eng. Commun. Comput. **18**(6) 583–601 (2007)
7. H.L. Resnikoff, J. Tian, R.O. Wells Jr., Biorthogonal wavelet space: parametrization and factorization. SIAM J. Math. Anal. **33**(1), 194–215 (2001)
8. D.W. Roach, The Parameterization of the Length Eight Orthogonal Wavelets with No Parameter Constraints, in *Approximation Theory XII: San Antonio 2007*, ed. by M. Neamtu, L. Schumaker (Nashboro Press, 2008), pp. 332–347
9. D.W. Roach, Frequency selective parameterized wavelets of length ten. J. Concrete Appl. Math. **8**(1), 1675–1679 (2010)
10. D.W. Roach, A subclass of the length 12 orthogonal wavelets, in *Approximation Theory XIII: San Antonio 2010*, ed. by M. Neamtu, L. Schumaker. Proceedings in Mathematics, vol. 13 (Springer, Berlin, 2012), pp 263–275
11. A. Said, W.A. Pearlman, A new fast and efficient image codec based on set partitioning in hierarchical trees. IEEE Trans. Circuits Syst. Video Technol. **6**, 243–250 (1996)
12. J. Schneid, S. Pittner, On the parametrization of the coefficients of dilation equations for compactly supported wavelets. Computing **51**, 165–173 (1993)
13. J.M. Shapiro, Embedded image coding using zerotrees of wavelet coefficients. IEEE Trans. Signal Process. **41**, 3445–3462 (1993)
14. R.O. Wells Jr., Parametrizing smooth compactly supported wavelets. Trans. Am. Math. Soc. **338**, 919–931 (1993)
15. M.V. Wickerhauser, *Comparison of Picture Compression Methods: Wavelet, Wavelet Packet, and Local Cosine Transform Coding, Wavelets: Theory, Algorithms, and Applications (Taormina, 1993)* (Academic Press, San Diego, CA, 1994), pp. 585–621

Potential Theoretic Approach to Design of Accurate Numerical Integration Formulas in Weighted Hardy Spaces

Ken'ichiro Tanaka, Tomoaki Okayama and Masaaki Sugihara

Abstract We propose a method for designing accurate numerical integration formulas on weighted Hardy spaces, which are regarded as spaces of transformed integrands by some useful variable transformations such as the double-exponential transformation. We begin with formulating an optimality of numerical integration formulas in the space by using the norms of the error operators corresponding to those formulas. Then, we derive an expression of the minimum value of the norms, which gives a criterion for an optimal sequence of sampling points for numerical integration. Based on the expression, we propose an algorithm designing accurate numerical integration formulas on the space by a potential theoretic approach. The effectiveness of the designed formulas is supported by some numerical examples.

Keywords Weighted Hardy space · Numerical integration · Potential theory · Green potential

K. Tanaka (✉)
Department of Mathematical Informatics, Graduate School of Information Science and Technology, The University of Tokyo, 7-3-1, Hongo, Bunkyo-ku, Tokyo 113-8656, Japan
e-mail: kenichiro@mist.i.u-tokyo.ac.jp

T. Okayama
Department of Systems Engineering, Graduate School of Information Sciences, Hiroshima City University, 3-4-1, Ozuka-higashi, Asaminami-ku, Hiroshima 731-3194, Japan
e-mail: okayama@hiroshima-cu.ac.jp

M. Sugihara
Department of Physics and Mathematics, College of Science and Engineering, Aoyama Gakuin University, 5-10-1, Fuchinobe, Chuo-ku, Sagamihara-shi, Kanagawa 252-5258, Japan
e-mail: sugihara@gem.aoyama.ac.jp

G.E. Fasshauer and L.L. Schumaker (eds.), *Approximation Theory XV: San Antonio 2016*, Springer Proceedings in Mathematics & Statistics 201,
DOI 10.1007/978-3-319-59912-0_17

1 Introduction

This paper is concerned with numerical integration formulas of the form

$$\int_{-\infty}^{\infty} f(x)\,\mathrm{d}x \approx \sum_{p=1}^{\ell} \sum_{q=0}^{m_p-1} c_{pq} f^{(q)}(a_p), \quad 1 \le \ell \le 2N+1, \quad m_1 + \cdots + m_\ell = 2N+1, \tag{1}$$

which generally represents a $(2N+1)$-point quadrature formula on $\mathbf{R}$. A typical formula of the form is the (truncated) trapezoidal rule

$$\int_{-\infty}^{\infty} f(x)\,\mathrm{d}x \approx h \sum_{k=-N}^{N} f(kh), \tag{2}$$

which is obtained by choosing $c_{ij} = h$, $\ell = 2N+1$, $m_1 = \cdots = m_{2N+1} = 1$, and $a_i = (i - N - 1)h$ in (1).

The interval of integration in (1) is $(-\infty, \infty)$, but even for other interval I, we may apply the formula of the form (1) by using a variable transformation $t = \psi(x)$ as

$$\int_I F(t)\,\mathrm{d}t = \int_{-\infty}^{\infty} F(\psi(x))\psi'(x)\,\mathrm{d}x,$$

and putting $f(x) = F(\psi(x))\psi'(x)$. In fact, the following variable transformations for $I = (-1, 1)$ are well known:

$$\psi_1(x) = \tanh\left(\frac{x}{2}\right) \qquad \text{(TANH transformation [3, 6, 9 -- -11])}, \tag{3}$$

$$\psi_2(x) = \operatorname{erf}(x) = \frac{2}{\sqrt{\pi}} \int_0^x \exp(-s^2)\,\mathrm{d}s \qquad \text{(ERF transformation [11])}, \tag{4}$$

$$\psi_3(x) = \tanh\left(\frac{\pi}{2}\sinh x\right) \qquad \text{(DE transformation [12])} \tag{5}$$

where "DE" stands for "double-exponential." Combination of each transformation and the trapezoidal rule yields a quadrature formula as

$$\int_{-1}^{1} F(t)\,\mathrm{d}t = \int_{-\infty}^{\infty} F(\psi_i(x))\psi_i'(x)\,\mathrm{d}x \approx h \sum_{k=-N}^{N} F(\psi_i(kh))\psi_i'(kh),$$

which is called the TANH formula, the ERF formula, and the DE formula, corresponding to each transformation. A significant difference between those transformations is a decay rate of $\psi_i'(x)$ as $|x| \to \infty$; $\psi_1'(x) = \mathrm{O}(\mathrm{e}^{-|x|})$ (single-exponential decay), $\psi_2'(x) = \mathrm{O}(\mathrm{e}^{-x^2})$ (Gaussian decay), and $\psi_3'(x) = \mathrm{O}(\mathrm{e}^{|x|-(\pi/2)\exp(|x|)})$ (double-exponential decay). There are several studies about how the decay rate affects the performance of the trapezoidal rule [4, 13, 15]. This is a discussion of optimality from a viewpoint of a variable transformation.

There is another discussion of optimality from a viewpoint of an approximation formula. The trapezoidal rule has been frequently used with the transformations above because of its high accuracy, but it is not obvious that the trapezoidal rule is an optimal formula among the possible formulas of the form (1). In order to discuss this point, Sugihara [7] introduced the weighted Hardy space given by

$$\boldsymbol{H}^{\infty}(\mathscr{D}_d, w) := \left\{ f : \mathscr{D}_d \to \mathbf{C} \;\middle|\; f \text{ is analytic on } \mathscr{D}_d,\ \|f\| := \sup_{z \in \mathscr{D}_d} \left| \frac{f(z)}{w(z)} \right| < \infty \right\}$$

as a set of the transformed functions f where d is a positive number, $\mathscr{D}_d := \{z \in \mathbf{C} \mid |\operatorname{Im} z| < d\}$, and w is a weight function which indicates the decay rate of $\psi'(x)$ (specified in Sect. 2). Sugihara considered the norm of the error operator of a quadrature formula for $f \in \boldsymbol{H}^{\infty}(\mathscr{D}_d, w)$ as a criterion of the accuracy of the formula. Then, he defined an optimal quadrature formula for $\boldsymbol{H}^{\infty}(\mathscr{D}_d, w)$ by a formula that minimizes the norm. Based on this notion, he demonstrated that the trapezoidal rule is *nearly* optimal by showing

(A lower bound of the minimum error norm)
$\approx$ (An upper bound of the error of the trapezoidal rule on $\mathbf{R}$)

in the case that w decays single-exponentially and double-exponentially. It should be emphasized here that the trapezoidal rule is not *exactly* optimal, and a better quadrature formula may exist. However, it is difficult to design an optimal formula. One of the reasons is the expression of the minimum error norm; the expression contains a two-stage optimization problem including an infinite dimensional constraint, and for this reason, the exact order of the minimum error norm is still not known except in the case that w decays single-exponentially [1, 2].

In this paper, we derive another convenient expression of the minimum error norm for a general weight function w. This expression gives an optimization problem determining sampling points for an optimal numerical integration formula for $\boldsymbol{H}^{\infty}(\mathscr{D}_d, w)$. Because this optimization problem is still difficult to solve exactly, we consider its approximate solution by using the method for designing accurate formulas for approximation of functions in $\boldsymbol{H}^{\infty}(\mathscr{D}_d, w)$, which is based on the characterization of the optimality of the approximation in that space. This method is proposed by the present authors [14], by using a technique in potential theory [5]. Then, we propose a method for designing accurate numerical integration formulas for $\boldsymbol{H}^{\infty}(\mathscr{D}_d, w)$ by using the approximate solution of the optimization problem.

The rest of this paper is organized as follows. In Sect. 2, we present some mathematical preliminaries concerning assumptions for weight functions w and the formulation of the optimality of a quadrature formula for $\boldsymbol{H}^{\infty}(\mathscr{D}_d, w)$. In Sect. 3, we derive an expression of the minimum error norm for a general weight function w. In Sect. 4, after reviewing our method for approximating functions in Sect. 4.1, we present the method to design accurate numerical integration formulas for $\boldsymbol{H}^{\infty}(\mathscr{D}_d, w)$ in Sect. 4.2. In Sect. 5, we present some numerical results showing the performance of the proposed quadrature formulas for three weight functions w. Finally, we conclude this paper in Sect. 6.

2 Mathematical Preliminaries

2.1 Weight Functions and Weighted Hardy Spaces

Let d be a positive real number, and let $\mathscr{D}_d$ be a strip region defined by $\mathscr{D}_d := \{z \in \mathbf{C} \mid |\operatorname{Im} z| < d\}$. In order to specify the analytical property of a weight function w on $\mathscr{D}_d$, we use a function space $\mathscr{B}(\mathscr{D}_d)$ of all functions ζ that are analytic on $\mathscr{D}_d$ such that

$$\lim_{x \to \pm\infty} \int_{-d}^{d} |\zeta(x + \mathrm{i}y)| \, \mathrm{d}y = 0$$

and

$$\lim_{y \to d-0} \int_{-\infty}^{\infty} (|\zeta(x + \mathrm{i}y)| + |\zeta(x - \mathrm{i}y)|) \, \mathrm{d}x < \infty.$$

Let w be a complex-valued function on $\mathscr{D}_d$. We regard w as a weight function on $\mathscr{D}_d$ if w satisfies the following assumption.

Assumption 1 The function w belongs to $\mathscr{B}(\mathscr{D}_d)$, does not vanish at any point in $\mathscr{D}_d$, and takes positive real values on the real axis.

For a weight function w that satisfies Assumption 1, we define a weighted Hardy space on $\mathscr{D}_d$ by

$$\boldsymbol{H}^{\infty}(\mathscr{D}_d, w) := \{f : \mathscr{D}_d \to \mathbf{C} \mid f \text{ is analytic on } \mathscr{D}_d \text{ and } \|f\| < \infty\}$$

where

$$\|f\| := \sup_{z \in \mathscr{D}_d} \left| \frac{f(z)}{w(z)} \right|.$$

In this paper, for the simplicity of our mathematical arguments, we apply the following additional assumption for a weight function w.

Assumption 2 The function w is even on $\mathbf{R}$.

2.2 Optimal Numerical Integration Formula

We provide a mathematical formulation for the optimality of numerical integration formulas for the weighted Hardy spaces $\boldsymbol{H}^\infty(\mathscr{D}_d, w)$ with weight functions satisfying Assumptions 1 and 2. In this regard, for a given positive integer N, we first consider all possible $(2N+1)$-point numerical integration formulas on $\mathbf{R}$ that can be applied to any functions $f \in \boldsymbol{H}^\infty(\mathscr{D}_d, w)$. Then, we choose one of the formulas such that it gives the minimum worst error in $\boldsymbol{H}^\infty(\mathscr{D}_d, w)$. The precise definition of the minimum worst error, denoted by $\mathscr{E}_N^{\min}(\boldsymbol{H}^\infty(\mathscr{D}_d, w))$, is given by

$$\mathscr{E}_N^{\min}(\boldsymbol{H}^\infty(\mathscr{D}_d, w)) := \inf_{1\le \ell \le 2N+1} \inf_{\substack{m_1,\ldots,m_\ell \\ m_1+\cdots+m_\ell=2N+1}} \inf_{a_p\in\mathscr{D}_d} \inf_{c_{pq}\in\mathbf{C}} \left\| E_{c_{pq},a_p}^{\ell,m_p} \right\| \tag{6}$$

where the error operator $E_{c_{pq},a_p}^{\ell,m_p} : \boldsymbol{H}^\infty(\mathscr{D}_d, w) \to \mathbf{C}$ is defined by

$$E_{c_{pq},a_p}^{\ell,m_p} f := \int_{-\infty}^{\infty} f(x)\,\mathrm{d}x - \sum_{p=1}^{\ell} \sum_{q=0}^{m_p-1} c_{pq} f^{(q)}(a_p).$$

That is, $\mathscr{E}_N^{\min}(\boldsymbol{H}^\infty(\mathscr{D}_d, w))$ is the norm of the error operator $E_{c_{pq},a_p}^{\ell,m_p}$. Therefore, we call it the minimum error norm. Here, we restrict ourselves without loss of generality to sequences of sampling points that are symmetric with respect to the origin. That is, they have the form $a_{-l}, a_{-l+1}, \ldots, a_{l-1}, a_l$ with $a_{-k} = -a_k$ $(k = 1, \ldots, l)$. This is because we consider even weight functions w according to Assumption 2. Owing to this reason, our definition (6) of $\mathscr{E}_N^{\min}(\boldsymbol{H}^\infty(\mathscr{D}_d, w))$ is slightly different from that given in [7].

The minimum error norm $\mathscr{E}_N^{\min}(\boldsymbol{H}^\infty(\mathscr{D}_d, w))$ can be expressed by the norm of a certain integral operator. For a sequence $\{a_{-N}, \ldots, a_N\} \subset \mathscr{D}_d$, let $B_N(x; \{a_i\}, \mathscr{D}_d)$ be defined by

$$B_N(x; \{a_i\}, \mathscr{D}_d) := \prod_{j=-N}^{N} \frac{T_d(x) - T_d(a_j)}{1 - \overline{T_d(a_j)}\, T_d(x)} \tag{7}$$

where

$$T_d(x) := \tanh\left(\frac{\pi}{4d}x\right).$$

The function given by (7) is called the transformed Blaschke product. According to Sugihara [7], we have

$$\mathscr{E}_N^{\min}(\boldsymbol{H}^\infty(\mathscr{D}_d, w)) = \inf_{a_i\in\mathscr{D}_d} \left[\sup_{\substack{g\in\boldsymbol{H}^\infty(\mathscr{D}_d,w) \\ \|g\|\le 1}} \left| \int_{-\infty}^{\infty} g(x)\, B_N(x; \{a_i\}, \mathscr{D}_d)\,\mathrm{d}x \right| \right]. \tag{8}$$

This expression gives an optimization problem that determines a sequence of sampling points for a numerical integration formula attaining the minimum error norm $\mathcal{E}_N^{\min}(\boldsymbol{H}^\infty(\mathcal{D}_d, w))$. However, it may be difficult to solve it directly because it is a two-stage optimization problem including an infinite dimensional constraint. Then, by using (8), we derive another convenient expression of the minimum error norm in Sect. 3 and propose an algorithm for designing accurate numerical integration formulas on the Hardy spaces $\boldsymbol{H}^\infty(\mathcal{D}_d, w)$ in Sect. 4.

3 Characterization of Optimal Numerical Integration Formulas

In this section, we prove the following theorem, which gives a new expression of the minimum error norm $\mathcal{E}_n^{\min}(\boldsymbol{H}^\infty(\mathcal{D}_d, w))$.

Theorem 1 *We have*

$$\begin{aligned}\mathcal{E}_N^{\min}(\boldsymbol{H}^\infty(\mathcal{D}_d, w)) &= \inf_{a_i \in \mathbf{R}} \int_{-\infty}^{\infty} w(x) \prod_{j=-N}^{N} \left|\tanh\left[\frac{\pi}{4d}(x - a_j)\right]\right|^2 \mathrm{d}x \\ &= \inf_{a_i \in \mathbf{R}} \int_{-\infty}^{\infty} w(x) \prod_{j=-N}^{N} |T_d(x - a_j)|^2 \, \mathrm{d}x \end{aligned} \tag{9}$$

where inf *denotes the infimum over any* $(2N+1)$ *sampling points* $a_i \in \mathbf{R}$.

As shown in [7], by setting

$$g(z) = w(z)\, \overline{B_N(\bar{z}; \{a_i\}, \mathcal{D}_d)}$$

in (8), we can obtain a lower bound of $\mathcal{E}_n^{\min}(\boldsymbol{H}^\infty(\mathcal{D}_d, w))$:

$$\mathcal{E}_N^{\min}(\boldsymbol{H}^\infty(\mathcal{D}_d, w)) \geq \inf_{a_i \in \mathcal{D}_d} \int_{-\infty}^{\infty} w(x)\, |B_N(x; \{a_i\}, \mathcal{D}_d)|^2 \, \mathrm{d}x. \tag{10}$$

Furthermore, by using the inequality shown in [8, Proof of Lemma 4.3],

$$\left|\frac{T_d(x) - T_d(a)}{1 - \overline{T_d(a)}\, T_d(x)}\right| \geq \left|\frac{T_d(x) - T_d(\operatorname{Re} a)}{1 - T_d(\operatorname{Re} a)\, T_d(x)}\right| = |T_d(x - \operatorname{Re} a)| \qquad (x \in \mathbf{R},\ a \in \mathcal{D}_d),$$

we restrict the range of the sampling points $\{a_i\}$ to $\mathbf{R}$ in (10) to deduce that

$$
\begin{aligned}
\mathscr{E}_n^{\min}(\boldsymbol{H}^\infty(\mathscr{D}_d, w)) &\geq \inf_{a_i \in \mathbf{R}} \int_{-\infty}^{\infty} w(x)\, |B_N(x; \{a_i\}, \mathscr{D}_d)|^2 \\
&= \inf_{a_i \in \mathbf{R}} \int_{-\infty}^{\infty} w(x) \prod_{j=-N}^{N} \left| T_d(x - a_j) \right|^2 \,\mathrm{d}x. \qquad (11)
\end{aligned}
$$

This inequality shows that the RHS of (9) is a lower bound of $\mathscr{E}_n^{\min}(\boldsymbol{H}^\infty(\mathscr{D}_d, w))$. Therefore, if we show the following proposition, we complete the proof of Theorem 1.

Proposition 4 *We have*

$$
\mathscr{E}_N^{\min}(\boldsymbol{H}^\infty(\mathscr{D}_d, w)) \leq \inf_{a_i \in \mathbf{R}} \int_{-\infty}^{\infty} w(x) \prod_{j=-N}^{N} \left| T_d(x - a_j) \right|^2 \,\mathrm{d}x. \qquad (12)
$$

We prove Proposition 4 in Sect. 3.2. To this end, for given d and w, we choose a numerical integration formula (i.e., $\{c_{pq}\}$ and $\{a_p\}$) and show that the RHS of (12) is an upper bound of the operator norm $E^{\ell, m_p}_{c_{pq}, a_p}$ given by (6). As such a $(2N + 1)$-point numerical integration formula, we consider the formula yielded by integrating the Hermite interpolation on $\mathbf{R}$ by the transformed Blaschke product $B_N(x; \{a_i\}; \mathscr{D}_d)$ in Sect. 3.1. Then, a proof of Proposition 4 follows in Sect. 3.2.

For the arguments in the following subsections, let $F(a)$ denote the objective function of $a = (a_{-N}, \ldots, a_N)$ in the RHS of (12), i.e.,

$$
\begin{aligned}
F(a) &:= \int_{-\infty}^{\infty} w(x) \prod_{j=-N}^{N} \left| T_d(x - a_j) \right|^2 \,\mathrm{d}x \\
&= \int_{-\infty}^{\infty} \exp\left(\log w(x) + 2 \sum_{i=-N}^{N} \log |T_d(x - a_i)| \right) \mathrm{d}x. \qquad (13)
\end{aligned}
$$

In the following subsections, we use the properties of F shown below.

Proposition 5 *The function F is continuous in $\mathbf{R}^{2N+1}$ and has a minimizer in $\mathbf{R}^{2N+1}$.*

Proposition 6 *The function F is differentiable with respect to each variable a_j in $\mathbf{R}^{2N+1}$ and*

$$
\frac{\partial F}{\partial a_j}(a) = \int_{-\infty}^{\infty} w(x) \prod_{j=-N}^{N} \left| T_d(x - a_j) \right|^2 \cdot \left(2 \, \frac{\partial}{\partial y} \log |T_d(x - y)| \bigg|_{y = a_j} \right) \mathrm{d}x. \quad (14)
$$

We show the sketch of the proofs of these propositions in Appendix 1 for readers' convenience, although they are fundamental results in calculus.

3.1 Hermite Interpolation on R by the Transformed Blaschke Product

In what follows, we use the constant C_d given by

$$C_d := \frac{\pi}{4d}. \tag{15}$$

Furthermore, we assume that a_is are mutually distinct because it follows from Proposition 5 that

$$\inf_{\substack{a_i \in \mathbf{R} \\ \text{distinct}}} F(a) = \inf_{a_i \in \mathbf{R}} F(a). \tag{16}$$

Remark 1 We have *not* proven that F has a minimizer whose components are mutually distinct. However, if F has a minimizer a^* with

$$a^*_{-N} < \cdots < a^*_j = \cdots = a^*_{j+k} < \cdots < a^*_N,$$

we can perturb these components as

$$\tilde{a}^*_{-N} < \cdots < \tilde{a}^*_j < \cdots < \tilde{a}^*_{j+k} < \cdots < \tilde{a}^*_N$$

so that $F(\tilde{a}^*)$ is arbitrary close to $F(a^*)$ because F is continuous. Therefore, we have (16).

In order to derive the explicit form of the Hermite interpolation, we start with the Lagrange interpolation by the transformed Blaschke product:

$$\begin{aligned} \tilde{f}^{\mathrm{L}}_{2N+1}(x) &:= \sum_{j=-N}^{N} f(a_j) \frac{w(x)}{w(a_j)} \frac{\displaystyle\prod_{\substack{-N \le i \le N \\ i \neq j}} \tanh\left[\frac{\pi}{4d}(x - a_i)\right]}{\displaystyle\prod_{\substack{-N \le i \le N \\ i \neq j}} \tanh\left[\frac{\pi}{4d}(a_j - a_i)\right]} \operatorname{sech}^2\left[\frac{\pi}{4d}(x - a_j)\right] \\ &= \sum_{j=-N}^{N} f(a_j) \frac{w(x)}{w(a_j)} \frac{\displaystyle\prod_{\substack{-N \le i \le N \\ i \neq j}} T_d(x - a_i)}{\displaystyle\prod_{\substack{-N \le i \le N \\ i \neq j}} T_d(a_j - a_i)} \frac{T_d'(x - a_j)}{C_d}, \end{aligned}$$

which is employed in [14] (see also (26) and (31) in this paper). Then, we alter the $(2N+1)$-point sequence $\{a_j\}$ to a $2(2N+1)$-point sequence

$$\{\alpha_1, \alpha_2, \ldots, \alpha_{2(2N+1)}\} = \{a_{-N}, \tilde{a}_{-N}, a_{-(N-1)}, \tilde{a}_{-(N-1)}, \ldots, a_N, \tilde{a}_N\},$$

and taking the limit

$$\tilde{a}_j \to a_j \qquad (j = -N, -(N-1), \dots, N) \tag{17}$$

to obtain the Hermite interpolation. If we take the limit $\tilde{a}_j \to a_j$ of the sum of the terms including $f(a_j)$ and $f(\tilde{a}_j)$, we have

$$\begin{aligned}
& f(a_j)\,\frac{w(x)}{w(a_j)}\,\frac{\prod_{k\neq 2(j+N)+1} T_d(x-\alpha_k)}{\prod_{k\neq 2(j+N)+1} T_d(a_j-\alpha_k)}\,\frac{T_d'(x-a_j)}{C_d} \\
& + f(\tilde{a}_j)\,\frac{w(x)}{w(\tilde{a}_j)}\,\frac{\prod_{k\neq 2(j+N)+2} T_d(x-\alpha_k)}{\prod_{k\neq 2(j+N)+2} T_d(\tilde{a}_j-\alpha_k)}\,\frac{T_d'(x-\tilde{a}_j)}{C_d} \\
= {} & f(a_j)\left[\frac{w(x)}{w(a_j)}\,\frac{\prod_{k\neq 2(j+N)+1} T_d(x-\alpha_k)}{\prod_{k\neq 2(j+N)+1} T_d(a_j-\alpha_k)}\,\frac{T_d'(x-a_j)}{C_d}\right. \\
& \qquad \left. + \frac{w(x)}{w(\tilde{a}_j)}\,\frac{\prod_{k\neq 2(j+N)+2} T_d(x-\alpha_k)}{\prod_{k\neq 2(j+N)+2} T_d(\tilde{a}_j-\alpha_k)}\,\frac{T_d'(x-\tilde{a}_j)}{C_d}\right] \\
& + (f(\tilde{a}_j)-f(a_j))\,\frac{w(x)}{w(\tilde{a}_j)}\,\frac{\prod_{k\neq 2(j+N)+2} T_d(x-\alpha_k)}{\prod_{k\neq 2(j+N)+2} T_d(\tilde{a}_j-\alpha_k)}\,\frac{T_d'(x-\tilde{a}_j)}{C_d} \\
= {} & f(a_j)\,w(x)\,\frac{\prod_{k\neq 2(j+N)+1,\,2(j+N)+2} T_d(x-\alpha_k)}{\prod_{k\neq 2(j+N)+1,\,2(j+N)+2} T_d(a_j-\alpha_k)}\,\frac{a_j-\tilde{a}_j}{T_d(a_j-\tilde{a}_j)} \\
& \times \frac{\dfrac{T_d(x-a_j)T_d'(x-a_j)}{C_d\,w(a_j)} - \dfrac{T_d(x-\tilde{a}_j)T_d'(x-\tilde{a}_j)}{C_d\,w(\tilde{a}_j)}}{a_j-\tilde{a}_j} \\
& + \frac{(f(\tilde{a}_j)-f(a_j))}{\tilde{a}_j-a_j}\,\frac{w(x)}{w(\tilde{a}_j)}\,\frac{\prod_{k=1}^{2(2N+1)} T_d(x-\alpha_k)}{\prod_{k\neq 2(j+N)+1,\,2(j+N)+2} T_d(\tilde{a}_j-\alpha_k)} \\
& \times \frac{\tilde{a}_j-a_j}{T_d(\tilde{a}_j-a_j)}\,\frac{T_d'(x-\tilde{a}_j)}{C_d\,T_d(x-\tilde{a}_j)} \\
\to {} & f(a_j)\,w(x)\,\frac{\prod_{k\neq 2(j+N)+1,\,2(j+N)+2} T_d(x-\alpha_k)}{\prod_{k\neq 2(j+N)+1,\,2(j+N)+2} T_d(a_j-\alpha_k)}\,\frac{\partial}{\partial a}\left[\frac{T_d(x-a)T_d'(x-a)}{C_d\,w(a)}\right]\Bigg|_{a=a_j} \\
& + f'(a_j)\,\frac{w(x)}{w(a_j)}\,\frac{T_d(x-a_j)^2\prod_{k\neq 2(j+N)+1,\,2(j+N)+2} T_d(x-\alpha_k)}{\prod_{k\neq 2(j+N)+1,\,2(j+N)+2} T_d(a_j-\alpha_k)}\,\frac{T_d'(x-a_j)}{T_d(x-a_j)}.
\end{aligned}$$

Then, by letting $\tilde{a}_i \to a_i$ for the remaining index i with $i \neq j$, we obtain

$$f(a_j)\, w(x) \left[\frac{\prod_{i \neq j} T_d(x - a_i)}{\prod_{i \neq j} T_d(a_j - a_i)} \right]^2 \frac{\partial}{\partial a} \left[\frac{T_d(x-a) T_d'(x-a)}{C_d\, w(a)} \right] \Bigg|_{a=a_j}$$

$$+ f'(a_j) \frac{w(x) \left[\prod_{i=-N}^{N} T_d(x - a_i) \right]^2}{w(a_j) \left[\prod_{i \neq j} T_d(a_j - a_i) \right]^2} \frac{\partial}{\partial x} \log \left| T_d(x - a_j) \right| . \tag{18}$$

Therefore, we have the Hermite interpolation by summing them for $j = -N, \ldots, N$. In the following, let $\tilde{f}_{2N+1}^{\mathrm{H}}$ denote the Hermite interpolation, i.e.,

$$\tilde{f}_{2N+1}^{\mathrm{H}}(x) := \sum_{j=-N}^{N} \left[f(a_j)\, w(x) \left[\frac{\prod_{i \neq j} T_d(x - a_i)}{\prod_{i \neq j} T_d(a_j - a_i)} \right]^2 \frac{\partial}{\partial a} \left[\frac{T_d(x-a) T_d'(x-a)}{C_d\, w(a)} \right] \Bigg|_{a=a_j} \right.$$
$$\left. + f'(a_j) \frac{w(x) \left[\prod_{i=-N}^{N} T_d(x - a_i) \right]^2}{w(a_j) \left[\prod_{i \neq j} T_d(a_j - a_i) \right]^2} \frac{\partial}{\partial x} \log \left| T_d(x - a_j) \right| \right] . \tag{19}$$

The upper bound of the error of $\tilde{f}_{2N+1}^{\mathrm{H}}$ is given by

$$\left| f(x) - \tilde{f}_{2N+1}^{\mathrm{H}}(x) \right| \leq w(x) \prod_{j=-N}^{N} \left| T_d(x - a_j) \right|^2 .$$

This inequality follows from the error estimate of the Lagrange interpolation shown in [8, Proof of Lemma 4.3]:

$$\left| f(x) - \tilde{f}_{2(2N+1)}^{\mathrm{L}}(x) \right| \leq w(x) \prod_{k=1}^{2(2N+1)} \left| T_d(x - \alpha_k) \right|$$

by taking the limit (17) under the conditions

$$f \in \boldsymbol{H}^{\infty}(\mathscr{D}_d, w), \quad \| f \| \leq 1.$$

3.2 *Proof of Proposition 4*

By integrating $\tilde{f}_{2N+1}^{\mathrm{H}}$ in (19) on $\mathbf{R}$, we obtain a $2(2N+1)$-point numerical integration formula. If the second term of (18) vanishes for all $j = -N, \ldots, N$, i.e.,

$$\int_{-\infty}^{\infty} w(x) \left[\prod_{i=-N}^{N} T_d(x - a_i) \right]^2 \frac{\partial}{\partial x} \log \left| T_d(x - a_j) \right| \mathrm{d}x = 0 \quad (j = -N, \ldots, N), \tag{20}$$

then the formula becomes an $(2N+1)$-point one without the values of the derivative of f. We can show the existence of a solution of (20). Actually, for the function F defined by (13), we have

$$\frac{\partial F}{\partial a_j}(a) = \int_{-\infty}^{\infty} w(x) \prod_{j=-N}^{N} |T_d(x-a_j)|^2 \cdot \left(2 \frac{\partial}{\partial y} \log |T_d(x-y)| \bigg|_{y=a_j} \right) \mathrm{d}x$$
$$= \int_{-\infty}^{\infty} w(x) \prod_{j=-N}^{N} |T_d(x-a_j)|^2 \cdot \left(-2 \frac{\partial}{\partial x} \log |T_d(x-y)| \bigg|_{y=a_j} \right) \mathrm{d}x$$

by Proposition 6. Therefore, Condition (20) is equivalent to

$$\frac{\partial F}{\partial a_j}(a) = 0 \qquad (j = -N, \ldots, N). \tag{21}$$

By Proposition 5, the function F has a minimizer, for which this condition is satisfied. Hence, we can prove Proposition 4 as follows:

$$\begin{aligned}
\mathscr{E}_N^{\min}(\boldsymbol{H}^\infty(\mathscr{D}_d, w)) &\leq \inf_{\substack{a_i \in \mathbf{R} \\ \{a_i\} \text{ satisfies (20)}}} \left[\sup_{\substack{f \in \boldsymbol{H}^\infty(\mathscr{D}_d, w) \\ \|f\| \leq 1}} \left| \int_{-\infty}^{\infty} f(x)\,\mathrm{d}x - \int_{-\infty}^{\infty} \tilde{f}_{2N+1}^{\mathrm{H}}(x)\,\mathrm{d}x \right| \right] \\
&\leq \inf_{\substack{a_i \in \mathbf{R} \\ \{a_i\} \text{ satisfies (21)}}} \left[\sup_{\substack{f \in \boldsymbol{H}^\infty(\mathscr{D}_d, w) \\ \|f\| \leq 1}} \int_{-\infty}^{\infty} \left| f(x) - \tilde{f}_{2N+1}^{\mathrm{H}}(x) \right| \mathrm{d}x \right] \\
&\leq \inf_{\substack{a_i \in \mathbf{R} \\ \{a_i\} \text{ satisfies (21)}}} \int_{-\infty}^{\infty} w(x) \prod_{j=-N}^{N} |T_d(x-a_j)|^2 \,\mathrm{d}x \\
&= \inf_{a_i \in \mathbf{R}} \int_{-\infty}^{\infty} w(x) \prod_{j=-N}^{N} |T_d(x-a_j)|^2 \,\mathrm{d}x.
\end{aligned} \tag{22}$$

Remark 2 According to (19) and (20), the integral of the Hermite interpolation $\tilde{f}_{2N+1}^{H}$ gives a numerical integration formula without the derivatives of f, which is written in the form

$$\int_{-\infty}^{\infty} \tilde{f}_{2N+1}^{H}(x)\,\mathrm{d}x = \sum_{j=-N}^{N} c_j\, f(a_j).$$

By combining (11) and the inequalities in (22), we know that the formula is optimal, i.e., attains the minimum error norm. Therefore, in the next section, we concentrate on finding $\{c_j\}$ and $\{a_j\}$ of the formula.

4 Design of Accurate Numerical Integration Formulas

For given d and w, we design a numerical integration formula for $\boldsymbol{H}^\infty(\mathscr{D}_d, w)$ by using Theorem 1. To this end, we consider the optimization problem given by (9) in order to obtain a sequence of sampling points $\{a_i\}$. However, because it may be difficult to obtain an exact solution of the problem, we consider the following approximate solution. By noting that

$$\begin{aligned}
&\int_{-\infty}^{\infty} w(x) \prod_{j=-N}^{N} \left|T_d(x-a_j)\right|^2 \mathrm{d}x \\
&\leq \int_{-\alpha}^{\alpha} w(x) \prod_{j=-N}^{N} \left|T_d(x-a_j)\right|^2 \mathrm{d}x + \int_{|x|>\alpha} w(x)\,\mathrm{d}x \\
&\leq \underline{2\alpha \sup_{x\in\mathbf{R}} \left\{ w(x) \prod_{j=-N}^{N} \left|T_d(x-a_j)\right|^2 \right\}} + \int_{|x|>\alpha} w(x)\,\mathrm{d}x \qquad (23)
\end{aligned}$$

for a sufficiently large $\alpha > 0$, we consider the minimization of the underlined value in (23) instead of the original problem. Because this minimization problem is equivalent to that given by

$$\inf_{a_i\in\mathbf{R}} \left[\sup_{x\in\mathbf{R}} \left\{ \log w(x)^{1/2} + \sum_{i=-N}^{N} \log\left|T_d(x-a_i)\right| \right\} \right], \qquad (24)$$

we obtain $\{a_i\}$ by this problem.

In fact, an approximate solution of (24) can be given by our method [14] for designing accurate formulas for approximating functions, which is based on potential theory. In the following, we first review this method in Sect. 4.1. Then, by using the method and (24), we propose a method for designing accurate formulas for numerical integration in Sect. 4.2.

4.1 *The Method for Designing Accurate Formulas for Approximating Functions*

In this section, we review the method for designing accurate formulas for approximating functions, which is detailed in Sects. 3 and 4 in [14].

Let $E_N^{\min}(\boldsymbol{H}^\infty(\mathscr{D}_d, w))$ be the minimum error norm of function approximation in $\boldsymbol{H}^\infty(\mathscr{D}_d, w)$ defined by

$$
\begin{aligned}
&E_N^{\min}(\boldsymbol{H}^\infty(\mathscr{D}_d, w)) \\
&:= \inf_{1 \leq l \leq N} \inf_{\substack{m_{-l},\ldots,m_l \\ m_{-l}+\cdots+m_l=2N+1}} \inf_{\substack{a_j \in \mathscr{D}_d \\ \text{distinct}}} \inf_{\phi_{jk}} \left[\sup_{\|f\| \leq 1} \sup_{x \in \mathbf{R}} \left| f(x) - \sum_{j=-l}^{l} \sum_{k=0}^{m_j-1} f^{(k)}(a_j)\, \phi_{jk}(x) \right| \right]
\end{aligned}
$$

where the ϕ_{jk}s are functions that are analytic on $\mathscr{D}_d$. The main idea is to pursue a formula achieving $E_N^{\min}(\boldsymbol{H}^\infty(\mathscr{D}_d, w))$. For this purpose, we begin with the following proposition, in which we restrict Lemma 4.3 in [8] to the case of even weight functions and sampling points that are symmetric with respect to the origin.

Proposition 7 ([8, Lemma 4.3]) *Let the weight function w satisfy Assumptions 1 and 2. Then, we have*

$$
\begin{aligned}
&E_N^{\min}(\boldsymbol{H}^\infty(\mathscr{D}_d, w)) \\
&= \inf_{a_\ell \in \mathbf{R}} \left[\sup_{\|f\| \leq 1} \sup_{x \in \mathbf{R}} \left| f(x) - \sum_{k=-N}^{N} f(a_k) \frac{w(x)\, B_{N;k}(x; \{a_\ell\}, \mathscr{D}_d)}{w(a_k)\, B_{N;k}(a_k; \{a_\ell\}, \mathscr{D}_d)} \frac{T_d'(a_k - x)}{C_d} \right| \right] \\
&= \inf_{a_\ell \in \mathbf{R}} \left[\sup_{x \in \mathbf{R}} |B_N(x; \{a_\ell\}, \mathscr{D}_d)\, w(x)| \right] \qquad (25)
\end{aligned}
$$

where

$$
B_{N;k}(x; \{a_\ell\}, \mathscr{D}_d) = \prod_{\substack{-N \leq m \leq N, \\ m \neq k}} T_d(x - a_m).
$$

If we can obtain sampling points $\{a_\ell\}$ that attain the infimum in (25), then the Lagrange interpolation formula

$$
\tilde{f}_{2N+1}^{\mathrm{L}}(x) = \sum_{k=-N}^{N} f(a_k) \frac{w(x)\, B_{N;k}(x; \{a_\ell\}, \mathscr{D}_d)}{w(a_k)\, B_{N;k}(a_k; \{a_\ell\}, \mathscr{D}_d)} \frac{T_d'(a_k - x)}{C_d} \qquad (26)
$$

gives an optimal approximation for $\boldsymbol{H}^\infty(\mathscr{D}_d, w)$. Then, we regard (25) as an optimization problem with respect to $\{a_\ell\}$ and consider its solution. In order to use the framework of potential theory, we consider the logarithm of the objective function of (25):

$$
\log |B_N(x; \{a_\ell\}, \mathscr{D}_d)\, w(x)| = V_{\mathscr{D}_d}^{\sigma_a}(x) + \log w(x) \qquad (27)
$$

where

$$
V_{\mathscr{D}_d}^{\sigma_a}(x) = \sum_{k=-N}^{N} \log |T_d(x - a_k)| = \sum_{k=-N}^{N} \log \left| \tanh \left[\frac{\pi}{4d}(x - a_k) \right] \right|.
$$

This is the negative of the Green potential with the measure σ_a given by

$$\sigma_a(z) := \sum_{k=-N}^{N} \delta(z - a_k)$$

where δ is Dirac's delta function. In the following, we use the notation $V_{\mathscr{D}_d}^{\mu} = -U_{\mathscr{D}_d}^{\mu}$ for a general measure μ on $\mathbf{R}$. By (27), the optimization problem given by (25) is equivalent to the following problem.

Problem 1 Find a sequence $\{a_\ell\}$ of sampling points that attain

$$\inf_{a_\ell \in \mathbf{R}} \left[\sup_{x \in \mathbf{R}} \left(V_{\mathscr{D}_d}^{\sigma_a}(x) + \log w(x) \right) \right].$$

This problem is rather difficult because it requires finding an optimal measure in the set of discrete ones. Then, we relax the constraint of Problem 1 and consider the following approximate problem, whose constraint covers general Borel measures $\mu_N \in \mathscr{M}(\mathbf{R}, 2(N+1))$ with $\mu(\mathbf{R}) = 2(N+1)$.

Problem 2 Find a positive real number α_N and a measure $\mu_N \in \mathscr{M}(\mathbf{R}, 2(N+1))$ that attain

$$\inf_{\substack{\mu_N \in \mathscr{M}(\mathbf{R}, 2(N+1)) \\ \operatorname{supp} \mu_N = [-\alpha_N, \alpha_N]}} \left[\sup_{x \in \mathbf{R}} \left(V_{\mathscr{D}_d}^{\mu_N}(x) + \log w(x) \right) \right].$$

Here, the condition $\mu_N(\mathbf{R}) = 2(N+1)$ corresponds to the constraint that the number of the sampling points $\{a_\ell\}$ is $2N+1$ in Problem 1.[1]

According to potential theory, Problem 2 gives a characterization of the solution of an energy minimization problem defined by $V_{\mathscr{D}_d}^{\mu_N}$, and the characterization can be rewritten in terms of a system of an integral equation and integral inequality. Then, we rewrite Problem 2 in terms of such a system. Furthermore, for the sake of analytical tractability, we restrict ourselves to measures μ_N with $\mathrm{d}\mu_N(x) = \nu_N(x)\mathrm{d}x$ and $\nu_N \in \mathscr{C}_{\mathrm{d}}^1(\mathbf{R}, 2(N+1))$ where

$$\mathscr{C}_{\mathrm{d}}^1(\mathbf{R}, 2(N+1)) := \Big\{ \nu_N : \mathbf{R} \to \mathbf{R}_{\geq 0} \mid \nu_N \text{ is continuously differentiable on } \operatorname{supp} \nu_N, \\ \int_{-\infty}^{\infty} \nu_N(x)\,\mathrm{d}x = 2(N+1) \Big\}.$$

The function ν_N is regarded as a density function. Then, we have the following problem.

---

[1]The difference between these values is owing to a technical treatment in estimating the error of the proposed formula based on Problem 2.

Problem 3 Find real numbers α_N and K_N, and a density function $\nu_N \in \mathscr{C}_{\mathrm{d}}^1(\mathbf{R}, 2(N+1))$ that satisfy $\operatorname{supp} \nu_N = [-\alpha_N, \alpha_N]$ and

$$V_{\mathscr{D}_d}^{\mu_N}(x) + \log w(x) = -K_N \quad \text{for any } x \in [-\alpha_N, \alpha_N], \tag{28}$$

$$V_{\mathscr{D}_d}^{\mu_N}(x) + \log w(x) \leq -K_N \quad \text{for any } x \in \mathbf{R} \setminus [-\alpha_N, \alpha_N] \tag{29}$$

where $\mathrm{d}\mu_N(x) = \nu_N(x)\mathrm{d}x$.

The conditions in Problem 3 do not give the entire form of $V_{\mathscr{D}_d}^{\mu_N}$ on $\mathbf{R}$. However, we can obtain it by solving a Laplace equation[2] based on the fact that the function $V_{\mathscr{D}_d}^{\mu_N}(x + \mathrm{i}\, y)$ is harmonic on the double-connected domain $\mathscr{D}_d \setminus [-\alpha_N, \alpha_N]$. Because it suffices to obtain the explicit form v^* of $V_{\mathscr{D}_d}^{\mu_N}$ on $\mathbf{R}$, we use an approximate analytical solution of the Laplace equation. By noting that the explicit form v^* contains the unknown parameters α_N and K_N, we first determine them and then solve the equation $V_{\mathscr{D}_d}^{\mu_N} = v^*$ with respect to μ_N (i.e., ν_N satisfying $\mathrm{d}\mu_N(x) = \nu_N(x)\mathrm{d}x$). By writing this procedure in the form of a problem, we obtain the following one.

Problem 4 Find solutions to the following two subproblems.

SP1 Determine the explicit forms of α_N, K_N, and $V_{\mathscr{D}_d}^{\mu_N}$ on $\mathbf{R} \setminus [-\alpha_N, \alpha_N]$ under the constraints $\nu_N \in \mathscr{C}_{\mathrm{d}}^1(\mathbf{R}, 2(N+1))$, $\operatorname{supp} \nu_N = [-\alpha_N, \alpha_N]$, (28), and (29) where $\mathrm{d}\mu_N(x) = \nu_N(x)\mathrm{d}x$.

SP2 Let the solutions for α_N, K_N, and $V_{\mathscr{D}_d}^{\mu_N}$ of SP1 be given by α_N^*, K_N^*, and v^*, respectively. Then, solve the equation $V_{\mathscr{D}_d}^{\mu_N} = v^*$ on $\mathbf{R}$, i.e.,

$$\int_{-\alpha_N^*}^{\alpha_N^*} \log \left| \tanh \left[\frac{\pi}{4d}(x - z) \right] \right| \nu_N(z)\, \mathrm{d}z = v^*(x) \qquad x \in \mathbf{R}, \tag{30}$$

and determine μ_N (i.e., ν_N satisfying $\mathrm{d}\mu_N(x) = \nu_N(x)\mathrm{d}x$).

In SP1 of Problem 4, we first obtain the solution of Eq. (30) with the unknown parameters α_N and K_N included and then determine them by using the condition $\mu_N(\mathbf{R}) = 2(N+1)$ and the smoothness of v^* on $\mathbf{R}$. We use the Fourier transform for obtaining the solution of Eq. (30) and use the fast Fourier transform (FFT) for computing it. See the diagram in Fig. 1 for the process we apply to reduce Problems 1–4.

According to the above arguments, we can establish an algorithm to design an accurate formula for each $\boldsymbol{H}^\infty(\mathscr{D}_d, w)$ as follows.

[2]The boundary conditions of the Laplace equation are as follows: $V_{\mathscr{D}_d}^{\mu_N} = 0$ for $\partial\mathscr{D}_d$ and $V_{\mathscr{D}_d}^{\mu_N} = -\log w - K_N$ for $[-\alpha_N, \alpha_N]$.

Problem 1 ⋯ Optimization problem to determine optimal sampling points a_ℓ

↓ Approximation of sampling points by a measure

Problem 2 ⋯ Optimization problem to determine an optimal measure μ_N

↓ Restriction of the range of the solution μ_N and its characterization by potential theory

Problem 3 ⋯ System of an equation and an inequality to determine a smooth measure μ_N

↓ Reformulation based on the harmonic property of a Green potential

Problem 4

SP1 ⋯ Problem to determine the unknown parameters and the Green potential in Problem 3

SP2 ⋯ Problem to determine the measure μ_N

Fig. 1 Diagram showing the relations of Problems 1–4

Algorithm 1

1. For given d, w, and N, determine $\tilde{\alpha}_N^*$ and $\tilde{K}_N^*$ by solving SP1 in Problem 4 numerically.
2. Compute a numerical solution $\tilde{\nu}_N(x)$ of $\nu_N(x)$ for $x \in [-\tilde{\alpha}_N^*, \tilde{\alpha}_N^*]$ by solving SP2 in Problem 4 numerically.
3. Compute the indefinite integral $I[\tilde{\nu}_N](x) := \int_0^x \tilde{\nu}_N(t)\,\mathrm{d}t$ for $x \in [-\tilde{\alpha}_N, \tilde{\alpha}_N]$.
4. Compute $I[\tilde{\nu}_N]^{-1}$, which is the inverse function of $I[\tilde{\nu}_N]$.
5. Generate sampling points a_i as $a_i = I[\tilde{\nu}_N]^{-1}(i)$ $(i = -N, \ldots, N)$.
6. Obtain an approximation formula of f for $x \in \mathbf{R}$ as

$$\tilde{f}^{\mathrm{L}}_{2N+1}(x) = \sum_{j=-N}^{N} f(a_j) \frac{w(x)\, B_{N;j}(x; \{a_i\}, \mathscr{D}_d)}{w(a_j)\, B_{N;j}(a_j; \{a_i\}, \mathscr{D}_d)} \frac{T'_d(a_j - x)}{C_d}$$

$$= \sum_{j=-N}^{N} f(a_j) \frac{w(x)}{w(a_j)} \frac{\displaystyle\prod_{\substack{-N \le i \le N,\\ i \ne j}} \tanh\left[\frac{\pi}{4d}(x - a_i)\right]}{\displaystyle\prod_{\substack{-N \le i \le N,\\ i \ne j}} \tanh\left[\frac{\pi}{4d}(a_j - a_i)\right]} \operatorname{sech}^2\left[\frac{\pi}{4d}(a_j - x)\right]. \tag{31}$$

Because several approximations are introduced in the process of reducing Problems 1–4, we cannot guarantee the rigorous optimality of Formula (31). Although derivation of a rigorously optimal formula is left as a theme of future work, we expect that Formula (31) is sufficiently close to an optimal formula. In fact, in the case that the weight w decays single-exponentially, Formula (31) achieves the optimal convergence rate for that case [14].

Table 1 The weight functions w and data for using Formula (33) for Examples 1–3

	Weight function w	n and $\tilde{h}$ for the trapezoidal rule in (34)
Example 1	$\mathrm{sech}(2x)$	$(n, \tilde{h}) = (500, 0.04)$
Example 2	$\exp(-x^2)$	$(n, \tilde{h}) = (500, 0.02)$
Example 3	$\mathrm{sech}((\pi/2)\sinh(2x))$	$(n, \tilde{h}) = (500, 0.005)$

4.2 *Accurate Formulas for Numerical Integration*

By using Algorithm 1 described in Sect. 4.1, we obtain an approximate solution of the optimization problem given by (24). Because this problem is deduced from Problem 1 by replacing w with $w^{1/2}$, we apply Algorithm 1 to (24) to obtain $\{a_j\}$.[3] Then, what remains is to determine the coefficients $\{c_j\}$. We may obtain them by integrating the Hermite interpolation $\tilde{f}^{\mathrm{H}}_{2N+1}(x)$ with respect to x from $-\infty$ to ∞. However, for easy tractability of numerical computation, we employ the Lagrange interpolation $\tilde{f}^{\mathrm{L}}_{2N+1}(x)$ to generate the coefficients.[4] That is, we set

$$\begin{aligned} c_j &= c_{N:j}(\{a_i\}, \mathscr{D}_d, w) \\ &:= \int_{-\infty}^{\infty} \frac{w(x)\, B_{N;j}(x; \{a_i\}, \mathscr{D}_d)}{w(a_j)\, B_{N;j}(a_j; \{a_i\}, \mathscr{D}_d)} \frac{T_d'(a_j - x)}{C_d}\, \mathrm{d}x. \end{aligned} \tag{32}$$

Consequently, we propose the following numerical integration formula:

$$\int_{-\infty}^{\infty} f(x)\, \mathrm{d}x \approx \sum_{j=-N}^{N} c_{N:j}(\{a_i\}, \mathscr{D}_d, w)\, f(a_j). \tag{33}$$

For the use of Formula (33), we need to compute the values of $c_{N:j}$ for given d, w, and N. However, we cannot always obtain the closed forms of the integrals given by (32). In such a case, we need to compute them by numerical integration. Therefore, we need an efficient and accurate method for this integration. We leave this issue as a theme of future work and employ the trapezoidal rule on equispaced grid to compute the integrals given by (32):

$$c_{N:j}(\{a_i\}, \mathscr{D}_d, w) \approx \tilde{h} \sum_{k=-n}^{n} \frac{w(k\tilde{h})\, B_{N:j}(k\tilde{h}; \{a_i\}, \mathscr{D}_d)}{w(a_j)\, B_{N:j}(a_j; \{a_i\}, \mathscr{D}_d)} \frac{T_d'(a_j - k\tilde{h})}{C_d}. \tag{34}$$

The actual values of n in (34) for the examples in Sect. 5 are shown in Table 1.

[3]The function $w^{1/2}$ may not satisfy Assumption 1. However, we can apply Algorithm 1 to $w^{1/2}$ in that case.

[4]We have not justified this treatment mathematically, although we have some numerical results implying its effectiveness. We leave its justification as a theme of future work.

5 Numerical Experiments

For numerical experiments of Formula (33), we used the following three functions as examples.[5] These functions are analytic on $\mathscr{D}_d$ for $d = \pi/4$.

Example 1 (*Function with single-exponential decay*)

$$f(x) = \mathrm{sech}(2x), \qquad \int_{-\infty}^{\infty} f(x)\,\mathrm{d}x = \frac{\pi}{2}.$$

Example 2 (*Function with Gaussian decay*)

$$f(x) = \frac{x^2}{(\pi/4)^2 + x^2}\,\mathrm{e}^{-x^2}, \qquad \int_{-\infty}^{\infty} f(x)\,\mathrm{d}x = \sqrt{\pi} - \frac{\pi^2 \mathrm{e}^{(\pi/4)^2}}{4}\left\{1 - \mathrm{erf}\left(\frac{\pi}{4}\right)\right\}.$$

Example 3 (*Function with double-exponential decay*)

$$f(x) = \frac{1}{\cosh((\pi/2)\sinh(2x))}, \qquad \int_{-\infty}^{\infty} f(x)\,\mathrm{d}x = 0.8119286510\cdots.$$

We applied Formula (33) to these functions to compute the approximate values of their integrals for $N = 5k$ $(k = 1, 2, \ldots, 10)$. Table 1 shows the weight functions of Examples 1–3 for which they belong to $\boldsymbol{H}^{\infty}(\mathscr{D}_d, w)$ and the parameters n and $\tilde{h}$ for the trapezoidal rule given by (34). We computed the errors of the approximate values and compared each of them with the error computed by the trapezoidal rule (2) adjusted to each function. As shown in Appendix 2, this adjustment is done by determining the grid spacing h according to the decay property of the weight function w corresponding to the integrand f. In particular, in the case that w decays double-exponentially like Example 3, the trapezoidal rule corresponds to the DE formula. Therefore, in the case of Example 3, we substantially compared Formula (33) and the DE formula. We performed all the computations by MATLAB R2015a programs with the floating point numbers. We used the double-precision numbers for Examples 1–2 and multiprecision numbers for Example 3, which are realized by the Multiprecision Computing Toolbox produced by Advanpix (http://www.advanpix.com).

[5]Because the integral of the function in Example 3 cannot be evaluated exactly, we computed it with 100 digits by using Mathematica and used that value as an alternative of its exact value.

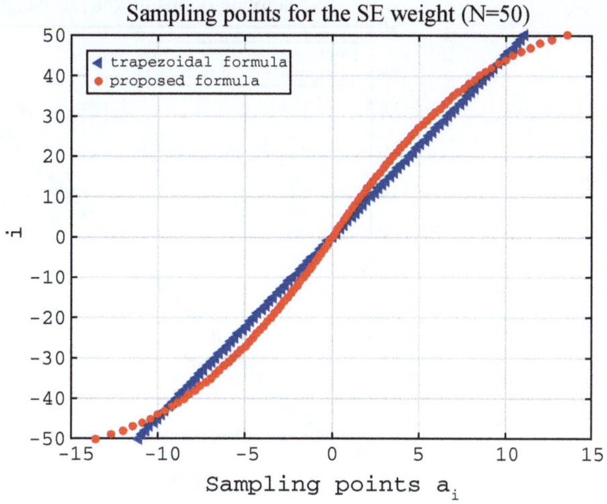

Fig. 2 Sampling points for Example 1 for $N = 50$. "Proposed formula" denotes the results of Formula (33)

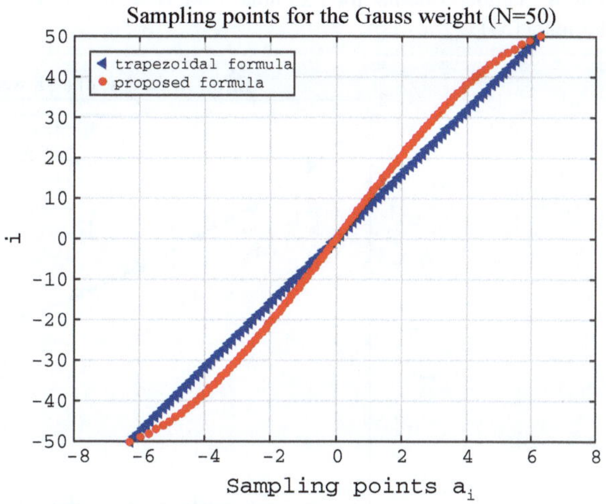

Fig. 3 Sampling points for Example 2 for $N = 50$. "Proposed formula" denotes the results of Formula (33)

We show the results of the sampling points and the errors in Figs. 2, 3, 4, 5, 6, and 7, respectively. In all the cases, Formula (33) outperforms the trapezoidal rule. By noting that we can make a formula by applying Formula (33) to a transformed integrand $f(x) = F(\psi(x))\psi'(x)$ with a transform ψ like (3)–(5), we can know from the result in the case of Example 3 that the formula made by that technique outperforms the DE formula.

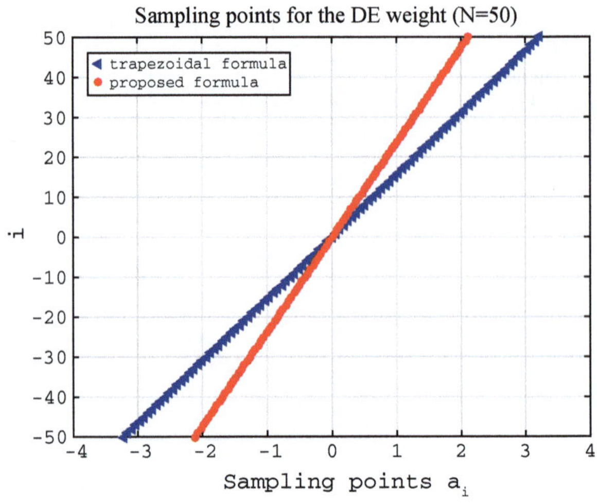

Fig. 4 Sampling points for Example 3 for $N = 50$. "Proposed formula" denotes the results of Formula (33)

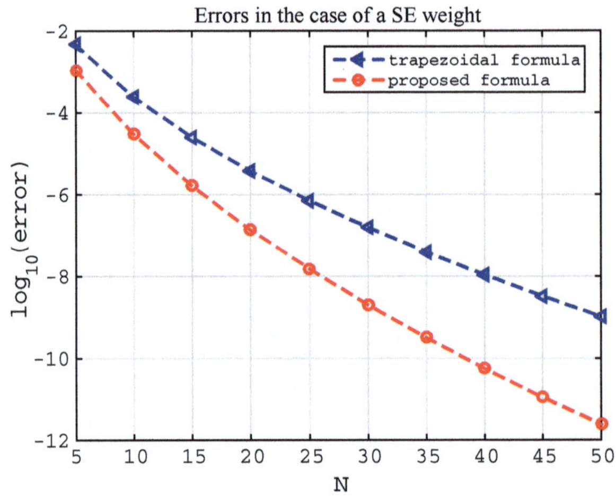

Fig. 5 Errors for Example 1. "Proposed formula" denotes the results of Formula (33)

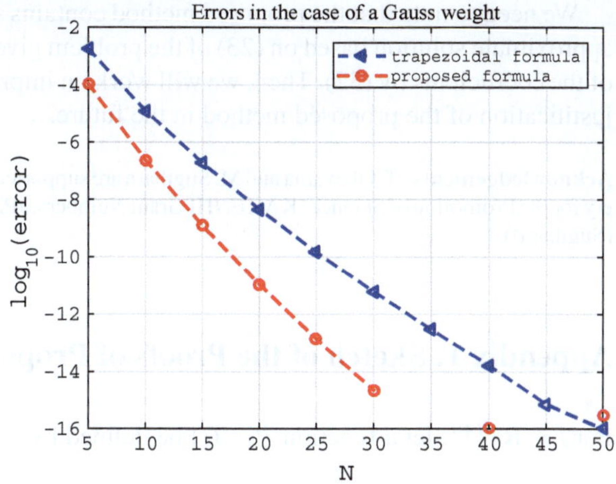

Fig. 6 Errors for Example 2. "Proposed formula" denotes the results of Formula (33)

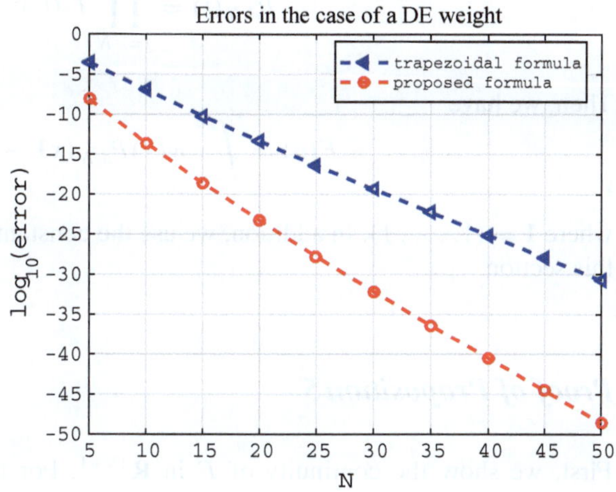

Fig. 7 Errors for Example 3. "Proposed formula" denotes the results of Formula (33)

6 Concluding Remarks

In this paper, we proposed the method for designing accurate numerical integration formulas in the weighted Hardy spaces $\boldsymbol{H}^\infty(\mathscr{D}_d, w)$. We proved Theorem 1 giving the explicit form of the minimum error norm $\mathscr{E}_N^{\min}(\boldsymbol{H}^\infty(\mathscr{D}_d, w))$ and proposed the approximate solution to the optimization problem in (9) that determines a sequence of sampling points of an optimal numerical integration formula for $\boldsymbol{H}^\infty(\mathscr{D}_d, w)$. In that solution, we employed the method for designing accurate formulas for approximating functions in $\boldsymbol{H}^\infty(\mathscr{D}_d, w)$, which was proposed in our paper [14]. Then, we obtained Formula (33) by using the sampling points and the coefficients corresponding to them. The numerical results showed the effectiveness of the formula.

We need to note that the proposed method contains some heuristic techniques: the approximate solution based on (23) of the problem given by (9), and the computation of the coefficients by (32). Then, we will work on improvement and/or mathematical justification of the proposed method in the future.

Acknowledgements T. Okayama and M. Sugihara are supported by the Grant-in-Aid of Japan Society for the Promotion of Science (KAKENHI Grant Numbers: JP24760060 (Okayama), JP25390146 (Sugihara)).

Appendix 1: Sketch of the Proofs of Propositions 5 and 6

For $t \in \mathbf{R}^{2N+1}$, let a function $P_{d,N}(t)$ be defined by

$$P_{d,N}(t) = \prod_{j=-N}^{N} T_d(t_j)^2.$$

Then, we have

$$F(a) = \int_{-\infty}^{\infty} w(x) P_{d,N}(x\mathbf{1} - a)\,\mathrm{d}x$$

where $\mathbf{1} = (1, \ldots, 1)$. In addition, we use the constant C_d given by (15) throughout this section.

Proof of Proposition 5

First, we show the continuity of F in $\mathbf{R}^{2N+1}$. For the partial derivative of $P_{d,N}(x\mathbf{1} - a)$:

$$\begin{aligned}\frac{\partial}{\partial a_k} P_{d,N}(x\mathbf{1} - a) &= \left(\prod_{j \neq k} T_d(x - a_j)^2\right) \cdot \left(-2T_d(x - a_k)T_d'(x - a_k)\right) \\ &= \left(\prod_{j \neq k} T_d(x - a_j)^2\right) \cdot T_d(x - a_k) \cdot \frac{-2C_d}{\cosh^2(C_d(x - a_k))},\end{aligned} \tag{35}$$

we have

$$\left| \frac{\partial}{\partial a_k} P_{d,N}(x\mathbf{1} - a) \right| \leq 2C_d. \tag{36}$$

Therefore, it follows from (36) and the mean value theorem that

$$\left|P_{d,N}(x\mathbf{1}-(a+\Delta a))-P_{d,N}(x\mathbf{1}-a)\right| \leq 2C_d \sum_{j=-N}^{N} |\Delta a_j|.$$

Consequently, we show the continuity of F in $\mathbf{R}^{2N+1}$ as follows:

$$\begin{aligned}|F(a+\Delta a)-F(a)| &\leq \int_{-\infty}^{\infty} w(x)\left|P_{d,N}(x\mathbf{1}-(a+\Delta a))-P_{d,N}(x\mathbf{1}-a)\right| \mathrm{d}x \\ &\leq 2C_d \left(\int_{-\infty}^{\infty} w(x)\,\mathrm{d}x\right) \sum_{j=-N}^{N} |\Delta a_j|.\end{aligned}$$

Next, we show that F has a minimizer in $\mathbf{R}^{2N+1}$. For this purpose, we introduce the variable transformations:

$$t = T_d(x), \quad b_j = T_d(a_j).$$

Then, by the addition theorem of tanh and integration by substitution, we have

$$F(a) = \int_{-1}^{1} \frac{\tilde{w}(t)}{C_d(1-t^2)} \left[\prod_{j=-N}^{N} \left(\frac{t-b_j}{1-b_j t}\right)^2\right] \mathrm{d}t \tag{37}$$

where $\tilde{w}(t) = w(C_d^{-1} \operatorname{arctanh} t)$. Let $G(b)$ and $g(b;t)$ denote the RHS of (37) and its integrand, respectively. Because F is continuous in $\mathbf{R}^{2N+1}$, so is the function G in $(-1,1)^{2N+1}$. In order to show the continuity of G in $[-1,1]^{2N+1}$, we consider b on the boundary of $[-1,1]^{2N+1}$, i.e., $b \in [-1,1]^{2N+1}$ with $b_j = \pm 1$ for some j. By letting J be the set of the indexes j for which $b_j = \pm 1$, we have

$$g(b;t) = \frac{\tilde{w}(t)}{C_d(1-t^2)} \left[\prod_{j \notin J} \left(\frac{t-b_j}{1-b_j t}\right)^2\right]. \tag{38}$$

Let $\{b^{(n)}\}$ be a sequence in $(-1,1)^{2N+1}$ with $b^{(n)} \to b$. By noting the continuity of $((t-b_j)/(1-b_j t))^2$ as a function of (t,b_j) in $[-1,1]^2$, we have

$$g(b^{(n)};t) \to g(b;t) \qquad (\forall t \in [-1,1]).$$

Furthermore, by noting that

$$\left(\frac{t-b_j}{1-b_j t}\right)^2 \leq 1 \qquad (\forall (t,b_j) \in [-1,1]^2), \tag{39}$$

we have

$$|g(b^{(n)};t)| \leq \frac{\tilde{w}(t)}{C_d(1-t^2)} \qquad (\forall t \in (-1,1)).$$

Because $\tilde{w}(t)/(C_d(1-t^2))$ is integrable on $(-1, 1)$, it follows from Lebesgue's dominated convergence theorem that

$$G(b^{(n)}) \to G(b),$$

which indicates the continuity of G in $[-1, 1]^{2N+1}$. Therefore, G has a minimizer in $[-1, 1]^{2N+1}$. In fact, we can show that a minimizer of G exists in $(-1, 1)^{2N+1}$. For any $b \in [-1, 1]^{2N+1}$ with $b_j = \pm 1$ for some j, we take $\tilde{b} \in (-1, 1)^{2N+1}$

$$\tilde{b}_j = \begin{cases} b_j & (|b_j| < 1) \\ 1-\varepsilon & (b_j = 1) \\ -1+\varepsilon & (b_j = -1) \end{cases}$$

by a sufficiently small $\varepsilon > 0$. Then, because of (38) and (39), we have

$$g(\tilde{b};t) = g(b;t)\left[\prod_{j\in J}\left(\frac{t-\tilde{b}_j}{1-\tilde{b}_j t}\right)^2\right] \leq g(b;t) \qquad (\forall t \in [-1,1]),$$

which implies that $G(\tilde{b}) \leq G(b)$. Hence, G has a minimizer in $(-1, 1)^{2N+1}$, which implies that F has a minimizer in $\mathbf{R}^{2N+1}$.

Proof of Proposition 6

The function $w(x)P_{d,N}(x\mathbf{1} - a)$ is integrable with respect to x on $\mathbf{R}$ and continuously differentiable with respect to a_j. According to (35) and (36), we have

$$\left|\frac{\partial}{\partial a_j}\left(w(x)P_{d,N}(x\mathbf{1}-a)\right)\right| \leq 2C_d w(x).$$

Therefore, by Lebesgue's dominated convergence theorem for derivatives, we have (14).

Appendix 2: Determination of the Grid Spacing h of the Midpoint Rule

We show how to determine the grid spacing h of the trapezoidal rule (2). For its error, we have

$$\left|\int_{-\infty}^{\infty} f(x)\,\mathrm{d}x - h\sum_{k=-N}^{N} f(kh)\right| = \left|\int_{-\infty}^{\infty} f(x)\,\mathrm{d}x - h\sum_{k=-\infty}^{\infty} f(kh) + h\sum_{|k|>N} f(kh)\right|$$
$$\leq E_D(f,h) + E_T(f,h,N)$$

where the discretization error $E_D(f, h)$ and truncation error $E_T(f, h, N)$ are given by

$$E_D(f,h) = \left|\int_{-\infty}^{\infty} f(x)\,\mathrm{d}x - h\sum_{k=-\infty}^{\infty} f(kh)\right|,$$
$$E_T(f,h,N) = \left|h\sum_{|k|>N} f(kh)\right|.$$

For the error $E_D(f, h)$ it is well known that

$$E_D(f,h) \leq C\exp\left(-\frac{2\pi d}{h}\right) \tag{40}$$

for $f \in \boldsymbol{H}^{\infty}(\mathscr{D}_d, w)$ where C is a positive number independent of h. On the other hand, for $f \in \boldsymbol{H}^{\infty}(\mathscr{D}_d, w)$ we have

$$E_T(f,h,N) \leq \|f\|\, h\sum_{|k|\geq N} w(kh). \tag{41}$$

Then, by evaluating the RHS in (41) according to the weight function w for each of Examples 1–3, we have

$$\text{Example 1: } E_T(f,h,N) \leq C_1\exp(-2Nh), \tag{42}$$
$$\text{Example 2: } E_T(f,h,N) \leq C_2\exp(-(Nh)^2), \tag{43}$$
$$\text{Example 3: } E_T(f,h,N) \leq C_3\exp(-(\pi/4)\exp(2Nh)) \tag{44}$$

where C_1, C_2, and C_3 are positive numbers independent of N and h.

According to Inequalities (40) and (42)–(44), we determine h so that the order of the upper bound of $E_D(f, h)$ is equal to that of $E_T(f, h, N)$ for each of Examples 1–3. In the case of Example 1, by equating the exponent of the RHS of (40) with that of (42) and substituting $d = \pi/4$ into the former, we have

$$-\frac{2\pi d}{h} = -2Nh \Rightarrow h = \sqrt{\frac{2\pi d}{2N}} = \frac{\pi}{2\sqrt{N}}.$$

In the cases of Examples 2–3, in a similar manner we have

$$-\frac{2\pi d}{h} = -(Nh)^2 \qquad \Rightarrow \quad h = \left(\frac{2\pi d}{N^2}\right)^{1/3} = \left(\frac{\pi^2}{2N^2}\right)^{1/3},$$
$$-\frac{2\pi d}{h} = -\frac{\pi}{4}\exp(2Nh) \quad \Rightarrow \quad h = \frac{W(16dN)}{2N} = \frac{W(4\pi N)}{2N} \approx \frac{\log(4\pi N)}{2N}$$

where W is the inverse of the function $g(x) = x\,e^x$, which is called Lambert's W function.

References

1. J.-E. Andersson, Optimal quadrature of H^p functions. Math. Z. **172**, 55–62 (1980)
2. J.-E. Andersson, B.D. Bojanov, A note on the optimal quadrature in H^p. Numer. Math. **44**, 301–308 (1984)
3. S. Haber, The tanh rule for numerical integration. SIAM J. Numer. Anal. **14**, 668–685 (1977)
4. M. Mori, Discovery of the double exponential transformation and its developments. Publ. RIMS Kyoto Univ. **41**, 897–935 (2005)
5. E.B. Saff, V. Totik, *Logarithmic Potentials with External Fields* (Springer, Berlin, 1997)
6. C. Schwartz, Numerical integration of analytic functions. J. Comput. Phys. **4**, 19–29 (1969)
7. M. Sugihara, Optimality of the double exponential formula – functional analysis approach. Numer. Math. **75**, 379–395 (1997)
8. M. Sugihara, Near optimality of the sinc approximation. Math. Comput. **72**, 767–786 (2003)
9. F. Stenger, *Numerical Methods Based on Sinc and Analytic Functions* (Springer, New York, 1993)
10. F. Stenger, *Handbook of Sinc Numerical Methods* (CRC Press, Boca Raton, 2011)
11. H. Takahasi, M. Mori, Quadrature formulas obtained by variable transformation. Numer. Math. **21**, 206–219 (1973)
12. H. Takahasi, M. Mori, Double exponential formulas for numerical integration. Publ. RIMS Kyoto Univ. **9**, 721–741 (1974)
13. K. Tanaka, M. Sugihara, K. Murota, M. Mori, Function classes for double exponential integration formulas. Numer. Math. **111**, 631–655 (2009)
14. K. Tanaka, T. Okayama, M. Sugihara, Potential theoretic approach to design of accurate formulas for function approximation in symmetric weighted Hardy spaces. IMA J. Numer. Anal. (2016). doi:10.1093/imanum/drw022
15. L.N. Trefethen, J.A.C. Weideman, The exponentially convergent trapezoidal rule. SIAM Rev. **56**, 385–458 (2014)

A Class of Intrinsic Trigonometric Mode Polynomials

Vesselin Vatchev

Abstract In this paper, we study a class of trigonometric polynomials that exhibit properties expected from intrinsic mode functions. In a series of lemmas, we provide sufficient conditions for a positiveness of the instantaneous frequency, number of zeros and extrema, and the proximity of upper and lower envelopes. The question of necessity of each of the conditions is discussed in numerical examples. We also introduce an orthonormal basis in L_2 of weak intrinsic mode functions.

Keywords Intrinsic modes · Instantaneous frequency · Trigonometric polynomials

1 Introduction

The empirical mode decomposition (EMD) method and the Hilbert–Huang transform were introduced in [8] to analyze nonstationary and nonlinear signals. The EMD method is a decomposition process to reduce any given data into a collection of intrinsic mode functions (IMFs) to which the Hilbert spectral analysis can be applied.

An IMF represents a simple oscillatory mode as a counterpart to the simple harmonic function but with variable amplitude and frequency along the time axis. In [8], IMF is defined as a function for which the cubic spline envelope through the maxima coincides with the negative of the cubic spline envelope through the minima.

The EMD has proven to be very effective in great variety of applications. The convergence of the method, its properties, and alternative characterizations of IMF inspired intense research in the area. Among the major results are the B-spline EMD, introduced in [3], and the synchrosqueezed wavelet transform, introduced in [4].

The EMD is a procedure to decompose a signal into intrinsic mode functions with presumably well-defined instantaneous frequencies. The analytic signal for a real f is $F = f + iHf$ where $Hf(t) = \text{p.v.} \int_{-\infty}^{\infty} \frac{f(x)}{t-x}\,dx$ is the Hilbert transform of f. In the case of a periodic f, the Hilbert transform is replaced by the conjugate operator $Hf(t) = \frac{1}{2\pi}\int_{-\pi}^{\pi} f(x)\tan\frac{t-x}{2}\,dx$. The *frequency* of f is denoted by $\theta(t)$ and

V. Vatchev (✉)
School of Mathematical and Statistical Sciences, The University of Texas Rio Grande Valley, One West University Blvd., Brownsville, TX 78520, USA
e-mail: vesselin.vatchev@utrgv.edu

G.E. Fasshauer and L.L. Schumaker (eds.), *Approximation Theory XV: San Antonio 2016*, Springer Proceedings in Mathematics & Statistics 201,
DOI 10.1007/978-3-319-59912-0_18

is defined through the relation $\tan\theta(t) = \frac{Hf(t)}{f(t)}$ and its derivative is the instantaneous frequency, that is,

$$\theta'(t) = \frac{Hf'(t)f(t) - Hf(t)f'(t)}{f^2(t) + Hf^2(t)}.$$

In order to have a meaningful definition from a physical point of view, θ' has to be positive.

In this paper, we study relations between the Fourier coefficients of a periodic function and the properties expected from IMF. In [10], we discussed that the function $\psi(t) = e^{\varepsilon\cos t}\sin(\varepsilon\sin t)$ has upper and lower envelopes differing only by sign for any $\varepsilon > 0$, but the instantaneous frequency is negative for $\varepsilon = 2.97$. The Fourier spectrum of ψ has a peak at 3 and is spread over all frequencies. Our interest in this paper is to provide sufficient conditions on the Fourier coefficients such that the resulting function is an IMF with positive frequency. The functions that we consider have dominant frequency and we are interested in how the Fourier coefficients can spread from that frequency while preserving the properties expected from IMF. In the next section, we provide sufficient conditions for positiveness of the instantaneous frequency, number of zeros and extrema, and the proximity of upper and lower envelopes. In Sect. 3 in a series of numerical examples, we discuss the necessity of the conditions. Functions for which the number of zeros and the number of extrema differ by at most 1 are called weak IMF. A special class of variable frequencies θ that generate an orthogonal basis of weak IMF in L_2 is considered in Sect. 4.

2 Intrinsic Trigonometric Mode Polynomials

In this section, we consider functions with dominant frequency. Without loss of generality, we assume that n is the dominant frequency and the corresponding harmonic is $\cos nt$. The function $\cos nt$ is an ideal IMF with a derivative $-n\sin nt$ and interlacing extrema

$$x_j = \frac{2j}{n}\pi, \quad y_j = \frac{2j+1}{n}\pi, \quad j = 0, 1, \dots, n-1 \tag{1}$$

where x_j are the zeros of $\sin\frac{nt}{2}$ and y_j are the zeros of $\cos\frac{nt}{2}$. The main problem we study is how to add harmonics and preserve the positive instantaneous frequency and the IMF properties. Since $\cos nt$ is our starting function, we define upper and lower envelopes as the interpolating trigonometric polynomials to f at the two sets of points x and y, defined in (1).

Definition 1 For positive integers n and $l (l < n)$, a trigonometric polynomial with real coefficients c_j and real β_j, $j = 1, \dots,$

$$f(t) = \cos nt + \sum_{j=n-l, j\neq n}^{n+l} c_j \cos(jt + \beta_j), \tag{2}$$

is an intrinsic trigonometric mode polynomial (ITMP) if it has positive instantaneous frequency and the interpolating trigonometric polynomials at each of the two collections x_j, $j = 0, \dots, n-1$ and y_j, $j = 0, \dots, n-1$ differ only by sign.

The coefficients c_j are the spectrum of f. The example from the introduction is the reason why we consider the spectrum of candidate IMF bounded from above by twice the dominant frequency. The points x and y are not extrema of the function, but if

$$A(n, l) = \sum_{j=n-l,\ j \neq n}^{n+l} |c_j| \tag{3}$$

is small, then f has maxima close either to the points x or to y, depending on oddity of n, and minima close to the other set, more details are considered in Lemma 2. In the next three lemmas, we provide upper bounds for $A(n, l)$ in order that the corresponding polynomial (2) to be an ITMP. Our first result is a sufficient condition for a positive instantaneous frequency.

Lemma 1 *A polynomial f of the form (2) has a positive instantaneous frequency if*

$$A(n, l) < \frac{n}{n + \frac{l}{2} + \sqrt{\left(n + \frac{l}{2}\right)^2 + n(n+l)}}, \quad l = 1, \dots, n-1.$$

Proof For the periodic function f, the analytic signal is $F(t) = f(t) + iHf(t) = r(t)e^{i\theta(t)}$ where $Hf(t) = \sin nt + \sum_{j=n-l,\ j\neq n}^{n+l} c_j \sin(jt + \beta_j)$ is the conjugate function of f. The amplitude of f is defined as $r(t) = \sqrt{f^2(t) + Hf^2(t)}$ and the instantaneous frequency is defined as $\theta'(t) = \frac{Hf'(t)f(t) - Hf(t)f'(t)}{f^2(t) + Hf^2(t)}$. Let $c_n = 1$ and $\beta_n = 0$, then in terms of the coefficients c_j, we have that

$$\begin{aligned} r^2(t)\theta'(t) &= \left(\sum_{j=n-l}^{n+l} jc_j \cos(jt + \beta_j) \right) \left(\sum_{j=n-l}^{n+l} c_j \cos(jt + \beta_j) \right) \\ &+ \left(\sum_{j=n-l}^{n+l} jc_j \sin(jt + \beta_j \right) \left(\sum_{j=n-l}^{n+l} c_j \sin(jt + \beta_j)) \right) \\ &= \sum_{j=n-l}^{n+l} jc_j f_j(t) \end{aligned}$$

where $f_j(t) = \sum_{k=n-l}^{n+l} c_k \cos((k-j)t + \beta_k - \beta_j)$. Since n and l are fixed, to simplify the notations, we set $A = A(n, l)$. By using the estimates $f_n > 1 - A$ and $|f_j| \leq A + 1$ for $j \neq n$, we have that

$$r^2(t)\theta'(t) = nf_n(t) + \sum_{j=n-l,\ j\neq n}^{n+l} jc_j f_j(t) \geq n(1-A) - \sum_{j=n-l, j\neq n}^{n+l} j|c_j|(1+A)$$

$$\geq n(1-A) - (1+A) \sum_{j=n-l, j\neq n}^{n+l} j|c_j| \geq n(1-A) - (n+l)(1+A)A$$

$$= -(n+l)A^2 - (2n+l)A + n.$$

From the choice of $A < \frac{n}{n+\frac{l}{2}+\sqrt{\left(n+\frac{l}{2}\right)^2+n(n+l)}}$, it follows that the quadratic polynomial above is positive for $0 < l < n$. To finish the proof, we need to show that $r^2 > 0$. Similarly to the above, we have for $r^2 = f^2 + Hf^2$ that

$$f^2 + Hf^2 = \sum_{j=n-l}^{n+l} c_j f_j = f_n + \sum_{j=n-l,\ j\neq n}^{n+l} c_j f_j \geq 1 - A - A(1+A) = 2 - (A+1)^2$$

and is positive for $A < \sqrt{2} - 1$. Since the estimate for $A(n,l)$ is decreasing as a function of l and for $l = 0$ equals $\sqrt{2} - 1$, we complete the proof of the lemma. □

In [10], we defined a weak IMF as a function whose number of zeros and number of extrema differ at most by one. The next lemma provides a sufficient condition in terms of $A(n,l)$, defined in (3), for a polynomial of type (2) to be a weak IMF.

Lemma 2 *A polynomial f of type (1) is a weak IMF if*

$$A(n,l) < \frac{n^2}{(n+l)\sqrt{n^2 + (n+l)^2}}, \quad l = 1, \ldots, n-1.$$

Proof Since we consider n and l fixed throughout the proof, we set $A = A(n,l)$. We prove that under the assumption for A the polynomial f has exactly $2n$ zeros and $2n$ extrema on $[0, 2\pi)$. The function $P(t) = f(t) - \cos nt$ is a trigonometric polynomial of degree at most $n + l$ and as such from Bernstein's inequality for trigonometric polynomials it follows that $|P^{(k)}| \leq (n+l)^k \|P\|_\infty \leq (n+l)^k A$. The polynomial f has a zero between any two consecutive $\xi_j = \frac{j}{n}\pi$ for $j = 0, \ldots, 2n - 1$. Indeed, $f(\xi_j) = \cos n\xi_j + P(\xi_j) = (-1)^j + P(\xi_j)$ and, since $A < 1$, we have that $f(\xi_j)$ has sign $(-1)^j$. If we assume that there are two zeros, say $z_1 < z_2$, between ξ_j and ξ_{j+1}, then from Rolle's theorem it follows that $f'(e) = 0$ for some point $e \in (z_1, z_2)$. Furthermore, from $0 = f(z_1) = \cos nz_1 + P(z_1)$ we have that $|\cos(nz_1)| = |P(z_1)| \leq A$. The same estimate holds at z_2, and since $\cos nt$ is strictly monotone on (ξ_j, ξ_{j+1}), it follows that $|\cos nt| < A$ on (z_1, z_2) and hence $|\sin nt| > \sqrt{1 - A^2}$ on (z_1, z_2). On the other hand, for the derivative $f'(t) = -n \sin nt + P'(t)$ on the interval (z_1, z_2) we have that $|f'| \geq n\sqrt{1 - A^2} - (n+l)A > 0$ for $A < \frac{n}{\sqrt{n^2+(n+l)^2}}$. We arrived at contradiction with the assumption that there are two zeros between ξ_j and ξ_{j+1}.

From $f'(t) = n(-\sin nt + \frac{P'(t)}{n})$ and $\frac{|P'|}{n} < \frac{n+l}{n}A < 1$, we have that f' has a zero between any two consecutive $z_j = \frac{2j+1}{2n}\pi$, $j = 0, \ldots, 2n-1$. If we assume that there are at least two, say e_1 and e_2, on one of those intervals, then from Rolle's theorem it follows that there is an inflection point on that interval. Similarly to the case for the count of the zeros of f, we estimate $|\cos nt| > \sqrt{1 - \frac{(n+l)^2}{n^2}A^2}$ and $|f''(t)| > n^2\sqrt{1 - \frac{(n+l)^2}{n^2}A^2} - P''(t) > n^2\sqrt{1 - \frac{(n+l)^2}{n^2}A^2} - (n+l)^2 A > 0$ for the choice of A. Again, we arrived at contradiction and hence f has exactly $2n$ zeros and $2n$ extrema on $[0, 2\pi)$. The proof of the lemma is complete. □

For polynomials of type (2) with symmetric spectrum, i.e., $c_{n+k} = c_{n-k}$, and $\beta_{n+k} = -\beta_{n-k}$, $k = 1, \ldots, n-1$ the condition on $A(n, l)$ from Lemma 1 can be relaxed.

Corollary 1 *Let $R_{n-1}(t) = 1 + \sum_{j=1}^{n-1} c_j \cos(jt + \beta_j)$; then, the function $f(t) = R_{n-1}(t)\cos nt$ has symmetric spectrum with respect to the dominant frequency n, $\theta' = n$ if $\sum_{j=1}^{n-1} |c_j| < 1$, and is a weak IMF if $\sum_{j=1}^{n-1} |c_j| < \frac{1}{2\sqrt{5}}$.*

Proof It is a straightforward computation to check that f has symmetric spectrum. For $\sum_{j=1}^{n-1} |c_j| < 1$ the polynomial R_{n-1} is positive and from [9], Theorem 3.1, it follows that $R_{n-1}(t)\cos nt$ has instantaneous frequency n. In the case $\sum_{j=1}^{n-1} |c_j| < \frac{1}{2\sqrt{5}}$, from Lemma 2, with $l = n-1$, it follows that f is a weak IMF. □

Our last lemma in the section holds for any trigonometric polynomial of degree less than $2n$ and is similar in nature to the results in [13].

Lemma 3 *For $T_n(t) = \frac{a_0}{2} + \sum_{j=1}^{2n-1}(a_j \cos jt + b_j \sin jt)$, the trigonometric interpolant at x_j, $j = 0, \ldots, n-1$, defined by (1), is*

$$U_n(t) = \frac{A_0}{2} + \sum_{j=0}^{\frac{n}{2}-1}(A_j \cos jt + B_j \sin jt) + \frac{A_{\frac{n}{2}}}{2}\cos\frac{n}{2}t,$$

$$A_j = a_j + a_{n-j} + a_{n+j} + a_{2n-j},\ j = 0, \ldots, \frac{n}{2}$$

$$B_j = b_j - b_{n-j} + b_{n+j} - b_{2n-j},\ j = 1, \ldots, \frac{n}{2} - 1,$$

and the trigonometric interpolant at y_j, $j = 0, \ldots, n-1$, defined by (1), is

$$L_n(t) = \frac{C_0}{2} + \sum_{j=0}^{\frac{n}{2}-1}(C_j \cos jt + D_j \sin jt) + \frac{C_{\frac{n}{2}}}{2}\cos\frac{n}{2}t,$$

$$C_j = a_j - a_{n-j} - a_{n+j} + a_{2n-j},\ j = 0, \ldots, \frac{n}{2}$$

$$D_j = b_j + b_{n-j} - b_{n+j} - b_{2n-j},\ j = 1, \ldots, \frac{n}{2} - 1.$$

Proof The proof follows from [13]. Briefly, the result for U_n is a direct corollary from [2]. The interpolant L_n can be considered as the interpolant, say S_n, at the points x to $T_n(t+\frac{\pi}{n})$. From $L_n(t)=S_n(t-\frac{\pi}{n})$, the proof follows from simple trigonometry. □

Summing up we obtain a sufficient condition on $A(n,l)$, defined in (3), for a function of type (2) to be an ITMP.

Theorem 1 *For any* $0 \le l < \frac{n}{2}$ *and* $0 \le A(n,l) < \frac{1}{3}$, *the trigonometric polynomial* f *of the form (2) is an ITMP.*

Proof Since the estimates for $A(n,l)$ in Lemmas 1 and 2 are decreasing functions of l, we need to compute them only for $l=\frac{n}{2}$. From Lemma 1, we have that $A(n,l) < A(n,\frac{n}{2}) < \frac{1}{3}$ for any $l < \frac{n}{2}$ and hence f has positive instantaneous frequency. The estimate $A(n,l) < \frac{1}{3}$ is stronger than $A(n,\frac{n}{2}) < \frac{4}{3\sqrt{13}}$ that is required in Lemma 2 with $l=\frac{n}{2}$ and hence f is a weak IMF for any $l < \frac{n}{2}$.

From Lemma 3 with $a_j=b_j=0$ for $j \ge \frac{n}{2}$, we get that

$$U_n(t) = 1 + \sum_{j=0}^{\frac{n}{2}-1} \big((a_{n-j}+a_{n+j})\cos jt - (b_{n-j}-b_{n+j})\sin(jt)\big)$$

where $a_j = c_j\cos\beta_j$, $b_j = c_j \sin\beta_j$ and $L_n = -U_n$. The proof is complete. □

The condition on the Fourier coefficients of f provided in Theorem 1 is only sufficient, and in the next section, we will consider a series of numerical examples to discuss its necessity. We also illustrate the application of EMD with trigonometric envelopes.

The above result suggests a modification of the pure greedy algorithm, for details see [11], in the trigonometric basis, when one picks the largest in absolute value Fourier coefficients and adds the adjacent terms compared to $\frac{1}{6^k}$ of the largest coefficient.

3 Numerical Examples

The extreme polynomials in all of the lemmas from Sect. 2 are of the form

$$f(t) = b\cos mt + (1+a\cos kt)\cos nt. \tag{4}$$

Next, we study their analytic properties and the performance of EMD on them. In [4], the polynomials of type (4) with $a=\frac{1}{5}$ and $b=0$ were considered as examples for different interpretations of the IMF depending on the Fourier coefficients. In the examples of the section, we discuss the relation of their Fourier coefficients of polynomials of type (4) and EMD and the resulting IMF.

The instantaneous frequency for real parameters a and b of the model signals (4) is computed explicitly in the next proposition.

Proposition 1 *The instantaneous frequency of f is $\theta'(t) = n$ if $b = 0$ and $k < n$. If $b = 0$ and $k \geq n$, then*

$$\theta'(t) = \frac{2n + 2a^2k\cos^2 nt + (2n+k)a\cos kt + ak\cos(2n-k)t}{2 + 2a^2\cos^2 nt + 2a\cos kt + 2a\cos(2n-k)t},$$

and if $a = 0$, then

$$\theta'(t) = \frac{n + b(m+n)\cos(n-m)t + b^2m}{1 + 2b\cos(n-m)t + b^2}. \tag{5}$$

Proof The proof follows from Theorem 1. The statement for $b = 0$ and $k < n$ also follows from the Bedrosian theorem, for details see [1, 9].

Our first example is an ideal IMF with a spectrum spreading outside $[0, 2n-1]$. This example shows that the spectral band conditions of Lemmas 1 and 2 are not necessary since the function still has positive instantaneous frequency (conclusion of Lemma 1) and is also a weak IMF (conclusion of Lemma 2). Further in each lemma, the corresponding bound conditions on the absolute sums $A(n, l)$, defined in (3), of the example are also violated and therefore not necessary conditions.

Example 1 Bounded spectrum and aliasing
For $b = 0$, $n = 12$, and $k = 24$,

$$f(t) = 1.4\left(\cos 12t + \frac{2}{7}\cos 36t\right) = (1 + 0.8\cos 24t)\cos 12t$$

and the derivative $f'(t) = -12\sin 12t(3a\cos 24t + 2a + 1)$ has extrema at the zeros of $\sin 12t$ where f has alternating values $\pm(1+a)$ and hence for $|a| < 1$ the function f is an IMF. The function θ' does not have real zeros, but the real part of the complex zeros, $z = \frac{1}{2n}\left(\pi - \arccos\frac{1+a+a^2}{a(a+2)}\right)$, could be arbitrary close to 0 as $a \to 1$. Figure 1 shows the graph of f and its widely varying instantaneous frequency. The effect is due to aliasing, for more details see [2].

In the next example, the support of the spectrum satisfies the condition of Theorem 1, but the coefficients have larger magnitude.

Example 2 Magnitude of the coefficients
For $a = 0$, $n = 15$, $m = 28 < 2n - 1 = 29$, and $b = -\frac{1}{3}$, the function $f(t) = \cos 15t - \frac{1}{3}\cos(28t - \frac{\pi}{7})$ does not fit the form in (4) due to the shift $\frac{\pi}{7}$, so the Proposition 1 cannot be used. However, a direct computation shows that the function still has positive instantaneous frequency. Direct computations also show that on the interval $[2.92, 2.96]$ the function has a positive local minimum and hence it is not

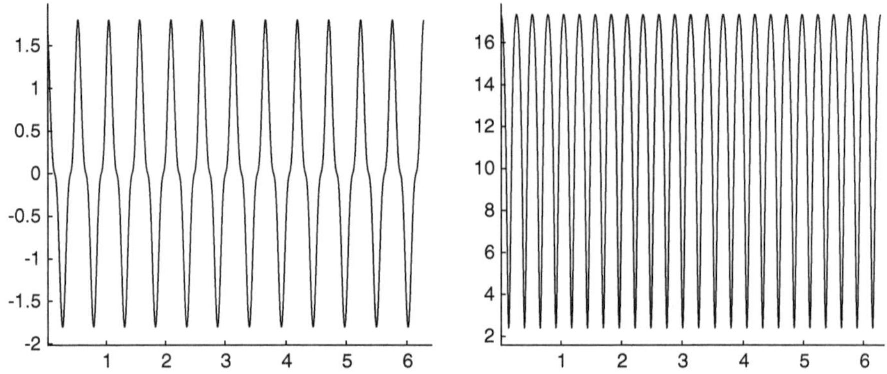

Fig. 1 $f(t) = (1 + 0.8\cos 24t)\cos 12t$ and its θ'

a weak IMF. The numerical results are illustrated in Fig. 2. Clearly, $A(15, 13) = \frac{1}{3}$ in that particular case while the estimates provided by Lemmas 1, 2 are 0.2930 and 0.2530. $A(15, 13)$ exceeds both upper bounds which agrees with the observation that f is not a weak IMF.

The final two examples demonstrate the effectiveness of the trigonometric envelopes replacing the usual cubic spline envelopes. The next example is an ITMP with symmetric spectrum in $[\frac{n}{2} + 1, \frac{3n}{2} - 1]$ to which the results from Theorem 1 can be applied.

Example 3 Controlled spectrum and coefficients
For $b = 0, a = 0.8, n = 12$, and $k = 4$, the function $f = (1 + 0.8\cos 4t)\cos 12t = 0.4\cos 8t + \cos 12t + 0.4\cos 16t$ has spectrum as in Theorem 1 and $A(12, 4) = 0.8$. In that case, the spline envelopes are very close to the trigonometric envelopes, see Fig. 3, and f is an IMF in the classical sense with $\theta' = 12$.

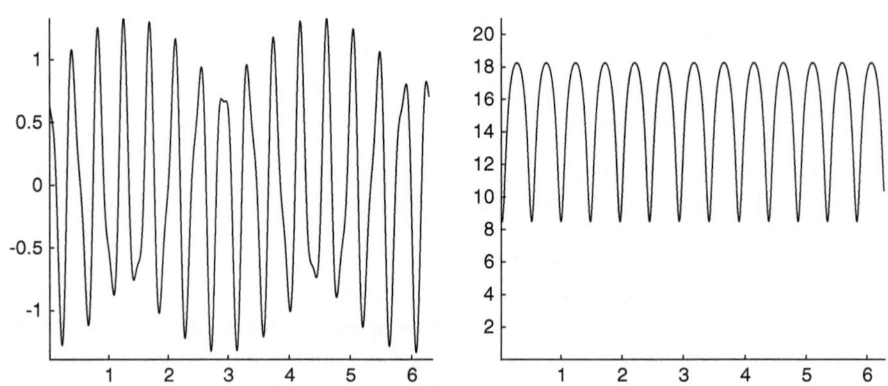

Fig. 2 $f(t) = \cos 15t - \frac{1}{3}\cos(28t - \frac{\pi}{7})$ and its θ'

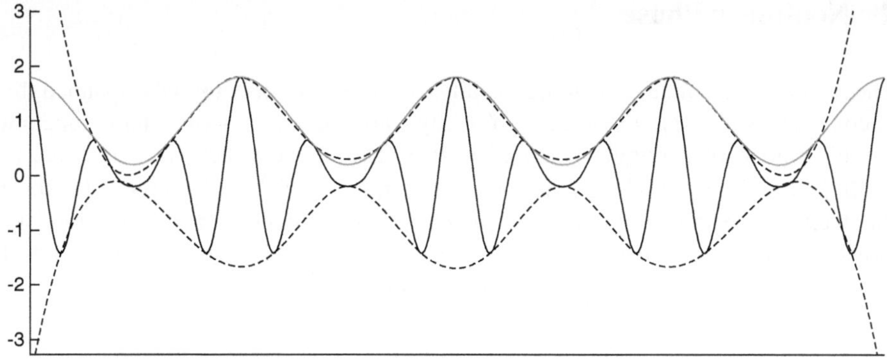

Fig. 3 $f = (1 + 0.8 \cos 4t) \cos 12t$ and its envelopes

In the next example, we apply the EMD with trigonometric envelopes to a signal of type (4).

Example 4 New frequency introduction and aliasing

For $b = -0.1, a = -0.2, n = 12, k = 9$, and $m = 3$, the polynomial is $f(t) = -0.1 \cos 3t + (1 - .2 \cos 9t) \cos 12t = \cos 12t + 0.1 \cos 21t$. From (5), we have that $\theta' = \frac{12.21+3.3 \cos 9t}{1.01+0.1 \cos 9t} > 0$. On the other hand, f can be considered as a superposition of two signals with frequencies 12 and 21. The EMD applied with cubic spline envelopes through the extrema produces mean very close to $-0.1 \cos 3t$. The envelopes are close to $U = 1 - 0.1 \cos 3t$, $L = -1 - 0.1 \cos 3t$ and the resulting IMF is $\psi = (1 - 0.2 \cos 9t) \cos 12t$ which has instantaneous frequency 12. The mean is an IMF with instantaneous frequency 3. Since $21 - 12 > 12 - 6$, we can conclude that the introduction of a component with instantaneous frequency 3 is due to aliasing (Fig. 4).

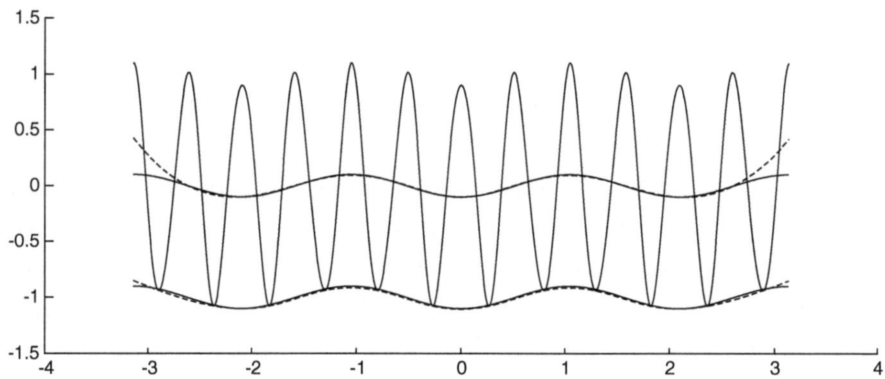

Fig. 4 $f = \cos 12t + 0.1 \cos 21t$ and its means

4 Nonlinear Phase

The result in Lemma 2 can be readily extended to polynomials spanned by $1, \cos j\theta(t), \sin j\theta(t), j = 1, 2, \ldots$ for any variable 2π-periodic phase function θ with a positive continuous first derivative. The above system is orthogonal in $L_2[0, 2\pi]$ with the weighted inner product $< f, g >= \int_0^{2\pi} \theta'(t) f(t) g(t)\, dt$. Since the weight is variable, Lemma 3 cannot be used directly. In this section, we consider orthonormal basis in the Hilbert space $L_2[0, 2\pi]$ with an inner product $< f, g >= \int_0^{2\pi} f(t) g(t)\, dt$, namely the functions

$$\sqrt{\frac{\theta'(t)}{\pi}}, \sqrt{\frac{\theta'(t)}{\pi}} \cos j\theta(t), \sqrt{\frac{\theta'(t)}{\pi}} \sin j\theta(t), \;\; j = 1, 2, \ldots$$

for a sufficiently smooth increasing function θ s.t. $\theta(0) = 0, \theta(2\pi) = 2\pi$. The corresponding Fourier series for a function f is

$$f(t) = \sqrt{\frac{\theta'(t)}{\pi}} \left(\frac{c_0}{2} + \sum_{j=1}^{\infty} (c_j \cos j\theta(t) + d_j \sin j\theta(t)) \right) \tag{6}$$

where

$$c_j = \int_0^{2\pi} \sqrt{\frac{\theta'(t)}{\pi}} f(t) \cos j\theta(t)\, dt, d_j = \int_0^{2\pi} \sqrt{\frac{\theta'(t)}{\pi}} f(t) \sin j\theta(t)\, dt.$$

In [12], we considered a method to construct phase functions of the type $\theta(t) = t + T_M(t)$ with $\min_{t\in[0,2\pi]} \theta'(t) = \delta > 0$, for a trigonometric polynomial T_M of degree M. Those functions can be further modified to satisfy particular conditions.

Proposition 2 *Let $k \geq 2$ be an integer and $p > 0$ then for any trigonometric polynomial T_M of degree M such that $\theta(t) = t + T_M(t)$ satisfies $\min_{t\in[0,2\pi]} \theta'(t) = \delta > 0$, the functions $\theta_\alpha(t) = \frac{1}{2\alpha} \int_{-\alpha}^{\alpha} \theta(t + x)\, dx$ are such that*

$$\lim_{\alpha\to\pi} \frac{\theta_\alpha^{(k)}(t)}{(\theta'_\alpha(t))^p} \to 0.$$

Proof Let $\theta(t) = t + \sum_{j=1}^{M} c_j \cos(jt + \beta_j)$, for some real β_j. Then for $k > 1$, we have that $\theta^{(k)}(t) = \sum_{j=1}^{M} \varepsilon_k j^k c_j \cos(jt + \tau_{j,k}), \varepsilon_k = \pm 1, \tau_{j,k}$ real. It follows that $\theta_\alpha^{(k)}(t) = \sum_{j=1}^{M} \frac{\sin(j\alpha)}{j\alpha} \varepsilon_k j^k c_j \cos(jt + \tau_{j,k})$. Hence, $\lim_{\alpha\to\pi} \theta_\alpha^{(k)}(t) = 0$. On the other hand, since $\theta' \geq \delta$, we have that $\theta'_\alpha(t) = \frac{1}{2\alpha} \int_{-\alpha}^{\alpha} \theta'(t + x)\, dx \geq \delta$ for any α. From continuity argument, we complete the proof. □

In [12], the proximity of the analytic signal, $H\cos n\theta(t)$, for $\cos n\theta(t)$, and the quadrature one, $\sin n\theta(t)$, was estimated in terms of the ratios as in the above proposition.

Next, we consider a sufficient condition for the basis functions to be weak IMF.

Lemma 4 *For a smooth increasing θ, $\theta(0)=0$, $\theta(2\pi)=2\pi$, with a 2π periodic derivative θ' and an integer n such that $\frac{\theta'''}{(\theta')^3}<2n^2$, the function $v(t)=\sqrt{\theta'(t)}\cos N\theta(t)$ is a weak IMF for any $N\geq n$.*

Proof It is enough to prove the statement only for $N=n$. In that case, the function v has $2n$ zeros. Its derivative is

$$v'(t)=\frac{\theta''(t)}{2\sqrt{\theta'(t)}}\cos n\theta(t)-n(\theta'(t))^{\frac{3}{2}}\sin n\theta(t)=R(t)\sin n\phi(t)$$

where $R(t)=-\sqrt{n^2(\theta'(t))^3+\frac{(\theta''(t))^2}{4\theta'(t)}}$, $\phi(t)=\theta(t)-\frac{\alpha(t)}{n}$, and $\tan\alpha(t)=\frac{\theta''(t)}{2n(\theta'(t))^2}$. The derivative $\phi'(t)=\theta'(t)\frac{4n^2(\theta'(t))^4+5(\theta''(t))^2-2\theta'''(t)\theta'(t)}{4n^2(\theta'(t))^4+(\theta''(t))^2}$ is positive if $\frac{\theta'''}{(\theta')^3}<2n^2$. Since α is an inverse tangent of a bounded function, we have that $-\frac{\pi}{2}<\alpha(t)<\frac{\pi}{2}$. It follows that $n(\phi(2\pi)-\phi(0))<(2n+1)\pi$ and hence v' has no more than $2n$ zeros. The proof of the lemma is complete. □

The statement of the lemma is trivial for $\theta(t)=t+at(t-2\pi)$, $|a|<\frac{1}{2\pi}$ or any concave instantaneous frequency θ', i.e., $\theta'''<0$. For any other function, it is asymptotically true as $n\to\infty$. In the case of splines or trigonometric polynomials from the Bernstein inequality for positive functions, for details see [5],

$$\theta'''\leq\|\theta'''\|_\infty\leq M\|\theta''\|_\infty\leq M\sqrt{Mr}\|\theta'\|_\infty$$

for some $r\leq M$. The last inequality is due to the fact that $\theta'>0$.

We conclude the section with a comment on the Fourier series approximation with the system considered above in the space $L_2[0,2\pi]$ with inner product $<f,g>=\int_0^{2\pi}f(t)g(t)\,dt$. For a fixed θ, the corresponding series expansion (6) has the same coefficients as the expansion of the function $g(x)=\frac{f(\theta^{-1}(x))}{\sqrt{\theta'(\theta^{-1}(x))}}$ where θ^{-1} is the inverse of θ, in the classical Fourier series, and hence, for a periodic function $f\in C^1$ we have that

$$\left\|f-\sqrt{\frac{\theta'}{\pi}}\sum_{j=0}^{n-1}(c_j\cos j\theta+d_j\sin j\theta)\right\|_\infty\leq\frac{C}{\sqrt{n}}\left\|\frac{1}{(\theta')^{\frac{3}{2}}}\left(f'-\frac{\theta''}{2\theta'}f\right)\right\|_2$$

where C is an absolute constant. This follows directly from the corresponding classical approximation estimate for absolutely converging Fourier series where the right-hand side involves the L_2 norm of the derivative of g.

5 Conclusion

In this paper, we considered two types of functions that exhibit properties expected from IMF. The class of ITMP, which is an extension from the simple harmonics, also illustrates the relation between the positive instantaneous frequency and the interlacing of zeros and extrema. The second class considered in Sect. 4 has the advantage both of a variable instantaneous phase and a convergent approximation method.

A further direction for investigations is to vary the phase function θ with the variation of n. This can be achieved by considering a greedy-type approximation, for details see [11], and matching pursuit-type algorithms, for details see [6, 7], with the dictionary $D = \left\{\sqrt{\frac{\theta'(t)}{\pi}} \cos n\theta(t) \middle| \theta' > 0\right\}$ and selecting a candidate weak IMF by maximizing

$$\left|\int_0^{2\pi} \sqrt{\theta'(t)} \cos n\theta(t) f(t)\, dt\right| =$$
$$\frac{1}{n}\left|\int_0^{2\pi} \frac{1}{\sqrt{\theta'(t)}}\left(f'(t) - \frac{\theta''(t)}{2\theta'(t)} f(t)\right) \sin n\theta(t)\, dt\right|.$$

The dictionary D for a particular choice of the smoothness on θ is related to the dictionary considered in [6, 7] while the optimization process is different and requires further study.

Acknowledgements The author is thankful for all the suggestions of the anonymous reviewers that greatly improved the exposition as well as for bringing to our attention the references [6, 7] about an effective matching pursuit process for EMD.

References

1. E. Bedrosian, A product theorem for Hilbert transform. Proc. IEEE **51**, 868–869 (1963)
2. J. Boyd, *Chebyshev and Fourier Spectral Methods*, 2nd edn. (Dover, New York, 2001)
3. Q. Chen, N. Huang, S. Riemenschneider, Y. Xu, A B-spline approach for empirical mode decomposition. Adv. Comput. Math. **24**, 171–195 (2006)
4. I. Daubechies, J. Lu, A. Wu, Synchrosqueezed wavelet transforms: an empirical mode decomposition-like tool. Appl. Comput. Harmon. Anal. **30**(2), 243–261 (2011)
5. T. Erdélyi, Markov-Bernstein Type Inequalities for Polynomials under Erdős-type Constraints, *Paul Erdős and His Mathematics I* (Springer, New York, 2002), pp. 219–239
6. T.Y. Hou, Z. Shi, P. Tavallali, Convergence of a data-driven timefrequency analysis method. Appl. Comput. Harmon. Anal. **37**(2), 235–270 (2014)
7. B. Huang, A. Kunoth, An optimization based empirical mode decomposition scheme. J. Comput. Appl. Math. **240**, 174–183 (2013)
8. N.E. Huang, Z. Shen, S.R. Long, M.C. Wu, H.H. Shih, Q. Zheng, N.C. Yen, C.C. Tung, H.H. Liu, The empirical mode decomposition and the Hilbert spectrum for nonlinear and nonstationary time series analysis. Proc. R. Soc. A **454**(1971), 903–995 (1998)

9. T. Qian, C. Qiuhui, L. Luoqing, Analytic unit quadrature signals with nonlinear phase. Physica D **203**, 80–87 (2005)
10. R.C. Sharpley, V. Vatchev, Analysis of the intrinsic mode functions. Constr. Approx. **24**(1), 17–47 (2006)
11. V.N. Temlyakov, *Greedy Approximation* (Cambridge University Press, Cambridge, 2011)
12. V. Vatchev, Analytic monotone pseudospectral interpolation. J. Fourier Anal. Appl. **21**, 715–733 (2015)
13. V. Vatchev, J. Del Castillo, Approximation of Fejér partial sums by interpolating functions. BIT Numer. Math. **53**(3), 779–790 (2013)

Kernel-Based Approximation Methods for Partial Differential Equations: Deterministic or Stochastic Problems?

Qi Ye

Abstract In this article, we present the kernel-based approximation methods to solve the partial differential equations using the Gaussian process regressions defined on the kernel-based probability spaces induced by the positive definite kernels. We focus on the kernel-based regression solutions of the multiple Poisson equations. Under the kernel-based probability measures, we show many properties of the kernel-based regression solutions including approximate formulas, convergence, acceptable errors, and optimal initialization. The numerical experiments show good results for the kernel-based regression solutions for the large-scale data.

Keywords Kernel-based approximation method · Gaussian process regression · Positive definite kernel · Kernel-based probability space · Hermite–Birkhoff interpolation · Multiple Poisson equation

1 Introduction

The kernel-based approximation methods form the basis of a recent fast growing research area with many applications in high-dimensional approximation, statistical learning, and engineering design. The kernel-based approximation methods cover the interdisciplinary fields of numerical analysis [2, 6, 21] and regression analysis [1, 19, 20]. The kernel-based approximation methods are also known by a variety of names in the monographs including radial basis functions, scattered data approximation, meshfree methods, and kriging.

In the paper [18] and the book [7], we already know that the kernel-based approximation methods have the mathematically equivalent formulas for the deterministic and stochastic interpolations. In this article, we present an idea to solve the deterministic partial differential equations numerically by the kernel-based approximation

Q. Ye (✉)
School of Mathematical Sciences, South China Normal University, Guangzhou, Guangdong 510631, China
e-mail: yeqi@m.scnu.edu.cn

G.E. Fasshauer and L.L. Schumaker (eds.), *Approximation Theory XV: San Antonio 2016*, Springer Proceedings in Mathematics & Statistics 201,
DOI 10.1007/978-3-319-59912-0_19

methods in the stochastic frameworks defined on the kernel-based probability spaces in [25]. Generally, we combine the knowledge and techniques of numerical analysis, regression analysis, and stochastic analysis to investigate the new theory of kernel-based approximation methods.

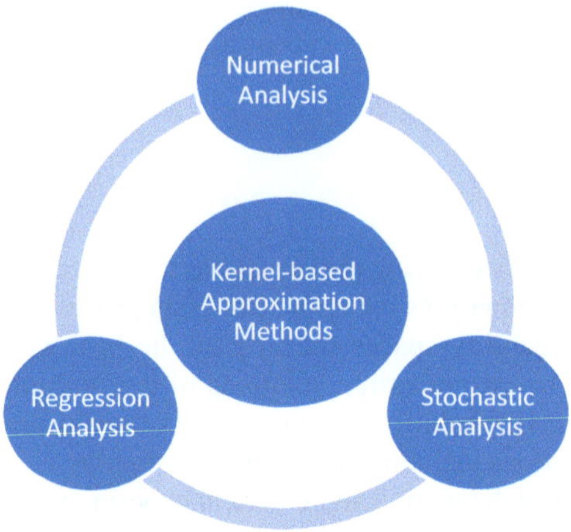

Now we review our previous related works. In the papers [4, 8, 9, 22–24], we firstly developed the kernel-based collocation algorithms to solve the stochastic partial differential equations extended by the survey in [14, 15, 18]. Next, by using previous results in [4], we introduced a new concept of kernel-based probability spaces in [25] to investigate the optimal Gaussian kernels for scattered data approximation. In [16], we compute the quasi-optimal meshfree numerical differentiation by the kernel-based probability measures. In [26], we generalize the simple kriging in the kernel-based probability spaces.

Moreover, we compare the differences of the original [4] and this article. The initial idea is inspired by [4]. Lemma 1 can be seen as a generalization of [4, Lemma 2.2]. Lemma 1 covers the general positive definite kernel while [4, Lemma 2.2] only works for the integral-type kernel. The concept of the kernel-based probability space has not been introduced in [4]. Lemma 1 guarantees that the kernel-based probability space is well defined in Definition 2. We will develop the approximate algorithms in the kernel-based probability space. Reference [4, Theorem 3.1] only shows that the normal random variables can be constructed by the special differential and boundary operators. In Theorem 1, we show that the multivariate normal random vector can be constructed by the vector operator composed of the general bounded linear functionals. So, we can compute the kernel-based estimators using the general Hermite–Birkhoff data. In [4], the kernel-based estimators are obtained by the maximum likelihood estimation. In this article, we compute the kernel-based estimators by the techniques of the conditional mean such as the Bayesian estimation. Since all random variables have the normal distributions, the formulas of the kernel-based estimators are consistent for both methods. The theory and algorithms discussed here can be viewed as the accomplished and strengthened forms of [4].

We show the main idea of the approximation in the kernel-based probability space by a simple example, that is, solving the multiple Poisson equation in Eq. (35) composed of the Laplace differential operator Δ and the gradient differential operator ∇ with the Dirichlet boundary condition. Given the scattered data points $\mathbf{x}_1, \ldots, \mathbf{x}_{n_1}$ in the domain $\mathscr{D}$ and $\mathbf{z}_1, \ldots, \mathbf{z}_{n_2}$ on the boundary $\partial\mathscr{D}$, we discretize the multiple Poisson equation by the related differential and boundary operators

$$\delta_{\mathbf{x}_1} \circ \Delta, \ldots, \delta_{\mathbf{x}_{n_1}} \circ \Delta,\ \delta_{\mathbf{x}_1} \circ \nabla, \ldots, \delta_{\mathbf{x}_{n_1}} \circ \nabla,\ \delta_{\mathbf{z}_1}, \ldots, \delta_{\mathbf{z}_{n_2}}. \tag{1}$$

According to Theorem 1, the kernel-based probability space gives a strong connection of the differential and boundary operators in Eq. (1) and the multivariate normal random variables

$$\delta_{\mathbf{x}_1} \circ \Delta S, \ldots, \delta_{\mathbf{x}_{n_1}} \circ \Delta S,\ \delta_{\mathbf{x}_1} \circ \nabla S, \ldots, \delta_{\mathbf{x}_{n_1}} \circ \nabla S,\ \delta_{\mathbf{z}_1} S, \ldots, \delta_{\mathbf{z}_{n_2}} S. \tag{2}$$

By the Gaussian process regression, we use the normal random variables in Eq. (2) to obtain the kernel-based regression solution of the multiple Poisson equation in Eq. (37), which is a linear combination of the kernel basis

$$\Delta_{\mathbf{y}} K(\mathbf{x}, \mathbf{x}_1), \ldots, \Delta_{\mathbf{y}} K(\mathbf{x}, \mathbf{x}_{n_1}), \nabla_{\mathbf{y}} K(\mathbf{x}, \mathbf{x}_1), \ldots, \nabla_{\mathbf{y}} K(\mathbf{x}, \mathbf{x}_{n_1}), K(\mathbf{x}, \mathbf{z}_1), \ldots, K(\mathbf{x}, \mathbf{z}_{n_2}), \tag{3}$$

where $\Delta_{\mathbf{y}}$ and $\nabla_{\mathbf{y}}$ represent the Laplace differentiation and the gradient differentiation with respect to the second argument of the kernel $K(\mathbf{x}, \mathbf{y})$, respectively. The linear system of the coefficients of the kernel-based regression solution is not required to have an exact solution in Eq. (38). For the classical collocation methods, we require that the collocation matrix is nonsingular to obtain the well-posed linear system by checking whether the linear functionals are linearly independent on the reproducing kernel Hilbert space of the positive definite kernel in Remark 1. Or, we remove the elements to obtain the solvable partial linear system. But, the important data may be missed at the removal process. Here, the coefficients of the kernel-based regression solution can be solved by the least squared methods. So, the kernel-based regression solution may not be a collocation solution for all Hermite–Birkhoff data. For the noncollocation solution, we can still verify the convergence of the kernel-based regression solutions by the kernel-based probability measures in Theorem 6. In particular, the kernel-based regression solution is consistent with the kernel-based collocation solution if the linear system of the coefficients of the kernel-based regression solution is well-posed. This shows that the kernel-based regression solution is the generalization of the kernel-based collocation solution.

In Sect. 2, we introduce the theorems of approximation for the general Hermite–Birkhoff data in the kernel-based probability spaces. Next, in Sect. 3, we discuss the kernel-based regression solutions of the multiple Poisson equation under the kernel-based probability measures. Finally, in Sect. 4, we show the consistences and differences of the kernel-based approximation methods for the partial differential equations in the deterministic and stochastic frameworks.

2 Approximation in Kernel-Based Probability Space

Now we present a new idea of approximation for observation data (general Hermite–Birkhoff data) using the Gaussian process regressions defined on a kernel-based probability space induced by a positive definite kernel.

Definition 1 A symmetric kernel $K : \mathscr{D} \times \mathscr{D} \to \mathbb{R}$ is called positive definite if, for all $n \in \mathbb{N}$, all sets of distinct points $X := \{\mathbf{x}_1, \ldots, \mathbf{x}_n\} \subseteq \mathscr{D}$, and all $\mathbf{c} := (c_1, \ldots, c_n)^T \in \mathbb{R}^N \setminus \{\mathbf{0}\}$, the quadratic form $\sum_{i=1}^n \sum_{j=1}^n c_i c_j K(\mathbf{x}_i, \mathbf{x}_j) > 0$.

In this article, we suppose that the kernel K is always symmetric. Clearly K is a positive definite kernel if and only if $\left(K(\mathbf{x}_i, \mathbf{x}_j)\right)_{i,j=1}^{n,n}$ is always a strictly positive definite matrix for all distinct data points $X \subseteq \mathscr{D}$.

For convenience, if it is not specially illustrated, then we suppose that the domain $\mathscr{D} \subseteq \mathbb{R}^d$ is always *regular* and *compact* in this article.

In this section, we construct a kernel-based probability space by a positive definite kernel using the theory and algorithm in [16, 24–26].

2.1 *Kernel-Based Probability Space*

Now we introduce the concept of kernel-based probability space. A Sobolev space is seen as a sample space of the probability space, and a kernel-based probability measure is defined on the Sobolev space. The normal random variables and the bounded (continuous) linear functionals are connected by the kernel-based probability space. Firstly we review the theorem in [25].

Lemma 1 ([25, Theorem 6.1]) *Let $\mathscr{H}^m(\mathscr{D})$ be an L_2-based Sobolev space of order $m > d/2$. Suppose that L is a bounded linear functional on $\mathscr{H}^m(\mathscr{D})$. If a function $\mu \in \mathscr{H}^m(\mathscr{D})$ and a positive definite kernel $K \in \mathrm{C}^{2m,1}(\mathscr{D} \times \mathscr{D})$, then there exists a probability measure $\mathbb{P}_K^\mu$ well defined on $\mathscr{H}^m(\mathscr{D})$ such that a random variable*

$$LS(\omega) := L\omega, \quad \text{for } \omega \in \mathscr{H}^m(\mathscr{D}), \tag{4}$$

has a normal distribution with mean $L\mu$ and variance $L_{\mathbf{x}}L_{\mathbf{y}}K(\mathbf{x}, \mathbf{y})$. Moreover, $\mathbb{P}_K^\mu$ is independent of L.

To clarify the notations, the above $L_{\mathbf{x}}$ and $L_{\mathbf{y}}$ denote an operator L with respect to the first and second arguments of the kernel $K(\mathbf{x}, \mathbf{y})$, respectively. Moreover, the space $\mathrm{C}^{2m,1}(\mathscr{D} \times \mathscr{D})$ consists of all functions which have the continuous derivatives up to order $2m$ and of which the $2m$th partial derivatives satisfy the Lipschitz condition. For example, the min kernel $K(x, y) := \min\{x, y\}$ belongs to $\mathrm{C}^{0,1}([0, 1] \times [0, 1])$. The functional $L : \mathscr{H}^m(\mathscr{D}) \to \mathbb{R}$ is always the real scalar in this article.

In Lemma 1, the Sobolev space $\mathscr{H}^m(\mathscr{D})$ is equipped with its Borel σ-algebra such that $\mathscr{H}^m(\mathscr{D})$ becomes a measurable space. So, we can view $\mathscr{H}^m(\mathscr{D})$ as a sample space, and $\omega \in \mathscr{H}^m(\mathscr{D})$ represents a sample path (trajectory). Lemma 1 assures the existence of the probability measure $\mathbb{P}_K^\mu$ called a kernel-based probability measure.

Definition 2 The Sobolev space $\mathscr{H}^m(\mathscr{D})$ endowed with the probability measure $\mathbb{P}_K^\mu$, which has the same properties as in Lemma 1, is called a kernel-based probability space, and $\mathbb{P}_K^\mu$ is called a kernel-based probability measure.

The kernel-based probability measure $\mathbb{P}_K^\mu$ can be seen as the generalization of the Wiener measure defined on the continuous function space which is used to construct the Brownian motion W_t by the coordinate mapping process $W_t(\omega) := \omega(t)$ in [11, Chap. 2].

When L is equal to a point evaluation function $\delta_{\mathbf{x}}$ at $\mathbf{x} \in \mathscr{D}$, then the collection of all $\delta_{\mathbf{x}} S$ is identical to the Gaussian process S with the mean μ and the covariance kernel K. Since the Sobolev space $\mathscr{H}^m(\mathscr{D})$ is a Hilbert space, the functional L given in Lemma 1 is isometrically equivalent to the unique element of $\mathscr{H}^m(\mathscr{D})$. Therefore, the collection of all LS is identical to the Gaussian process $L \mapsto LS$ indexed by the Hilbert space $\mathscr{H}^m(\mathscr{D})$ (see [10, Definition 1.19]). This shows that $\mathbb{P}_K^\mu$ can be seen as a Gaussian measure centered at μ.

Lemma 1 guarantees that the kernel-based probability measure $\mathbb{P}_K^\mu$ is independent of any functional L. So, we can use a vector operator to obtain a random vector defined on the kernel-based probability space. Now we give a generalization of Lemma 1 to a vector operator in Theorem 1.

The motivation of the theorems discussed here is that Theorem 1 can be seen as the extension of [4, Theorem 3.1] such that the kernel-based estimators are computed and coded easily in the complex domain by the Gaussian process regression. To be more precise, we will replace the integral-type kernel to the general positive definite kernel. Thus, the computational process of the integral can be avoided to accelerate the computational time. [4, Theorem 3.1] only covers the special differential and boundary operators. Theorem 1 shows that the general vector operators can be connected to the multivariate normal variables.

Theorem 1 *Let $\mathscr{H}^m(\mathscr{D})$ be an L_2-based Sobolev space of order $m > d/2$. Suppose that a vector operator $\boldsymbol{L} := (L_1, \ldots, L_n)^T$ is composed of finitely many bounded linear functionals $L_1, \ldots, L_n$ on $\mathscr{H}^m(\mathscr{D})$. If a function $\mu \in \mathscr{H}^m(\mathscr{D})$ and a positive definite kernel $K \in \mathrm{C}^{2m,1}(\mathscr{D} \times \mathscr{D})$, then there exists a probability measure $\mathbb{P}_K^\mu$ well defined on $\mathscr{H}^m(\mathscr{D})$ such that a random vector*

$$\boldsymbol{L}S(\omega) := \boldsymbol{L}\omega, \quad \textit{for } \omega \in \mathscr{H}^m(\mathscr{D}), \tag{5}$$

has a multivariate normal distribution with a vector mean $\boldsymbol{L}\mu$ and a covariance matrix $\boldsymbol{L}_{\mathbf{x}}\boldsymbol{L}_{\mathbf{y}}^T K(\mathbf{x}, \mathbf{y})$.

Proof Since $\boldsymbol{L}S$ is composed of $L_1 S, \ldots, L_n S$, Lemma 1 shows that $\boldsymbol{L}S$ is a multivariate normal vector defined by the kernel-based probability measure $\mathbb{P}_K^\mu$.

Since $L_j S$ has a mean $L_j \mu$ for any $j = 1, \dots, n$, the mean of $\boldsymbol{L}S$ is equal to $\boldsymbol{L}\mu$. Next, we will compute the covariance matrix of $\boldsymbol{L}S$. Let $L_j S^0 := L_j S - L_j \mu$ for any $j = 1, \dots, n$. Then $L_j S^0$ is a normal random variable with the mean 0 and the variance $L_{j,\mathbf{x}} L_{j,\mathbf{y}} K(\mathbf{x}, \mathbf{y})$ same as the normal random variable given in [25, Theorem 2.1]. Thus, [25, Corollary 2.7] guarantees that $\mathrm{E}\left(L_i S^0 L_j S^0\right) = L_{i,\mathbf{x}} L_{j,\mathbf{y}} K(\mathbf{x}, \mathbf{y})$ for any $i, j = 1, \dots, n$. Therefore, $\mathrm{Cov}(L_i S, L_j S) = \mathrm{E}\left(L_i S^0 L_j S^0\right) = L_{i,\mathbf{x}} L_{j,\mathbf{y}} K(\mathbf{x}, \mathbf{y})$. This shows that the covariance matrix $\left(\mathrm{Cov}(L_i S, L_j S)\right)_{i,j=1}^{n,n}$ is equal to $(L_{i,\mathbf{x}} L_{j,\mathbf{y}} K(\mathbf{x}, \mathbf{y}))_{i,j=1}^{n,n}$.

In Theorem 1, the matrix operator $\boldsymbol{L}_{\mathbf{x}} \boldsymbol{L}_{\mathbf{y}}^T$ is composed of the multiple functional $L_{i,\mathbf{x}} L_{j,\mathbf{y}}$ for $i, j = 1, \dots, n$, that is, $\boldsymbol{L}_{\mathbf{x}} \boldsymbol{L}_{\mathbf{y}}^T := \left(L_{i,\mathbf{x}} L_{j,\mathbf{y}}\right)_{i,j=1}^{n,n}$. Obviously, the covariance matrix $\boldsymbol{L}_{\mathbf{x}} \boldsymbol{L}_{\mathbf{y}}^T K(\mathbf{x}, \mathbf{y})$ can be written as

$$
\mathsf{A}_{\boldsymbol{L}} := \begin{pmatrix} L_{1,\mathbf{x}} L_{1,\mathbf{y}} K(\mathbf{x}, \mathbf{y}) & \cdots & L_{1,\mathbf{x}} L_{n,\mathbf{y}} K(\mathbf{x}, \mathbf{y}) \\ & \ddots & \\ L_{n,\mathbf{x}} L_{1,\mathbf{y}} K(\mathbf{x}, \mathbf{y}) & \cdots & L_{n,\mathbf{x}} L_{n,\mathbf{y}} K(\mathbf{x}, \mathbf{y}) \end{pmatrix}.
$$

For convenience, we will replace $\boldsymbol{L}_{\mathbf{x}} \boldsymbol{L}_{\mathbf{y}}^T K(\mathbf{x}, \mathbf{y})$ by $\mathsf{A}_{\boldsymbol{L}}$. Since the kernel K is always fixed in the computational process, the kernel K does not appear at the index of the matrix. The symmetry of K guarantees that $\mathsf{A}_{\boldsymbol{L}}$ is a symmetric matrix. For any choice of $\boldsymbol{L}$, the covariance matrix $\mathsf{A}_{\boldsymbol{L}}$ is always a symmetric positive definite matrix. Thus, the (Moore–Penrose) *pseudo-inverse* $\mathsf{A}_{\boldsymbol{L}}^{\dagger}$ of $\mathsf{A}_{\boldsymbol{L}}$ is well defined by the eigen decomposition of $\mathsf{A}_{\boldsymbol{L}}$ even if $\mathsf{A}_{\boldsymbol{L}}$ has zero eigenvalue (see [13, Sect. 5.4]). In particular, we have that $\mathsf{A}_{\boldsymbol{L}}^{\dagger} = \mathsf{A}_{\boldsymbol{L}}^{-1}$ if $\mathsf{A}_{\boldsymbol{L}}$ is nonsingular. To be more precise, the covariance matrix $\mathsf{A}_{\boldsymbol{L}}$ always has the eigen decomposition $\mathsf{A}_{\boldsymbol{L}} = \mathsf{V}_{\boldsymbol{L}} \mathsf{D}_{\boldsymbol{L}} \mathsf{V}_{\boldsymbol{L}}^T$, where $\mathsf{D}_{\boldsymbol{L}}$ and $\mathsf{V}_{\boldsymbol{L}}$ are composed of the nonnegative eigenvalues and orthonormal eigenvectors of $\mathsf{A}_{\boldsymbol{L}}$, respectively. Thus, the pseudo-inverse $\mathsf{A}_{\boldsymbol{L}}^{\dagger}$ has the form $\mathsf{A}_{\boldsymbol{L}}^{\dagger} = \mathsf{V}_{\boldsymbol{L}} \mathsf{D}_{\boldsymbol{L}}^{\dagger} \mathsf{V}_{\boldsymbol{L}}^T$, where $\mathsf{D}_{\boldsymbol{L}}^{\dagger}$ is taken by the reciprocal of each nonzero element on the diagonal of $\mathsf{D}_{\boldsymbol{L}}$.

Remark 1 The bounded linear functionals $\boldsymbol{L}$ discussed here are defined on the Sobolev space $\mathscr{H}^m(\mathscr{D})$. For the classical kernel-based approximation methods in [6, 7, 21], we always assume that the bounded linear functionals $\boldsymbol{L}$ are linearly independent on the reproducing kernel Hilbert space (native space) $\mathscr{H}_K(\mathscr{D})$ such that $\mathsf{A}_{\boldsymbol{L}}$ is a strictly positive definite matrix by the reproducing properties. If $\mathsf{A}_{\boldsymbol{L}}$ is a strictly positive definite matrix, then $\mathsf{A}_{\boldsymbol{L}}$ has the inverse such that the collocation solution is constructed well by any general Hermite–Birkhoff data. But, the covariance matrix $\mathsf{A}_{\boldsymbol{L}}$ in Theorem 1 may be singular. Next, we will show that the singular matrix $\mathsf{A}_{\boldsymbol{L}}$ can be also used to obtain the good approximate solution in the kernel-based probability space.

As follows we will construct the kernel-based estimator by the multivariate normal vector $\boldsymbol{L}S$.

2.2 Kernel-Based Approximation by Gaussian Process Regressions

Now we show the construction of kernel-based estimator and kernel-based regression function in the kernel-based probability space including approximate formulas, convergence, acceptable errors, and optimal initialization. Different from [4], we will compute the kernel-based estimators by the conditional mean.

Let $\mathscr{H}^m(\mathscr{D})$ be an L_2-based Sobolev space of order $m > d/2$. Suppose that $P : \mathscr{H}^m(\mathscr{D}) \to \mathbb{R}$ is a bounded linear functional on $\mathscr{H}^m(\mathscr{D})$ and $\boldsymbol{L} : \mathscr{H}^m(\mathscr{D}) \to \mathbb{R}^n$ is a vector operator composed of bounded linear functionals on $\mathscr{H}^m(\mathscr{D})$.

Given observation $\boldsymbol{f} \in \mathbb{R}^n$ corresponding to $\boldsymbol{L}$, we will estimate an unknown value $v \in \mathbb{R}$ with respect to P. Furthermore, we suppose that the values v and $\boldsymbol{f}$ are observed by some function $u \in \mathscr{H}^m(\mathscr{D})$ such as $v = Pu$ and $\boldsymbol{f} = \boldsymbol{L}u$. This shows that we can catch the exact value v if we can choose the collocation solution u correctly. The observation data $(\boldsymbol{L}, \boldsymbol{f})$ can be viewed as the general Hermite–Birkhoff data, and the kernel-based approximation discussed here can be roughly thought of as generalized Hermite–Birkhoff interpolation.

Remark 2 The classical kernel-based approximation methods usually assume that the observed data $\boldsymbol{f}$ are evaluated by some function u in $\mathscr{H}_K(\mathscr{D})$ such that the error analysis can be developed well by the reproducing properties. By [21, Theorem 10.45], we can only verify that $\mathscr{H}_K(\mathscr{D}) \subseteq C^m(\mathscr{D})$ if $K \in C^{2m}(\mathscr{D} \times \mathscr{D})$. But, we weaken the condition to $u \in \mathscr{H}^m(\mathscr{D})$ here. Thus, the new kernel-based approach is more feasible to solve the partial differential equations.

Kernel-based Approximation. According to Theorem 1, given the function $\mu \in \mathscr{H}^m(\mathscr{D})$ and the positive definite kernel $K \in C^{2m,1}(\mathscr{D} \times \mathscr{D})$, the normal random variable PS and the multivariate normal random vector $\boldsymbol{L}S$ are well defined on the Sobolev space $\mathscr{H}^m(\mathscr{D})$ endowed with the kernel-based probability measure $\mathbb{P}_K^\mu$ such as Eqs. (4) and (5) hold. Therefore, PS and $\boldsymbol{L}S$ have the joint normal distribution with the vector mean

$$\begin{pmatrix} P\mu \\ \boldsymbol{L}\mu \end{pmatrix}, \tag{6}$$

and the covariance matrix

$$\begin{pmatrix} P_{\mathbf{x}}P_{\mathbf{y}}K(\mathbf{x},\mathbf{y}) & P_{\mathbf{x}}\boldsymbol{k}_L(\mathbf{x})^T \\ P_{\mathbf{x}}\boldsymbol{k}_L(\mathbf{x}) & \mathsf{A}_L \end{pmatrix}, \tag{7}$$

where the vector function $\boldsymbol{k}_L$ is composed of the kernel basis $L_{1,\mathbf{y}}K(\cdot,\mathbf{y}), \ldots, L_{n,\mathbf{y}}K(\cdot,\mathbf{y})$, that is, $\boldsymbol{k}_L := \boldsymbol{L}_{\mathbf{y}}K(\cdot,\mathbf{y})$.

Let a collocation set $\mathscr{I}_L(\boldsymbol{f}) := \{\omega \in \mathscr{H}^m(\mathscr{D}) : \boldsymbol{L}\omega = \boldsymbol{f}\}$. Then $u \in \mathscr{I}_L(\boldsymbol{f})$ and $\mathscr{I}_L(\boldsymbol{f})$ covers all collocation solutions in $\mathscr{H}^m(\mathscr{D})$. By the construction of $\boldsymbol{L}S$, we have the connection of $\boldsymbol{L}\omega = \boldsymbol{f}$ and $\boldsymbol{L}S = \boldsymbol{f}$. This shows that the collocation condition can be transferred into the given event of conditional expectation such as the Gaussian process regression. Moreover, let a set $\mathscr{I}_P(v) :=$

$\{\omega \in \mathscr{H}^m(\mathscr{D}) : P\omega = v\}$. If $Pu = v$, then $u \in \mathscr{I}_P(v)$. The construction of PS indicates that v could be thought as the observed value of PS. Therefore, we can use the Bayesian methods in [3] to compute the optimal estimator as the conditional mean of PS given $\boldsymbol{L}S = \boldsymbol{f}$, that is,

$$\hat{s}_{P|\boldsymbol{L},\boldsymbol{f}} := \mathrm{E}\left(PS|\boldsymbol{L}S = \boldsymbol{f}\right). \tag{8}$$

Usually, we will compare the importance of estimators to choose the "best" estimator. The kernel-based probability measure $\mathbb{P}_K^\mu$ gives a tool to compute the weight of different value v. Since P and $\boldsymbol{L}$ are bounded, the sets $\mathscr{I}_P(v)$ and $\mathscr{I}_{\boldsymbol{L}}(\boldsymbol{f})$ are closed. This assures that $\mathscr{I}_P(v)$ and $\mathscr{I}_{\boldsymbol{L}}(\boldsymbol{f})$ belong to the Borel σ-algebra of $\mathscr{H}^m(\mathscr{D})$. So, the events $\mathscr{I}_P(v)$ and $\mathscr{I}_{\boldsymbol{L}}(\boldsymbol{f})$ can be measured by $\mathbb{P}_K^\mu$. Since the intersection $\mathscr{I}_P(v) \cap \mathscr{I}_{\boldsymbol{L}}(\boldsymbol{f})$ covers all collocation solutions whose values are equal to v with respect to P, the importance of v can be weighted by the conditional probability

$$\mathbb{P}_K^\mu\left(\mathscr{I}_P(v)\middle|\mathscr{I}_{\boldsymbol{L}}(\boldsymbol{f})\right) = \frac{\mathbb{P}_K^\mu\left(\mathscr{I}_P(v) \cap \mathscr{I}_{\boldsymbol{L}}(\boldsymbol{f})\right)}{\mathbb{P}_K^\mu\left(\mathscr{I}_{\boldsymbol{L}}(\boldsymbol{f})\right)}.$$

The conditional probability $\mathbb{P}_K^\mu\left(\mathscr{I}_P(v)\middle|\mathscr{I}_{\boldsymbol{L}}(\boldsymbol{f})\right)$ measures the possibility of v observed by $\boldsymbol{f}$. The estimator $\hat{s}_{P|\boldsymbol{L},\boldsymbol{f}}$ can be thought as the average of all v measured by $\mathbb{P}_K^\mu$, that is,

$$\hat{s}_{P|\boldsymbol{L},\boldsymbol{f}} = \int_{\mathbb{R}} v p_K^\mu(v|\boldsymbol{f})\mathrm{d}v = \sum_{v\in\mathbb{R}} v\mathbb{P}_K^\mu\left(\mathscr{I}_P(v)\middle|\mathscr{I}_{\boldsymbol{L}}(\boldsymbol{f})\right),$$

where $p_K^\mu(\cdot|\cdot)$ is the conditional probability density function of PS given $\boldsymbol{L}S$ under $\mathbb{P}_K^\mu$. This shows that $\hat{s}_{P|\boldsymbol{L},\boldsymbol{f}}$ is the average of all collocation solutions with respect to P weighted by the kernel-based probabilities.

Definition 3 The conditional mean $\hat{s}_{P|\boldsymbol{L},\boldsymbol{f}}$ in Eq. (8) is called a kernel-based estimator with respect to P for the data $(\boldsymbol{L}, \boldsymbol{f})$ under the kernel-based probability measure $\mathbb{P}_K^\mu$.

Moreover, we look at the properties of $\hat{s}_{P|\boldsymbol{L},\boldsymbol{f}}$.

Theorem 2 *The kernel-based estimator $\hat{s}_{P|\boldsymbol{L},\boldsymbol{f}}$ is evaluated by P acting on a function $\hat{s}_{\boldsymbol{L},\boldsymbol{f}}$, that is,*

$$\hat{s}_{P|\boldsymbol{L},\boldsymbol{f}} = P\hat{s}_{\boldsymbol{L},\boldsymbol{f}}. \tag{9}$$

Moreover $\hat{s}_{\boldsymbol{L},\boldsymbol{f}}$ has the form

$$\hat{s}_{\boldsymbol{L},\boldsymbol{f}}(\mathbf{x}) := \mu(\mathbf{x}) + \boldsymbol{k}_{\boldsymbol{L}}(\mathbf{x})^T\mathbf{c}, \tag{10}$$

where $\mathbf{c}$ is a least-squared solution of the linear system

$$\mathsf{A}_{\boldsymbol{L}}\mathbf{c} = \boldsymbol{f} - \boldsymbol{L}\mu. \tag{11}$$

Proof According to the normal distributions of PS and $\boldsymbol{L}S$ in Eqs. (6) and (7), we have that

$$\mathrm{E}\left(PS|\boldsymbol{L}S=\boldsymbol{f}\right)=P_{\mathbf{x}}\mu(\mathbf{x})+P_{\mathbf{x}}\boldsymbol{k}_{\boldsymbol{L}}(\mathbf{x})^{T}\mathsf{A}_{\boldsymbol{L}}^{\dagger}\left(\boldsymbol{f}-\boldsymbol{L}\mu\right), \tag{12}$$

by the approach of multivariate statistics in [5, Proposition 3.13]. Let a function

$$\eta_{\boldsymbol{L},\boldsymbol{f}}(\mathbf{x}):=\mu(\mathbf{x})+\boldsymbol{k}_{\boldsymbol{L}}(\mathbf{x})^{T}\mathsf{A}_{\boldsymbol{L}}^{\dagger}\left(\boldsymbol{f}-\boldsymbol{L}\mu\right). \tag{13}$$

Since the least-squared solution $\mathbf{c}$ of Eq. (11) has the form $\mathbf{c}=\mathsf{A}_{\boldsymbol{L}}^{\dagger}\left(\boldsymbol{f}-\boldsymbol{L}\mu\right)$ in [13, Theorem 2 in Sect. 5.4], we have that

$$\eta_{\boldsymbol{L},\boldsymbol{f}}(\mathbf{x})=\hat{s}_{\boldsymbol{L},\boldsymbol{f}}(\mathbf{x}). \tag{14}$$

The Eqs. (13) and (14) assure that

$$P_{\mathbf{x}}\mu(\mathbf{x})+P_{\mathbf{x}}\boldsymbol{k}_{\boldsymbol{L}}(\mathbf{x})^{T}\mathsf{A}_{\boldsymbol{L}}^{\dagger}\left(\boldsymbol{f}-\boldsymbol{L}\mu\right)=P_{\mathbf{x}}\hat{s}_{\boldsymbol{L},\boldsymbol{f}}(\mathbf{x}). \tag{15}$$

Combining Eqs. (8), (12), and (15), we verify equality (9).

It is clear that $\hat{s}_{\boldsymbol{L},\boldsymbol{f}}$ is independent of P. In particular, if $\mu=0$, then the coefficients $\mathbf{c}$ of $\hat{s}_{\boldsymbol{L},\boldsymbol{f}}$ are the least-squared solution of $\mathsf{A}_{\boldsymbol{L}}\mathbf{c}=\boldsymbol{f}$. In this article, the kernel-based regression function is given by the fixed μ and K if it is not specified. To simplify the notations, the μ and K are not indexed at the kernel-based regression function.

Definition 4 The approximate function $\hat{s}_{\boldsymbol{L},\boldsymbol{f}}$ in Eqs. (10) and (11) is called a kernel-based regression solution for the data $(\boldsymbol{L},\boldsymbol{f})$ under the kernel-based probability measure $\mathbb{P}_{K}^{\mu}$.

For convenience, we rewrite the kernel-based estimator $\hat{s}_{P|\boldsymbol{L},\boldsymbol{f}}$ as $P\hat{s}_{\boldsymbol{L},\boldsymbol{f}}$.

Remark 3 By the classical kernel-based approximation methods, if $\boldsymbol{L}$ are linearly independent on $\mathscr{H}_K(\mathscr{D})$, then $\mathsf{A}_{\boldsymbol{L}}$ is nonsingular such that the *kernel-based collocation function* $s_{\boldsymbol{L},\boldsymbol{f}}$ can be written as

$$s_{\boldsymbol{L},\boldsymbol{f}}(\mathbf{x})=\boldsymbol{k}_{\boldsymbol{L}}(\mathbf{x})^{T}\mathbf{c}, \tag{16}$$

where the coefficients $\mathbf{c}$ are solved by a linear system

$$\mathsf{A}_{\boldsymbol{L}}\mathbf{c}=\boldsymbol{f}. \tag{17}$$

It is clear that $\boldsymbol{L}s_{\boldsymbol{L},\boldsymbol{f}}=\boldsymbol{f}$.

Different from $s_{\boldsymbol{L},\boldsymbol{f}}$, the kernel-based regression solution $\hat{s}_{\boldsymbol{L},\boldsymbol{f}}$ may not be a collocation solution. However, the construction of $\hat{s}_{\boldsymbol{L},\boldsymbol{f}}$ is dependent of a collection of collocation solutions. So, we only call $\hat{s}_{\boldsymbol{L},\boldsymbol{f}}$ a regression solution. This indicates that a noncollocation solution could be also a good approximation of the general Hermite–Birkhoff data. Moreover, comparing with $s_{\boldsymbol{L},\boldsymbol{f}}$, if $\mu=0$ and $\mathsf{A}_{\boldsymbol{L}}$ is nonsingular, then

$\hat{s}_{L,f} = s_{L,f}$. Thus, the additional term μ of $\hat{s}_{L,f}$ can be seen as the *initial conjecture* of the approximation.

The regression analysis of Gaussian processes also shows that

$$P_{\mathbf{x}}\mu(\mathbf{x}) + P_{\mathbf{x}}\boldsymbol{b}_L(\mathbf{x})^T(\boldsymbol{L}S - \boldsymbol{L}\mu) = \underset{U\in\text{span}\{\boldsymbol{L}S\}}{\text{argmin}}\ \mathrm{E}\,|U - PS|^2,$$

where $\boldsymbol{b}_L(\mathbf{x}) := \mathsf{A}_L^{\dagger}\boldsymbol{k}_L(\mathbf{x})$ is a generalized Lagrange basis. It is clear that $\hat{s}_{L,f} = \mu + \boldsymbol{b}_L^T(\boldsymbol{f} - \boldsymbol{L}\mu)$. This shows that the kernel-based estimator $P\hat{s}_{L,f}$ can be viewed as the observation of the optimal estimator of PS over the linear span of $\boldsymbol{L}S$.

Remark 4 The kernel-based collocation solution $s_{L,f}$ in Eqs. (16) and (17) can be rewritten by the new basis $\boldsymbol{b}_L$ such as $s_{L,f} = \boldsymbol{b}_L^T\boldsymbol{f}$ if A_L is nonsingular. The Riesz' representation [21, Theorem 16.1] guarantees the optimal recovery

$$s_{L,f} = \underset{u\in\mathcal{H}_K(\mathcal{D})}{\text{argmin}}\ \|u\|_{\mathcal{H}_K(\mathcal{D})} \text{ subjected to } \boldsymbol{L}u = \boldsymbol{f}.$$

This shows that the kernel-based approximation has the mathematically equivalent formulas by the deterministic and stochastic optimizations.

Convergence. Next, we look at the convergence of the kernel-based estimator by the stochastic approach. Assume that the infinite countable observations $f_1, \ldots, f_n, \ldots$ are evaluated by some $u \in \mathcal{H}^m(\mathcal{D})$ with respect to the infinite countable bounded linear functionals $L_1, \ldots, L_n, \ldots$. This shows that $L_1u = f_1, \ldots, L_nu = f_n, \ldots$ Let $\boldsymbol{f}_n$ and $\boldsymbol{L}_n$ be composed of finitely many $f_1, \ldots, f_n$ and $L_1, \ldots, L_n$, respectively, that is, $\boldsymbol{f}_n := (f_1, \ldots, f_n)^T$ and $\boldsymbol{L}_n := (L_1, \ldots, L_n)^T$. Let $\boldsymbol{f}_\infty$ and $\boldsymbol{L}_\infty$ be composed of infinitely many $f_1, \ldots, f_n, \ldots$ and $L_1, \ldots, L_n, \ldots$, respectively, $\boldsymbol{f}_\infty := (f_1, \ldots, f_n, \ldots)^T$ and $\boldsymbol{L}_\infty := (L_1, \ldots, L_n, \ldots)^T$. By the same techniques of [26, Proposition 1], we can obtain the convergence of the sequence of kernel-based estimators $P\hat{s}_{\boldsymbol{L}_n,\boldsymbol{f}_n}$ for $n \in \mathbb{N}$.

Theorem 3 *If there exists a unique solution u in $\mathcal{H}^m(\mathcal{D})$ such that $\boldsymbol{L}_\infty u = \boldsymbol{f}_\infty$, then the kernel-based estimator $P\hat{s}_{\boldsymbol{L}_n,\boldsymbol{f}_n}$ converges to the value Pu and the kernel-based regression solution $\hat{s}_{\boldsymbol{L}_n,\boldsymbol{f}_n}$ converges to the exact solution u uniformly.*

Proof The uniqueness of u shows that $\mathscr{I}_{\boldsymbol{L}_\infty}(\boldsymbol{f}_\infty) = \{u\}$. Thus,

$$\mathrm{E}\left(PS \middle| \mathscr{I}_{\boldsymbol{L}_\infty}(\boldsymbol{f}_\infty)\right) = Lu. \tag{18}$$

Moreover, since $\mathscr{I}_{\boldsymbol{L}_n}(\boldsymbol{f}_n) \supseteq \mathscr{I}_{\boldsymbol{L}_{n+1}}(\boldsymbol{f}_{n+1})$ for all $n \in \mathbb{N}$ and $\bigcap_{n\in\mathbb{N}}\mathscr{I}_{\boldsymbol{L}_n}(\boldsymbol{f}_n) = \mathscr{I}_{\boldsymbol{L}_\infty}(\boldsymbol{f}_\infty)$, we have the convergence

$$\lim_{n\to\infty} \mathrm{E}\left(PS \middle| \mathscr{I}_{\boldsymbol{L}_n}(\boldsymbol{f}_n)\right) = \mathrm{E}\left(PS \middle| \mathscr{I}_{\boldsymbol{L}_\infty}(\boldsymbol{f}_\infty)\right). \tag{19}$$

The construction of $\boldsymbol{L}_nS$ guarantees that

$$\mathrm{E}\left(PS \middle| \mathscr{I}_{L_n}(f_n)\right) = \mathrm{E}\left(PS \middle| L_n S = f_n\right). \tag{20}$$

According to Eqs. (8) and (12), we have that

$$\mathrm{E}\left(PS \middle| L_n S = f_n\right) = P\hat{s}_{L_n, f_n}. \tag{21}$$

Combining Eqs. (18)–(21), we conclude that

$$\lim_{n\to\infty} P\hat{s}_{L_n, f_n} = Pu. \tag{22}$$

Take any $\mathbf{x} \in \mathscr{D}$. According to the Sobolev imbedding theorem, the point evaluation function $\delta_{\mathbf{x}}$ is a bounded linear functional on $\mathscr{H}^m(\mathscr{D})$. Replacing P to $\delta_{\mathbf{x}}$, Eq. (22) shows that

$$\lim_{n\to\infty} \hat{s}_{L_n, f_n}(\mathbf{x}) = u(\mathbf{x}).$$

Therefore, the compactness of $\mathscr{D}$ guarantees that the convergence of $\hat{s}_{L_n, f_n}$ is uniform.

Remark 5 In Theorem 3, the uniqueness is an important condition for the convergence. If we have two distinct solutions u_1 and u_2 such that $L_\infty u_1 = L_\infty u_2 = f_\infty$, then the kernel-based regression function $\hat{s}_{L_n, f_n}$ may not converge to any solution u_1, u_2. For example, if $L_\infty := \delta_{X_\infty}$ for a nondense set X_∞ composed of infinitely countable data points in $\mathscr{D}$, then the equation $L_\infty u = f_\infty$ has infinitely many solutions such that $\hat{s}_{L_n, f_n}$ may not converge to any solution u at the nondense location in $\mathscr{D}$ for X_∞. This shows that the convergence of $\hat{s}_{L_n, f_n}$ cannot be guaranteed without the uniqueness condition.

In the classical kernel-based approximation methods, the convergence of the kernel-based collocation solution of the partial differential equations is dependent of the fill distances of the domain and the boundary in [21, Theorem 16.15], for example, the fill distance $h_{X,\mathscr{D}} := \sup_{\mathbf{x}\in\mathscr{D}} \min_{\mathbf{x}_k \in X} \|\mathbf{x} - \mathbf{x}_k\|_2$ for the data $X \subseteq \mathscr{D}$. In particular, if there exists a sequence of the set X_n of the data points in $\mathscr{D}$ such that $h_{X_n,\mathscr{D}} \to 0$ when $n \to \infty$, then the collection $X_\infty := \cup_{n=1}^\infty X_n$ is dense in $\mathscr{D}$; hence the solution of $L_\infty u = f_\infty$ for $L_\infty := \delta_{X_\infty}$ is unique in $\mathrm{C}(\mathscr{D})$. This shows that Theorem 3 covers the convergence of the kernel-based approximation by the fill distance.

Acceptable Error. In statistics, we can check the errors by the probability measures. Now we will use the same idea to measure the errors of the kernel-based estimator $P\hat{s}_{L,f}$ by conditional mean squared error

$$\mathrm{E}\left(\left|PS - P\hat{s}_{L,f}\right|^2 \middle| LS = f\right) = \sigma^2_{P|L}, \tag{23}$$

where $\sigma^2_{P|L}$ is the conditional variance of PS given $LS = f$, that is,

$$\sigma^2_{P|L} := P_{\mathbf{x}} P_{\mathbf{y}} K(\mathbf{x}, \mathbf{y}) - P_{\mathbf{x}} k_L(\mathbf{x})^T \mathsf{A}_L^\dagger P_{\mathbf{y}} k_L(\mathbf{y}).$$

The standard deviation $\sigma_{P|L}$ is the square root of the conditional variance $\sigma_{P|L}^2$.

Different from the deterministic error analysis, we will determine whether the kernel-based estimator matches the error ε at the confidence level using the techniques of the confidence interval in statistical hypothesis testing in [12].

Let a set $\mathscr{E}_P(\varepsilon) := \left\{\omega \in \mathscr{H}^m(\mathscr{D}) : \left|P\omega - P\hat{s}_{L,f}\right| \leq \varepsilon\right\}$ for any $\varepsilon > 0$. If the correct collocation solution u satisfies $\left|Pu - P\hat{s}_{L,f}\right| \leq \varepsilon$, then $u \in \mathscr{E}_P(\varepsilon)$. This shows that $\mathscr{E}_P(\varepsilon) \cap \mathscr{I}_L(\boldsymbol{f})$ covers all collocation solutions accepting the error ε. So, the error ε is rejected if $u \notin \mathscr{E}_P(\varepsilon)$. Moreover, since P is bounded, the set $\mathscr{E}_P(\varepsilon)$ is closed such that $\mathscr{E}_P(\varepsilon)$ belongs to the Borel σ-algebra of $\mathscr{H}^m(\mathscr{D})$. Moreover, we know that $u \in \mathscr{I}_L(\boldsymbol{f})$ surely. If $\mathscr{E}_P(\varepsilon) \cap \mathscr{I}_L(\boldsymbol{f})$ is measured as a large part of $\mathscr{I}_L(\boldsymbol{f})$, then we believe that $P\hat{s}_{L,f}$ would be a good approximation accepting the error ε in high probability. By the kernel-based probability measure of $\mathscr{E}_P(\varepsilon)$ conditioned on $\mathscr{I}_L(\boldsymbol{f})$, we can determine whether $P\hat{s}_{L,f}$ accepts or rejects the error ε. In statistical hypothesis testing, an event expresses small probability in happening if the confidence is more than 99%. So, we define the acceptable errors under the kernel-based probability measures as follows.

Definition 5 We call the kernel-based estimator $P\hat{s}_{L,f}$ accepting the error $\varepsilon > 0$ under the kernel-based probability measure $\mathbb{P}_K^{\mu}$ if the conditional probability $\mathbb{P}_K^{\mu}\left(\mathscr{E}_P(\varepsilon)\middle|\mathscr{I}_L(\boldsymbol{f})\right) \geq 99\%$.

If $P\hat{s}_{L,f}$ accepts the error ε under $\mathbb{P}_K^{\mu}$, then we are 99% confident that the difference of Pu and $P\hat{s}_{L,f}$ is at most ε under $\mathbb{P}_K^{\mu}$.

Theorem 4 *If $\varepsilon \geq 10\sigma_{P|L}$, then the kernel-based estimator $P\hat{s}_{L,f}$ accepts the error ε under the kernel-based probability measure $\mathbb{P}_K^{\mu}$.*

Proof The construction of PS and $\boldsymbol{L}S$ shows that

$$\mathbb{P}_K^{\mu}\left(\mathscr{E}_P(\varepsilon)\middle|\mathscr{I}_L(\boldsymbol{f})\right) = \mathbb{P}_K^{\mu}\left(\left|PS - P\hat{s}_{L,f}\right| \leq \varepsilon \middle| \boldsymbol{L}S = \boldsymbol{f}\right). \tag{24}$$

Moreover, by the conditional mean squared error in Eq. (23), the Chebyshev's inequality guarantees that

$$\mathbb{P}_K^{\mu}\left(\left|PS - P\hat{s}_{L,f}\right| \leq \varepsilon \middle| \boldsymbol{L}S = \boldsymbol{f}\right) \leq 1 - \frac{\sigma_{P|L}^2}{\varepsilon^2}. \tag{25}$$

Combining Eq. (24) and inequality (25), we have that

$$\mathbb{P}_K^{\mu}\left(\mathscr{E}_P(\varepsilon)\middle|\mathscr{I}_L(\boldsymbol{f})\right) \geq 1 - \frac{\sigma_{P|L}^2}{\varepsilon^2}. \tag{26}$$

Since $\varepsilon^2 \geq 100\sigma_{P|L}^2$, inequality (26) assures that $\mathbb{P}_K^{\mu}\left(\mathscr{E}_P(\varepsilon)\middle|\mathscr{I}_L(\boldsymbol{f})\right) \geq 99\%$. Therefore, the proof is completed.

Remark 6 If A_L is nonsingular, then the standard deviation $\sigma_{P|L}$ has the same form as the generalized power function

$$\tau_L(P) := \sqrt{P_{\mathbf{x}} P_{\mathbf{y}} K(\mathbf{x}, \mathbf{y}) - P_{\mathbf{x}} \boldsymbol{k}_L(\mathbf{x})^T \mathsf{A}_L^{-1} P_{\mathbf{y}} \boldsymbol{k}_L(\mathbf{y})}.$$

Moreover, by [21, Theorem 16.3], we can obtain the upper bound of $s_{L,f}$ such as

$$\left| P s_{L,f} - P u \right| \leq \tau_L(P) \left\| u \right\|_{\mathscr{H}_K(\mathscr{D})}, \tag{27}$$

if $u \in \mathscr{H}_K(\mathscr{D})$. By Theorem 4, we find that the standard deviation $\sigma_{P|L}$ determines the lower bound of the accepted error ε. This shows that the lower bounds of the acceptable errors under the kernel-based probability measures are strongly correlated to the upper bounds of the deterministic errors given in Eq. (27).

Optimal Initialization. Finally, we will look at how to choose an optimal initial conjecture $\hat{\mu}$ by the observed values $\boldsymbol{f}$. The kernel-based probability measure $\mathbb{P}_K^{\mu}$ is constructed by the initial conjecture μ and the positive definite kernel K in Theorem 1. The initial conjecture μ can be viewed as the center of $\mathbb{P}_K^{\mu}$ defined on $\mathscr{H}^m(\mathscr{D})$. Subjected to the collocation conditions, the event $\mathscr{I}_L(\boldsymbol{f})$ happens absolutely when the data $(\boldsymbol{L}, \boldsymbol{f})$ are given. So, by the similar technique of maximum likelihood estimation, the optimal initial conjecture $\hat{\mu}$ is chosen to maximize the kernel-based probability measure of $\mathscr{I}_L(\boldsymbol{f})$, to be more precise, the optimal initial conjecture $\hat{\mu}$ is solved by the maximization

$$\max_{\mu \in \mathscr{H}^m(\mathscr{D})} \mathbb{P}_K^{\mu}\left(\mathscr{I}_L(\boldsymbol{f})\right). \tag{28}$$

Since $\mathbb{P}_K^{\mu}$ masses the largest probability at the center μ, there exists the optimal solution $\hat{\mu} := u$ to the optimization (28). By the construction of $\boldsymbol{LS}$, we have that

$$\mathbb{P}_K^{\mu}\left(\mathscr{I}_L(\boldsymbol{f})\right) = \mathbb{P}_K^{\mu}\left(\boldsymbol{LS} = \boldsymbol{f}\right) \backsimeq p_K^{\mu}(\boldsymbol{f}), \tag{29}$$

where p_K^{μ} is the joint probability density function of $\boldsymbol{LS}$ under $\mathbb{P}_K^{\mu}$. Here, the notation $\backsimeq$ represents that the probability density function equivalently specifies the relative likelihood of the observed values of the random variables. According to Eq. (29), the optimization (28) can be equivalently transferred into

$$\min_{\mu \in \mathscr{H}^m(\mathscr{D})} \left(\boldsymbol{f} - \boldsymbol{L}\mu\right)^T \mathsf{A}_L^{\dagger} \left(\boldsymbol{f} - \boldsymbol{L}\mu\right). \tag{30}$$

Usually, it is so difficult to solve the optimization (30), and we replace the optimization (30) by finding the optimal initial conjecture $\hat{\mu}$ in a subspace $\mathscr{M}$ spanned by finite linearly independent well-computable functions $p_1, \ldots, p_l \in \mathscr{H}^m(\mathscr{D})$, that is, $\mathscr{M} := \mathrm{span}\{p_1, \ldots, p_l\}$. Usually, l is far less than n.

To simplify the complexity of the optimization (30), we will solve the optimal initial conjecture $\hat{\mu}$ of another simple optimization problem

$$\min_{\mu \in \mathscr{M}} \left(\boldsymbol{f} - \boldsymbol{L}\mu\right)^T \mathsf{A}_L^{\dagger} \left(\boldsymbol{f} - \boldsymbol{L}\mu\right). \tag{31}$$

Let $\boldsymbol{p} := (p_1, \ldots, p_l)^T$ and $\mathsf{P}_L := \left(L_j p_k\right)_{j,k=1}^{n,l}$. Thus, the optimal initial conjecture $\hat{\mu}$ is an optimal solution of the optimization (31), to be more precise,

$$\hat{\mu}(\mathbf{x}) := \boldsymbol{p}(\mathbf{x})^T \boldsymbol{\beta}, \tag{32}$$

where the coefficients $\boldsymbol{\beta}$ are an optimal solution of the optimization problem

$$\min_{\boldsymbol{b}\in\mathbb{R}^l} \left(\boldsymbol{f} - \mathsf{P}_L \boldsymbol{b}\right)^T \mathsf{A}_L^{\dagger} \left(\boldsymbol{f} - \mathsf{P}_L \boldsymbol{b}\right). \tag{33}$$

Definition 6 We call the optimal solution $\hat{\mu}$ in Eqs. (32) and (33) an optimal initial conjecture for the kernel-based regression solution $\hat{s}_{L,f}$ under the class of the kernel-based probability measure $\mathbb{P}_K^{\mu}$ for $\mu \in \mathscr{M}$.

Now we show that the optimization problem (33) can be solved by a linear system.

Theorem 5 *The optimization problem* (33) *is equivalent to the linear system*

$$\mathsf{P}_L^T \mathsf{A}_L^{\dagger} \mathsf{P}_L \boldsymbol{\beta} = \mathsf{P}_L^T \mathsf{A}_L^{\dagger} \boldsymbol{f}. \tag{34}$$

Proof Let F be a quadratic objective function defined by

$$F(\boldsymbol{b}) := \left(\boldsymbol{f} - \mathsf{P}_L \boldsymbol{b}\right)^T \mathsf{A}_L^{\dagger} \left(\boldsymbol{f} - \mathsf{P}_L \boldsymbol{b}\right).$$

Then the vector expression for the gradient of F is

$$\nabla F(\boldsymbol{b}) := \mathsf{P}_L^T \mathsf{A}_L^{\dagger} \boldsymbol{f} - \mathsf{P}_L^T \mathsf{A}_L^{\dagger} \mathsf{P}_L \boldsymbol{b}.$$

Thus, the critical point of F is the solution of Eq. (34). Moreover, since the minimizer of F over $\mathbb{R}^l$ is the critical point of F, the optimal solution of the optimization problem (33) is the solution of Eq. (34).

Next, since A_L is a symmetric positive definite matrix, its pseudo-inverse $\mathsf{A}_L^{\dagger}$ is a symmetric positive definite matrix. This assures that $\mathsf{P}_L^T \mathsf{A}_L^{\dagger} \mathsf{P}_L$ is a symmetric positive definite matrix. Therefore, the solution of Eq. (34) is the optimal solution of the optimization problem (33).

In the following section, we will solve the multiple Poisson equation using the kernel-based probability measure $\mathbb{P}_K^{\mu}$.

3 Solving Poisson Equation in Kernel-Based Probability Space

In this section, we will show how to obtain the kernel-based regression solution of the multiple Poisson equation under the kernel-based probability measure using the approximation methods in Sect. 2.

Let $\mathscr{D}$ be a ball of $\mathbb{R}^d$ centered at origin with the radius $\frac{1}{2}$ and let $\partial\mathscr{D}$ be the boundary of $\mathscr{D}$. Given a function $f \in \mathrm{C}(\mathscr{D})$, a vector function $\boldsymbol{\eta}$ composed of functions $\eta_1, \ldots, \eta_d \in \mathrm{C}(\mathscr{D})$, and a function $g \in \mathrm{C}(\partial\mathscr{D})$, we look at the multiple Poisson equation with the Dirichlet boundary condition

$$\Delta u = f \text{ in } \mathscr{D}, \quad \nabla u = \boldsymbol{\eta} \text{ in } \mathscr{D}, \quad u = g \text{ on } \partial\mathscr{D}, \tag{35}$$

where Δ is a Laplace differential operator and ∇ is a gradient operator. We always suppose that Poisson equation (35) has a solution u in the Sobolev space $\mathscr{H}^m(\mathscr{D})$ for $m > d/2$ and $m \geq 2$. By the maximum principle, the solution u is unique.

Remark 7 In this article, we look at the multiple Poisson equation as a simple example to illustrate how to solve the complex system of differential equations by the kernel-based approximation methods. Comparing with the regular Poisson equation, the multiple Poisson equation has the additional term of the gradient part. Actually, the regular Poisson equation with the Dirichlet boundary condition always exists the unique solution. But, the ill-posed gradient part may cause that the multiple Poisson equation would not have the solution. To be more precise, the good $\boldsymbol{\eta}$ is necessary to dependent of f such that $\nabla^T\boldsymbol{\eta} = f$. For convenience, we only discuss the well-posed multiple Poisson equation in this article. For the continuous case, the gradient part is a unnecessary term of the Poisson equation because $\nabla^T\nabla = \Delta$. For the discrete case, the well-posed gradient part will provide the additionally useful information to construct the good kernel-based regression solutions. For coding the program simply, we derive the special system of the Poisson equation adding the gradient part to discuss the kernel-based regression solutions. Generally, we have the same results of another elliptic partial differential equations with complex differential and boundary operators.

Firstly, we choose the distinct points

$$X := \left\{\mathbf{x}_1, \ldots, \mathbf{x}_{n_1}\right\} \subseteq \mathscr{D} \text{ and } Z := \left\{\mathbf{z}_1, \ldots, \mathbf{z}_{n_2}\right\} \subseteq \partial\mathscr{D}, \tag{36}$$

to obtain the observation $\boldsymbol{f}$, to be more precise, the observed values $\boldsymbol{f}$ composed of the elements $f(\mathbf{x}_j)$, $\boldsymbol{\eta}(\mathbf{x}_j)$, and $g(\mathbf{z}_k)$ for all $j = 1, \ldots, n_1$ and $k = 1, \ldots, n_2$. The corresponding vector operator $\boldsymbol{L}$ is set up by the linear functionals $\delta_{\mathbf{x}_j} \circ \Delta$, $\delta_{\mathbf{x}_j} \circ \nabla$, and $\delta_{\mathbf{z}_k}$ for all $j = 1, \ldots, n_1$ and $k = 1, \ldots, n_2$. Let $n := (1 + d)n_1 + n_2$. Then the vectors $\boldsymbol{L}$ and $\boldsymbol{f}$ include n elements. Obviously, the functionals $\delta_{\mathbf{x}_j} \circ \Delta$, $\delta_{\mathbf{x}_j} \circ \nabla$, and $\delta_{\mathbf{z}_k}$ are bounded on $\mathscr{H}^m(\mathscr{D})$ for all $j = 1, \ldots, n_1$ and $k = 1, \ldots, n_2$.

Next, we choose the positive definite kernel $K \in \mathrm{C}^{2m,1}(\mathscr{D} \times \mathscr{D})$ and the initial conjecture μ in the function space $\mathscr{M}$ spanned by a finite polynomial basis $p_1, \ldots, p_l$, to be more precise, $\mu := \boldsymbol{p}^T\boldsymbol{b}$ where the vector function $\boldsymbol{p}$ is composed of $p_1, \ldots, p_l$ and $\boldsymbol{b} \in \mathbb{R}^l$. It is obvious that $\mathscr{M} \subseteq \mathscr{H}^m(\mathscr{D})$. Let $\mathsf{P}_L \in \mathbb{R}^{n\times l}$ be composed of the elements $\Delta p_i(\mathbf{x}_j)$, $\nabla p_i(\mathbf{x}_j)$, and $p_i(\mathbf{z}_k)$ for all $i = 1, \ldots, l$, $j = 1, \ldots, n_1$ and $k = 1, \ldots, n_2$, more precisely,

$$\mathsf{P}_L^T := \begin{pmatrix} \Delta p_1(\mathbf{x}_1) & \cdots & \Delta p_1(\mathbf{x}_{n_1}) & \nabla^T p_1(\mathbf{x}_1) & \cdots & \nabla^T p_1(\mathbf{x}_{n_1}) & p_1(\mathbf{z}_1) & \cdots & p_1(\mathbf{z}_{n_2}) \\ \vdots & \ddots & \vdots & \vdots & \ddots & \vdots & \vdots & \ddots & \vdots \\ \Delta p_l(\mathbf{x}_1) & \cdots & \Delta p_l(\mathbf{x}_{n_1}) & \nabla^T p_l(\mathbf{x}_1) & \cdots & \nabla^T p_l(\mathbf{x}_{n_1}) & p_l(\mathbf{z}_1) & \cdots & p_l(\mathbf{z}_{n_2}) \end{pmatrix}.$$

Then $\boldsymbol{L}\mu = \mathsf{P}_L \boldsymbol{b}$.

According to Theorem 1, the kernel-based probability measure $\mathbb{P}_K^\mu$ is well defined on the Sobolev space $\mathscr{H}^m(\mathscr{D})$. Therefore, we can construct the multivariate normal random variables $\boldsymbol{L}S$ on the kernel-based probability space such that the covariance matrix A_L of $\boldsymbol{L}S$ is composed of the elements $\Delta_{\mathbf{x}}\Delta_{\mathbf{y}}K(\mathbf{x}_i,\mathbf{x}_j) = \Delta_{\mathbf{x}}\Delta_{\mathbf{y}}K(\mathbf{x}_j,\mathbf{x}_i)$, $\Delta_{\mathbf{x}}\nabla_{\mathbf{y}}K(\mathbf{x}_i,\mathbf{x}_j) = \nabla_{\mathbf{x}}\Delta_{\mathbf{y}}K(\mathbf{x}_j,\mathbf{x}_i)$, $\nabla_{\mathbf{x}}\nabla_{\mathbf{y}}^T K(\mathbf{x}_i,\mathbf{x}_j) = \nabla_{\mathbf{x}}\nabla_{\mathbf{y}}^T K(\mathbf{x}_j,\mathbf{x}_i)$, $\Delta_{\mathbf{x}}K(\mathbf{x}_i,\mathbf{z}_k)$ $= \Delta_{\mathbf{y}}K(\mathbf{z}_k,\mathbf{x}_i)$, $\nabla_{\mathbf{x}}K(\mathbf{x}_i,\mathbf{z}_k) = \nabla_{\mathbf{y}}K(\mathbf{z}_k,\mathbf{x}_i)$, and $K(\mathbf{z}_k,\mathbf{z}_t) = K(\mathbf{z}_t,\mathbf{z}_k)$ for $i,j = 1,\ldots,n_1$ and $k,t = 1,\ldots,n_2$. To be more precise, we have that

$$\mathsf{A}_L := \begin{pmatrix} \mathsf{A}_{11} & \mathsf{A}_{12} & \mathsf{A}_{13} \\ \mathsf{A}_{21} & \mathsf{A}_{22} & \mathsf{A}_{23} \\ \mathsf{A}_{31} & \mathsf{A}_{32} & \mathsf{A}_{33} \end{pmatrix},$$

where

$$\mathsf{A}_{11} := \begin{pmatrix} \Delta_{\mathbf{x}}\Delta_{\mathbf{y}}K(\mathbf{x}_1,\mathbf{x}_1) & \cdots & \Delta_{\mathbf{x}}\Delta_{\mathbf{y}}K(\mathbf{x}_1,\mathbf{x}_{n_1}) \\ \vdots & \ddots & \vdots \\ \Delta_{\mathbf{x}}\Delta_{\mathbf{y}}K(\mathbf{x}_{n_1},\mathbf{x}_1) & \cdots & \Delta_{\mathbf{x}}\Delta_{\mathbf{y}}K(\mathbf{x}_{n_1},\mathbf{x}_{n_1}) \end{pmatrix},$$

$$\mathsf{A}_{21} = \mathsf{A}_{12}^T := \begin{pmatrix} \nabla_{\mathbf{x}}\Delta_{\mathbf{y}}K(\mathbf{x}_1,\mathbf{x}_1) & \cdots & \nabla_{\mathbf{x}}\Delta_{\mathbf{y}}K(\mathbf{x}_1,\mathbf{x}_{n_1}) \\ \vdots & \ddots & \vdots \\ \nabla_{\mathbf{x}}\Delta_{\mathbf{y}}K(\mathbf{x}_{n_1},\mathbf{x}_1) & \cdots & \nabla_{\mathbf{x}}\Delta_{\mathbf{y}}K(\mathbf{x}_{n_1},\mathbf{x}_{n_1}) \end{pmatrix},$$

$$\mathsf{A}_{22} := \begin{pmatrix} \nabla_{\mathbf{x}}\nabla_{\mathbf{y}}^T K(\mathbf{x}_1,\mathbf{x}_1) & \cdots & \nabla_{\mathbf{x}}\nabla_{\mathbf{y}}^T K(\mathbf{x}_1,\mathbf{x}_{n_1}) \\ \vdots & \ddots & \vdots \\ \nabla_{\mathbf{x}}\nabla_{\mathbf{y}}^T K(\mathbf{x}_{n_1},\mathbf{x}_1) & \cdots & \nabla_{\mathbf{x}}\nabla_{\mathbf{y}}^T K(\mathbf{x}_{n_1},\mathbf{x}_{n_1}) \end{pmatrix},$$

$$\mathsf{A}_{31} = \mathsf{A}_{13}^T := \begin{pmatrix} \Delta_{\mathbf{y}}K(\mathbf{z}_1,\mathbf{x}_1) & \cdots & \Delta_{\mathbf{y}}K(\mathbf{z}_{n_2},\mathbf{x}_1) \\ \vdots & \ddots & \vdots \\ \Delta_{\mathbf{y}}K(\mathbf{z}_1,\mathbf{x}_{n_1}) & \cdots & \Delta_{\mathbf{y}}K(\mathbf{z}_{n_2},\mathbf{x}_{n_1}) \end{pmatrix},$$

$$\mathsf{A}_{32} = \mathsf{A}_{23}^T := \begin{pmatrix} \nabla_{\mathbf{y}}K(\mathbf{z}_1,\mathbf{x}_1) & \cdots & \nabla_{\mathbf{y}}K(\mathbf{z}_{n_2},\mathbf{x}_1) \\ \vdots & \ddots & \vdots \\ \nabla_{\mathbf{y}}K(\mathbf{z}_1,\mathbf{x}_{n_1}) & \cdots & \nabla_{\mathbf{y}}K(\mathbf{z}_{n_2},\mathbf{x}_{n_1}) \end{pmatrix},$$

and

$$\mathsf{A}_{33} := \begin{pmatrix} K(\mathbf{z}_1, \mathbf{z}_1) & \cdots & K(\mathbf{z}_{n_2}, \mathbf{z}_1) \\ \vdots & \ddots & \vdots \\ K(\mathbf{z}_1, \mathbf{z}_{n_2}) & \cdots & K(\mathbf{z}_{n_2}, \mathbf{z}_{n_2}) \end{pmatrix}.$$

Moreover, by Theorem 2, the kernel basis $\boldsymbol{k}_L(\mathbf{x})$ is constructed by the functions $\Delta_\mathbf{y} K(\mathbf{x}, \mathbf{x}_j)$, $\nabla_\mathbf{y} K(\mathbf{x}, \mathbf{x}_j)$, and $K(\mathbf{x}, \mathbf{z}_k)$ for all $j = 1, \dots, n_1$ and $k = 1, \dots, n_2$. To be more precise, we have that

$$\boldsymbol{k}_L(\mathbf{x}) := \begin{pmatrix} \boldsymbol{k}_1(\mathbf{x}) \\ \boldsymbol{k}_2(\mathbf{x}) \\ \boldsymbol{k}_3(\mathbf{x}) \end{pmatrix},$$

where

$$\boldsymbol{k}_1(\mathbf{x}) := \begin{pmatrix} \Delta_\mathbf{y} K(\mathbf{x}, \mathbf{x}_1) \\ \vdots \\ \Delta_\mathbf{y} K(\mathbf{x}, \mathbf{x}_{n_1}) \end{pmatrix}, \quad \boldsymbol{k}_2(\mathbf{x}) := \begin{pmatrix} \nabla_\mathbf{y} K(\mathbf{x}, \mathbf{x}_1) \\ \vdots \\ \nabla_\mathbf{y} K(\mathbf{x}, \mathbf{x}_{n_1}) \end{pmatrix}, \quad \boldsymbol{k}_3(\mathbf{x}) := \begin{pmatrix} K(\mathbf{x}, \mathbf{z}_1) \\ \vdots \\ K(\mathbf{x}, \mathbf{z}_{n_2}) \end{pmatrix}.$$

According to Eqs. (10) and (11) in Theorem 2, the *kernel-based regression solution* $\hat{s}_{L,f}$ has the form

$$\hat{s}_{L,f}(\mathbf{x}) := \boldsymbol{p}(\mathbf{x})^T \boldsymbol{b} + \boldsymbol{k}_L(\mathbf{x})^T \mathbf{c}, \tag{37}$$

where the coefficients $\mathbf{c}$ are the least-squared solution of

$$\mathsf{A}_L \mathbf{c} = \boldsymbol{f} - \mathsf{P}_L \boldsymbol{b}. \tag{38}$$

Now we show the convergence of $\hat{s}_{L,f}$.

Theorem 6 *If the data points X and Z are the part of the countable data points X_∞ and Z_∞ which are dense in $\mathscr{D}$ and $\partial\mathscr{D}$, respectively, then the kernel-based regression solution $\hat{s}_{L,f}$ uniformly converges to the exact solution u of Poisson equation* (35) *when $n_1, n_2 \to \infty$.*

Proof Let $\boldsymbol{L}_\infty$ be a vector operator composed of $\delta_{\mathbf{x}_j} \circ \Delta$, $\delta_{\mathbf{x}_j} \circ \nabla$, and $\delta_{\mathbf{z}_k}$ for all $\mathbf{x}_j \in X_\infty$ and $\mathbf{z}_k \in Z_\infty$ and let $\boldsymbol{f}_\infty$ be a relative observation composed of $f(\mathbf{x}_j)$, $\boldsymbol{\eta}(\mathbf{x}_j)$, and $g(\mathbf{z}_k)$ for all $\mathbf{x}_j \in X_\infty$ and $\mathbf{z}_k \in Z_\infty$. Then $\boldsymbol{L}$ and $\boldsymbol{f}$ are the parts of $\boldsymbol{L}_\infty$ and $\boldsymbol{f}_\infty$, respectively. Since there exists the unique solution $u \in \mathscr{H}^m(\mathscr{D})$ of Poisson equation (35), the maximum principle guarantees that u is the unique solution in $\mathscr{H}^m(\mathscr{D})$ such that $\boldsymbol{L}_\infty u = \boldsymbol{f}_\infty$. Therefore, Theorem 3 assures the uniform convergence of $\hat{s}_{L,f}$.

Moreover, we look at the acceptable error ε of the kernel-based estimator $\Delta\hat{s}_{L,f}(\mathbf{x})$ under the kernel-based probability measure $\mathbb{P}_K^\mu$ for any $\mathbf{x} \in \mathscr{D}$. Theorem 4 assures that the lower bound of ε is dependent of the standard deviation $\sigma_{\delta_\mathbf{x} \circ \Delta | L}$. To be more precise, when $\sigma_{\delta_\mathbf{x} \circ \Delta | L}$ is small enough, then ε can be tiny. Let $\boldsymbol{L}_1$ be a vector operator composed of the functionals $\delta_{\mathbf{x}_j} \circ \Delta$ for $j = 1, \ldots, n_1$. Then $\boldsymbol{L}_1$ is a subpart of $\boldsymbol{L}$. By the construction of $\sigma_{\delta_{\mathbf{x}} \circ \Delta | \boldsymbol{L}}$ and $\sigma_{\delta_{\mathbf{x}} \circ \Delta | \boldsymbol{L}_1}$, we have that

$$\sigma_{\delta_{\mathbf{x}} \circ \Delta | \boldsymbol{L}} \leq \sigma_{\delta_{\mathbf{x}} \circ \Delta | \boldsymbol{L}_1}. \tag{39}$$

If the matrix A_{11} is nonsingular, then we have that

$$\sigma_{\delta_{\mathbf{x}} \circ \Delta | \boldsymbol{L}} = \tau_{\boldsymbol{L}_1}(\delta_{\mathbf{x}} \circ \Delta). \tag{40}$$

According to the proof of [21, Theorem 11.13], we can obtain the upper bound of the power function $\tau_{\boldsymbol{L}_1}(\delta_{\mathbf{x}} \circ \Delta)$ in terms of the fill distance

$$h_{X,\mathscr{D}} := \sup_{\mathbf{x} \in \mathscr{D}} \min_{j=1,\ldots,n_1} \left\| \mathbf{x} - \mathbf{x}_j \right\|_2,$$

that is,

$$\tau_{\boldsymbol{L}_1}(\delta_{\mathbf{x}} \circ \Delta) \leq C_1 h_{X,\mathscr{D}}^{m-2}, \tag{41}$$

where the positive constant C_1 is independent of the differential operators and the data points. Combining Eqs. (39), (40), and (41), we have that

$$\sigma_{\delta_{\mathbf{x}} \circ \Delta | \boldsymbol{L}} \leq C_1 h_{X,\mathscr{D}}^{m-2}. \tag{42}$$

Therefore, Eq. (42) shows that the acceptable error ε of $\Delta \hat{s}_{\boldsymbol{L},f}(\mathbf{x})$ under $\mathbb{P}_K^{\mu}$ can be tiny when the fill distance $h_{X,\mathscr{D}}$ is small enough. By the same ways, for $\mathbf{x} \in \mathscr{D}$ and $\mathbf{z} \in \partial\mathscr{D}$, we also have that

$$\left\| \sigma_{\delta_{\mathbf{x}} \circ \nabla | \boldsymbol{L}} \right\|_2 \leq C_2 h_{X,\mathscr{D}}^{m-1}, \quad \sigma_{\delta_{\mathbf{z}} | \boldsymbol{L}} \leq C_3 h_{Z,\partial\mathscr{D}}^{m}, \tag{43}$$

where the positive constants C_1 and C_2 are independent of the differential operators and the data points, and $h_{Z,\partial\mathscr{D}}$ represents the fill distance of the data points Z for the boundary $\partial\mathscr{D}$. Thus, Eq. (43) shows that the acceptable errors ε of $\nabla \hat{s}_{\boldsymbol{L},f}(\mathbf{x})$ and $\hat{s}_{\boldsymbol{L},f}(\mathbf{z})$ under $\mathbb{P}_K^{\mu}$ can be tiny when the fill distances $h_{X,\mathscr{D}}$ and $h_{Z,\partial\mathscr{D}}$ are small enough, respectively.

Remark 8 The classical kriging is mainly used to obtain the geostatistical estimation such as a method of standard interpolation (see [19]). The recent paper [17] shows how to generalize the universal kriging to solve the uncertainty propagation using the observation of gradients. By the construction of kernel-based probability space, the kernel-based estimation discussed here can be viewed as a wider extension of kriging to solve the partial differential equations. The initial thought is given by the uncertainty differential equations sampling with Bayesian formulations in [14, 15]. But, the stochastic approach here is different from the uncertainty quantification. We solve the deterministic partial differential equations in the kernel-based probability spaces.

3.1 Numerical Example of Trivial Initial Conjecture μ

Suppose that the circle $\mathscr{D} \subseteq \mathbb{R}^2$ and that the right-hand sides of Poisson equation (35) are given by

$$\begin{aligned} f(\mathbf{x}) &:= -2\pi^2\psi_1(\mathbf{x}) - 8\pi^2\psi_2(\mathbf{x}), \quad g(\mathbf{x}) := \psi_1(\mathbf{x}) + \psi_2(\mathbf{x}), \\ \eta_1(\mathbf{x}) &:= \pi\varphi_1(\mathbf{x}) + \pi\varphi_2(\mathbf{x}), \quad \eta_2(\mathbf{x}) := \pi\varphi_3(\mathbf{x}) + \pi\varphi_4(\mathbf{x}), \end{aligned} \tag{44}$$

for $\mathbf{x} := (x_1, x_2) \in \mathbb{R}^2$, where

$$\begin{aligned} \psi_1(\mathbf{x}) &:= \sin(\pi x_1 + \pi/2)\sin(\pi x_2 + \pi/2), \quad \psi_2(\mathbf{x}) := \sin(2\pi x_1 + \pi)\sin(\pi x_2 + \pi), \\ \varphi_1(\mathbf{x}) &:= \cos(\pi x_1 + \pi/2)\sin(\pi x_2 + \pi/2), \quad \varphi_2(\mathbf{x}) := \cos(2\pi x_1 + \pi)\sin(\pi x_2 + \pi), \\ \varphi_3(\mathbf{x}) &:= \sin(\pi x_1 + \pi/2)\cos(\pi x_2 + \pi/2), \quad \varphi_4(\mathbf{x}) := \sin(2\pi x_1 + \pi)\cos(\pi x_2 + \pi). \end{aligned}$$

Thus, the exact solution of Poisson equation (35) has the form

$$u(\mathbf{x}) := \psi_1(\mathbf{x}) + \psi_2(\mathbf{x}).$$

Here, we choose the trivial initial conjecture

$$\mu(\mathbf{x}) := 0,$$

and the positive definite kernel (compactly supported kernel)

$$K(\mathbf{x}, \mathbf{y}) := \phi(\|\mathbf{x} - \mathbf{y}\|_2), \tag{45}$$

set up by a compactly supported radial basis function

$$\phi(r) := (3 + 18\theta r + 35\theta^2 r^2)(1 - \theta r)_+^6, \tag{46}$$

for the shape parameter $\theta > 0$ (see [21, Table 9.1]). So, we can obtain the kernel basis

$$\Delta_{\mathbf{y}} K(\mathbf{x}, \mathbf{x}_j) := 112\theta^2(-1 - 4\theta\left\|\mathbf{x} - \mathbf{x}_j\right\|_2 + 20\theta^2\left\|\mathbf{x} - \mathbf{x}_j\right\|_2^2)(1 - \theta\left\|\mathbf{x} - \mathbf{x}_j\right\|_2)_+^4,$$

$$\nabla_{\mathbf{y}} K(\mathbf{x}, \mathbf{x}_j) := \begin{pmatrix} 56\theta^2(x_1 - x_{1j})(1 + 5\theta\left\|\mathbf{x} - \mathbf{x}_j\right\|_2)(1 - \theta\left\|\mathbf{x} - \mathbf{x}_j\right\|_2)_+^5 \\ 56\theta^2(x_2 - x_{2j})(1 + 5\theta\left\|\mathbf{x} - \mathbf{x}_j\right\|_2)(1 - \theta\left\|\mathbf{x} - \mathbf{x}_j\right\|_2)_+^5 \end{pmatrix},$$

and

$$K(\mathbf{x}, \mathbf{z}_k) := (3 + 18\theta\|\mathbf{x} - \mathbf{z}_k\|_2 + 35\theta^2\|\mathbf{x} - \mathbf{z}_k\|_2^2)(1 - \theta\|\mathbf{x} - \mathbf{z}_k\|_2)_+^6,$$

for all $j = 1, \ldots, n_1$ and $k = 1, \ldots, n_2$. Moreover $\mathsf{P}_L = 0$.

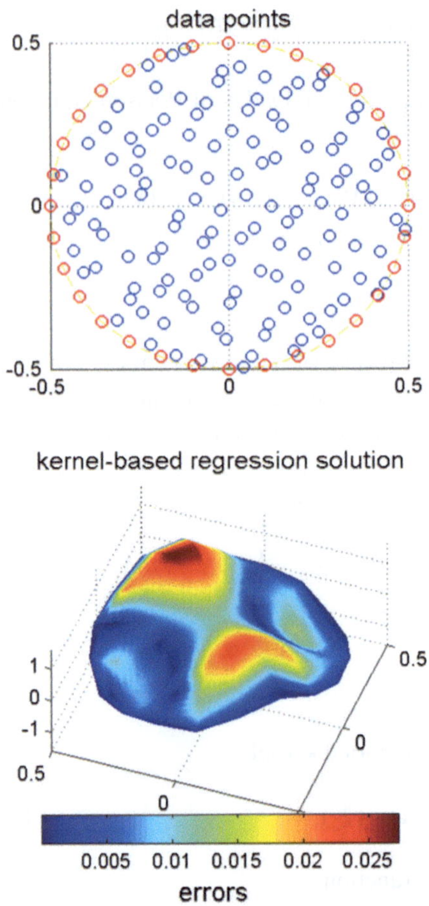

Fig. 1 The numerical examples of the kernel-based regression solution $\hat{s}_{L,f}$ of Poisson equation (35). The $\hat{s}_{L,f}$ is constructed by the compactly supported kernel for $\theta := 0.6$. The *top panels* represent the Halton points $n_1 := 204$ (*left*), 787 (*right*) and the evenly spaced points $n_2 := 72$ (*left*), 108 (*right*). The *bottom panels* represent the numerical solutions for the different data points. The color bars represent the values of absolute errors. The fill distances $h_{X,\mathscr{D}} = 0.0941$ (*left*), $h_{Z,\partial\mathscr{D}} = 0.1013$ (*left*) and $h_{X,\mathscr{D}} = 0.0440$ (*right*), $h_{Z,\partial\mathscr{D}} = 0.0294$ (*right*)

Therefore, we can compute the kernel-based regression solution $\hat{s}_{L,f}$ in Eqs. (37) and (38). In this article, all numerical examples are coded by the Matlab programs. The numerical experiments in Fig. 1 show that $\hat{s}_{L,f}$ is still computable well by the ill-posed matrix A_L for the large data site. The main reason is that the coefficients of $\hat{s}_{L,f}$ can be solved by the least-squared method.

Remark 9 In this numerical example, the exact solution u belongs to $\mathrm{C}^\infty(\mathscr{D})$. So, we can use any smooth kernel to construct the kernel-based regression function. Usually, the conditional variance $\sigma^2_{P|L}$ is smaller when the kernel becomes smoother such that we can obtain the better acceptable error. But, the acceptable error is only discussed for the fixed kernel in this article. Here, we do not compare the kernel-based approximate algorithms induced by the different kernels. Now we look at a special example. Let $m_1 > m_2$. If a positive definite kernel $K \in \mathrm{C}^{2m_1,1}(\mathscr{D} \times \mathscr{D})$, then the kernel-based probability measure $\mathbb{P}_K^0$ defined on $\mathscr{H}^{m_2}(\mathscr{D})$ can be viewed as the extension of the kernel-based probability measure $\mathbb{P}_K^0$ defined on $\mathscr{H}^{m_1}(\mathscr{D})$. This shows that $\mathbb{P}_K^0$ vanishes out of $\mathscr{H}^{m_1}(\mathscr{D})$ in $\mathscr{H}^{m_2}(\mathscr{D})$ such that $\mathbb{P}_K^0$ does not well measure the function $\omega \in \mathscr{H}^{m_2}(\mathscr{D})\backslash\mathscr{H}^{m_1}(\mathscr{D})$. So, we conjecture that the smoothness of K would be strongly related to the exact smoothness of u. We will investigate this topic in our next research paper.

3.2 *Numerical Example of Optimal Initial Conjecture μ*

Suppose that the circle $\mathscr{D} \subseteq \mathbb{R}^2$ and that the right-hand sides of Poisson equation (35) are given by Eq. (44). Let $p_1(\mathbf{x}) := 1$, $p_2(\mathbf{x}) := x_1$, and $p_3(\mathbf{x}) := x_2$. Then we choose the initial conjecture $\mu \in \mathrm{span}\{p_1, p_2, p_3\}$ and the positive definite kernel $K(\mathbf{x}, \mathbf{y})$ given in Eqs. (45) and (46). Thus, we have that

$$\mathsf{P}_L^T := \begin{pmatrix} 0 \cdots 0\; 0\; 0 \cdots 0\; 0 & 1 & \cdots & 1 \\ 0 \cdots 0\; 1\; 0 \cdots 1\; 0 & z_{11} & \cdots & z_{1n_2} \\ 0 \cdots 0\; 0\; 1 \cdots 0\; 1 & z_{21} & \cdots & z_{2n_2} \end{pmatrix}.$$

Moreover, the coefficients $\boldsymbol{\beta}$ of the optimal initial conjecture are solved by the linear system such as Eq. (34) in Theorem 5.

Therefore, we can compute the kernel-based regression solution $\hat{s}_{L,f}$ in Eqs. (37) and (38) by replacing $\boldsymbol{b}$ to $\boldsymbol{\beta}$. Figure 2 shows that the kernel-based regression solution $\hat{s}_{L,f}$ set up by the optimal initial conjecture $\hat{\mu}$ performs the nonworse approximation than the random initial conjecture μ as many as possible. Actually, the optimal initial conjecture $\hat{\mu}$ is not the truly optimal initialization to minimize the exact errors. The optimal initial conjecture is dependent of the kernel-based probability measures. It is usually hard to obtain the truly optimal initialization. But, we can compute the optimal initial conjecture by the observed data. The numerical experiments show that the optimal conjecture is still a good choice for the construction of the kernel-based regression solution.

4 Final Remarks of Comparisons

Finally, we compare the consistence and differences of the kernel-based approximate solutions $s_{L,f}$ and $\hat{s}_{L,f}$ of the partial differential equations induced by the deterministic and stochastic frameworks.

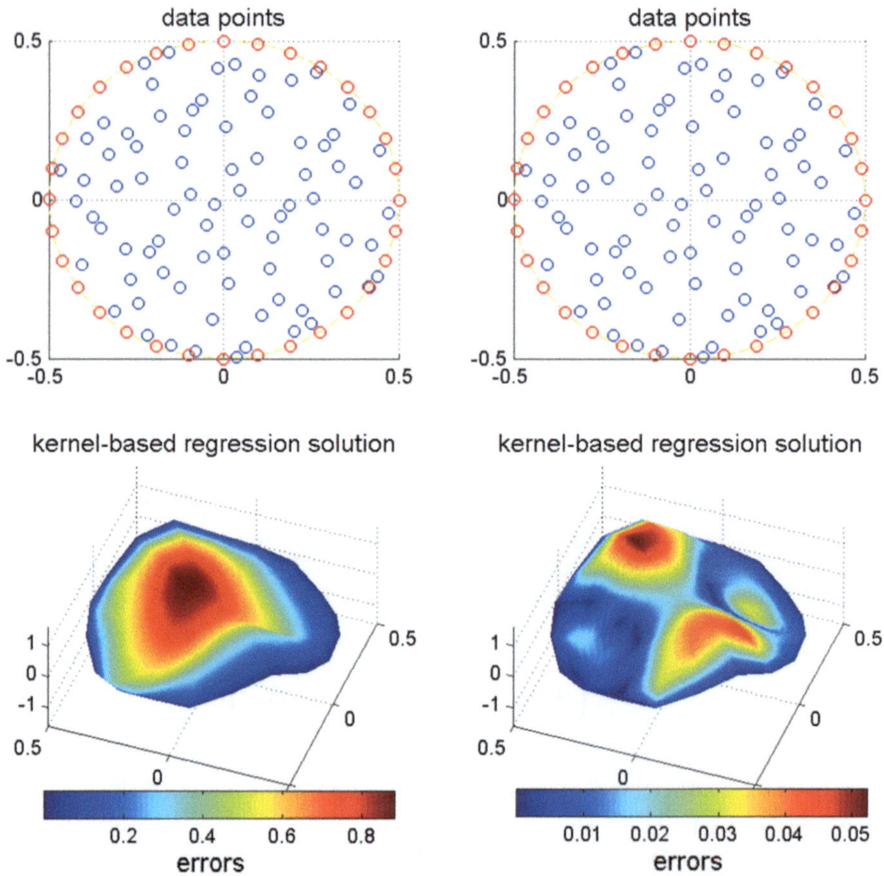

Fig. 2 The numerical examples of the kernel-based regression solution $\hat{s}_{L,f}$ of Poisson equation (35). The $\hat{s}_{L,f}$ is constructed by the compactly supported kernel for $\theta := 0.6$. The *top panels* represent the Halton points $n_1 := 80$ and the evenly spaced points $n_2 := 32$. The *bottom panels* represent the numerical solutions for the different data points. The color bars represent the values of absolute errors. The *left panels* represent the uniformly random coefficients $\boldsymbol{b}$, and the *right panels* represent the optimal coefficients $\boldsymbol{\beta} = (-0.1788, -0.0860, 0.2098)^T$. The fill distances $h_{X,\mathscr{D}} = 0.1272$ and $h_{Z,\partial\mathscr{D}} = 0.1013$

- The kernel-based collocation solution $s_{L,f}$ is given by a deterministic approach, that is, collocation methods. But, the kernel-based regression solution $\hat{s}_{L,f}$ is given by a stochastic approach, that is, Gaussian process regression.
- For the kernel-based collocation, we suppose that the exact solution of partial differential equations belongs to $\mathscr{H}_K(\mathscr{D})$. For the kernel-based regression, we suppose that the exact solution of partial differential equation belongs to $\mathscr{H}^m(\mathscr{D})$.
- The kernel-based collocation and the kernel-based regression are the meshfree methods. This shows that the kernel-based approximation methods do not require any construction of triangular meshes.

- The formulas of $s_{L,f}$ and $\hat{s}_{L,f}$ are almost the same. Both approximate solutions are constructed by the kernel basis $\boldsymbol{k}_L$, and the related coefficients are solved by the linear systems of the positive definite matrix A_L, that is, $\boldsymbol{k}_L(\mathbf{x}) = \boldsymbol{L}_\mathbf{y} K(\mathbf{x}, \mathbf{y})$ and $\mathsf{A}_L = \boldsymbol{L}_\mathbf{x}\boldsymbol{L}_\mathbf{y}^T K(\mathbf{x}, \mathbf{y})$. When A_L is nonsingular, then the forms of $s_{L,f}$ and $\hat{s}_{L,f}$ are consistent.
- The coefficients of $s_{L,f}$ are the exact solution of the linear system. The coefficients of $\hat{s}_{L,f}$ are the least-squared solution of the linear system. This shows that A_L of $s_{L,f}$ is needed to be a nonsingular matrix. But A_L of $\hat{s}_{L,f}$ can be a singular matrix. Generally, the linear functionals $\boldsymbol{L}$ of $s_{L,f}$ are required to be linearly independent on the reproducing kernel Hilbert space $\mathscr{H}_K(\mathscr{D})$. But, the linear functionals $\boldsymbol{L}$ of $\hat{s}_{L,f}$ have not any restriction on the Sobolev space $\mathscr{H}^m(\mathscr{D})$.
- The $s_{L,f}$ is a collocation solution interpolating the Hermite–Birkhoff data. The $\hat{s}_{L,f}$ may not be a collocation solution. But $\hat{s}_{L,f}$ is dependent of a collection of all collocation solutions. Roughly $\hat{s}_{L,f}$ is the average of all collocation solutions under the kernel-based probability measures.
- The $s_{L,f}$ is an optimal recovery interpolation to minimize the energies. The $\hat{s}_{L,f}$ is an optimal estimation to minimize the mean squared errors.
- The convergence of $s_{L,f}$ and $\hat{s}_{L,f}$ is guaranteed by the density of the Hermite–Birkhoff data.
- The convergence analysis of $s_{L,f}$ is verified by the reproducing properties. The convergence analysis of $\hat{s}_{L,f}$ is verified by the conditional kernel-based means.
- The upper bound of the deterministic errors of $s_{L,f}$ is controlled by the generalized power functions. The lower bounds of the acceptable errors of $\hat{s}_{L,f}$ are determined by the standard deviations under kernel-based probability measures. The forms of the generalized power functions and the standard deviations are consistent when A_L is nonsingular.
- The classical collocation solution $s_{L,f}$ is usually independent of the initial conjecture μ. However, the regression solution $\hat{s}_{L,f}$ can contain the information of the initial conjecture μ. Roughly the initial conjecture μ of $\hat{s}_{L,f}$ changes the kernel-based probabilities of the confidence intervals in the statistical hypothesis testing. Using the techniques of maximum likelihood estimation, we can also obtain the optimal initial conjecture $\hat{\mu}$ of $\hat{s}_{L,f}$ under the kernel-based probability measures.

In this article, we investigate the kernel-based approximation method for the partial differential equations by the stochastic approach but not the deterministic approach. Generalization of the survey of the kernel-based approximation methods for the deterministic and stochastic problems in [7, 18], we present a new idea to solve the deterministic problems in the stochastic frameworks such as the approximation in the kernel-based probability spaces using the theory and algorithms of numerical analysis, regression analysis, and stochastic analysis. We strongly believe that the approximation in the kernel-based probability spaces will open up another research area of approximation theory.

Acknowledgements I would like to express my gratitude to the grant of the "Thousand Talents Program" for junior scholars of China, the grant of the Natural Science Foundation of China (11601162), and the grant of South China Normal University (671082, S80835, and S81031).

References

1. A. Berlinet, C. Thomas-Agnan, *Reproducing Kernel Hilbert Spaces in Probability and Statistics* (Kluwer Academic Publishers, Boston, 2004)
2. M.D. Buhmann, *Radial Basis Functions: Theory and Implementations* (Cambridge University Press, Cambridge, 2003)
3. B.P. Carlin, T.A. Louis, *Bayesian Methods for Data Analysis*, 3rd edn. (Taylor & Francis Group CRC Press, New York, 2009)
4. I. Cialenco, G.E. Fasshauer, Q. Ye, Approximation of stochastic partial differential equations by a kernel-based collocation method. Int. J. Comput. Math. **89**, 2543–2561 (2012)
5. M.L. Eaton, *Multivariate Statistics: A Vector Space Approach* (Institute of Mathematical Statistics, Beachwood, 2007)
6. G. Fasshauer, *Meshfree Approximation Methods with MATLAB* (World Scientific Publishing, Hackensack, 2007)
7. G. Fasshauer, M.J. McCourt, *Kernel-Based Approximation Methods Using MATLAB* (World Scientific Publishing, Hackensack, 2015)
8. G.E. Fasshauer, Q. Ye, Kernel-based collocation methods versus Galerkin finite element methods for approximating elliptic stochastic partial differential equations, in *Meshfree Methods for Partial Differential Equations VI*, ed. by M. Griebel, M.A. Schweitzer (Springer, Berlin, 2013), pp. 155–170
9. G.E. Fasshauer, Q. Ye, A kernel-based collocation method for elliptic partial differential equations with random coefficients, in *Monte Carlo and Quasi-Monte Carlo Methods 2012*, ed. by J. Dick, F.Y. Kuo, G.W. Peters, I.H. Sloan (Springer, New York, 2013), pp. 331–348
10. S. Janson, *Gaussian Hilbert Spaces* (Cambridge University Press, Cambridge, 1997)
11. I. Karatzas, S.E. Shreve, *Brownian Motion and Stochastic Calculus*, 2nd edn. (Springer, New York, 1991)
12. J. Kiefer, Conditional confidence statements and confidence estimators. J. Am. Stat. Assoc. **72**, 789–827 (1977)
13. D. Kincaid, Ward Cheney, *Numerical Analysis*, 3rd edn. (Brook/Cole, Pacific Grove, 2002)
14. P.-S. Koutsourelakis, Accurate uncertainty quantification using inaccurate computational models. SIAM J. Sci. Comput. **31**, 3274–3300 (2009)
15. P.-S. Koutsourelakis, A multi-resolution, non-parametric, Bayesian framework for identification of spatially-varying model parameters. J. Comput. Phys. **228**, 6184–6211 (2009)
16. L. Ling, Q. Ye, Quasi-optimal meshfree numerical differentiation (2016), pp. 1–21 (submitted)
17. B.A. Lockwood, M. Anitescu, Gradient-enhanced universal kriging for uncertainty propagation. Nucl. Sci. Eng. **170**, 168–195 (2012)
18. M. Scheuerer, R. Schaback, M. Schlather, Interpolation of spatial data - a stochastic or a deterministic problem? Eur. J. Appl. Math. **24**, 601–629 (2013)
19. M.L. Stein, *Interpolation of Spatial Data: Some Theory for Kriging* (Springer, New York, 1999)
20. I. Steinwart, A. Christmann, *Support Vector Machines* (Springer, New York, 2008)
21. H. Wendland, *Scattered Data Approximation* (Cambridge University Press, Cambridge, 2005)
22. Q. Ye, Analyzing reproducing kernel approximation method via a Green function approach. Ph.D. Thesis, Illinois Institute of Technology, Chicago (2012)
23. Q. Ye, Approximation of nonlinear stochastic partial differential equations by a kernel-based collocation method. Int. J. Appl. Nonlinear Sci. **1**, 156–172 (2014)
24. Q. Ye, Kernel-based methods for stochastic partial differential equations (2015), pp. 1–54, arXiv:1303.5381v8
25. Q. Ye, Optimal designs of positive definite kernels for scattered data approximation. Appl. Comput. Harmonic Anal. **41**, 214–236 (2016)
26. Q. Ye, Generalizations of simple kriging methods in spatial data analysis (2016), pp. 1–20

MIX
Papier aus verantwortungsvollen Quellen
Paper from responsible sources
FSC® C105338

If you have any concerns about our products,
you can contact us on
ProductSafety@springernature.com

In case Publisher is established outside the EU,
the EU authorized representative is:
Springer Nature Customer Service Center GmbH
Europaplatz 3, 69115 Heidelberg, Germany

Printed by Libri Plureos GmbH
in Hamburg, Germany